D1288446

# SPECTRAL ANALYSIS OF ORGANIC COMPOUNDS

*An Introductory Programmed Text*

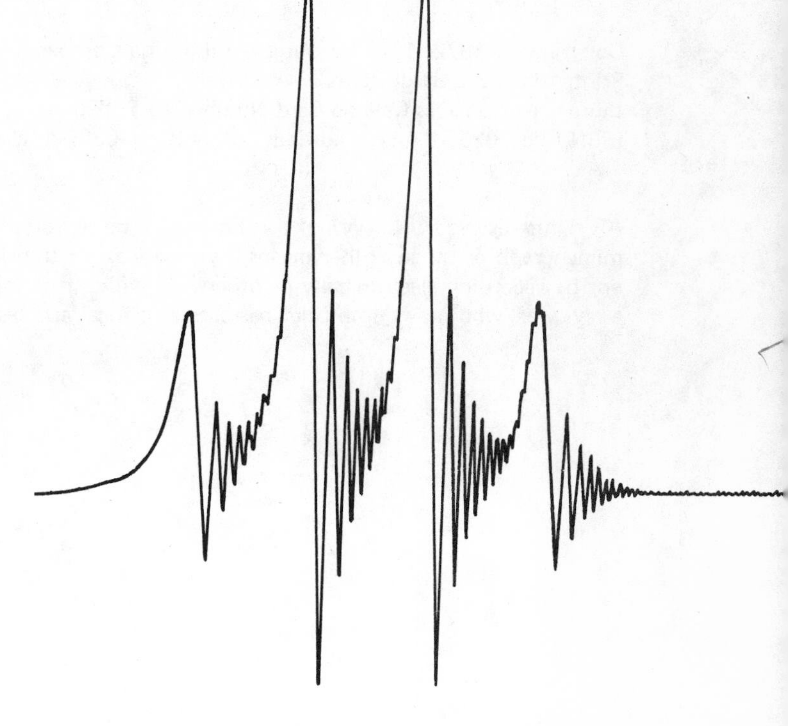

by

CLIFFORD J. CRESWELL

OLAF A. RUNQUIST

Hamline University
St. Paul, Minnesota

and

MALCOLM M. CAMPBELL

Heriot-Watt University
Edinburgh, Scotland

SECOND EDITION

Burgess Publishing Company

Minneapolis, Minn.

Copyright © 1972, 1970 by Burgess Publishing Company
Printed in the United States of America
Library of Congress Catalog Card Number 72-77099
ISBN 8087-0335-8, paper edition ISBN 8087-0337-4, cloth edition

All rights reserved. No part of this book may be reproduced in any form whatsoever, by photograph or mimeograph or by any other means, by broadcast or transmission, by translation into any kind of language, nor by recording electronically or otherwise, without permission in writing from the publisher, except by a reviewer, who may quote brief passages in critical articles and reviews.

8 9 0

QD
271
C76
1972
c.2

# PREFACE - To the Second Edition

Mass spectroscopy is a powerful tool in structure determination. The information which can be deduced from mass spectroscopic examination of a compound includes molecular weight, elemental composition, and often indicates the presence of functional groups. The data obtained from mass spectroscopic examination complements the information derived from nuclear magnetic resonance, infrared and ultraviolet spectroscopy. Use of the four major spectroscopic techniques in conjunction has revolutionized structure elucidation in the last decade. Interest in mass spectroscopy is not solely confined to structure elucidation. The highly excited ions formed may undergo complex rearrangements and fragmentations, the study of which is accounting for more and more space in the chemical literature. The primary purpose of a second edition is to include Chapter 7 to assist in developing the ability to interpret mass spectra and use the information in deducing possible structures for organic compounds.

Although mass spectroscopy has not been included in the chapters on Synthesis of Spectral Data and Practice Problems (Chapters 5 and 6), the molecular weights or formulas of compounds given in these chapters could have been obtained from mass spectral data. The technique of generating formulas from molecular ion masses is illustrated in Chapter 7.

This new edition is a programmed book. An explanation of how to use this text has been given in the preface of the first edition. S I units have been used throughout this book but cm and g have been retained.

This book is meant to be used as a supplement in the undergraduate organic chemistry course and/or a text in the undergraduate advanced organic course (organic qualitative analysis).

We gratefully acknowledge the financial assistance of the Louis W. and Maud Hill Family Foundation.

C.J.C.
O.A.R.
M.M.C.

April, 1972

# PREFACE - To the First Edition

Spectroscopy has provided the chemist with powerful tools for the determination of structures and conformations of organic compounds. Competence in interpreting and using spectral data is essential to any serious student of chemistry. The primary purpose of this book is to assist in developing the ability to interpret spectra and use the information in determining the structure of organic compounds.

This is a programmed book. It is organized into a logical set of questions which, when answered correctly, will lead the student to the desired learning goals. Each question is accompanied by a correct response so that proposed answers may be immediately checked. This book differs from many programmed texts in that topics are covered in more depth, and the questions require a greater degree of independent thought. Some questions may be answered quickly. Other questions are more involved and will require plotting of graphs, solving equations, or deriving mathematical relationships.

This programmed text is divided into six chapters, each of which has a short introduction indicating what should be accomplished. Following the introduction are statements (marked S), questions (marked Q), answers (marked A), and review statements (marked R). The statements give some basic information and are followed by a series of graded questions and answers. *This manual will be a useful learning aid only if each question is answered by the student before the correct response is consulted.* Reading the question and then reading the answer *is not* satisfactory. The following steps are recommended for using this book:

1. Cover each page of the book with a piece of scratch paper in such a way that you can see only a statement. Read the statement carefully.
2. Uncover the first question. *Write out* the answer to the question on the piece of scratch paper.
3. Move the scratch paper down the page so that a correct response to the first question can be seen. Read the entire answer given and compare it with your answer.
4. If your answer is correct, proceed to the next question. If your answer is incorrect, restudy the previous statement and questions. DO NOT PROCEED UNTIL YOU UNDERSTAND AND CAN CORRECTLY ANSWER EACH QUESTION. To become expert in the interpretation of spectra and the art of structural determination requires considerable practice. There is no substitute for experience.

We are indebted to the helpful suggestions made by colleagues and the many students at Hamline University who have used this book. We wish particularly to thank Deborah Johnson for her effort in proofreading the manuscript.

C.J.C.
O.A.R.

March, 1970

# CONTENTS

CONTENTS

# Chapter 1 THE INTERACTION OF LIGHT WITH ATOMS AND MOLECULES

After completing this chapter, you should be able to:

a) determine the period, frequency, energy, and wave number for light of any wavelength
b) explain why atoms and molecules will absorb light of only certain wavelengths
c) understand the origin of vibrational, rotational, and electronic spectra
d) determine the difference between rotational, vibrational, and electronic energy levels of a molecule from a knowledge of the wavelengths of light absorbed by the molecule
e) diagram the bonding and anti-bonding orbitals for simple molecules and determine if the orbitals are sigma, pi, or nonbonding.

---

S-1 Spectroscopy is the study of interactions of light with atoms and molecules. Light or electromagnetic radiation can be regarded as being wavelike or corpuscular. Some physical properties of light are best explained by its wave characteristics while other properties are best explained by its particle nature. Thus, light is said to be dualistic in nature. The wave nature of light will be considered first.

A diagram of a wave, with its essential features labeled, is shown below:

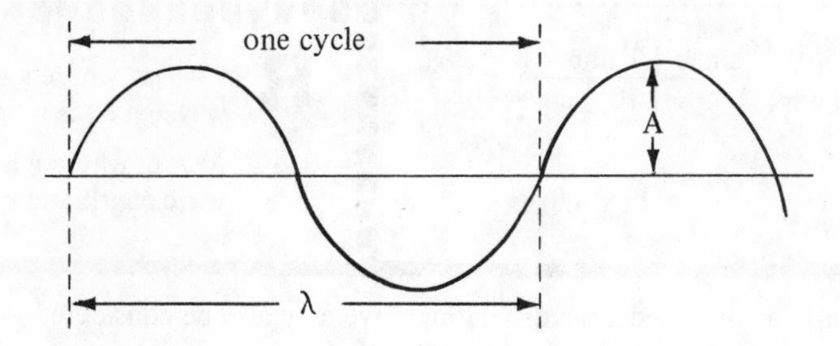


$\lambda$ = wavelength: distance that the wave moves during one cycle.
unit: unit of length/cycle.

A = wave amplitude: the maximum disturbance from the horizontal axis.
unit: unit of length.

$\tau$ = period: the time for one complete cycle.
unit: sec/cycle or s/cycle.

$\nu$ = frequency of oscillation: the number of cycles that occur in every second.
unit: cycle/sec or Hertz.

When referring to $\lambda$, $\tau$, and $\nu$, the word cycle is understood and is not usually used. The relationship between wavelength ($\lambda$) and frequency ($\nu$) for a light wave is

$$\nu \lambda = c,$$

where $c$ is the velocity of light ($3.0 \times 10^8$ m/s or $3.0 \times 10^{10}$ cm/s).

---

**Q-1** Calculate the period and frequency of light with a wavelength of $2.0 \times 10^5$ cm.

---

**A-1**

$\nu = c/\lambda$

$\nu = \frac{3.0 \times 10^{10} \text{ cm/s}}{2.0 \times 10^5 \text{ cm}} = 1.5 \times 10^5 \text{ s}^{-1}$

$\tau = 1/\nu$

$\tau = \text{s}/1.5 \times 10^5 = 6.7 \times 10^{-6} \text{ s}$

**Q-2** If the period of a light wave is $2.0 \times 10^{-17}$ s, what is the length of this wave in nanometer units?

(1 nm = $10^{-9}$ m = $10^{-7}$ cm)

---

**A-2**

$\nu = 1/\tau = 1/2.0 \times 10^{-17}\text{s}$

$= 5.0 \times 10^{16} \text{ s}^{-1}$

$\lambda = c/\nu = \frac{3.0 \times 10^{10} \text{ cm/s}}{5.0 \times 10^{16} \text{ s}^{-1}}$

$= 6.0 \times 10^{-7} \text{ cm} = 6.0 \text{ nm}$

**Q-3** Calculate the period of electromagnetic radiation with

$\lambda = 4.0 \times 10^3$ cm.

---

**A-3**

$\nu = c/\lambda = \frac{3.0 \times 10^{10} \text{ cm/s}}{4.0 \times 10^3 \text{ cm}}$

$= 7.5 \times 10^6 \text{ s}^{-1}$

$\tau = 1/\nu = 1/7.5 \times 10^6 \text{ s}^{-1}$

$= 1.3 \times 10^{-7} \text{ s}$

**Q-4** What is the frequency of green light with a wavelength of 500 nm?

---

**A-4**

$\nu = c/\lambda = \frac{3.0 \times 10^{10} \text{ cm/s}}{500 \text{ nm}} \cdot \frac{1 \text{ nm}}{10^{-7} \text{ cm}}$

$= 6.0 \times 10^{14} \text{ s}^{-1}$

**R**

$\nu = 1/\tau$, where $\nu$ is the frequency of the wave and $\tau$ is its period.

$\lambda\nu = c$, where $\nu$ is the frequency, $\lambda$ is the wavelength and $c$ is the velocity of light.

---

**S-2** Light which can be described as an oscillating wave may also be considered as a stream of energy packets or particles traveling with a high velocity ($3.0 \times 10^{10}$ cm/s). These energy packets are called photons. The frequency ($\nu$) of the wave theory can be related to the energy of the photons ($E$) of the particle theory through Planck's equation.

$E = h\nu$

where $h$ is Planck's constant, a proportionality factor whose value is $6.63 \times 10^{-34}$ joule second (J s).

Wave number, $\bar{\nu}$, is a wave characteristic that is proportional to energy and is defined as the number of waves per centimeter.

$\bar{\nu} = 1/\lambda$

---

**Q-5** Find the energy of the photons which correspond to light of frequency $3.0 \times 10^{15} \text{ s}^{-1}$.

---

**A-5**

$E = h\nu = (6.63 \times 10^{-34} \text{ J s})(3.0 \times 10^{15} \text{ s}^{-1})$

$= 2.0 \times 10^{-18} \text{ J}$

**Q-6** Find the frequency of light which corresponds to photons of energy $5.0 \times 10^{-12}$ J.

A-6 $\nu = \frac{E}{h} = \frac{5.0 \times 10^{-12}\ \text{J}}{6.63 \times 10^{-34}\ \text{J s}}$

$= 7.5 \times 10^{21}\ \text{s}^{-1}$

Q-7 From the equation $E = h\nu$ and $\lambda\nu = c$, derive an equation which relates energy to wavelength.

---

A-7 $E = h\nu \quad \nu = c/\lambda$

$E = hc/\lambda$

Q-8 What is the energy of photons with a wavelength equal to 0.05 nm?

---

A-8 $E = h\nu = hc/\lambda$

$E = \frac{6.63 \times 10^{-34}\ \text{J s}}{5 \times 10^{-9}\ \text{cm}} \left| \frac{3.0 \times 10^{10}\ \text{cm}}{\text{s}} \right.$

$E = 4.0 \times 10^{-15}\ \text{J}$

Q-9 Show that $\bar{\nu}$ is proportional to energy.

---

A-9 $E = h\nu$ and $\nu = c/\lambda$

Therefore, $E = hc/\lambda$

$1/\lambda = \bar{\nu}$

Therefore, $E = hc\bar{\nu}$

$hc$ = constant

Q-10 What is the wave number for light with a wavelength of 400 nm?

---

A-10 $\bar{\nu} = \frac{1}{\lambda} = \frac{1}{400\ \text{nm}} \left| \frac{1\ \text{nm}}{10^{-7}\ \text{cm}} \right.$

$= 2.5 \times 10^{4}\ \text{cm}^{-1}$

Q-11 What is the energy of photons which have a wave number of $2.5 \times 10^{-5}\ \text{cm}^{-1}$?

---

A-11 $E = hc\bar{\nu}$

$= 6.63 \times 10^{-34}\ \text{J s} \left| \frac{3 \times 10^{10}\ \text{cm}}{\text{s}} \right| \frac{2.5 \times 10^{-5}}{\text{cm}}$

$= 5.0 \times 10^{-28}\ \text{J}$

**R** $E = h\nu$ where $E$ is the energy of the photon, $\nu$ is the frequency of light, and $h$ is a proportionality constant with a value of $6.63 \times 10^{-34}$ J s. $\bar{\nu} = \frac{1}{\lambda}$ where $\bar{\nu}$ is the wave number.

---

S-3 Electromagnetic radiation is divided into several regions called radio wave, microwave, infrared, visible, ultraviolet, and x-ray. The wavelengths, wave numbers, frequencies, and energies associated with each of these regions are given on the diagram below.

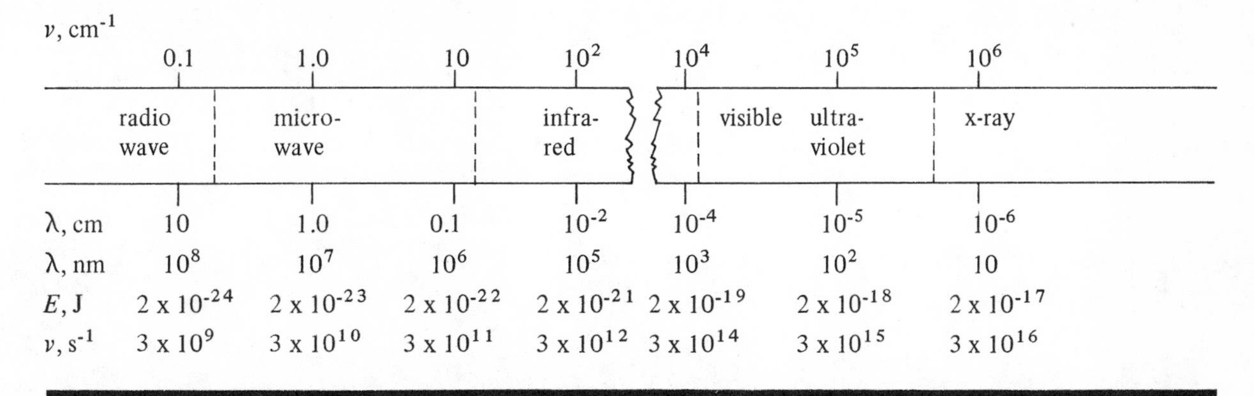

Q-12 In what region of the electromagnetic radiation will the following wavelengths be located?

1 cm

0.8 $\mu$m (micrometer) (1 $\mu$m = $10^{-6}$m)

10 $\mu$m

100 nm

10 nm

A-12

| | |
|---|---|
| 1 cm | microwave |
| 0.8 $\mu$m | visible |
| 10 $\mu$m | infrared |
| 100 nm | ultraviolet |
| 10 nm | x-ray |

Q-13 In what region of the electromagnetic radiation will the following wave numbers be located?

983 $cm^{-1}$

3.0 x $10^4$ $cm^{-1}$

5.0 $cm^{-1}$

8.7 x $10^4$ $cm^{-1}$

A-13

| | |
|---|---|
| 983 $cm^{-1}$ | infrared |
| 3.0 x $10^4$ $cm^{-1}$ | visible |
| 5.0 $cm^{-1}$ | microwave |
| 8.7 x $10^4$ $cm^{-1}$ | ultraviolet |

Q-14 Arrange the following regions of light in order of increasing energy.

microwave

x-ray

visible

ultraviolet

infrared

A-14 microwave < infrared < visible < ultraviolet < x-ray

**R** Electromagnetic radiation is grouped into regions called radio wave, microwave, infrared, visible, ultraviolet, and x-ray.

S-4 When continuous light (i.e., light which consists of all possible wavelengths within a given range) is passed through a prism, the light is dispersed into its component wavelengths. When these dispersed wavelengths are passed through cells containing samples of atoms or molecules, the emergent light is no longer continuous. Some of the light waves have interacted with and have been absorbed by the atoms or molecules in the cells. The missing wavelengths can be detected by permitting the light emerging from the sample cell to fall on a photographic plate or some other detecting device. This procedure is called *absorption spectroscopy* and the recorded image is called a spectrum. A spectral "line" is the wavelength at which light has been absorbed. Absorption spectroscopy is illustrated by the following diagram:

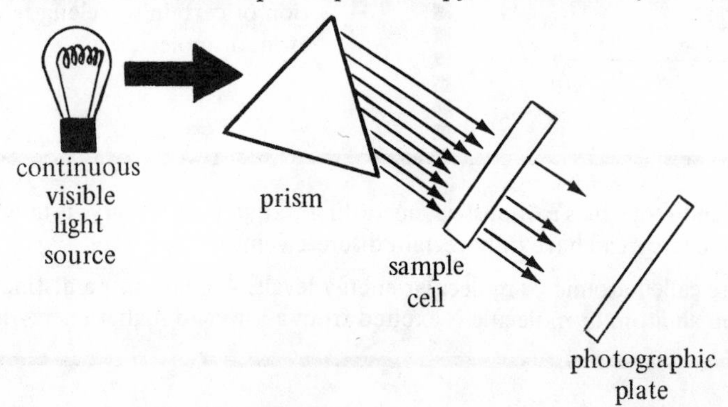


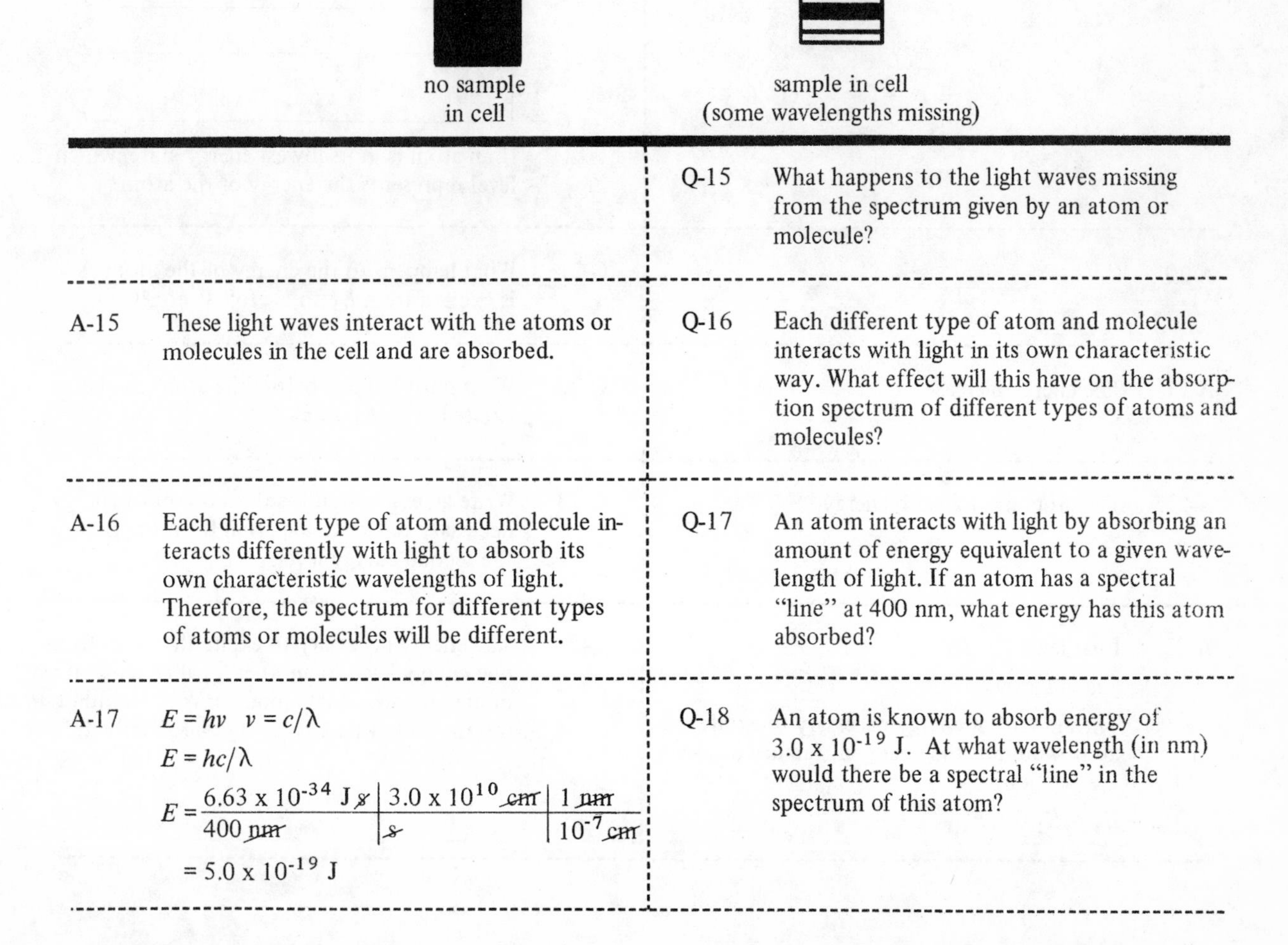


Q-15 What happens to the light waves missing from the spectrum given by an atom or molecule?

A-15 These light waves interact with the atoms or molecules in the cell and are absorbed.

Q-16 Each different type of atom and molecule interacts with light in its own characteristic way. What effect will this have on the absorption spectrum of different types of atoms and molecules?

A-16 Each different type of atom and molecule interacts differently with light to absorb its own characteristic wavelengths of light. Therefore, the spectrum for different types of atoms or molecules will be different.

Q-17 An atom interacts with light by absorbing an amount of energy equivalent to a given wavelength of light. If an atom has a spectral "line" at 400 nm, what energy has this atom absorbed?

A-17 $E = h\nu \quad \nu = c/\lambda$

$E = hc/\lambda$

$$E = \frac{6.63 \times 10^{-34}\ \text{J}\,\cancel{\text{s}}}{400\ \cancel{\text{nm}}} \left| \frac{3.0 \times 10^{10}\ \cancel{\text{cm}}}{\cancel{\text{s}}} \right| \frac{1\ \cancel{\text{nm}}}{10^{-7}\ \cancel{\text{cm}}}$$

$= 5.0 \times 10^{-19}$ J

Q-18 An atom is known to absorb energy of $3.0 \times 10^{-19}$ J. At what wavelength (in nm) would there be a spectral "line" in the spectrum of this atom?

A-18 $E = \dfrac{hc}{\lambda}$ $\lambda = \dfrac{hc}{E}$

$$\lambda = \frac{6.63 \times 10^{-34}\ \text{J s}}{3.0 \times 10^{-19}\ \text{J}} \left| \frac{3.0 \times 10^{10}\ \text{cm}}{\text{s}} \right| \frac{1\ \text{nm}}{10^{-7}\ \text{cm}}$$

$= 660$ nm

Q-19 An atom is known to absorb energy equal to $5.0 \times 10^{-19}$ J. At what wave number would there be an absorption line in the spectrum of this atom?

A-19 $E = hc/\lambda = hc\bar{\nu}$ $\bar{\nu} = \dfrac{E}{hc}$

$$\bar{\nu} = \frac{5.0 \times 10^{-19}\ \text{J}}{6.63 \times 10^{-34}\ \text{J s}} \left| \frac{\text{s}}{3.0 \times 10^{10}\ \text{cm}} \right.$$

$= 2.5 \times 10^{4}\ \text{cm}^{-1}$

**R** An absorption spectrum arises from the absorption of certain wavelengths of light by an atom or molecule.

S-5 The energy that atoms and molecules can possess, according to quantum theory, is quantized. That is, in an atom or molecule, energy can have only certain discrete values.

The allowed energies are called atomic or molecular energy levels. A quantum (a distinct amount) of energy is absorbed when an atom or molecule is excited from a lower to higher energy level.

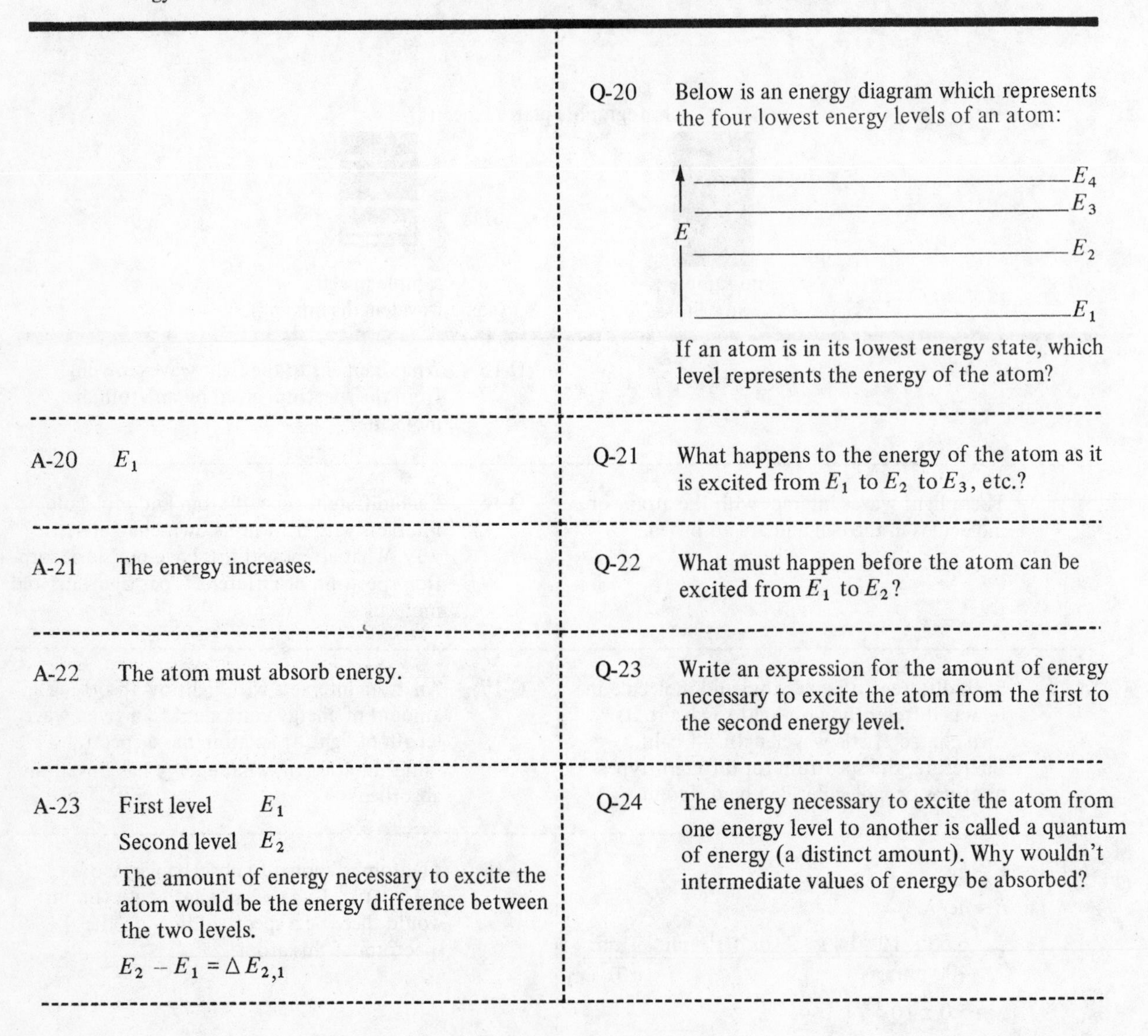

Q-20 Below is an energy diagram which represents the four lowest energy levels of an atom:



If an atom is in its lowest energy state, which level represents the energy of the atom?

A-20 $E_1$

Q-21 What happens to the energy of the atom as it is excited from $E_1$ to $E_2$ to $E_3$, etc.?

A-21 The energy increases.

Q-22 What must happen before the atom can be excited from $E_1$ to $E_2$?

A-22 The atom must absorb energy.

Q-23 Write an expression for the amount of energy necessary to excite the atom from the first to the second energy level.

A-23 First level $E_1$

Second level $E_2$

The amount of energy necessary to excite the atom would be the energy difference between the two levels.

$E_2 - E_1 = \Delta E_{2,1}$

Q-24 The energy necessary to excite the atom from one energy level to another is called a quantum of energy (a distinct amount). Why wouldn't intermediate values of energy be absorbed?

A-24 The atom has no energy levels available at intermediate energies.

Q-25 A transition from one energy level to another is indicated by use of an arrow,

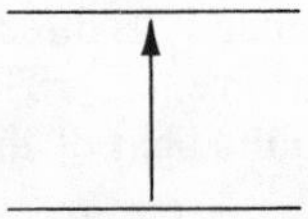

Draw on the diagram of Q-20 arrows to represent the three transitions that this atom could undergo starting from its lowest energy level.

A-25

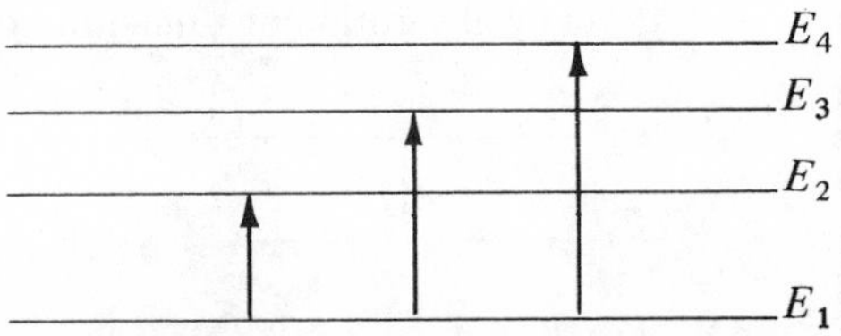

$E_4$

$E_3$

$E_2$

$E_1$

Q-26 Electromagnetic radiation can interact with atoms and molecules to supply the quantum of energy necessary to cause transitions to higher energy levels. Account for the missing wavelengths of light present in the absorption spectrum of an atom or molecule (see S-4).

A-26 An absorption spectrum is produced when an atom or molecule absorbs wavelengths of light with energy equal to the difference between two energy levels.

Q-27 Give an expression which relates the energy of absorption to the wavelength of light absorbed in the process

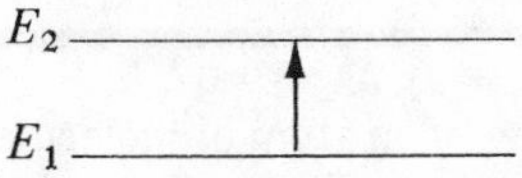

$E_2$

$E_1$

A-27

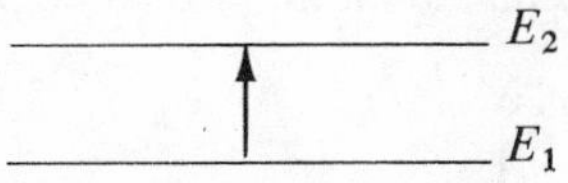

$E_2$

$E_1$

$\Delta E_{2,1} = E_2 - E_1$

$E = hc/\lambda$

Therefore,

$\Delta E_{2,1} = hc/\lambda$

Q-28 What factor determines the wavelength of light absorbed in a spectrum?

A-28 The differences between the various energy levels in the atom or molecule.

Q-29 What explanation would account for the fact that the absorption spectrum for each kind of atom and molecule is unique?

A-29 Each atom or molecule has unique differences between its energy levels.

Q-30 Consider the transitions represented in A-25. Which transition would give rise to the absorption of light of *longest* wavelength?

A-30 $\Delta E = hc/\lambda$

Wavelength is inversely proportional to transition energy. The longest wavelength absorption would be due to the $E_2 - E_1$ transition.

Q-31 For the diagram shown in A-25, which transition would cause absorption of light with the largest wave number?

A-31 $1/\lambda = \bar{\nu}$

$\Delta E = hc\bar{\nu}$

Wave number, $\bar{\nu}$, is directly proportional to the transition energy. The $E_4 - E_1$ transition would require light of the largest wave number.

Q-32 In the diagram of Q-27, assume that the difference in energy between the $E_2$ and $E_1$ levels is $6.0 \times 10^{-18}$ J.

What is the wavelength of light in Å needed to accomplish this transition?

A-32 $E_2 - E_1 = hc/\lambda \quad \lambda = hc/\Delta E_{2,1}$

$$\lambda = \frac{6.63 \times 10^{-34}\ \text{J s}}{6.0 \times 10^{-18}\ \text{J}} \left| \frac{3.0 \times 10^{10}\ \text{cm}}{\text{s}} \right| \frac{1\ \text{nm}}{10^{-7}\ \text{cm}}$$

$= 33.0$ nm

**R** The energy of atoms and molecules is quantized (only certain allowed energy levels). An absorption spectrum is the result of an atom or molecule being excited to a higher energy level by absorption of a quantum of energy.

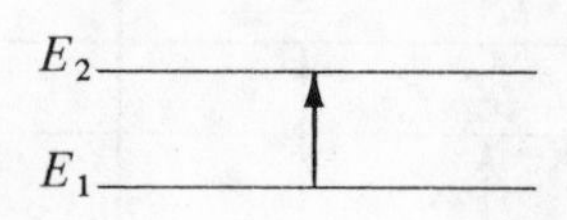


$E_2 - E_1 = \Delta E_{2,1} = hc/\lambda = hc\bar{\nu}$

$E_2 - E_1$ = transition energy.

Each different kind of atom or molecule gives its own characteristic absorption spectrum.

S-6 The energy of an atom or molecule is described in terms of its translational, rotational, vibrational, and electronic energy. Energy levels exist for each of these different types of energies.

Translational energy is the kinetic energy atoms or molecules possess due to motion from one place in space to another.

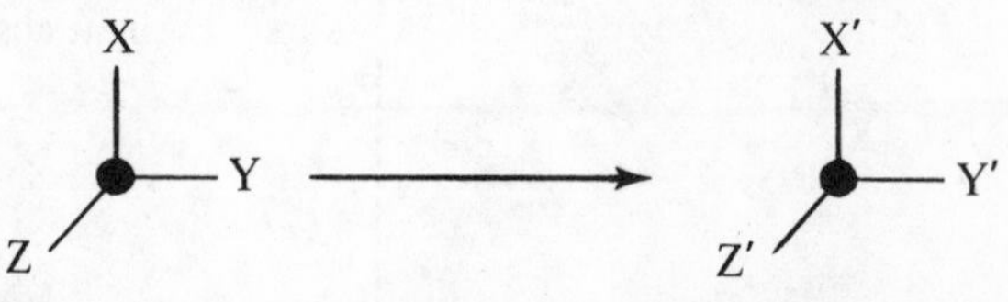


Q-33 Kinetic energy is given by the equation

$E_k = 1/2\ mv^2$

where $m$ is the mass of the moving body and $v$ is its velocity. As the velocity of a molecule increases, what happens to its translational energy?

A-33 Increases. Energy is proportional to the square of the velocity.

Q-34 As temperature increases, what happens to the average velocity of atoms and molecules?

A-34 The average velocity increases with temperature.

Q-35 What happens to the kinetic energy as temperature increases?

A-35 The average kinetic energy increases as temperature increases.

Q-36 The relationship between average translational kinetic energy, $E_k$, for an atom or molecule and the temperature on the absolute scale, T, is given by the equation

$$E_k = \frac{3}{2} k T$$

where $k$ is a constant equal to $1.38 \times 10^{-23}$ J/K (where K is the symbol used for the Kelvin units of temperature). At what temperature will the atoms or molecules be in their lowest translational energy level?

A-36 $E_k = \frac{3}{2} k T$

$E_k = 0$ at T = 0 K

Therefore, at absolute zero the atoms and molecules will be in their lowest translational energy level.

Q-37 Calculate the average translational kinetic energy of an atom or molecule at 25°C (298 K).

A-37 $E_k = \frac{3}{2} k T$

$E_k = \frac{3}{2} (1.38 \times 10^{-23} \text{ J/K})(298 \text{ K})$

$= 6.2 \times 10^{-21}$ J

Q-38 Translational energy levels are so close together that they are considered to be continuous.

Draw a diagram to represent translational energy levels for an atom or molecule.

A-38

E

continuous

E = 0

Q-39 As the temperature increases, thermal energy is absorbed to excite the atoms or molecules into higher translational energy levels. How much energy is required to excite an atom or molecule from a given translational energy level to the next higher level?

A-39 Since the translational energy levels are very nearly continuous, a very small (infinitesimally small) amount of energy is required. It is for this reason that absorption spectra due to translational transitions are not observed.

**R** The energy of an atom or molecule is made up of translational, rotational, vibrational, and electronic energy.

Translational energy is the kinetic energy atoms or molecules possess due to motion in space.

Quantum restrictions are not important in the consideration of translational energy levels (any energy is possible).

S-7 Rotational energy is the kinetic energy molecules possess due to rotation about an axis through their center of gravity.

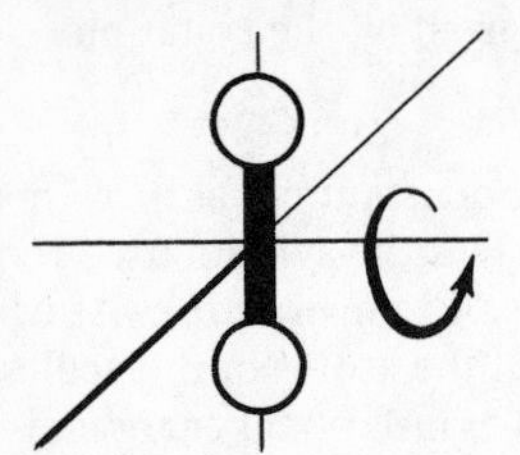

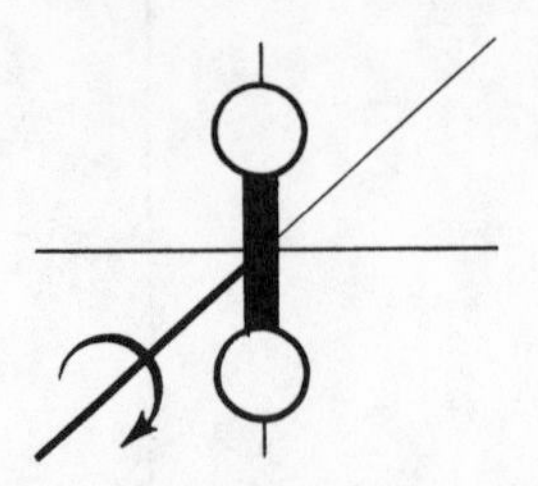

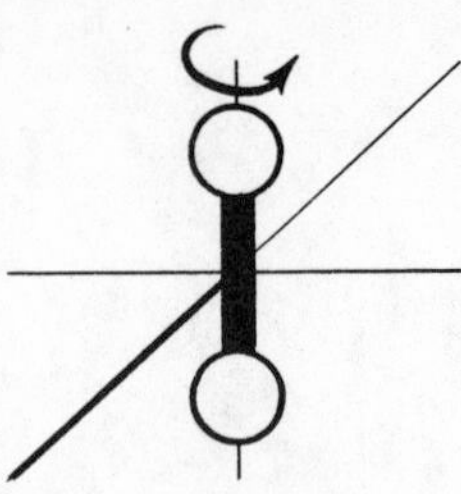

Q-40 As temperature increases, what should happen to the speed of rotation of a molecule?

A-40 As the temperature increases, the molecule should rotate faster.

Q-41 As the molecule rotates faster, what happens to its rotational kinetic energy?

A-41 As the molecule rotates faster, its rotational kinetic energy increases.

Q-42 The rotational energy of a molecule is quantized. Using the energy model account for the increase in rotational kinetic energy when a molecule is heated.

A-42 Thermal energy is absorbed in the amount necessary to excite the molecule to higher rotational energy levels.

Q-43 How else might a molecule be excited to higher rotational energy levels?

A-43 The molecule could absorb a quantum of light of just the right energy to excite it to higher energy levels.

Q-44 The energy difference between rotational energy levels corresponds to light in the microwave region of electromagnetic radiation. A molecule absorbs light with $\lambda = 1.0$ cm. What is the difference in energy between the rotational energy levels which gives rise to this absorption?

A-44 $\Delta E = hc/\lambda$

$$\Delta E = \frac{6.63 \times 10^{-34}\ \text{J}\,\cancel{\text{s}} \;\Big|\; 3.0 \times 10^{10}\ \cancel{\text{cm}}}{\Big|\ \cancel{\text{s}} \;\Big|\; 1.0\ \cancel{\text{cm}}}$$

$= 2.0 \times 10^{-23}$ J

Q-45 Will there be any rotational energy at absolute zero?

A-45 Kinetic energy is proportional to absolute temperature. Therefore, at T equal to zero, $E_k$ equals zero.

Q-46 Below is a typical rotational energy level diagram for a molecule. What does the arrow indicate? (Describe in terms of what happens to the molecule.)

$E_3$ ______

$E_2$ ______

↑

$E_1$ ______

$E_0$ ______

A-46 The arrow indicates the excitation of the molecule from the $E_1$ to the $E_2$ rotational energy level. This requires absorption of energy. Therefore, at the $E_2$ energy level, the molecule will have more rotational energy, causing it to rotate faster.

Q-47 The first three rotational levels of the CO molecule have energies of 0, 7.6 x $10^{-23}$, and 22.9 x $10^{-23}$ J/molecule. What is the wavelength of light necessary to excite the CO molecule from the $E_1$ to the $E_2$ rotational level?

A-47 $\Delta E = E_2 - E_1 = 22.9 \times 10^{-23} - 7.6 \times 10^{-23}$

$= 15.3 \times 10^{-23}$ J/molecule

$\Delta E = hc/\lambda \quad \lambda = hc/\Delta E$

$$\lambda = \frac{6.63 \times 10^{-34}\ \text{J s} \quad 3.0 \times 10^{10}\ \text{cm}}{\text{s} \quad 15.3 \times 10^{-23}\ \text{J}}$$

$\lambda = 0.13$ cm

Q-48 The spacing between rotational energy levels is inversely proportional to the moment of inertia of a molecule. The moment of inertia is defined as

$I = \mu r^2$

where $\mu$ is the reduced mass $\frac{m_1 m_2}{m_1 + m_2}$ and $r$ is the distance between the masses. Which of the following two compounds would be expected to have the larger spacing between rotational energy levels?

| Compound | $\mu$ | Bond Length |
|---|---|---|
| HBr | $1.65 \times 10^{-24}$g | 0.141 nm |
| HI | $1.66 \times 10^{-24}$g | 0.160 nm |

A-48 HBr

$\Delta E \propto 1/I$

$I = \mu r^2 \quad \mu_{HI} \cong \mu_{HBr}$

$r^2_{HI} > r^2_{HBr}$

$I_{HI} > I_{HBr}$

Therefore, $\Delta E_{HBr} > \Delta E_{HI}$

Q-49 Which of the following two compounds would be expected to have the larger spacing between rotational energy levels?

| Compound | $\mu$ | Bond Length |
|---|---|---|
| NO | $12.4 \times 10^{-24}$g | 0.115 nm |
| NaCl | $23.2 \times 10^{-24}$g | 0.236 nm |

A-49 NO $\quad \Delta E \propto 1/I$

$r^2_{NaCl} \cong 4r^2_{NO}$

$\mu_{NaCl} \cong 2\mu_{NO}$

$I_{NaCl} > I_{NO}$

Therefore, $\Delta E_{NO} > \Delta E_{NaCl}$

Q-50 Given two compounds, A-B and C-D, with similar reduced masses. The absorption spectra for an $E_1$ to $E_2$ transition for the two compounds gives a peak for A-B at 1 cm and for C-D at 1.5 cm. Which compound has the greater bond distance?

A-50 $E = hc/\lambda$

$\lambda_{CD} > \lambda_{AB}$

$\Delta E_{AB} > \Delta E_{CD}$

$\Delta E \propto 1/I$

$I_{CD} > I_{AB}$

$I = \mu r^2$

$r^2_{CD} > r^2_{AB}$

Therefore, compound CD would have the greater bond distance.

**R** Rotational energy is the kinetic energy molecules possess due to rotation about an axis through their center of gravity. Rotational energy is quantized and gives rise to absorption spectra in the microwave region of the electromagnetic spectrum.

The difference between rotational energy levels is inversely proportional to the moment of inertia ($I$) of the molecule.

$\Delta E \propto 1/I$

$I = \mu r^2$

where $\mu = \dfrac{m_1 m_2}{m_1 + m_2}$ and $r$ is the separation between masses.

S-8 Vibrational energy is the potential and kinetic energy molecules possess due to vibrational motion. The atoms in a molecule can be considered as point masses held together by bonds acting like springs. Because molecules are not rigid, their flexibility results in vibrational motion.

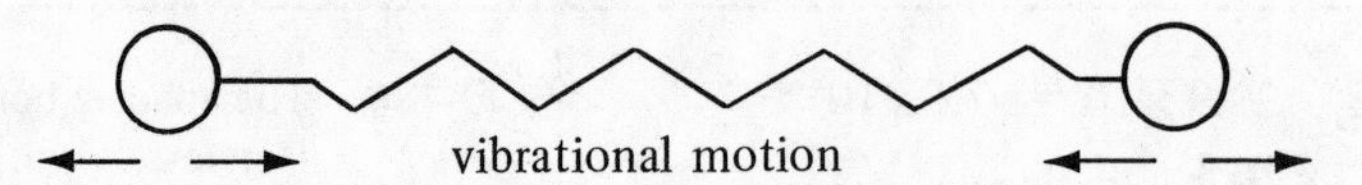


The *force constant* ($f$) of a spring is a measure of the energy required to stretch the spring.

Q-51 What is the relationship expected between vibrational energy and temperature?

A-51 As the temperature increases, the amount of vibrational energy would be expected to increase.

Q-52 Using the ball and spring model (see S-8), predict what would happen to a vibrating molecule as its energy increases.

A-52 The magnitude of the displacement would increase.

Q-53 Draw a diagram to indicate the difference in the magnitude of the displacement of a ball and spring model at $T_1$ and $T_2$.

($T_1 < T_2$)

A-53

Stretching magnitude $T_1$

Stretching magnitude $T_2$

Q-54 What does the force or energy necessary to displace the balls depend upon?

A-54 The strength of the spring (the magnitude of the force constant).

Q-55 The magnitude of the force constant for a spring is analogous to the strength of a chemical bond. As the force constant increases, what happens to the strength of the chemical bond?

A-55 The strength of the bond increases.

A large force constant results from a strong chemical bond.

Q-56 If a stronger chemical bond requires more energy to excite a molecule from one vibrational energy level to another, what is the qualitative relationship between bond strength and vibrational energy level spacings?

---

A-56 As the bond strength increases, the spacing between vibrational energy levels increases.

Q-57 A typical vibrational energy level spacing is $2.0 \times 10^{-20}$ J. What wave number of light is necessary to cause a transition of $2.0 \times 10^{-20}$ J?

---

A-57 $\Delta E = hc\bar{\nu}$

$$\bar{\nu} = \frac{\Delta E}{hc}$$

$$\bar{\nu} = \frac{2.0 \times 10^{-20}\ \text{J}}{6.63 \times 10^{-34}\ \text{J s}} \cdot \frac{\text{s}}{3.0 \times 10^{10}\ \text{cm}}$$

$$= 1.0 \times 10^{3}\ \text{cm}^{-1}$$

Q-58 In what region of the electromagnetic spectrum will vibrational energy absorption take place (refer to S-3)?

---

A-58 Infrared.

Q-59 Which of the three carbon to carbon bonds, C–C, C=C, or C≡C, would be expected to have the largest force constant? Why?

---

A-59 –C≡C–

Strengths of bonds increase single < double < triple.

Q-60 In which of the three bonds, C–C, C=C, or C≡C, would the vibrational energy level spacings be expected to be closest together? Why?

---

A-60 C–C. The bond strength is weakest for this interaction, thus, the vibrational energy levels are closest together.

Q-61 Draw energy level diagrams to represent the first two vibrational energy levels for the C–C, C=C, and C≡C bonds.

---

A-61

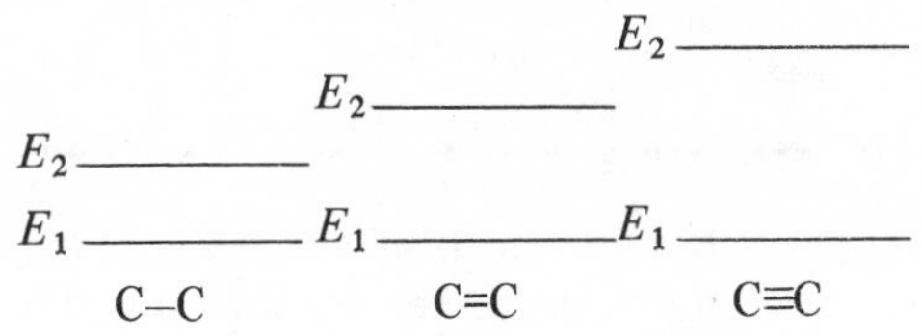

$E_2$ ——— $E_2$ ——— $E_2$ ———

$E_1$ ——— $E_1$ ——— $E_1$ ———

C–C C=C C≡C

Q-62 On the diagram below, label the vibrational, rotational, and translational energy levels.

$E$

A-62

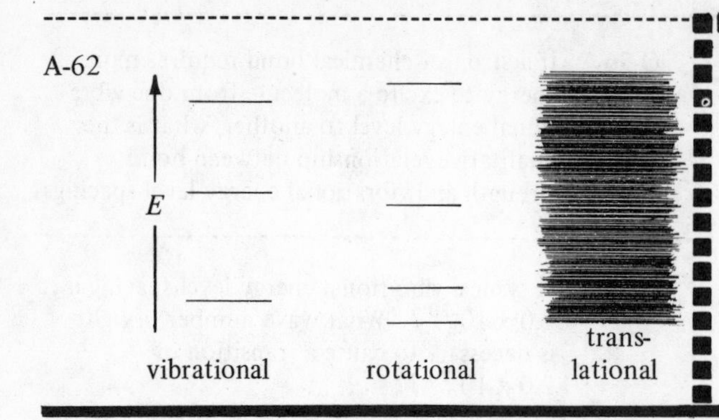


**R** Vibrational energy is the potential and kinetic energy molecules possess due to vibrational motion.

Vibrational energy is quantized and gives rise to absorption spectra in the infrared region of the electromagnetic spectrum.

The relative spacing between vibrational energy levels increases with increasing strength of the chemical bond between atoms.

S-9 Electronic energy is the energy molecules and atoms possess due to the potential and kinetic energy of their electrons. The kinetic energy of the electron is the result of motion, and potential energy arises from the interaction of the electron with nuclei and other electrons.

Q-63 Within an atom there are many energy levels available for the electrons. These energy levels are designated by four quantum numbers. What names are given these four quantum numbers? What values are allowed for each?

A-63 $n$ = principal

1,2,3, — — —

$\ell$ = angular momentum,

0,1,2, — — — — $n$-1

$m$ = magnetic,

$-\ell$ — — — 0 — — $+\ell$

$s$ = spin, ± 1/2

Q-64 The $\ell$ quantum number values are usually designated by letters. Give the letters that indicate the quantum values 0,1,2, and 3.

A-64

| $\ell$ | letter |
|---|---|
| 0 | *s* |
| 1 | *p* |
| 2 | *d* |
| 3 | *f* |

Q-65 What two quantum numbers determine the energy of an electron?

A-65 $n$, $\ell$

Q-66 Draw an electronic energy level diagram for $n$ = 1. Include energy level notations.

A66 $n$ = 1 $\ell$ = 0

$n$ =1 — — — — — 1*s*

Q-67 Draw an electronic energy level diagram for $n$ = 1 and $n$ = 2. Include energy level notations.

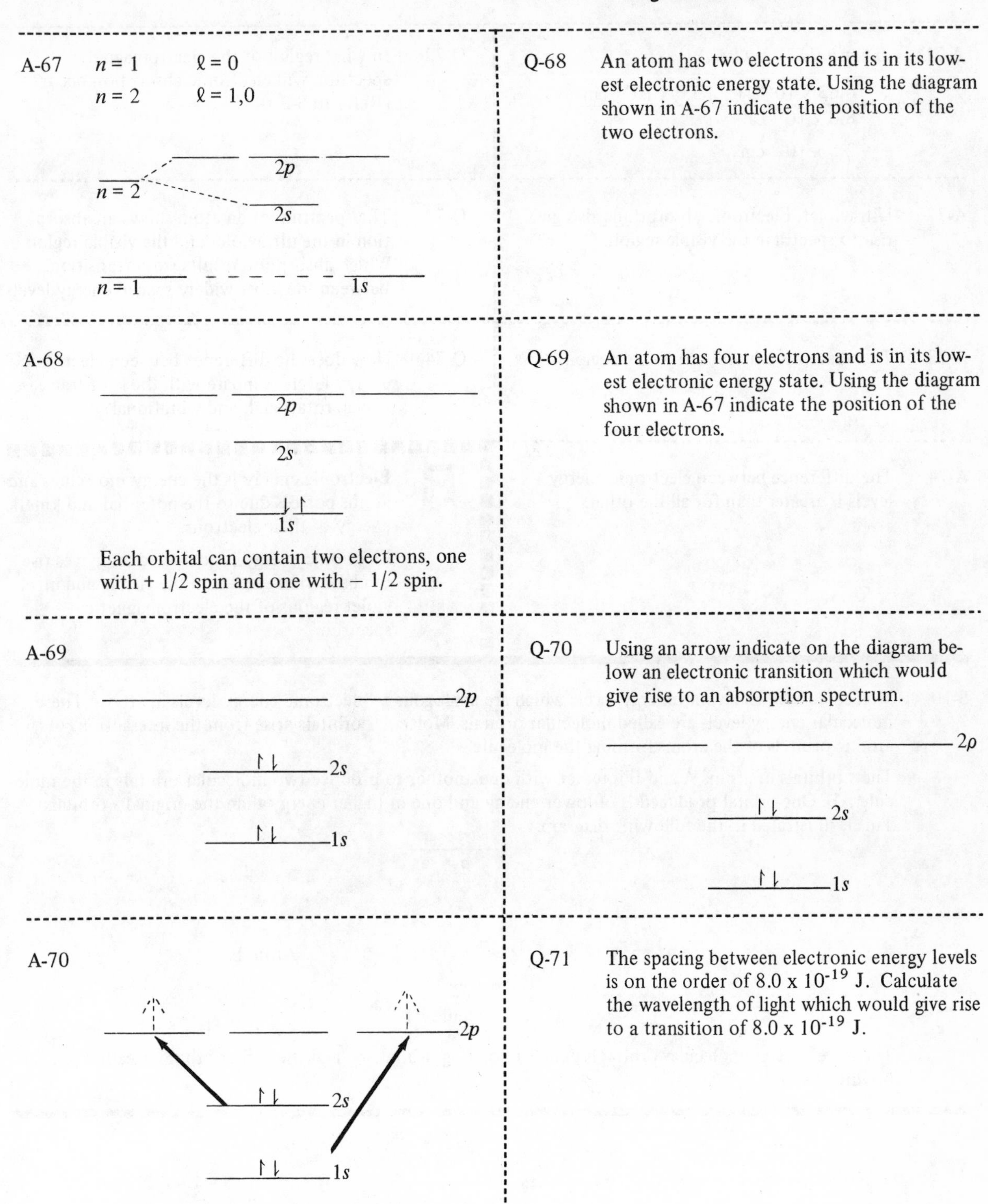

A-67 $n = 1$ $\ell = 0$

$n = 2$ $\ell = 1,0$



Q-68 An atom has two electrons and is in its lowest electronic energy state. Using the diagram shown in A-67 indicate the position of the two electrons.

A-68



Each orbital can contain two electrons, one with + 1/2 spin and one with – 1/2 spin.

Q-69 An atom has four electrons and is in its lowest electronic energy state. Using the diagram shown in A-67 indicate the position of the four electrons.

A-69



Q-70 Using an arrow indicate on the diagram below an electronic transition which would give rise to an absorption spectrum.



A-70



Either transition could be drawn. An arrow to any one of the 2*p* orbitals would be the same since each *p* orbital has the same energy.

Q-71 The spacing between electronic energy levels is on the order of $8.0 \times 10^{-19}$ J. Calculate the wavelength of light which would give rise to a transition of $8.0 \times 10^{-19}$ J.

A-71 $\Delta E = hc/\lambda \quad \lambda = hc/\Delta E$

$$\lambda = \frac{6.63 \times 10^{-34} \text{ J s} \mid 3.0 \times 10^{10} \text{ cm}}{8.0 \times 10^{-19} \text{ J} \mid \text{ s}}$$

$= 2.5 \times 10^{-5}$ cm

Q-72 In what region of the electromagnetic spectrum will electronic absorption occur? (Refer to S-3.)

A-72 Ultraviolet. Electronic absorptions also give rise to spectra in the visible region.

Q-73 The spectrum of an atom shows an absorption in the ultraviolet and the visible region. Which absorption results from transition between the most widely spaced energy levels?

A-73 The absorption in the ultraviolet region.

Q-74 How does the difference between electronic energy levels compare with those of translational, rotational, and vibrational?

A-74 The difference between electronic energy levels is greater than for all the others.

**R** Electronic energy is the energy molecules and atoms possess due to the potential and kinetic energy of their electrons.

Electronic energy is quantized and gives rise to absorption spectra in the visible and ultraviolet regions of the electromagnetic spectrum.

S-10 Molecules have electronic energy levels which are analogous to electronic energy levels in atoms. These molecular energy levels are called molecular orbitals. Molecular orbitals arise from the interactions of atomic orbitals of the atoms forming the molecule.

The *s* orbitals of atoms A and B interact with one another to produce two molecular orbitals in the molecule AB. One orbital produced is of lower energy and one of higher energy than the original *s* orbitals. This is illustrated in the following diagram:

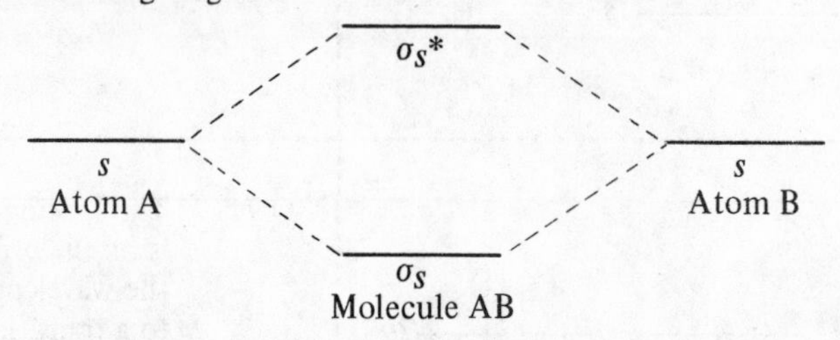


The lower energy molecular orbital is called a bonding orbital, $\sigma_s$, and the higher orbital is called an antibonding orbital, $\sigma_s$*.

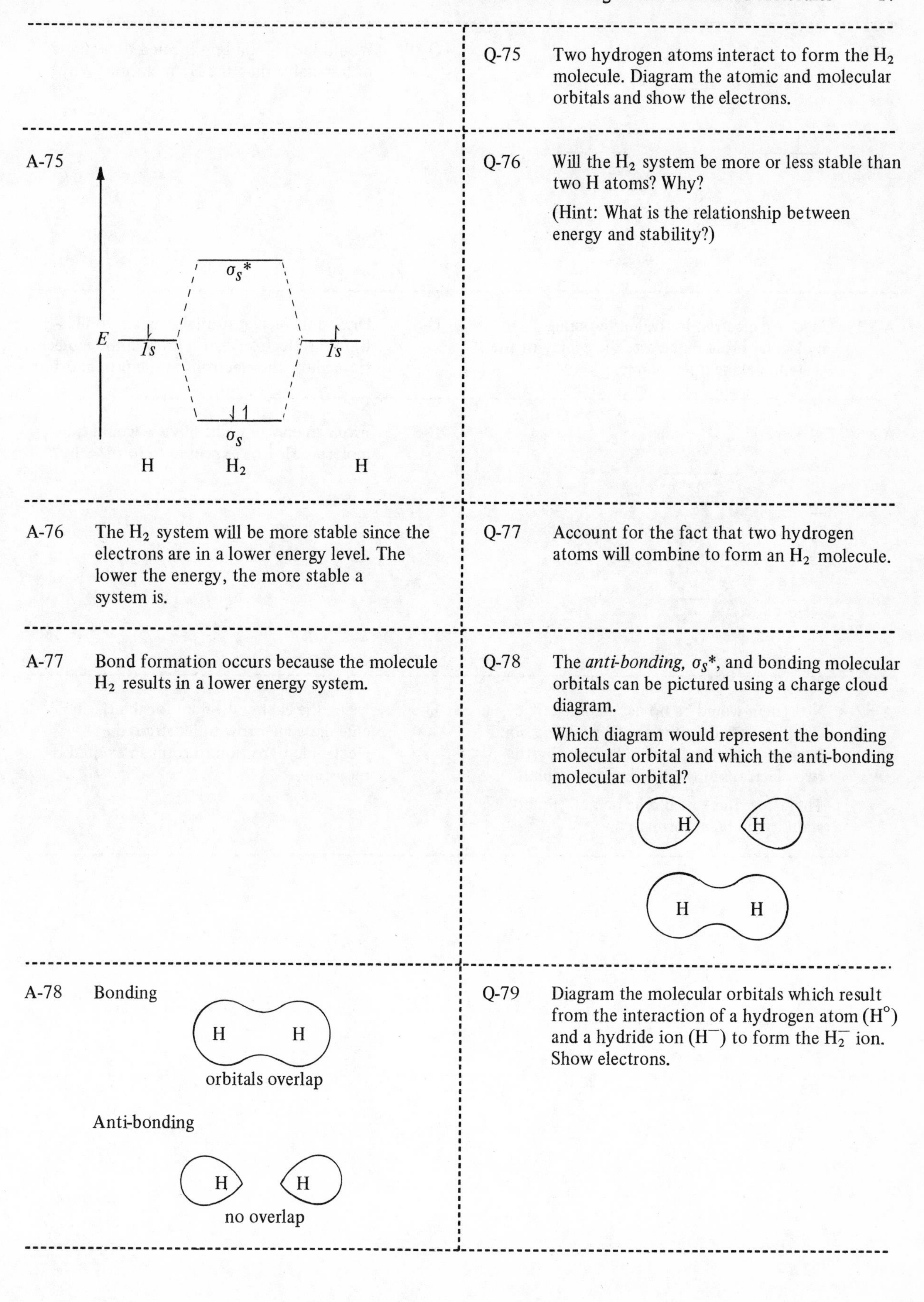

Q-75 Two hydrogen atoms interact to form the $H_2$ molecule. Diagram the atomic and molecular orbitals and show the electrons.

A-75



Q-76 Will the $H_2$ system be more or less stable than two H atoms? Why?

(Hint: What is the relationship between energy and stability?)

A-76 The $H_2$ system will be more stable since the electrons are in a lower energy level. The lower the energy, the more stable a system is.

Q-77 Account for the fact that two hydrogen atoms will combine to form an $H_2$ molecule.

A-77 Bond formation occurs because the molecule $H_2$ results in a lower energy system.

Q-78 The *anti-bonding,* $\sigma_s$*, and bonding molecular orbitals can be pictured using a charge cloud diagram.

Which diagram would represent the bonding molecular orbital and which the anti-bonding molecular orbital?



A-78 Bonding



orbitals overlap

Anti-bonding



no overlap

Q-79 Diagram the molecular orbitals which result from the interaction of a hydrogen atom (H°) and a hydride ion ($H^-$) to form the $H_2^-$ ion. Show electrons.

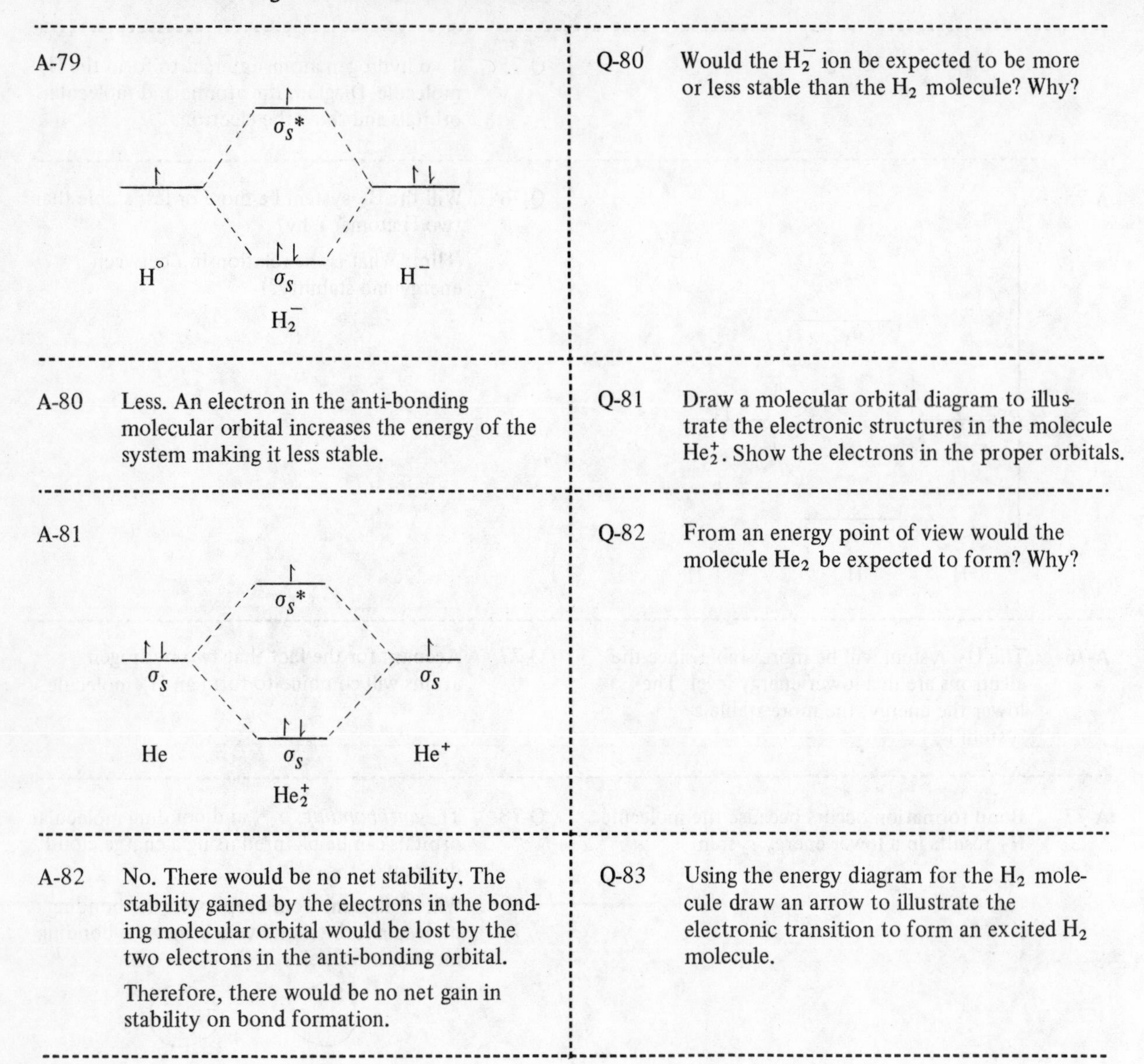


A-79

Q-80 Would the $H_2^-$ ion be expected to be more or less stable than the $H_2$ molecule? Why?

A-80 Less. An electron in the anti-bonding molecular orbital increases the energy of the system making it less stable.

Q-81 Draw a molecular orbital diagram to illustrate the electronic structures in the molecule $He_2^+$. Show the electrons in the proper orbitals.

A-81

Q-82 From an energy point of view would the molecule $He_2$ be expected to form? Why?

A-82 No. There would be no net stability. The stability gained by the electrons in the bonding molecular orbital would be lost by the two electrons in the anti-bonding orbital. Therefore, there would be no net gain in stability on bond formation.

Q-83 Using the energy diagram for the $H_2$ molecule draw an arrow to illustrate the electronic transition to form an excited $H_2$ molecule.

A-83

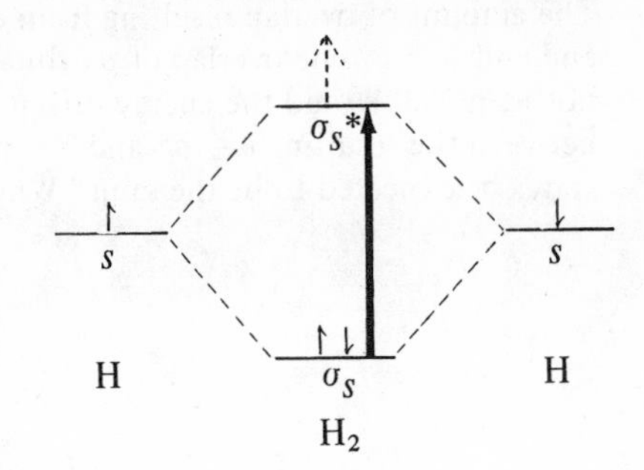


Q-84 Molecular orbitals that are formed which are symmetrical to rotation around the bond are called $\sigma$ orbitals. Molecular orbitals which are not symmetrical to rotation around the bond are called $\pi$ orbitals. Below are some contour diagrams for molecular orbitals. Label them as being $\sigma$ or $\pi$ molecular orbitals.

Atomic orbitals — Molecular orbitals

1. A (s) + B (s) → A B

2. A (p) + B (p) → A B

3. A (p) + B (p) → A B

A-84 1. and 3. would be $\sigma$ molecular orbitals.
2. would be a $\pi$ molecular orbital.

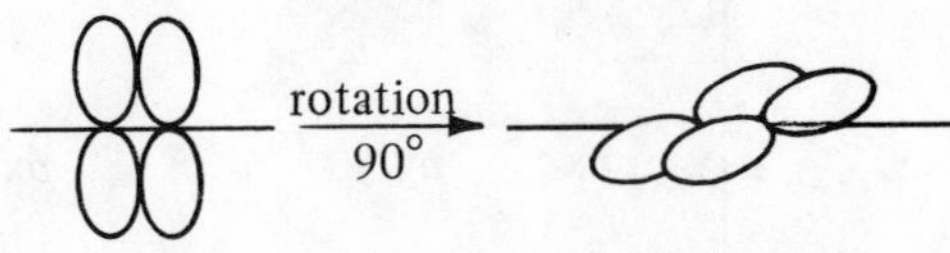


This rotation is not symmetrical about the bond axis.

Q-85 Would the overlap of an $s$ and a $p$ orbital, as described below, give rise to a $\sigma$ or $\pi$ molecular orbital?

A (s) + B (p) → A B molecular orbital

A-85 $\sigma$

Q-86 Draw an energy diagram to show the anti-bonding and bonding molecular orbitals which arise from the interaction of two $p$ orbitals end to end. Are the molecular orbitals $\sigma$ or $\pi$ orbitals?

A-86

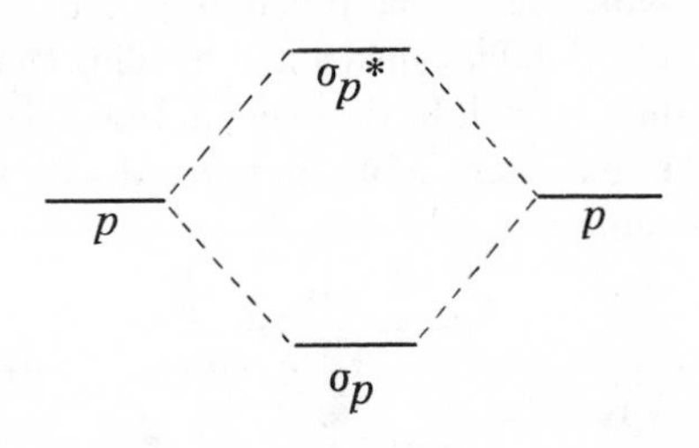


$\sigma$ molecular orbitals

Q-87 Draw an energy diagram to show the molecular orbitals arising from side-to-side interaction of $p$ orbitals. Are the molecular orbitals $\sigma$ or $\pi$ orbitals?

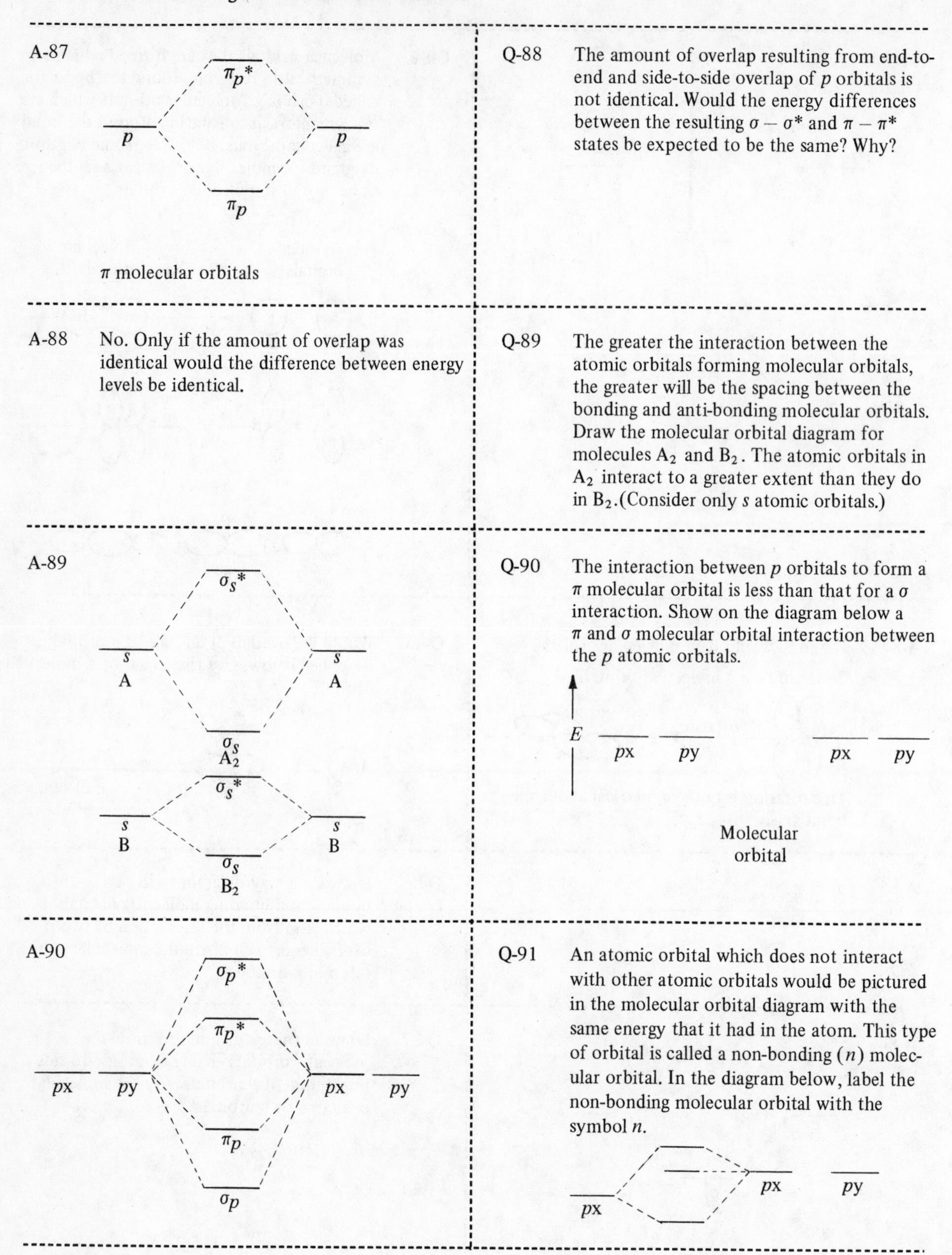


A-87

$\pi$ molecular orbitals

Q-88 The amount of overlap resulting from end-to-end and side-to-side overlap of $p$ orbitals is not identical. Would the energy differences between the resulting $\sigma - \sigma^*$ and $\pi - \pi^*$ states be expected to be the same? Why?

A-88 No. Only if the amount of overlap was identical would the difference between energy levels be identical.

Q-89 The greater the interaction between the atomic orbitals forming molecular orbitals, the greater will be the spacing between the bonding and anti-bonding molecular orbitals. Draw the molecular orbital diagram for molecules $A_2$ and $B_2$. The atomic orbitals in $A_2$ interact to a greater extent than they do in $B_2$. (Consider only $s$ atomic orbitals.)

A-89

Q-90 The interaction between $p$ orbitals to form a $\pi$ molecular orbital is less than that for a $\sigma$ interaction. Show on the diagram below a $\pi$ and $\sigma$ molecular orbital interaction between the $p$ atomic orbitals.

A-90

Q-91 An atomic orbital which does not interact with other atomic orbitals would be pictured in the molecular orbital diagram with the same energy that it had in the atom. This type of orbital is called a non-bonding ($n$) molecular orbital. In the diagram below, label the non-bonding molecular orbital with the symbol $n$.

A-91

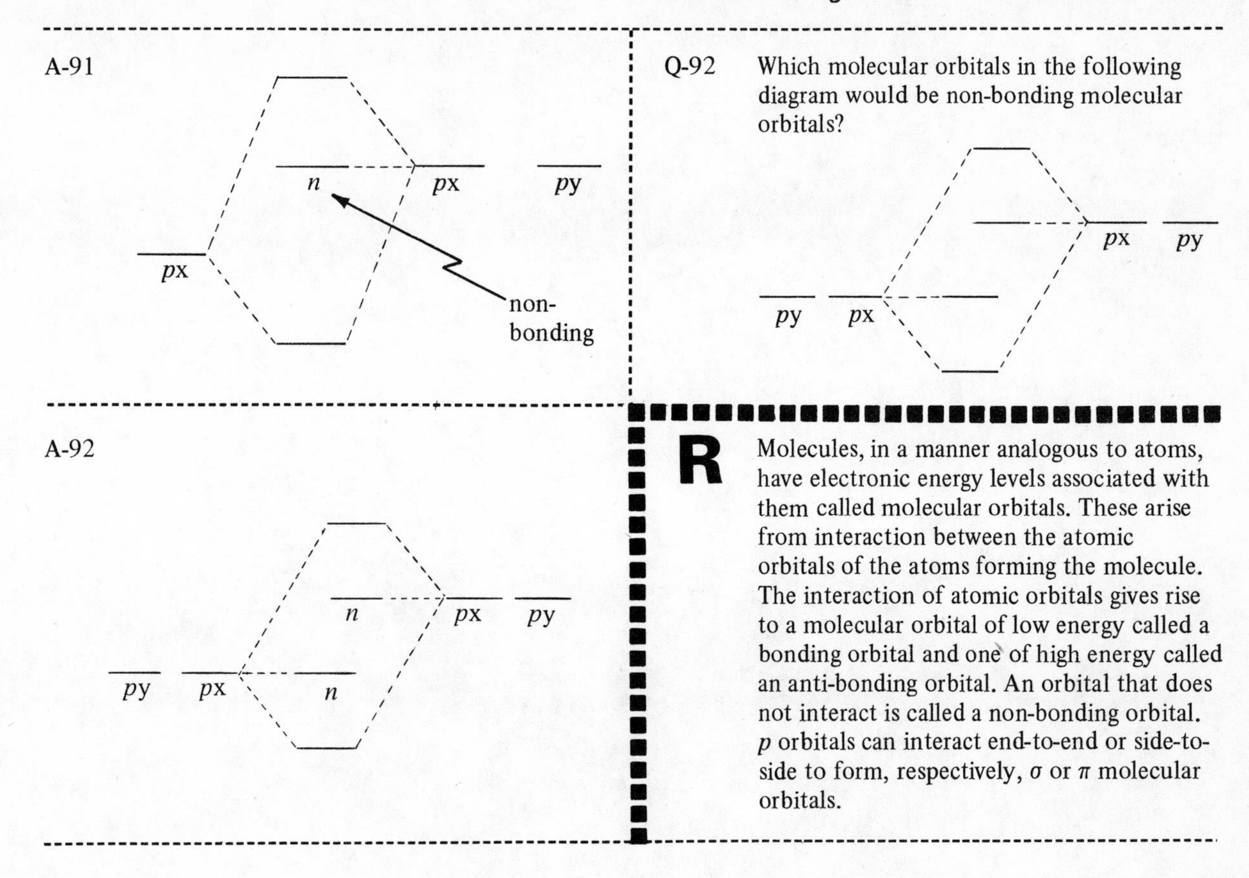


Q-92 Which molecular orbitals in the following diagram would be non-bonding molecular orbitals?

A-92

**R** Molecules, in a manner analogous to atoms, have electronic energy levels associated with them called molecular orbitals. These arise from interaction between the atomic orbitals of the atoms forming the molecule. The interaction of atomic orbitals gives rise to a molecular orbital of low energy called a bonding orbital and one of high energy called an anti-bonding orbital. An orbital that does not interact is called a non-bonding orbital. $p$ orbitals can interact end-to-end or side-to-side to form, respectively, $\sigma$ or $\pi$ molecular orbitals.

## REFERENCES

**Introductory**:

Barrow, Gordon M. 1964. *The Structure of Molecules.* New York: W. A. Benjamin.
Companion, Audrey L. 1964. *Chemical Bonding.* New York: McGraw-Hill.
Gray, Harry B. 1964. *Electrons and Chemical Bonding.* New York: W. A. Benjamin.

**Intermediate**:

Barrow, Gordon M. 1962. *Introduction to Molecular Spectroscopy.* New York: McGraw-Hill.
Hanna, Melvin W. 1969. Quantum Mechanics in Chemistry, 2nd edition. New York: W. A. Benjamin.
Roberts, John D. 1962. *Notes on Molecular Orbital Calculation.* New York: W. A. Benjamin.

# Chapter 2 ULTRAVIOLET AND VISIBLE SPECTRA

Ultraviolet and visible spectroscopy is useful in determining structures of organic molecules and in quantitative analysis. This chapter will deal mainly with the theory of electronic transitions and with the elucidation of structure from ultraviolet spectral data.

**PART I: Electronic Transitions**

After completing this section you should

a) be able to determine the types of chromophores present in a molecule and determine which chromophore gives rise to the lowest energy transition
b) be able to predict approximate wavelength regions for different types of transitions
c) understand why conjugated systems absorb at longer wavelength
d) be able to predict solvent effects on $\pi \rightarrow \pi^*$ and $n \rightarrow \pi^*$ transitions
e) be able to calculate $\mathcal{E}_{max}$ from the Lambert-Beer relationship.

**PART II: Use of Ultraviolet Spectra in Structural Determinations**

After completing this section you should

a) be able to predict $\lambda_{max}^{hexane}$ for conjugated dienes
b) be able to predict $\lambda_{max}^{ethanol}$ for $\alpha$, $\beta$-unsaturated ketones
c) be able to predict structures of compounds with the aid of ultraviolet spectra.

## Ultraviolet and Visible Spectra
## PART I: Electronic Transitions

S-1 The electronic spectrum of a molecule results from a transition between two different molecular electronic energy levels. A transition between two states will be denoted by using the notation illustrated below:

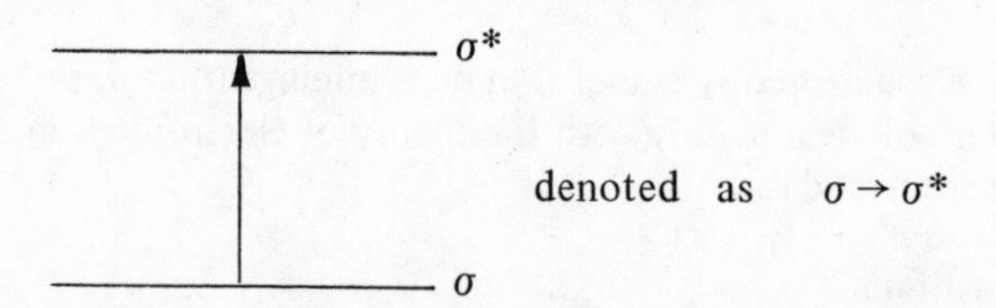


Q-1 Two atoms, each containing a single half-filled *s* orbital, are allowed to interact so as to form a $\sigma$ bond. Draw an energy diagram for this process and label the bonding and anti-bonding molecular orbitals. Show the electrons in the lowest energy state.

A-1

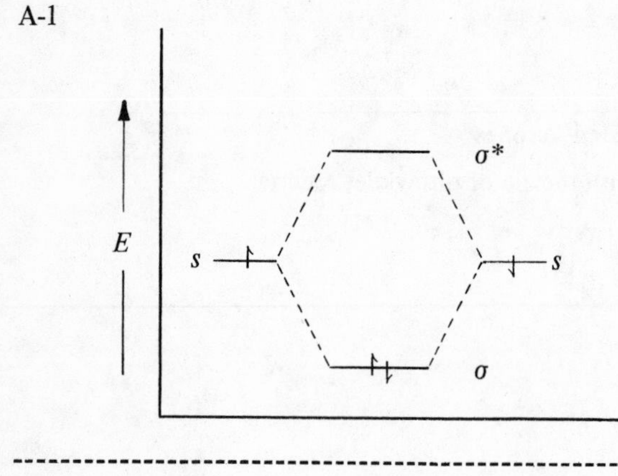


Q-2 Two carbon atoms interact so as to form a $\pi$ bond. Draw an energy diagram for this process showing only *p* orbital interactions. Label the bonding ( $\pi$ ) and anti-bonding ( $\pi^*$ ) molecular orbitals.

A-2

E

$\pi^*$

p p

$\pi$

Q-3 Two atoms interact to form a bond. A *p* orbital of atom A interacts side-to-side with a *p* orbital of atom B. A second *p* orbital of atom B does not interact. Sketch an energy level diagram for interaction between A and B and label all the orbitals.

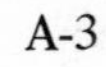

A-3

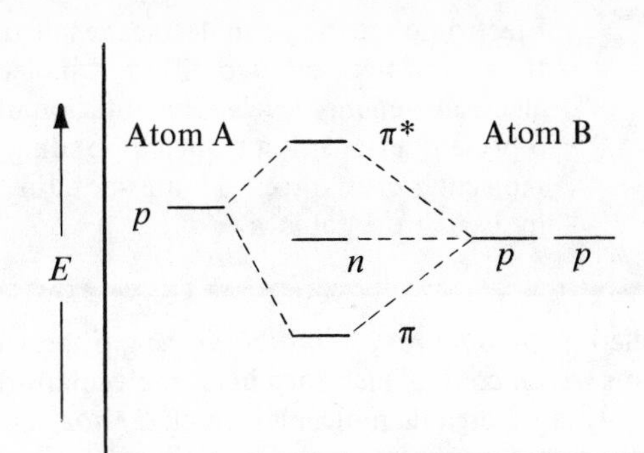


The second $p$ orbital of atom B does not interact; therefore, its energy remains the same.

Q-4 A total of four electrons must be added to the molecular orbitals formed by the interaction described in Q-3. Show where these four electrons should be placed when the molecule is in the ground state (lowest energy state).

A-4

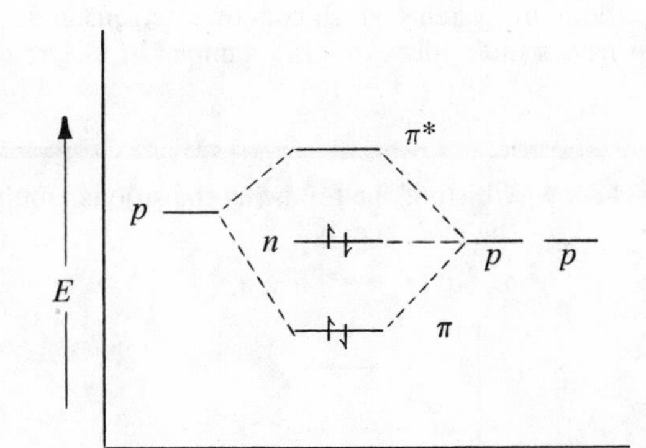


Q-5 The energy level diagram for ethylene is given below. What kind of electronic transitions can the highest energy electrons undergo?

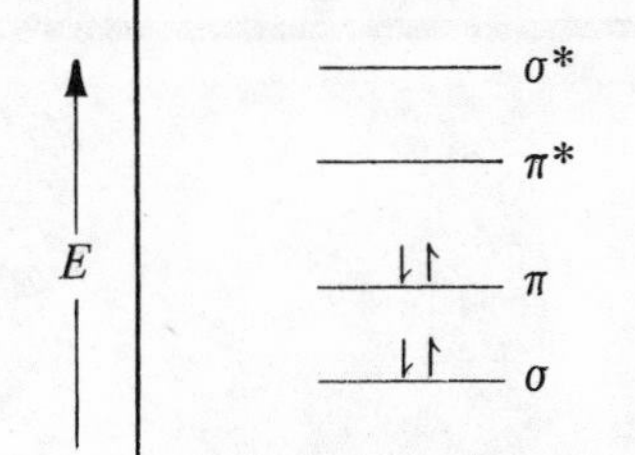


A-5 $\pi \rightarrow \pi^*$

$\pi \rightarrow \sigma^*$

Q-6 The energy level diagram for the carbonyl group is given below. What kind of electronic transitions can the highest energy electrons undergo?

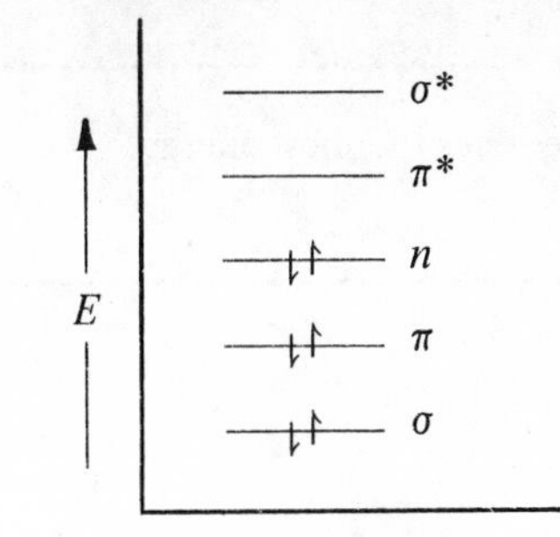


A-6 $n \rightarrow \pi^*$

$n \rightarrow \sigma^*$

Q-7 Which of the two possible transitions for the non-bonding electrons in the carbonyl group (see Q-6) will require the greatest energy? Which transition will require the longest wavelength of light?

A-7 The $n \rightarrow \sigma^*$ transition will require the most energy. Because wavelength is inversely proportional to energy, the $n \rightarrow \pi^*$ transition will require the longer wavelength of light (lower energy).

**R** Electronic spectra of molecules result from transitions between two different molecular electronic energy levels. The notation used to represent a transition from a $\sigma$ bonding molecular orbital to a $\sigma^*$ anti-bonding molecular orbital is $\sigma \rightarrow \sigma^*$.

---

S-2 Systems responsible for the absorption of light are called *chromophores or chromophoric* groups. Chromophores which give rise to $\sigma \rightarrow \sigma^*$ transitions are systems which contain electrons in $\sigma$ molecular orbitals. Compounds containing only $\sigma$ molecular orbitals are saturated organic molecules which *do not* contain atoms with lone pair electrons. Examples of $\sigma \rightarrow \sigma^*$ type chromophores are $\geq$C – C$\leq$ and $\geq$C – H.

Chromophores which give rise to $n \rightarrow \sigma^*$ transitions are systems which contain electrons in non-bonding and $\sigma$ molecular orbitals. Compounds containing only $n$ and $\sigma$ molecular orbitals are saturated organic molecules which contain one or more atoms with lone pair electrons. Examples of $n \rightarrow \sigma^*$ type chromophores are $\geq$C – Ö –, $\geq$C – S̈ –, $\geq$C – N̈ –, and $\geq$C – C̈l: .

Chromophores which give rise to $\pi \rightarrow \pi^*$ type transitions are systems which contain electrons in $\pi$ molecular orbitals. Unsaturated organic compounds have $\pi$ molecular orbitals. Examples of $\pi \rightarrow \pi^*$ type chromophores are >C = C< and – C ≡ C –.

---

Q-8 Which of the following transitions requires the most energy?

$\sigma \rightarrow \sigma^*$, $n \rightarrow \sigma^*$, $\pi \rightarrow \pi^*$

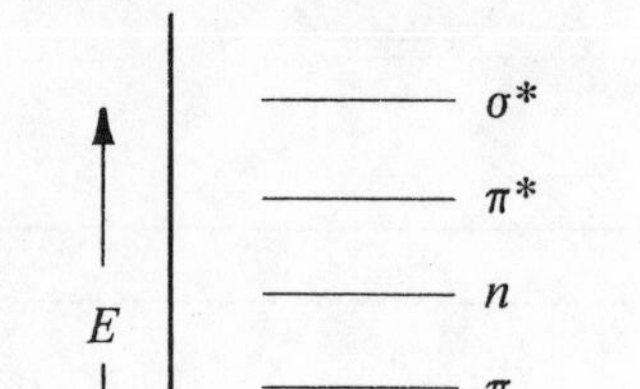


---

A-8 $\sigma \rightarrow \sigma^*$ requires the most energy.

Q-9 What types of transitions are possible in cyclopentene ($C_5H_8$)?

---

A-9
$\sigma \rightarrow \sigma^*$
$\sigma \rightarrow \pi^*$
$\pi \rightarrow \pi^*$
$\pi \rightarrow \sigma^*$

Q-10 What *chromophore* in cyclopentene is responsible for the lowest energy transition?

---

A-10 >C = C<
The $\pi \rightarrow \pi^*$ is the lowest energy transition.

Q-11 Cyclopentene absorbs light near 190 nm. Absorption at higher wavelength does not occur. What type of transition is responsible for the observed absorption? Illustrate by use of an energy level diagram.

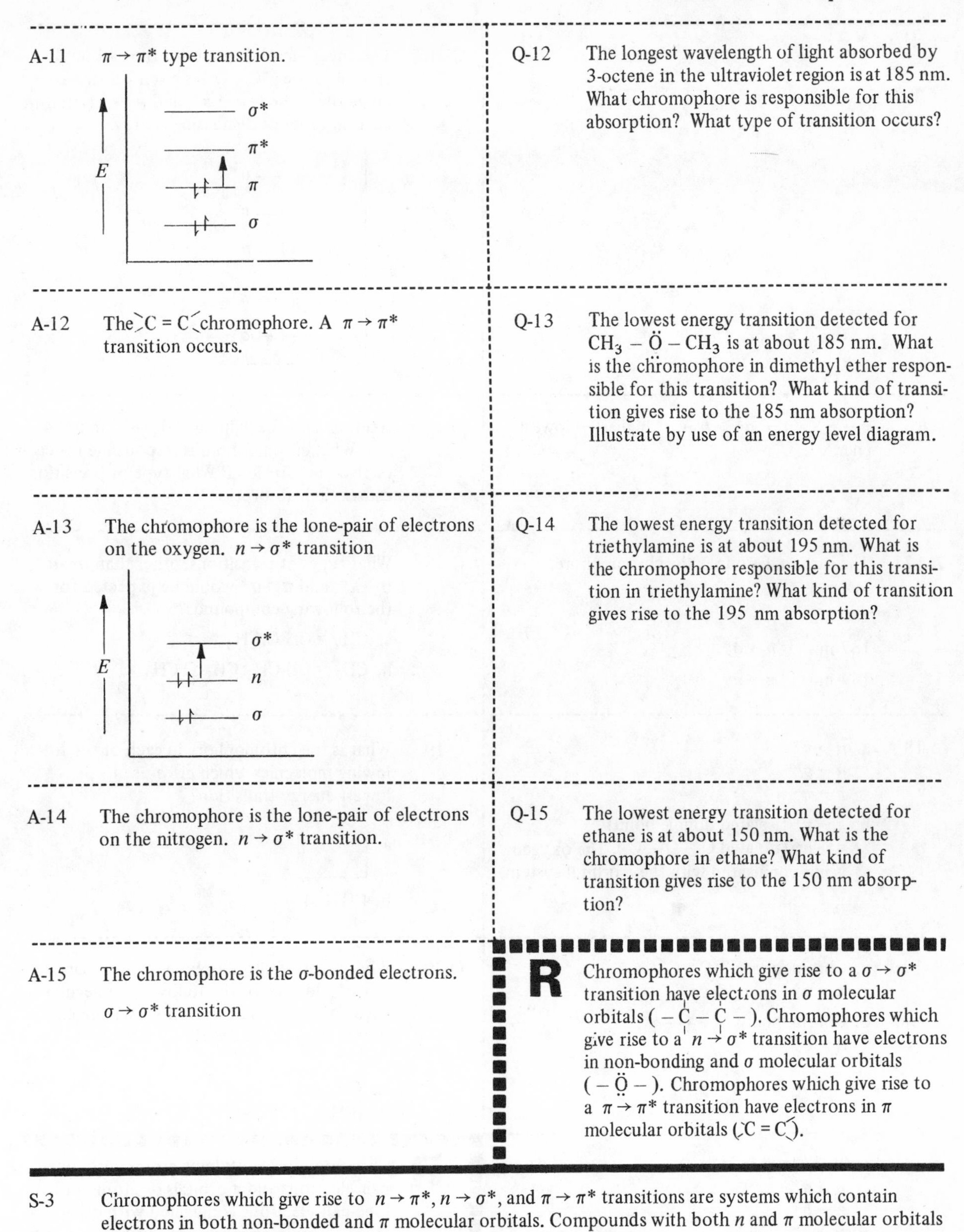

A-11 $\pi \rightarrow \pi^*$ type transition.



Q-12 The longest wavelength of light absorbed by 3-octene in the ultraviolet region is at 185 nm. What chromophore is responsible for this absorption? What type of transition occurs?

A-12 The $\rangle C = C \langle$ chromophore. A $\pi \rightarrow \pi^*$ transition occurs.

Q-13 The lowest energy transition detected for $CH_3 - \ddot{O} - CH_3$ is at about 185 nm. What is the chromophore in dimethyl ether responsible for this transition? What kind of transition gives rise to the 185 nm absorption? Illustrate by use of an energy level diagram.

A-13 The chromophore is the lone-pair of electrons on the oxygen. $n \rightarrow \sigma^*$ transition



Q-14 The lowest energy transition detected for triethylamine is at about 195 nm. What is the chromophore responsible for this transition in triethylamine? What kind of transition gives rise to the 195 nm absorption?

A-14 The chromophore is the lone-pair of electrons on the nitrogen. $n \rightarrow \sigma^*$ transition.

Q-15 The lowest energy transition detected for ethane is at about 150 nm. What is the chromophore in ethane? What kind of transition gives rise to the 150 nm absorption?

A-15 The chromophore is the $\sigma$-bonded electrons.

$\sigma \rightarrow \sigma^*$ transition

**R** Chromophores which give rise to a $\sigma \rightarrow \sigma^*$ transition have electrons in $\sigma$ molecular orbitals ($-\overset{|}{\underset{|}{C}}-\overset{|}{\underset{|}{C}}-$). Chromophores which give rise to a $n \rightarrow \sigma^*$ transition have electrons in non-bonding and $\sigma$ molecular orbitals ($-\ddot{\underset{..}{O}}-$). Chromophores which give rise to a $\pi \rightarrow \pi^*$ transition have electrons in $\pi$ molecular orbitals ($\rangle C = C \langle$).

S-3 Chromophores which give rise to $n \rightarrow \pi^*$, $n \rightarrow \sigma^*$, and $\pi \rightarrow \pi^*$ transitions are systems which contain electrons in both non-bonded and $\pi$ molecular orbitals. Compounds with both $n$ and $\pi$ molecular orbitals are those containing atoms with lone pair electrons and a $\pi$ orbital *or* atoms with lone-pair electrons conjugated with other atoms containing $\pi$ orbitals. Examples of these types of chromophores are $\rangle C = \ddot{\underset{..}{O}}$ and $\rangle C = C \langle - \ddot{\underset{..}{O}} -$.

Q-16 The energy level diagram for the carbonyl chromophore ($\gt C = \ddot{O}$ ) is given below. Arrange the $n \rightarrow \sigma^*$, $\pi \rightarrow \pi^*$, and $n \rightarrow \pi^*$ transitions in order of decreasing energy.

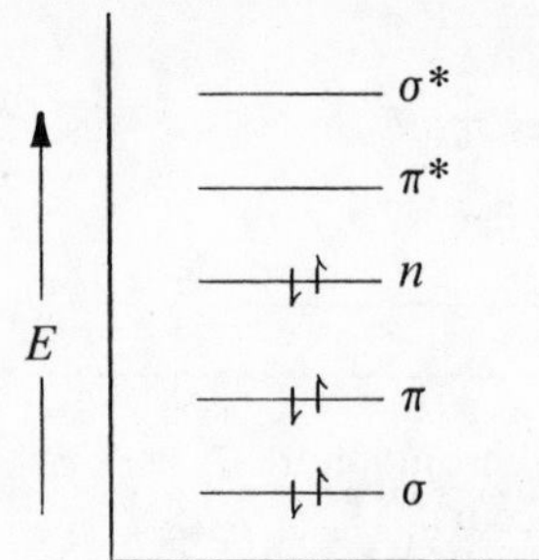


A-16 $\pi \rightarrow \pi^* > n \rightarrow \sigma^* > n \rightarrow \pi^*$ highest → lowest energy.

Q-17 Acetone absorbs light at 280, 187, and 154 nm. What chromophore is responsible for each of these absorptions? What type of transition causes each absorption?

A-17 The carbonyl group is the chromophore.

280 nm $n \rightarrow \pi^*$

187 nm $n \rightarrow \sigma^*$

154 nm $\pi \rightarrow \pi^*$

Q-18 What types of transitions, other than $\sigma \rightarrow \sigma^*$, $\sigma \rightarrow \pi^*$, and $\pi \rightarrow \sigma^*$ would be expected for the following compounds?

a. $CH_2 = CH\ \ddot{O}\ CH_3$

b. $CH_2 = CH\ CH_2\ CH_2\ \ddot{O}\ CH_3$

A-18 a. $n \rightarrow \pi^*$
$n \rightarrow \sigma^*$
$\pi \rightarrow \pi^*$

b. $\pi \rightarrow \pi^*$ (isolated double bond)
$n \rightarrow \sigma^*$ (isolated $C - \ddot{O} - C$ ). The oxygen is not conjugated with the $\pi$ orbital system.

Q-19 What is the chromophore in each of the following molecules which gives rise to the lowest energy transition?

a. 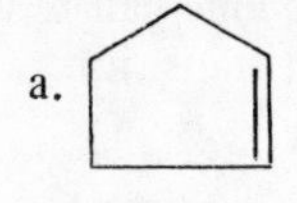

b. $CH_3OH$

A-19 a. Chromophore $\gt C = C \lt$.
$\pi \rightarrow \pi^*$

b. Chromophore $-\ddot{O}-$.
$n \rightarrow \sigma^*$

Q-20 What types of chromophores, other than $\sigma \rightarrow \sigma^*$, do each of the following molecules have? What types of transitions will be expected to occur in each?

a. $CH_3\overset{O}{\overset{\|}{C}}H$

b. $(CH_3)_2\ \ddot{N}\ CH{=}CH_2$

A-20 a. Chromophore $\gt C = O$.
$n \rightarrow \pi^*$
$n \rightarrow \sigma^*$
$\pi \rightarrow \pi^*$

b. Chromophore $-\ddot{N} - C = C \lt$ system.
$n \rightarrow \pi^*$
$n \rightarrow \sigma^*$
$\pi \rightarrow \pi^*$

**R** Chromophores containing atoms with lone-pair electrons and $\pi$ orbitals *or* atoms with lone-pair electrons conjugated with other atoms containing $\pi$ orbitals give rise to $n \rightarrow \pi^*$, $n \rightarrow \sigma^*$, and $\pi \rightarrow \pi^*$ transitions.

S-4 In general compounds which have only $\sigma \rightarrow \sigma^*$ transitions absorb light near 150 nm while compounds with $n \rightarrow \sigma^*$ and $\pi \rightarrow \pi^*$ transitions (due to unconjugated chromophores) absorb light close to 200 nm. Compounds with $n \rightarrow \pi^*$ transitions absorb light in the quartz ultraviolet (200-400 nm). The ultraviolet region below 200 nm (vacuum ultraviolet) will not be considered in detail since it is a difficult area in which to obtain spectra and yields little information regarding the structure of organic molecules.

---

Q-21 A compound whose energy diagram appears below showed absorptions at 190 nm and 300 nm. What type of transition is associated with each absorption?

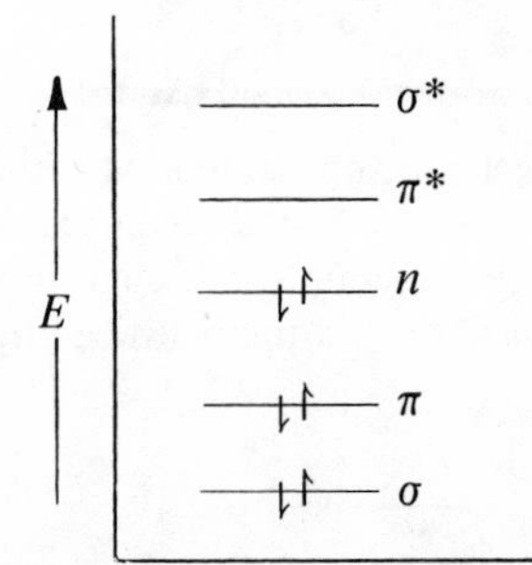


A-21 300 nm $n \rightarrow \pi^*$

190 nm $n \rightarrow \sigma^*$

Q-22 A compound whose energy diagram is shown below absorbed light at 185 nm but had no absorption bands in the quartz ultraviolet. What type of transition was the absorption due to?

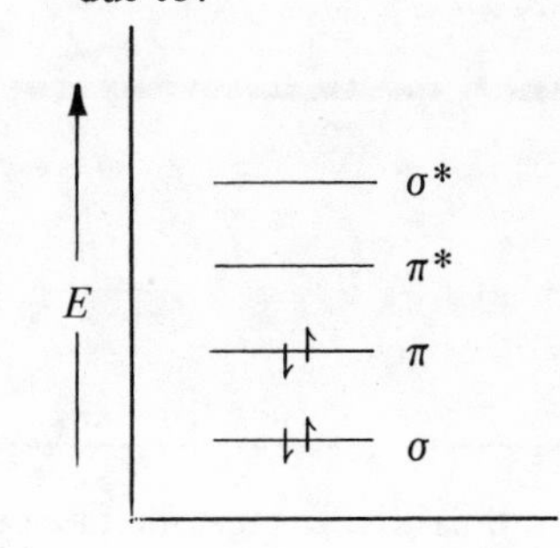


A-22 $\pi \rightarrow \pi^*$

Q-23 Acetaldehyde ($CH_3COH$) has absorption peaks at 160, 180, and 290 nm. What type of transition is responsible for each of these absorptions?

A-23 290 $n \rightarrow \pi^*$

180 $n \rightarrow \sigma^*$

160 $\pi \rightarrow \pi^*$

Q-24 What transition is responsible for each of the three absorptions in the spectrum of acetone ($CH_3COCH_3$)? Will these absorptions occur in the quartz or in the vacuum region of the ultraviolet?

A-24 $n \rightarrow \pi^*$ Quartz (actual 279)
$n \rightarrow \sigma^*$ Vacuum (actual 189)
$\pi \rightarrow \pi^*$ Vacuum (actual 166)

**R**

| Transition | Approximate region of absorption |
|---|---|
| $\sigma \rightarrow \sigma^*$ | 150 nm |
| $n \rightarrow \sigma^*$ | less than 200 nm |
| $\pi \rightarrow \pi^*$ (unconjugated) | less than 200 nm |
| $n \rightarrow \pi^*$ | 300 nm |

Vacuum ultraviolet region is below 200 nm. Quartz ultraviolet region is between 200 and 400 nm.

S-5 Different electronic environments within molecules will affect the degree of interaction between orbitals. Consequently, the energy differences between similar electronic levels in two different molecules will not be identical. For example, the environments about the chromophores in compounds A and B are not the same because of the nature of the groups attached. The energy level diagrams below illustrate this effect.

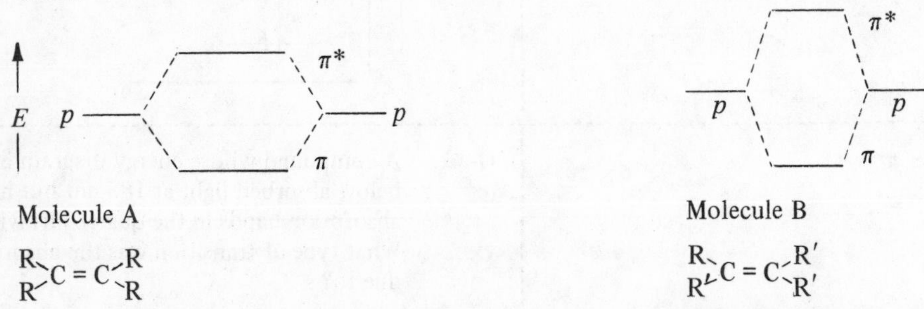


Molecule A

$R_2C{=}CR_2$

Molecule B

$R_2C{=}CR'_2$

Q-25 The energy diagrams for two different carbon-carbon double bonds are shown in S-5. In which compound is there the greatest interaction between $p$ orbitals? Which bond is the longer?

A-25 Compound B has the greatest interaction between $p$ orbitals. The $\pi$ bond in compound B will be the shorter bond (see Q-89, Chapter I).

Q-26 Which of the two compounds described in S-5 will have the lower energy $\pi \rightarrow \pi^*$ transition?

A-26 The $\pi \rightarrow \pi^*$ transition in compound A will require the lower energy.

Q-27 The absorptions due to the $\pi \rightarrow \pi^*$ transitions in ethylene and 3-octene occur at 163 and 185 nm, respectively. Why aren't the absorptions at the same wavelength?

A-27 Both absorptions are due to $\pi \rightarrow \pi^*$ transitions. Because the double bonds have different groups attached, the electronic environments of the chromophores are different. Thus, the energy difference between the $\pi$ and $\pi^*$ states are not the same.

Q-28 Is the interaction of $p$ orbitals greater in ethylene or 3-octene?

A-28 The interaction is greater in ethylene. The lower wavelength absorption indicates the greater energy difference between $\pi$ and $\pi^*$ states.

Q-29 The $\pi \rightarrow \pi^*$ transitions for two compounds are given below. Why don't the two transitions absorb at the same wavelength?

| | |
|---|---|
| $CH_3C \equiv CH$ | 187 nm |
| $CH_3C(=O)CH_3$ | 154 nm |

A-29 While both absorptions are due to a $\pi \rightarrow \pi^*$ type transition, the electronic environment of the carbonyl group is not identical to the environment about the triple bond. Thus, the transitions require different energy.

Q-30 For the three compounds given below, can the observed absorption bands be due to the same type of transition? Explain.

| | |
|---|---|
| $CH_3-Cl$ | 172 nm |
| $CH_3-I$ | 258 nm |
| $CH_3-Br$ | 204 nm |

A-30 Yes. All the absorptions are due to $n \rightarrow \sigma^*$ transitions. Because the electronegativity of each halogen is different, the electronic environment in each molecule will be different; this results in a different degree of interaction of molecular orbitals and, consequently, a difference in energy between the $n$ and $\sigma^*$ states.

**R** Different electronic environments within molecules affect the degree of interaction between orbitals. As interactions between orbitals increase the length of the bond formed becomes less, and the difference in energy between the bonding and anti-bonding molecular orbitals increases.

S-6 The intensity of the absorption due to a $\pi \rightarrow \pi^*$ type transition is always 10 to 100 times more intense than $n \rightarrow \sigma^*$ or $n \rightarrow \pi^*$ type absorptions. The spectrum of a compound which has both $n \rightarrow \pi^*$ and $\pi \rightarrow \pi^*$ transitions is shown below. The position of maximum absorption of each band (called $\lambda_{max}$) corresponds to the wavelength of light necessary for the transitions. The width of the bands is, in part, due to instrumentation.

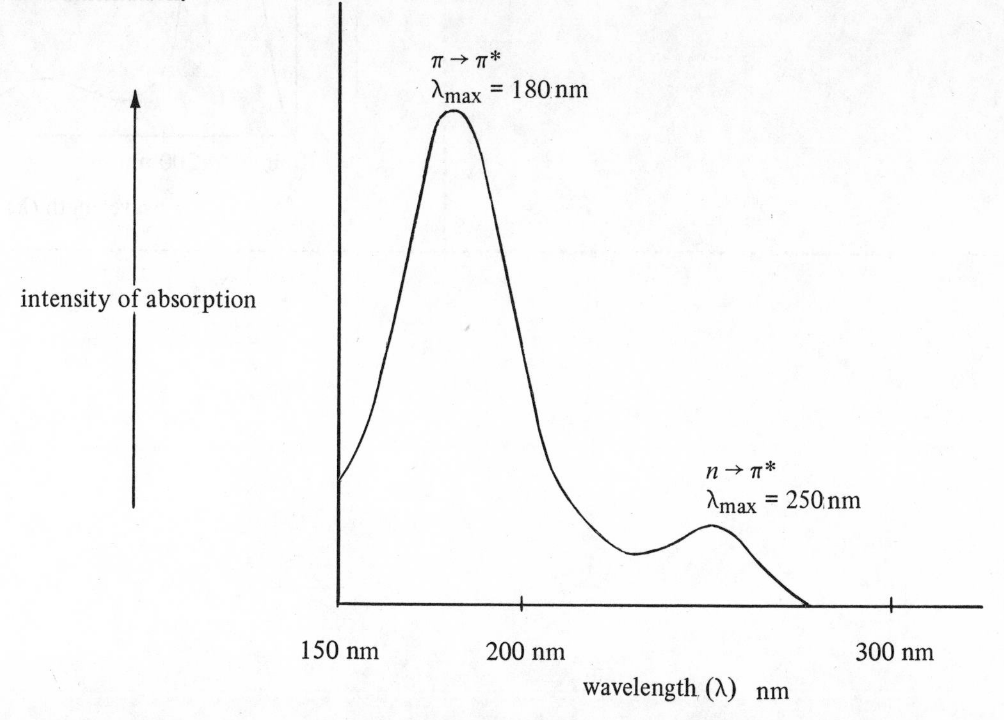

Q-31 For the following spectrum, indicate the $\lambda_{max}$ for each peak.

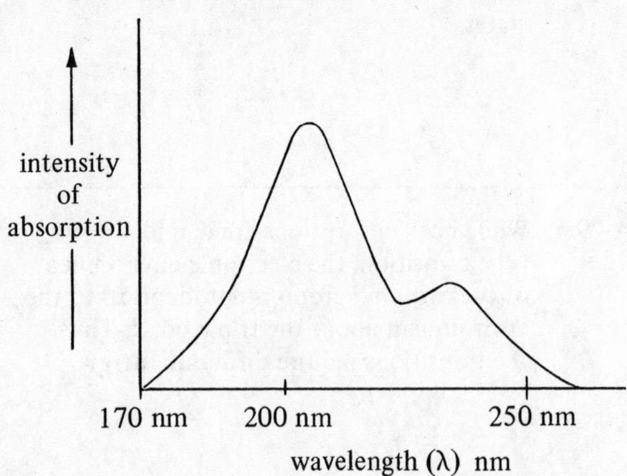


A-31 $\lambda_{max}$ = 205 nm.
$\lambda_{max}$ = 235 nm.

Q-32 The spectrum for a compound which has both $n \rightarrow \pi^*$ and $\pi \rightarrow \pi^*$ transitions is given below. Give $\lambda_{max}$ for the band resulting from each type of transition.

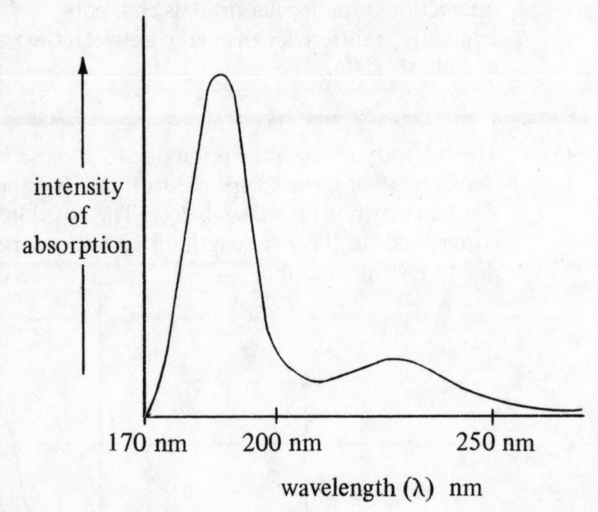

A-32 $n \to \pi^*$ $\lambda_{max}$ = 230 nm.
$\pi \to \pi^*$ $\lambda_{max}$ = 190 nm.

Q-33 A compound was known to be either a saturated amine $\rangle N - \overset{|}{\underset{|}{C}} - \overset{|}{\underset{|}{C}} - \overset{|}{\underset{|}{C}} -$ or an unsaturated amine $\rangle N - \overset{|}{\underset{|}{C}} - \overset{|}{C} = C \langle$. The spectrum of the compound is given below. What structure is indicated? Justify your answer.

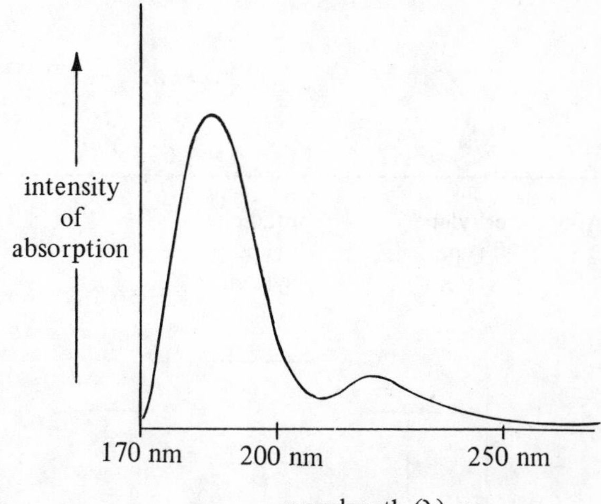


A-33 The structure $\rangle N - \overset{|}{\underset{|}{C}} - \overset{|}{C} = C \langle$ is indicated. The presence of two peaks is indicative of two types of transitions. The strong $\lambda_{max}$ of 185 nm is due to a $\pi \to \pi^*$ transition while the weak $\lambda_{max}$ at 220 nm is due to the $n \to \sigma^*$ transition.

**R** The intensity of $\pi \to \pi^*$ transitions is 10 to 100 times as great as the intensity of $n \to \sigma^*$ and $n \to \pi^*$ type absorptions. The position of maximum absorption of a band is denoted as $\lambda_{max}$.

S-7 In conjugated systems, such as $\rangle C = \overset{|}{C} - \overset{|}{C} = C \langle$, $\pi$ orbitals from each double bond interact to form a new set of bonding and anti-bonding orbitals. This interaction is illustrated below.

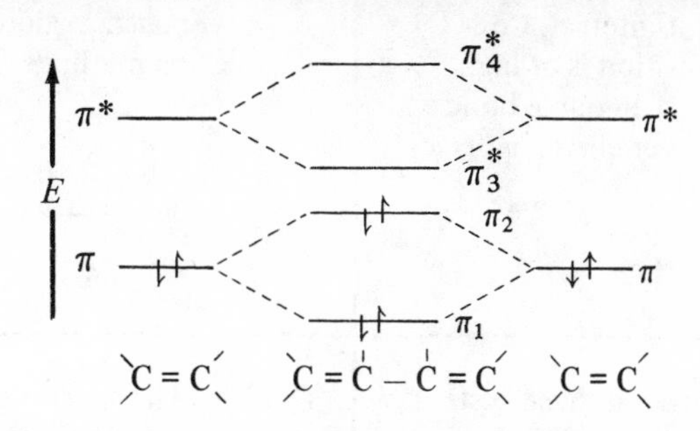


As the conjugated system in a molecule becomes longer (involves more atoms with $\pi$ bonds), the difference in energy between the ground states and the excited states for the $\pi \to \pi^*$ transitions becomes less. Consequently, as the conjugated system increases in length the energy required for a $\pi \to \pi^*$ transition becomes less and absorption will occur at longer wavelength.

**Q-34** For the conjugated system diagrammed in S-7 what is the lowest energy transition possible?

---

**A-34** $\pi_2 \rightarrow \pi_3{}^*$

**Q-35** The $\lambda_{max}$ for ethylene is about 185 nm while $\lambda_{max}$ for 1,3-butadiene ($CH_2 = CH - CH = CH_2$) is 217 nm.

Using the energy level diagram shown in S-7 explain why butadiene absorbs at longer wavelength.

---

**A35**

ethylene type system | butadiene type system

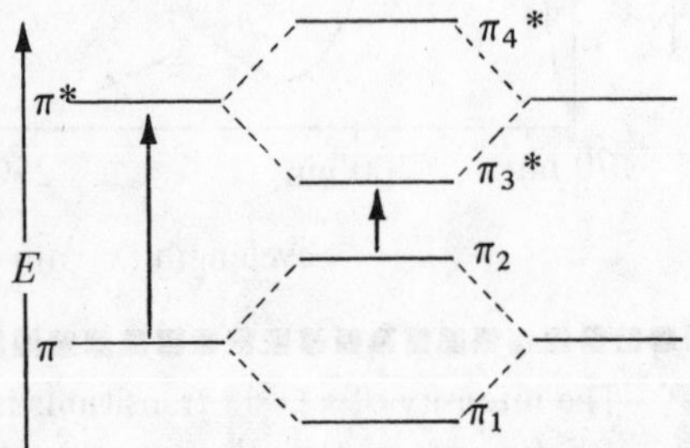

$\pi^*$, $\pi$, $E$ | $\pi_4{}^*$, $\pi_3{}^*$, $\pi_2$, $\pi_1$

The $\pi \rightarrow \pi^*$ transition in ethylene requires higher energy than the $\pi_2 \rightarrow \pi_3{}^*$ transition in butadiene.

**Q-36** Why doesn't 1,4-pentadiene absorb light above 200 nm in the ultraviolet region of the spectrum?

$CH_2{=}CHCH_2CH{=}CH_2$ $\lambda_{max}$ = 175 nm

$CH_2{=}CHCH{=}CH_2$ $\lambda_{max}$ = 217 nm

---

**A-36** In 1,4-pentadiene the two double bonds are *not* conjugated and do not interact. Consequently, the $\pi \rightarrow \pi^*$ transition is of high energy. In 1,3 butadiene the double bonds are conjugated and the lower energy $\pi_2 \rightarrow \pi_3{}^*$ transition is possible.

**Q-37** Predict the difference in the spectrum in the *quartz* region of the ultraviolet for the two compounds

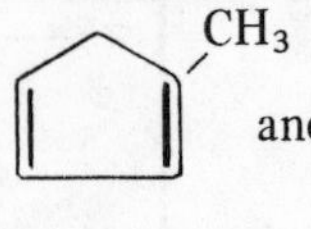

$CH_3$ (A) and $CH_2$ (B)

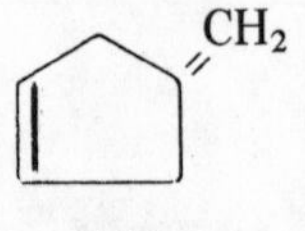

---

**A-37** Compound **A** has a conjugated diene system and would absorb light between 200 and 400 nm. $\lambda_{max}$ (actual) $\approx$ 235 nm.

Compound **B** has two isolated double bonds and would have *no* absorption in the 200 – 400 nm region.

**Q-38** The spectra of compounds **C** and **D** (below) are obtained in the region 200 – 400 nm. The concentration is such that only the high intensity $\pi \rightarrow \pi^*$ transition absorptions are detected. What difference in the spectra of the two compounds would be expected?

O (C) O (D)

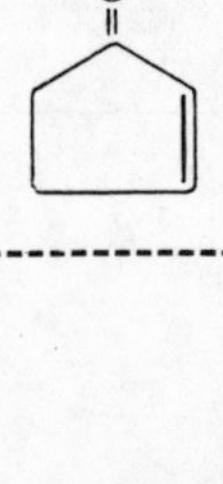

A-38 Compound **C** would have no $\pi \rightarrow \pi^*$ absorption between 200 and 400 nm. Compound **D** would show absorption in the 200–230 nm region. An absorption at 218 nm is actually observed for D.

Q-39 Which of the following would be expected to absorb light of the longest and shortest wavelength? Why?

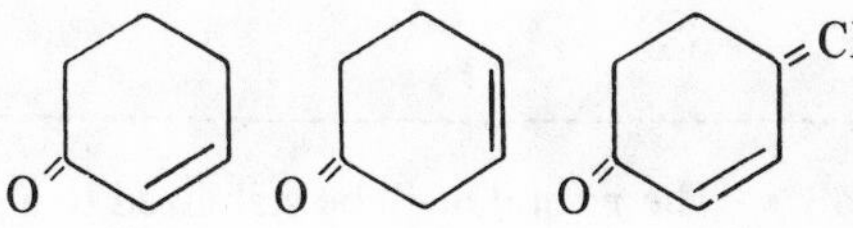

**A** **B** **C**

Consider only the $\pi \rightarrow \pi^*$ transition.

A-39 **B** the shortest, **C** the longest. The longer the conjugated system, the lower the energy for the $\pi \rightarrow \pi^*$ transition. **C** has the longest conjugated system. B has only isolated double bonds.

Q-40 Which of the following would be expected to absorb light of the longest and shortest wavelength? Why?

**A** $CH_3(CH_2)_5\ CH_3$
**B** $(CH_3)_2\ C = CHCH_2CH = C\ (CH_3)_2$
**C** $CH_2 = CH\ CH = CHCH_3$

A-40 **C** would absorb at longest wavelength, **A** at shortest. **C** has a conjugated system.

Q-41 The $\pi \rightarrow \pi^*$ transitions for compounds **E** and **F** (below) were obtained. One compound had a $\lambda_{max}$ = 303 nm while the other had $\lambda_{max}$ = 263 nm. Which compound has the $\lambda_{max}$ = 303 nm?

$CH_3CH{=}CHCH{=}CHC(=O)H$ **E**

$CH_3CH{=}CHCH{=}CHCH{=}CHC(=O)H$ **F**

A-41 Compound B has $\lambda_{max}$ 303. The longer conjugated system absorbs light of longer wavelength.

Q-42 The compound called $\beta$-carotene has eleven conjugated double bonds and absorbs light in the colored region of the spectrum. Explain.

A-42 As the length of the conjugated system increases, the energy difference between the highest occupied energy level and the lowest unoccupied energy level decreases. In $\beta$-carotene this energy difference must correspond to visible (low energy) light.

**R** In conjugated systems, $\pi$ orbitals interact to form a new set of bonding and anti-bonding orbitals in which the energy difference between the ground and excited states for the $\pi \rightarrow \pi^*$ transition is less. As the length of the conjugated system increases, the wavelength of absorption for the $\pi \rightarrow \pi^*$ transition becomes greater.

S-8 In most $\pi \rightarrow \pi^*$ transitions the excited states are more polar than the ground states. As a result, in the $\pi \rightarrow \pi^*$ transition, absorption will shift toward longer wavelength in polar solvents. This shift toward longer wavelength is referred to as a *bathochromic* effect (red shift).

Molecules with non-bonded electrons are able to interact with hydrogen bonding solvents to a greater extent in the ground state than in their excited state. As a result the $n \rightarrow \pi^*$ transition absorption will shift toward shorter wavelength as the hydrogen bonding ability of the solvent increases. A shift toward shorter wavelength is called a *hypsochromic* effect (blue shift).

Q-43 Polar interactions of solvents with molecules result in the lowering of energy states. Will the $\pi$ or $\pi^*$ state be stabilized to the greater extent by a polar solvent?

A-43 The $\pi^*$ state will be stabilized to a greater extent because it is more polar than the $\pi$ state. As solvent interaction increases the stability of the state increases.

Q-44 On the energy level diagram below show the relative changes in energy that a polar solvent has on both the $\pi$ and $\pi^*$ state.

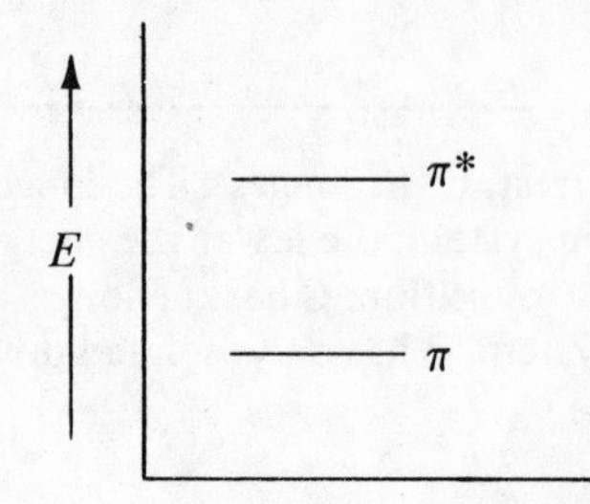


A-44

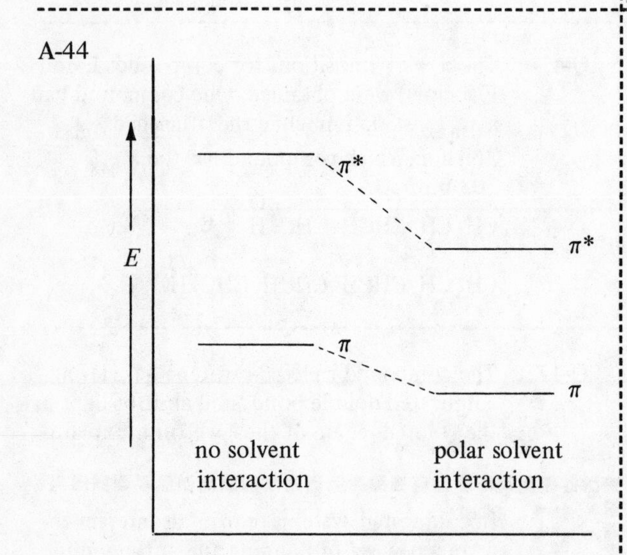


Q-45 Using the energy level diagram drawn for A-44 explain why a polar solvent causes a $\pi \rightarrow \pi^*$ transition to shift to longer wavelength.

A-45 The energy required for a $\pi \rightarrow \pi^*$ transition in a molecule interacting with a polar solvent is less than the energy required for the $\pi \rightarrow \pi^*$ transition for the molecule in a non-polar solvent.

Q-46 A conjugated diene has a $\lambda_{max}$ at 219 nm in hexane solvent $\lambda_{max}^{hexane} = 219$ nm. Will $\lambda_{max}$ be greater or smaller than 219 nm if the solvent is changed to ethanol? Explain.

A-46 $\lambda_{max}^{ethanol}$ is greater than 219 nm. Ethanol is a more polar solvent than hexane. Therefore, ethanol would stabilize the $\pi^*$ state to a greater extent than the $\pi$ state.

Q-47 Hydrogen bond interactions between lone-pair electrons and solvent result in the lowering of the energy state. For an $n \rightarrow \pi^*$ type transition will the $n$ or $\pi^*$ state be stabilized to the greater extent by a hydrogen bonding solvent? Why?

A-47 The $n$ state will be stabilized to the greater extent. Non-bonded electrons interact to a greater extent with hydrogen bonding solvent in their ground states than in their excited states.

Q-48 Below is the energy level diagram for a ketone. Show the effect on the $n \rightarrow \pi^*$ transition when the molecule is dissolved in a polar, hydrogen bonding solvent such as ethanol. Show the effect of the same solvent on the $\pi \rightarrow \pi^*$ transition. Explain.

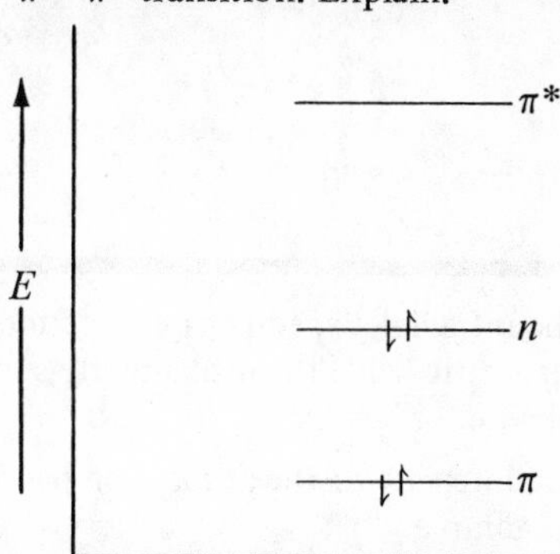


A-48

π* — π*
E
n — n
π — π

no solvent — solvated with ethanol

The solvent will hydrogen bond and stabilize the $n$ ground state to a greater extent than it will stabilize the excited ($\pi^*$) state. The energy difference for the $n \rightarrow \pi^*$ transition will become larger (blue shift). The polar solvent stabilizes the $\pi^*$ state to a greater extent than it stabilizes the $\pi$ state. Thus, the energy difference for the $\pi \rightarrow \pi^*$ transition becomes less (red shift).

Q-49 A certain compound dissolved in hexane had a $\lambda_{max}^{hexane}$ =305 nm. When the same compound was dissolved in ethanol the $\lambda_{max}^{ethanol}$ was 307 nm. Is the absorption due to an $n \rightarrow \pi^*$ or a $\pi \rightarrow \pi^*$ type transition? Give your reasoning.

A-49 $\pi \rightarrow \pi^*$

Ethanol is a more polar solvent and better hydrogen bonding solvent than hexane. The transition is of lower energy in ethanol than in hexane. Therefore, the excited state must be stabilized to a *greater* extent by ethanol than the ground state. The shift to longer wavelength with more polar solvents is characteristic of $\pi \rightarrow \pi^*$ transition.

Q-50 The following spectral data is available for a compound.

$\lambda_{max}^{ethanol}$ = 287 nm

$\lambda_{max}^{dioxane}$ = 295 nm

What type of transition gives rise to the observed peaks?

A-50 The transition must be an $n \rightarrow \pi^*$ or an $n \rightarrow \sigma^*$ since increasing the hydrogen bonding ability of the solvent causes a blue shift.

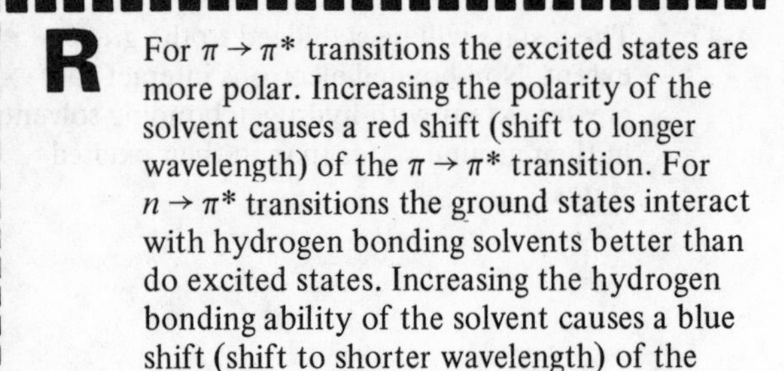

For $\pi \rightarrow \pi^*$ transitions the excited states are more polar. Increasing the polarity of the solvent causes a red shift (shift to longer wavelength) of the $\pi \rightarrow \pi^*$ transition. For $n \rightarrow \pi^*$ transitions the ground states interact with hydrogen bonding solvents better than do excited states. Increasing the hydrogen bonding ability of the solvent causes a blue shift (shift to shorter wavelength) of the $n \rightarrow \pi^*$ transition.

---

S-9 The ultraviolet spectra of compounds are usually obtained by passing light of a given wavelength (monochromatic light) through a dilute solution of the substance in a non-absorbing solvent such as water, ethanol, or hexane.

The intensity of the absorption band is measured by the percent of the incident light that passes through the sample.

$$\% \text{ transmittance} = (100)\, I/I_0$$

where $I_0$ is the intensity of incident light and $I$ is the intensity of the transmitted light.

Because light absorption is a function of the concentration of the absorbing molecules, a more precise way of reporting intensity of absorption is by use of the Lambert-Beer Law

$$\log I/I_0 = -\varepsilon c \ell = -\text{ Absorbance}$$

where $I$ and $I_0$ have the same meaning as before, $c$ is the concentration of the sample in mole/l, $\ell$ is the length of cell through which the light passes in cm, and $\varepsilon$ is a constant called the *molar absorptivity*. Molar absorptivity of absorption bands are reported as $\varepsilon_{max}$.

---

Q-51 A compound had an absorption maximum at 235 nm ($\lambda_{max}$ = 235). At 235 nm, 20% of the incident light was transmitted by a $2.0 \times 10^{-4}$ molar solution of the sample in a 1.0 cm cell. What was the molar absorptivity at 235 nm ($\varepsilon_{max}$)?

A-51 $\log I/I_0 = -\varepsilon c \ell$

$\ell = 1.0$

$c = 2.0 \times 10^{-4}$ mole/l

$\log I/I_0 = \log 0.20 = -0.70$

$-0.70 = -\varepsilon\,(2.0 \times 10^{-4})\,(1.0)$

$\varepsilon = 3.5 \times 10^3$

Q-52 A $1.0 \times 10^{-5}$ molar solution of a compound has a % transmittance of 50 at $\lambda_{max}$ = 280 nm when a 1.0 cm cell is used. Calculate $\varepsilon_{max}$ at 280 nm.

A-52 $\log I/I_0 = -\varepsilon c \ell$

$\ell = 1.0$ cm.

$c = 1.0 \times 10^{-5}$ M

$\log I/I_0 = \log 0.50 = -0.30$

$-0.30 = -\varepsilon\,(1.0 \times 10^{-5})\,(1.0)$

$\varepsilon = 3.0 \times 10^4$

Q-53 The $\varepsilon_{max}$ for aniline at $\lambda_{max}$ = 280 nm is 1 430. A solution of aniline in water is to be prepared such that the % transmission of the solution in a 1.0 cm cell will be 30. How many grams of aniline will be required to prepare 100 ml of solution?

A-53 $\log I/I_0 = -\varepsilon c \ell$

$\ell = 1.0$ cm

$\varepsilon = 1\ 430$

$\log I/I_0 = \log 0.30 = -0.52$

$-0.52 = -(1\ 430)(c)(1.0)$

$c = 3.6 \times 10^{-4}$ mole/l

$MW_{C_6H_7N} = 93$

$(3.6 \times 10^{-4})(93)$ g of aniline to prepare one liter of solution.

0.0034 g of aniline to prepare 100 ml of solution.

Q-54 2-cyclohexenone ($C_6H_8O$) has an $\varepsilon_{max}$ of 10 000 at $\lambda_{max} = 225$ nm. If a solution of this compound is to be prepared such that the absorbance is 0.30 when using 1.0 cm cells, how many grams of 2-cyclohexenone will be required per liter of solution? Remember that $\log I/I_0 = -$ absorbance.

A-54 $\log I/I_0 = -\varepsilon c \ell = -$ absorbance

$\varepsilon = 10\ 000$

$\ell = 1.0$ cm

Absorbance = 0.30

$(10\ 000)(1.0)(c) = 0.30$

$c = 3.0 \times 10^{-5}$ mole/l

$MW_{C_6H_8O} = 96$

Grams of 2-cyclohexenone/l of solution = $(96)(3.0)\ 10^{-5}$

$= 2.9 \times 10^{-3}$ g

Q-55 The $\lambda_{max}^{hexane}$ and $\varepsilon_{max}^{hexane}$ for butyl nitrite are given below. Calculate the % transmission at each $\lambda_{max}$ for a $1.0 \times 10^{-4}$ molar solution of the compound in hexane if a 1.0 cm cell is used.

| | $\lambda_{max}^{hexane}$ | $\varepsilon_{max}^{hexane}$ |
|---|---|---|
| butyl nitrite | 220 | 14 500 |
| | 356 | 87 |

A-55 $\log I/I_0 = -\varepsilon c \ell$

$\varepsilon_{220} = 14\ 500$

$c = 1.0 \times 10^{-4}$

$\ell = 1.0$

$\log I/I_0 = -(14\ 500)(1.0 \times 10^{-4})(1.0)$

$= -1.45 = -2.0 + 0.55$

$I/I_0 = .035$

% transmittance = 3.5%

$\varepsilon_{356} = 87$

$\log I/I_0 = -(87)(1.0 \times 10^{-4})(1.0)$

$= -0.0087$

$= -1.0000 + 0.9913$

$I/I_0 = 0.98$

% transmittance = 98%

Q-56 Below is the ultraviolet spectrum of compound **A**. Give $\lambda_{max}$ and $\varepsilon_{max}$ for the compound.

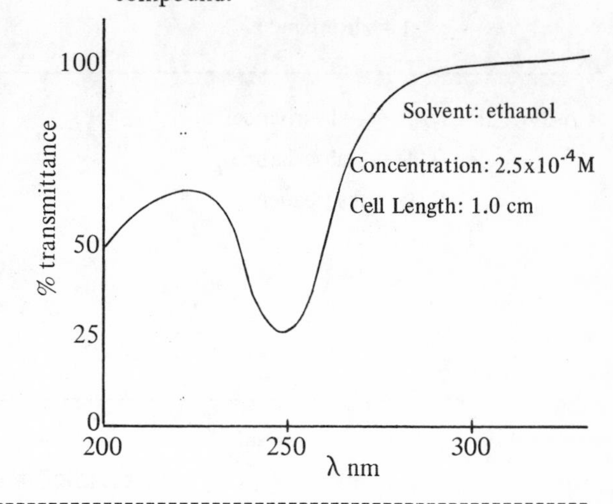

A-56 $\lambda_{max}^{ethanol} = 250$ nm

$\mathcal{E}_{max} = -\log 0.25/(2.5 \times 10^{-4})(1.0)$

$= 2\,400$

Q-57 Below is the ultraviolet spectrum of compound B (mole wt = 100). Give $\lambda_{max}$ and $\mathcal{E}_{max}$ for the compound.

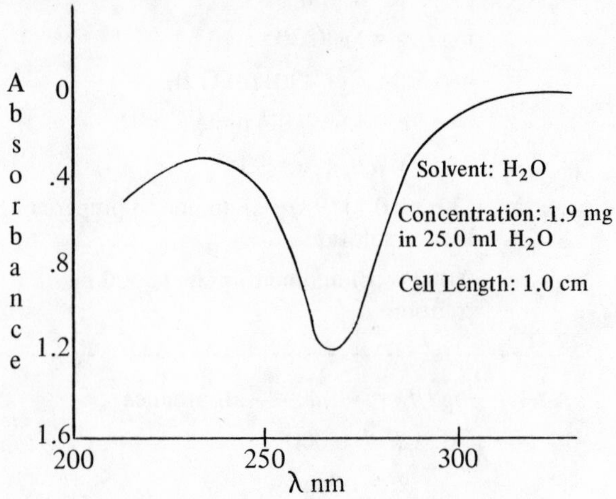

A-57 $\lambda_{max} = 270$ nm

Concentration:

$$\frac{1.9\text{mg}}{25.0\text{ml}} \left| \frac{1000\text{ml}}{1\ \text{l}} \right| \frac{1\text{g}}{1000\text{mg}} \left| \frac{1\text{ mole}}{100\text{g}} \right. = 7.6 \times 10^{-4}\text{M}$$

$\mathcal{E}_{max} = 1.2/(1.0)(7.6 \times 10^{-4})$

$= 1.58 \times 10^{3}$

Q-58 If the % transmission of a solution is 10% at 254 nm what is the absorbance at that wavelength?

A-58 $\log I/I_0 = -$ absorbance

$\log 0.10 = -$ absorbance

1 = absorbance

Q-59 If the % transmission of a solution is 90% at 310 nm what is the absorbance at that wavelength?

A-59 $\log I/I_0 = -$ absorbance

$\log 0.90 = -$ absorbance

0.05 = absorbance

**R** Lambert-Beer Law:

$\log I/I_0 = -\mathcal{E}c\ell = -$ absorbance

$I$ = intensity of transmitted light

$I_0$ = intensity of incident light

$\ell$ = length of cell through which light passes

$c$ = concentration in moles/l

$\mathcal{E}$ = molar absorptivity

## Ultraviolet and Visible Spectra

## PART II: Use of Ultraviolet Spectra in Structural Determinations

S-10 A bathochromic effect (red shift) is observed when the length of a conjugated system is increased. Bathochromic effects are also observed when conjugated systems are substituted with various groups. Some general rules have been formulated which enable $\lambda_{max}^{hexane}$ to be calculated for certain conjugated systems.

The first system to be considered will be that of conjugated dienes ( $>C=\overset{|}{C}-\overset{|}{C}=C<$ ). The unsubstituted diene, butadiene, with $\lambda_{max}^{hexane} = 217$ nm is used as the parent system. Each double bond which extends the conjugated system increases $\lambda_{max}$ by 30 nm. Each *alkyl* group attached to a carbon of the conjugated system increases $\lambda_{max}$ by 5 nm. If the diene system is within a ring (i.e., 1,3-cyclohexadiene) $\lambda_{max}$ is increased by 36 nm. For each double bond of the conjugated system which is exocyclic (i.e., attached to a ring, ) $\lambda_{max}$ is increased by 5 nm. The calculated values are reliable to within ±5 nm.

A summary of these rules and red shifts caused by other groups is listed below. These rules are *not* applicable for "crossed conjugated systems" (such as **A** below) or aromatic systems.

$=CH_2$

**A**

crossed
conjugated
system

| | |
|---|---|
| Value assigned to parent diene system | 217 nm |
| Increment added for: | |
| diene system within a ring (homoannular diene) | 36 nm |
| each alkyl substituent or ring residue | 5 nm |
| the exocyclic nature of any double bond | 5 nm |
| a double bond extension | 30 nm |
| auxochrome – OAcyl | 0 nm |
| – OAlkyl | 6 nm |
| – SAlkyl | 30 nm |
| – Cl, – Br | 5 nm |
| – $NAlkyl_2$ | 60 nm |

(Reprinted with permission from A.I. Scott, *Interpretation of the Ultraviolet Spectra of Natural Products*, Pergamon Press, Oxford, 1964.)

Q-60 How many substituents are attached to double bond carbons in the molecule below? Has the basic conjugated system been extended? How many exocyclic double bonds are in the molecule?

$CH_3$

A-60 Three substituents are attached to double bond carbons. The conjugated diene system has not been extended. There are no exocyclic double bonds.

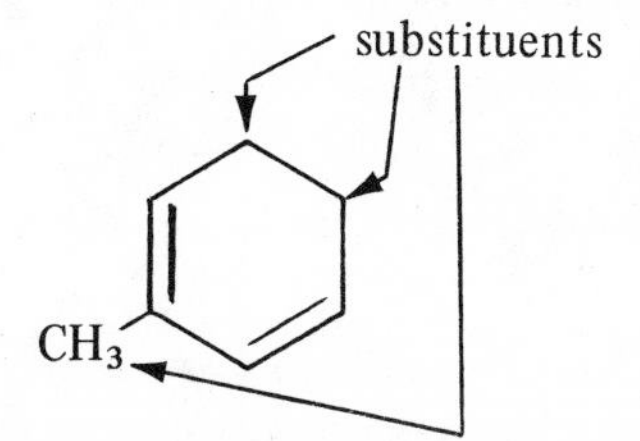


Q-61 For the following compound how many substituents are attached to double bond carbons of the conjugated system? Are there any exocyclic double bonds?

$CH_3$ A B

A-61 There are four substituents attached to double bond carbons of the conjugated system. There is one exocyclic double bond (the double bond is exocyclic to ring A).

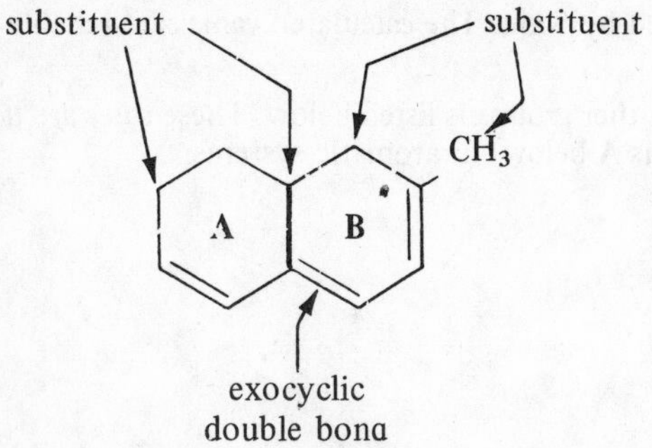

substituent — substituent — $CH_3$ — A — B — exocyclic double bond

Q-62 Using the values given in S-10 calculate $\lambda_{max}^{hexane}$ for the following compound.

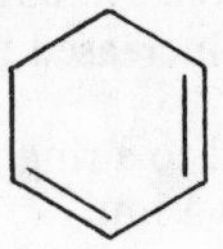

A-62

| | |
|---|---|
| Parent system | 217 nm |
| Homoannular diene | 36 nm |
| Alkyl substituents (2x5) | 10 nm |
| $\lambda_{max}^{hexane}$ (calc) = | 263 nm |
| $\lambda_{max}^{hexane}$ (actual) = | 258 nm |

Q-63 Using the values given in S-10 calculate $\lambda_{max}^{hexane}$ for

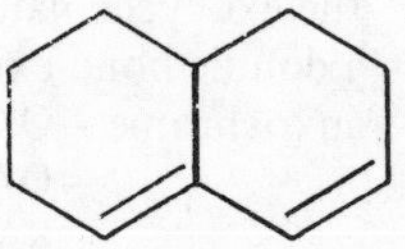

A-63

| | |
|---|---|
| Parent system | 217 nm |
| Exocyclic double bond | 5 nm |
| Alkyl substituents (3x3) | 15 nm |
| $\lambda_{max}^{hexane}$ (calc) = | 237 nm |
| $\lambda_{max}^{hexane}$ (actual) = | 235 nm |

Q-64 Calculate $\lambda_{max}^{hexane}$ for

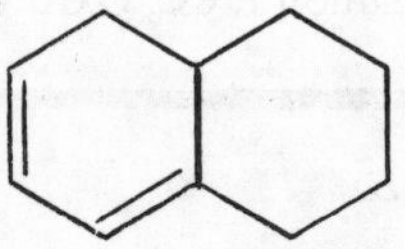

A-64

| | |
|---|---|
| Parent system | 217 nm |
| Homoannular ring diene | 36 nm |
| Exocyclic double bond | 5 nm |
| Alkyl substituents (3x5) | 15 nm |
| $\lambda_{max}^{hexane}$ (calc) = | 273 nm |
| $\lambda_{max}^{hexane}$ (actual) = | 275 nm |

Q-65 Calculate the $\lambda_{max}^{hexane}$ for

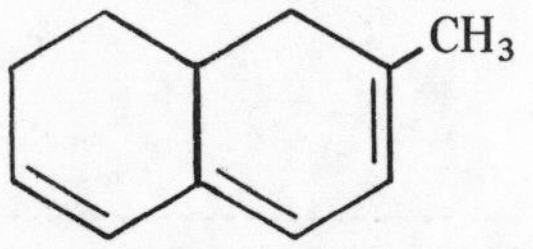

$CH_3$

A-65

| | |
|---|---|
| Parent system | 217 nm |
| Homoannular ring system | 36 nm |
| Alkyl substituents (4x5) | 20 nm |
| Exocyclic double bond (1x5) | 5 nm |
| Extension of conjugated system (1 x 30) | 30 nm |
| $\lambda_{max}^{hexane}$ (calc) = | 308 nm |

Q-66 Calculate $\lambda_{max}^{hexane}$ for

$C_9H_{19}$ $CH_3$ $CH_3$ $CH_3CO$ O

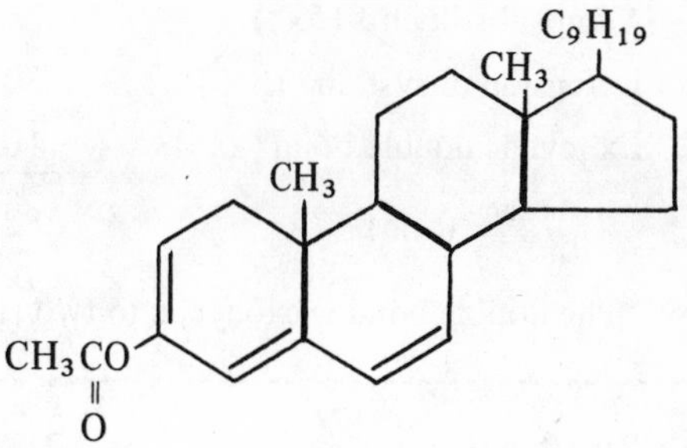

A-66

| | |
|---|---|
| Parent system | 217 nm |
| Homoannular ring system | 36 nm |
| Alkyl substituents (3x5) | 15 nm |
| Substituent ($OCOCH_3$) | 0 nm |
| Exocyclic double bond (1x5) | 5 nm |
| Extension of conjugated system (1 x 30) | 30 nm |
| $\lambda_{max}^{hexane}$ (calc) = | 303 nm |
| $\lambda_{max}^{hexane}$ (actual) = | 304 nm |

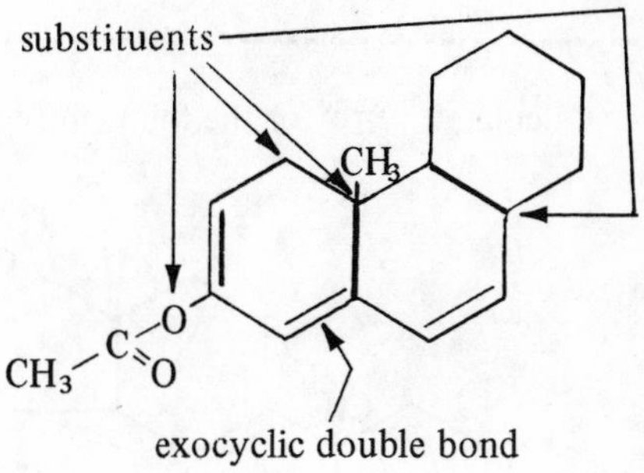

substituents $CH_3$ O C O $CH_3$

exocyclic double bond

Q-67 Calculate $\lambda_{max}^{hexane}$ for the following compound.

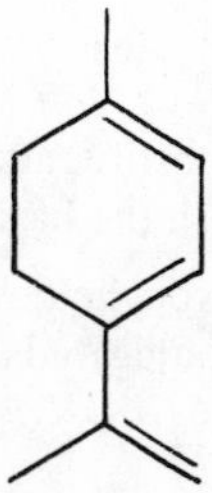

A-67

| | |
|---|---|
| Parent system | 217 nm |
| Homoannular ring | 36 nm |
| Alkyl substituents (4x5) | 20 nm |
| Exocyclic double bond (0x5) | 0 nm |
| Extension of conjugated system (1x30) | 30 nm |
| $\lambda_{max}^{hexane}$ (calc) = | 303 nm |

Q-68 Calculate $\lambda_{max}^{hexane}$ for the following compound.

$CH_3$

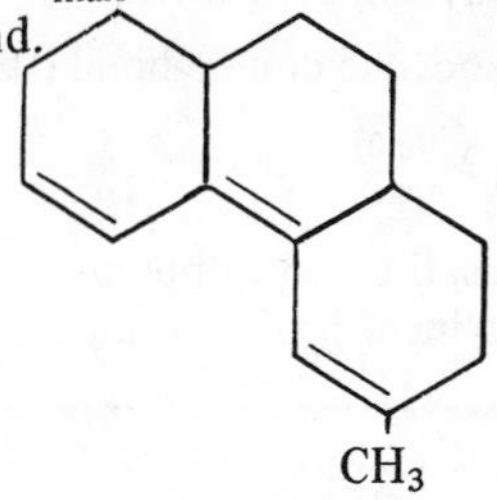

A-68

| | |
|---|---|
| Parent system | 217 nm |
| Alkyl substituents (5x5) | 25 nm |
| Extension of system (1x30) | 30 nm |
| Exocyclic double bond (2x5) | 10 nm* |
| $\lambda_{max}^{hexane}$ (calc) = | 282 nm |

*The double bond is exocyclic to two rings.

Q-69 Calculate $\lambda_{max}^{hexane}$ for the following compound.

$CH_2$

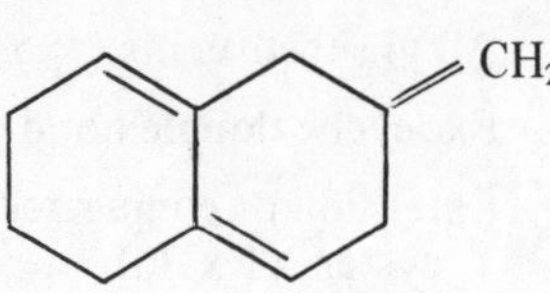

A-69

| | |
|---|---|
| Parent system | 217 nm |
| Alkyl substituents (4x5) | 20 nm* |
| Exocyclic double bond (2x5) | 10 nm‡ |
| Extension of system (0x30) | 0 nm |
| $\lambda_{max}^{hexane}$ (calc) = | 247 nm |

*Only substituents on double bonds conjugated with the base system are counted.

‡Only exocylic double bonds which are part of the conjugated system are counted.

Q-70 Calculate $\lambda_{max}^{hexane}$ for the two isomers

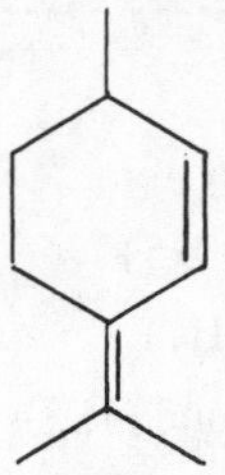

A and B

Assuming that $\lambda_{max}^{hexane}$ can be predicted with a certainty of 5 nm, could the compounds shown above be distinguished by using ultraviolet spectroscopy?

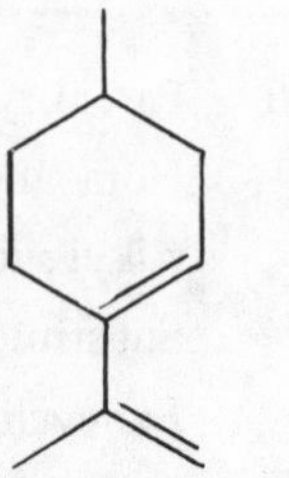

A-70 Compound **A**:

| | |
|---|---|
| Parent system | 217 nm |
| Alkyl substituents (4x5) | 20 nm |
| Exocyclic double bond (1x5) | 5 nm |
| $\lambda_{max}^{hexane}$ (calc) = | 242 nm |

Compound **B**:

| | |
|---|---|
| Base Value | 217 nm |
| Alkyl substituents (3x5) | 15 nm |
| Exocyclic double bond (0x5) | 0 nm |
| $\lambda_{max}^{hexane}$ (calc) = | 232 nm |

Yes, the two compounds could probably be distinguished.

Q-71 Calculate $\lambda_{max}^{hexane}$ for the following compound.

$CH_3$ $C_8H_{17}$ $CH_3$ HO

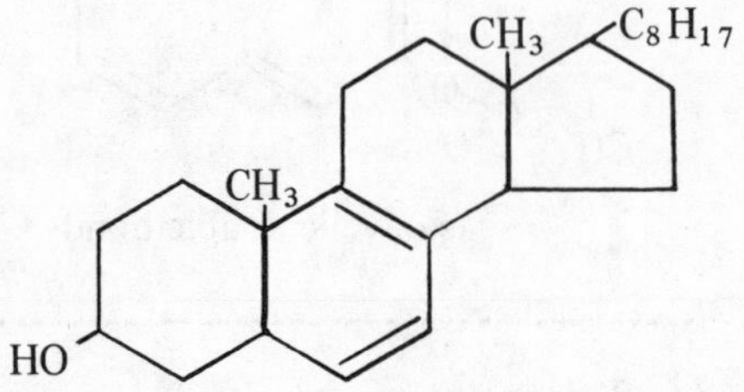

A-71

| | |
|---|---|
| Parent system | 217 nm |
| Homoannular diene system | 36 nm |
| Alkyl substituents (4 x 5) | 20 nm |
| $\lambda_{max}^{hexane}$ (calc) = | 273 nm |
| $\lambda_{max}^{hexane}$ (actual) = | 275 nm |

Q-72 Using values given in S-10 calculate $\lambda_{max}^{hexane}$ for the following compound.

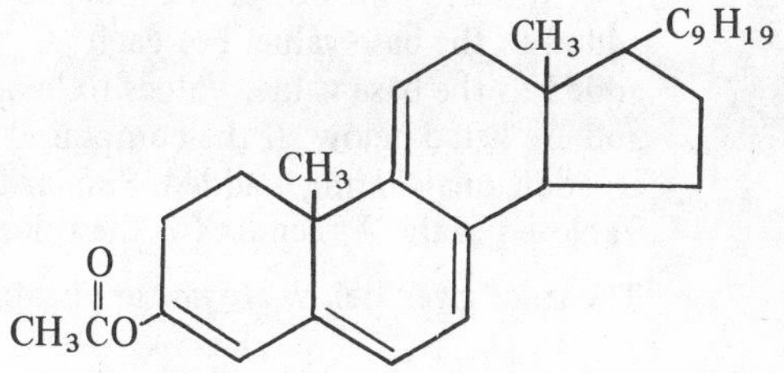

A-72

| | |
|---|---|
| Parent system | 217 nm |
| Homoannular diene system | 36 nm |
| Alkyl substituents (5 x 5) | 25 nm |
| OAcyl substituent | 0 nm |
| Extension of conjugated system (2 x 30) | 60 nm |
| Exocyclic double bonds (3 x 5) | 15 nm |
| $\lambda_{max}^{hexane}$ (calc) = | 353 nm |
| $\lambda_{max}^{hexane}$ (actual) = | 356 nm |

Q-73 Calculate $\lambda_{max}^{hexane}$ for the following compound.

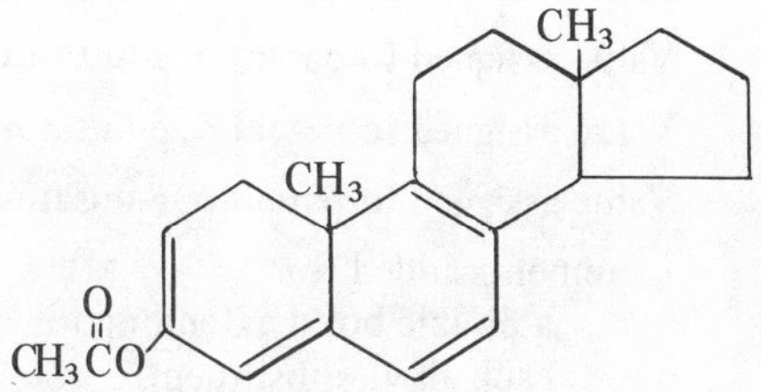

Compare the $\lambda_{max}^{hexane}$ (calc) for this compound and its isomer given in Q-72.

A-73

| | |
|---|---|
| Parent system | 217 nm |
| Homoannular diene system | 36 nm |
| Alkyl substituents (5 x 5) | 25 nm |
| OAcyl substituent | 0 nm |
| Extension of conjugated system (2 x 30) | 60 nm |
| Exocyclic double bonds (1 x 5) | 5 nm |
| $\lambda_{max}^{hexane}$ (calc) = | 343 nm |

**R** Rules used for predicting $\lambda_{max}^{hexane}$ of dienes are given in Table I of the Appendix.

S-11 The $\lambda_{max}^{ethanol}$ for the $\pi \rightarrow \pi^*$ transitions in $\alpha, \beta$-unsaturated acyclic and cyclic ketones ($\gt C = \overset{|}{C} - \underset{|}{C} = O$) may be calculated by a method analogous to that used for the dienes. The base value used for the system is 215 nm. For each exocyclic double bond which is part of the conjugated system, 5 nm is added to the base value. For each $\gt C = C \lt$ group extending the conjugated system, 30 nm is added to the base value. Values to be added for substituents vary with the position of the substituent and are listed below. If the compound has a homoannular diene system (both C = C in a single ring) an additional 39 nm is added. Similar trends are followed by $\alpha, \beta$-unsaturated acids and esters (see Table III in the Appendix for the rules regarding these compounds).

The rules given below are *not* applicable to "crossed" conjugated nor aromatic systems.

**Rules for $\alpha, \beta$-unsaturated Ketone and Aldehyde Absorption**

$$\overset{\delta}{C} = \overset{\gamma}{C} - \overset{\beta}{C} = \overset{\alpha}{C} - C = O$$

$\mathcal{E}$ values are usually above 10 000 and increase with the length of the conjugated system.

| | | |
|---|---|---|
| Value assigned to parent $\alpha, \beta$-unsaturated six-ring or acyclic ketone | | 215 nm |
| Value assigned to parent $\alpha, \beta$-unsaturated five-ring ketone | | 202 nm |
| Value assigned to parent $\alpha, \beta$-unsaturated aldehyde | | 207 nm |
| Increments added for: | | |
| a double bond extending the conjugation | | 30 nm |
| each alkyl substituent | $\alpha$ | 10 nm |
| | $\beta$ | 12 nm |
| | $\gamma$ and higher | 18 nm |
| each substituent | | |
| – OH | $\alpha$ | 35 nm |
| | $\beta$ | 30 nm |
| | $\delta$ | 50 nm |
| – OAc | $\alpha, \beta, \delta$ | 6 nm |
| – OMe | $\alpha$ | 35 nm |
| | $\beta$ | 30 nm |
| | $\gamma$ | 17 nm |
| | $\delta$ | 31 nm |
| – SAlk | $\beta$ | 85 nm |
| – Cl | $\alpha$ | 15 nm |
| | $\beta$ | 12 nm |
| – Br | $\alpha$ | 25 nm |
| | $\beta$ | 30 nm |
| – $NR_2$ | $\beta$ | 95 nm |
| each exocyclic double bond | | 5 nm |
| homoannular diene | | 39 nm |
| $\lambda_{max}^{EtOH}$ (calc) | Total | |

(Reprinted with permission from A.I. Scott, *Interpretation of the Ultraviolet Spectra of Natural Products*, Pergamon Press, Oxford, 1964.)

Q-74 At what positions does the following unsaturated ketone have substituents?

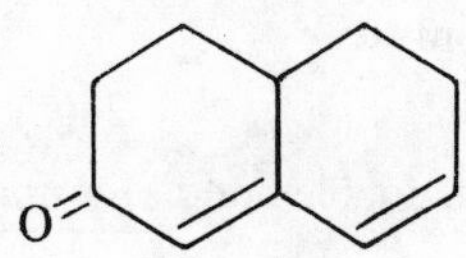

A-74 α : None
β : One
γ : None
δ : One

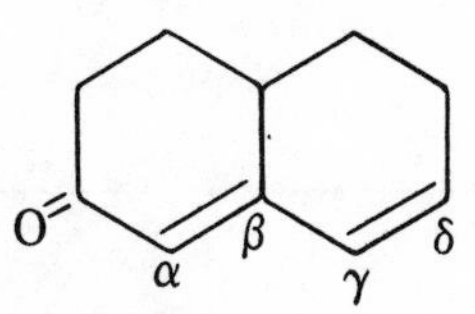

Q-75 Using the values given in S-12 calculate the $\lambda_{max}^{alcohol}$ for the compound shown in the previous question.

A-75

| | |
|---|---|
| Parent system | 215 nm |
| Substituents β (1 x 12) | 12 nm |
| δ (1 x 18) | 18 nm |
| Exocyclic double bond (1 x 5) | 5 nm |
| Extension of system (1 x 30) | 30 nm* |
| $\lambda_{max}^{alcohol}$ (calc) = | 280 nm |

*Remember that the basic system is

$$-\overset{O}{\overset{\|}{C}}-\overset{|}{C}=C\langle .$$

Q-76 Calculate the $\lambda_{max}^{alcohol}$ for

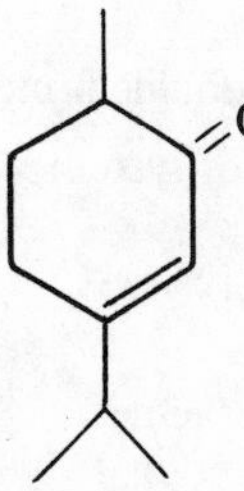

A-76

| | |
|---|---|
| Parent system | 215 nm |
| Substituents β (2 x 12) | 24 nm |
| $\lambda_{max}^{ethanol}$ (calc) = | 239 nm |

Q-77 Calculate $\lambda_{max}^{alcohol}$ for each of the compounds below. If the values predicted were within 5 nm of the observed $\lambda_{max}^{ethanol}$, could the two compounds be distinguished by their ultraviolet spectrum?

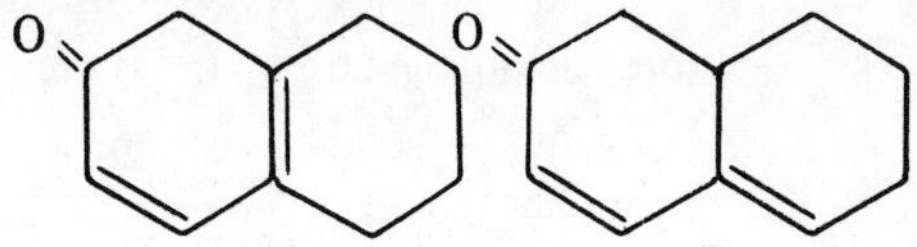

A B

A-77

| | A | B |
|---|---|---|
| Parent system | 215 nm | 215 nm |
| Substituents γ | 18(1) nm | 18(1) nm |
| δ | 18(1) nm | 36(2) nm |
| Homoannular diene system | 39 nm | 0 nm |
| Exocyclic double bond | 0 nm | 5 nm |
| Extension of system | 30 nm | 30 nm |
| $\lambda_{max}^{ethanol}$ (calc) = | 338 nm | 286 nm |

The two compounds could be distinguished from their ultraviolet spectrum.

Q-78 Calculate $\lambda_{max}^{alcohol}$ for each of the following isomers.

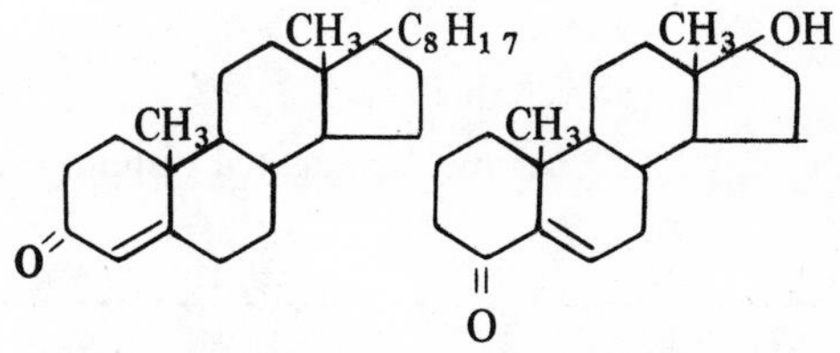

$CH_3$ $C_8H_{17}$ $CH_3$ $CH_3$ OH $CH_3$ O O

A B

A-78

| | A | B |
|---|---|---|
| Parent system | 215 nm | 215 nm |
| Substituents α | | 10(1) nm |
| β | 24(2) nm | 12(1) nm |
| Exocyclic double bond | 5 nm | 5 nm |
| $\lambda_{max}^{alcohol}$ (calc) = | 244 nm | 242 nm |
| $\lambda_{max}^{alcohol}$ (actual) = | 234 nm | 240 nm |

Q-79 Calculate $\lambda_{max}^{alcohol}$ for the following compounds.

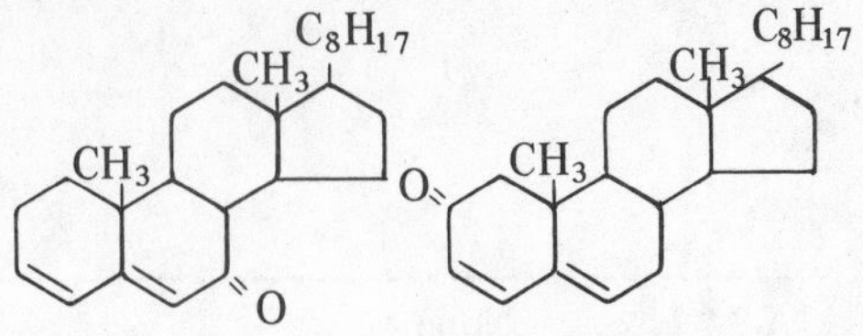

A-79

| | A | B |
|---|---|---|
| Parent system | 215 nm | 215 nm |
| Substituents β | 12(1) nm | |
| γ | | 18(1) nm |
| δ | 18(1) nm | 18(1) nm |
| Exocyclic double bond | 5 nm | 5 nm |
| Extension of system | 30 nm | 30 nm |
| $\lambda_{max}^{ethanol}$ (calc) = | 280 nm | 286 nm |
| $\lambda_{max}^{ethanol}$ (actual) = | 280 nm | 290 nm |

Q-80 A compound was believed to have either structure **A** or **B**. The spectrum of the compound had $\lambda_{max}^{alcohol}$ = 352 nm. What is the most likely structure for the compound?

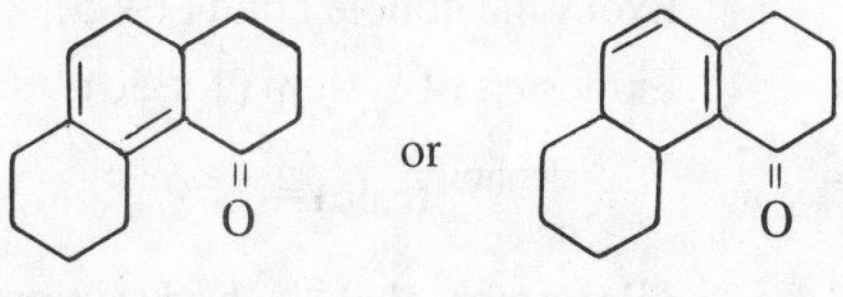

A-80

| | A | B |
|---|---|---|
| Base value | 215 nm | 215 nm |
| Homoannular ring | 39 nm | 39 nm |
| Substituents α | 10(1) nm | 10(1) nm |
| β | 12(1) nm | 12(1) nm |
| γ | 18(1) nm | |
| δ | 18(1) nm | 18(1) nm |
| Exocyclic double bond | 15(3) nm | |
| Extension of system | 30 nm | 30 nm |
| $\lambda_{max}^{alcohol}$ (calc) = | 357 nm | 324 nm |

The compound is more likely to have structure A.

Q-81 A compound known to be a substituted cyclohexanone derivative had a $\lambda_{max}^{ethanol}$ of 235 nm. Could the compound be a conjugated dieneone (i.e., $-\overset{|}{C}=\overset{|}{C}-\overset{|}{C}=\overset{|}{C}-\overset{O}{\overset{\|}{C}}-$)? Explain?

A-81 No. If the compound were a conjugated dieneone, $\lambda_{max}^{ethanol}$ (calc) would be at least 245 nm.

| | |
|---|---|
| Parent value | 215 nm |
| Extension of conjugated system | 30 nm |
| | 245 nm |

Q-82 For the compound described in Q-81 how many nm must be accounted for by substituents?

A-82 Actual $\lambda_{max}$ 235 nm
– Parent system 215 nm
20 nm must be accounted for. 20 nm

Q-83 Assume for the compound described in Q-81 that substituents attached to the conjugated system are alkyl. What are possible structures for the chromophoric system? Show alkyl substituents as R groups.

A-83

| | $\lambda_{max}^{ethanol}$ (calc) |
|---|---|
| One α and one β substituent A | 237 nm |
| Two β substituents B | 239 nm |
| One α substituent and an exocyclic double bond C. | 230 nm |

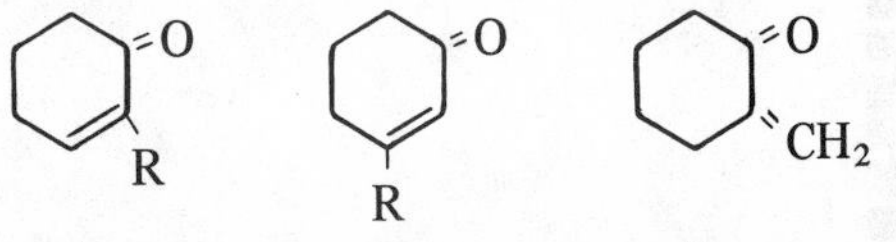

A B C

Q-84 A compound, $C_7H_{10}O$, known to be either a cyclohexenone derivative or an acyclic ketone had an observed $\lambda_{max}^{ethanol}$ of 257 nm. Is the compound likely to have a cyclic structure? How many carbon-carbon double bonds does the molecule have?

A-84 The compound is most likely an acyclic dienone, $>C=C-C=C-C(=O)-$. The only possible cyclohexenone structures that fit the molecular formula are

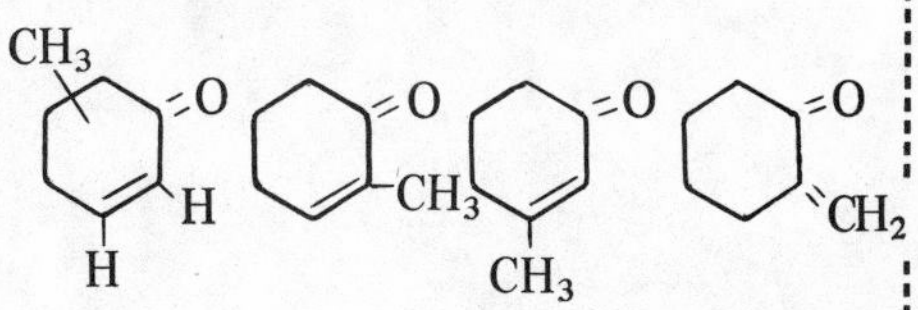

$\lambda_{max}^{ethanol}$ (calc) = 227 237 239 230

An acyclic dienone has a minimum value for $\lambda_{max}^{ethanol}$ (calc) of 245 nm. For the compound $CH_2=CHCH=C(CH_3)COCH_3$ $\lambda_{max}^{ethanol}$ (calc) = 255 nm.

Q-85 The compound $C_{10}H_{16}O$ was known to be a substituted derivative of cyclohexenone. The spectrum of the compound is shown below.

% transmittance

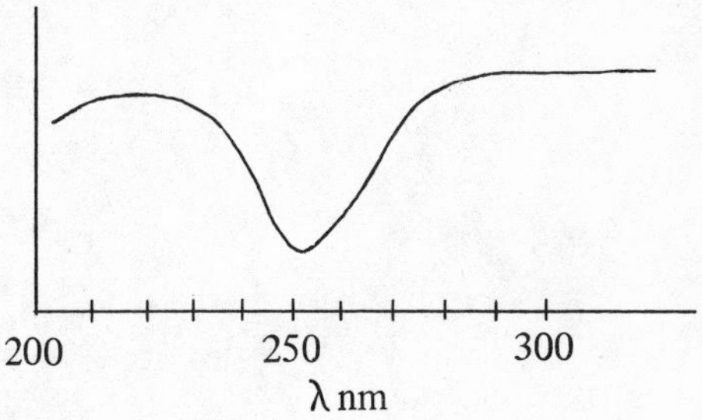

200 250 300
λ nm

Give two structures consistent with the data.

A-85 The basic structure must be

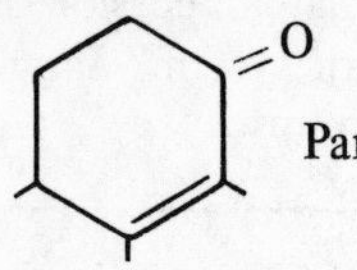

Parent $\lambda_{max}^{ethanol}$ (calc) = 227 or

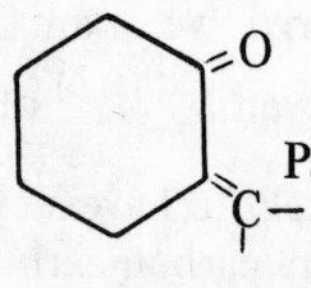

Parent $\lambda_{max}^{ethanol}$ (calc) = 230.

Extension of the conjugated system is not possible since the compound can have only two double bonds and one ring ($C_nH_{2n-4}O$).

Possible structures are

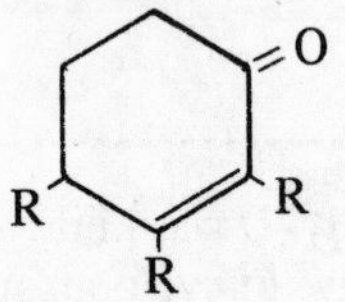

and

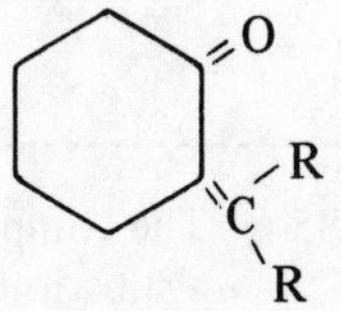

**R** Rules for predicting $\lambda_{max}^{ethanol}$ of α, β-unsaturated aldehydes and ketones are given in Table II of Appendix. Rules for predicting $\lambda_{max}^{ethanol}$ of α, β-unsaturated acids and esters are given in Table III of the Appendix.

S-12 The ultraviolet spectrum of benzene shows several distinct bands. Substitution of the ring with groups containing non-bonded electrons or $\pi$ electrons adjacent to the ring will cause shifts to longer wavelengths (*bathochromic* effect). For disubstituted benzene rings, prediction of $\lambda_{max}$ is not always possible but three general rules are useful.

1. When an electron withdrawing group (i.e., $-NO_2$, $-C{=}O$ ) and an electron donating group (i.e., $-OH$, $-OCH_3$, $-X$) are situated *para* to each other, there is a red shift (*bathochromic* effect).
2. When an electron withdrawing and a donating group are situated *meta* or *ortho* to each other, the spectrum differs little from that of the separate monosubstituted aromatic compounds.
3. When two electron withdrawing or two electron donating groups are situated *para* to each other, the spectrum differs little from that of the separate monosubstituted aromatic compounds.

**Absorption Maxima of the Substituted Benzene Rings Ph – R**

| R | $\lambda_{max}$ nm ($\varepsilon$) (solvent $H_2O$ or MeOH) | | | | | |
|---|---|---|---|---|---|---|
| – H | 203 | (7400) | 254 | (204) | | |
| $-NH_3^+$ | 203 | (7500) | 254 | (160) | | |
| – Me | 206 | (7000) | 261 | (225) | | |
| – I | 207 | (7000) | 257 | (700) | | |
| – Cl | 209 | (7400) | 263 | (190) | | |
| – Br | 210 | (7900) | 261 | (192) | | |
| – OH | 210 | (6200) | 270 | (1450) | | |
| – OMe | 217 | (6400) | 269 | (1480) | | |
| $-SO_2NH_2$ | 217 | (9700) | 264 | (740) | | |
| – CN | 224 | (13,000) | 271 | (1000) | | |
| $-CO_2^-$ | 224 | (8700) | 268 | (560) | | |
| $-CO_2H$ | 230 | (11,600) | 273 | (970) | | |
| $-NH_2$ | 230 | (8600) | 280 | (1430) | | |
| $-O^-$ | 235 | (9400) | 287 | (2600) | | |
| – NHAc | 238 | (10,500) | | | | |
| – COMe | 245 | (9800) | | | | |
| $-CH=CH_2$ | 248 | (14,000) | 282 | (750) | 291 | (500) |
| – CHO | 249 | (11,400) | | | | |
| – Ph | 251 | (18,300) | | | | |
| – OPh | 255 | (11,000) | 272 | (2000) | 278 | (1800) |
| $-NO_2$ | 268 | (7800) | | | | |
| $-CH\overset{t}{=}CHCO_2H$ | 273 | (21,000) | | | | |
| $-CH\overset{t}{=}CHPh$ | 295 | (29,000) | | | | |

(Most values taken with permission from H.H. Jaffé and M. Orchin, *Theory and Applications of Ultraviolet Spectroscopy*, Wiley, New York, 1962.)

Some rules for predicting $\lambda_{max}^{ethanol}$ for disubstituted compounds of the type $R-C_6H_4-C(=O)-Z$ where Z = H, OH, OAlkyl, or alkyl have been formulated.

These rules are listed in the table below and refer to the strongest band in the 200–400 nm region.

**Rules for the Principal Band of Substituted Benzene Derivatives** $R\,C_6H_4\,COZ$

| | Orientation | EtOH $\lambda_{calc}$ nm |
|---|---|---|
| Parent Chromophore: | | |
| Z = alkyl or ring residue | | 246 |
| Z = H | | 250 |
| Z = OH or OAlkyl | | 230 |
| Increment for each substituent: | | |
| R = alkyl or ring residue | *o-, m-* | 3 |
| | *p-* | 10 |
| R = OH, OMe, OAlkyl | *o-, m-* | 7 |
| | *p-* | 25 |
| R = O | *o-* | 11 |
| | *m-* | 20 |
| | *p-* | 78 |
| R = Cl | *o-, m-* | 0 |
| | *p-* | 10 |
| R = Br | *o-, m-* | 2 |
| | *p-* | 15 |
| R = $NH_2$ | *o-, m-* | 13 |
| | *p-* | 58 |
| R = NHAc | *o-, m-* | 20 |
| | *p-* | 45 |
| R = NHMe | *p-* | 73 |
| R = $NMe_2$ | *o-, m-* | 20 |
| | *p-* | 85 |

(Reprinted with permission from A.I. Scott, *Interpretation of the Ultraviolet Spectra of Natural Products,* Pergamon Press, Oxford, 1964.)

Q-86 Calculate $\lambda_{max}^{ethanol}$ for $m-NH_2-C_6H_4-C(=O)-OH$

A-86

| | |
|---|---|
| Parent value | 230 nm |
| Substituent | 58 nm |
| $\lambda_{max}^{ethanol}$ (calc) = | 288 nm |
| $\lambda_{max}^{ethanol}$ (actual) = | 288 nm |

Q-87 Calculate $\lambda_{max}^{ethanol}$ for $p-OCH_3-C_6H_4-C(=O)-CH_3$

A-87

| | |
|---|---|
| Parent value = | 246 nm |
| Substituent | 7 |
| $\lambda_{max}^{ethanol}$ (calc) = | 253 nm |

Q-88 A compound, known to be a bromobenzoic acid, had a $\lambda_{max}^{ethanol}$ (actual) = 244 nm.

What is the most likely structure for the compound?

A-88

(Structure: benzene ring bearing COOH (C=O with OH) and Br in the meta position — m-bromobenzoic acid)

$\lambda_{max}^{ethanol}$(calc) = 245 nm

**R** Rules for predicting $\lambda_{max}^{ethanol}$ of substituted aromatic aldehydes and acid derivatives are given in Table IV of the Appendix.

## REFERENCES

**Introductory:**

Barrow, Gordon M. 1964. *The Structure of Molecules,* Chapter 5. New York: W. A. Benjamin.

Bauman, Robert P. 1962. *Absorption Spectroscopy.* New York: John Wiley and Sons.

Dyer, John R. 1965. *Applications of Absorption Spectroscopy of Organic Compounds,* Chapter 2. Englewood Cliffs, N.J.: Prentice-Hall.

Gray, Harry B. 1964. *Electrons and Chemical Bonding,* Chapters VIII and IX. New York: W. A. Benjamin.

**Intermediate:**

Barrow, Gordon M. 1962. *Introduction to Molecular Spectroscopy,* Chapters 10 and 11. New York: McGraw-Hill.

Scott, A. I. 1964. *Interpretation of the Ultraviolet Spectra of Natural Products.* New York: Macmillan Co.

# Chapter 3 INFRARED SPECTROSCOPY

**PART I: Vibrational Energy Levels**

After completing this section, you should be able to

a) determine the kinetic and potential energy and the vibrational frequency of a harmonic oscillator
b) determine the force constants, vibrational frequencies, and spacings between vibrational energy levels for various chemical bonds
c) determine the number of rotational, vibrational, and translational degrees of freedom for molecules
d) determine which vibrational modes are infrared inactive.

**PART II: Interpretation of Infrared Spectra**

After completing this section, you should be able to

a) locate the seven major spectral regions and identify types of bond vibrations which cause absorption in these regions
b) distinguish between two different functional group isomers by use of infrared spectra
c) distinguish between position isomers of alkenes, alkynes, and derivatives of benzene by use of infrared spectra
d) determine the presence or absence of major functional groups by use of infrared spectra.

# INFRARED SPECTROSCOPY

## PART I: Vibrational Energy Levels

S-1 The infrared spectrum of a molecule is a result of transitions between two different *vibrational* energy levels. Vibrational motions of a molecule resemble the motions observed for a ball attached to a spring (i.e., harmonic oscillator). The ball and spring model will be used to develop the concepts for vibrational motion.

The stretching vibration of a simple ball and spring system is illustrated below. The potential energy, $E_p$, of the stretching vibration is a function of the force constant, *k*, of the spring and the displacement, *d*, of the ball, from the rest position.

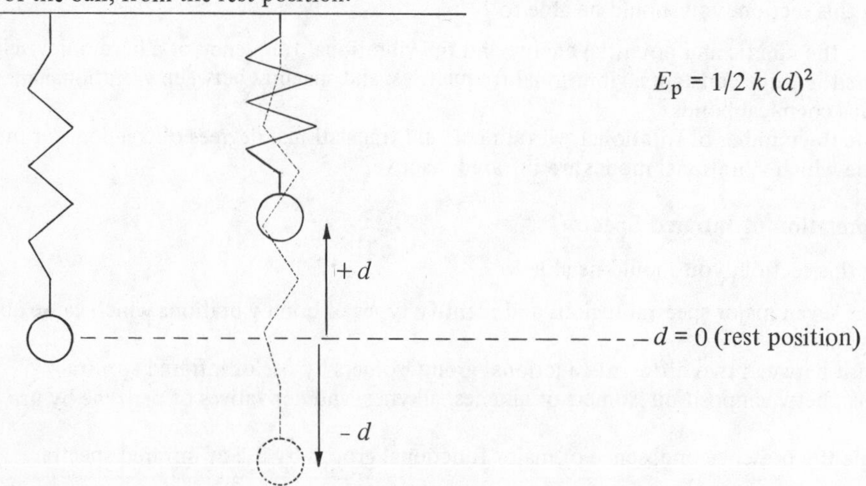


The kinetic energy, $E_k$, of the system shown above is given by the equation

$E_k = 1/2\, mv^2$

where *m* is the mass of the ball and *v* is its velocity.

---

Q-1 What is the potential energy of the spring system shown in S-1 when the ball is in the rest position?

A-1 $E_p = 0$. At rest position $d = 0$. Therefore, $E_p = 1/2\, k\, (d)^2 = 0$

Q-2 The ball of the spring system shown in S-1 is stretched to position $-d$ and released. At what positions will the potential energy of the system be a maximum and a minimum?

A-2 Maximum potential energy is at the position of greatest displacement, $+d$ and $-d$. Minimum potential energy is at the rest position, $d = 0$.

Q-3 At what positions will the *kinetic* energy of the ball be a maximum and a minimum for the system set in motion as described in Q-2?

A-3 Maximum kinetic energy will occur at maximum velocity which is attained at position $d = 0$. Minimum kinetic energy ($E_k = 0$) will occur at positions of maximum displacement, $+d$ and $-d$.

Q-4 Balls of equal mass are attached to two different springs. Spring A has a force constant twice as great as for spring B. Each ball is stretched 1 cm from the rest position. Which system has the greater potential energy?

A-4 Spring **A** has the greater potential energy.

$E_p = 1/2\, k\, (d)^2$

$k_A > k_B$

Q-5 Which system described in Q-4 requires the greater energy to displace the balls 5 cm?

A-5 System **A**, the system with the larger force constant.

**R** For a harmonic oscillator (one vibrating ball on a spring)

$E_p = 1/2\, k\, (d)^2$

$E_k = 1/2\, mv^2$

S-2 A spring with a ball attached to it is set in vibration. The *frequency* of vibration (number of oscillations or cycles per second), $\nu$, of the system is a function of the force constant of the spring, $k$, and the mass of the ball, $m$.

$$\nu = \frac{1}{2\pi}\sqrt{\frac{k}{m}}$$

Q-6 Spring **A** is attached to a ball with a mass of 1.0 g. The spring has a force constant of $4.0 \times 10^{-5}$ N/cm (newton/cm). What is the frequency of this system when set into vibration? What is the unit for the frequency?

(Hint: $1 \times 10^{-5}\ \text{N} = 1\frac{\text{g cm}}{\text{s}^2}$) ($1\ \text{N} = 1\frac{\text{kg m}}{\text{s}^2}$)

A-6

$$\nu = \frac{1}{2\pi}\sqrt{\frac{\frac{4.0\ \text{g cm}}{\text{s}^2\ \text{cm}}}{1.0\ \text{g}}}$$

$$\nu = \frac{2}{2\pi} = \frac{1}{\pi}\ \text{cycle/s or Hertz (Hz)}$$

Q-7 Spring **B** is attached to a ball with a mass of 1.0 g. The spring has a force constant of $9.0 \times 10^{-5}$ N/cm. What is the frequency of this ball when set into vibration?

A-7

$$\nu = \frac{1}{2\pi}\sqrt{\frac{\frac{9.0\ \text{g cm}}{\text{s}^2\ \text{cm}}}{1.0\ \text{g}}}$$

$\nu = 3/2\pi$ cycle/s or Hertz

Q-8 As the force constant of a spring increases does the frequency of vibration increase or decrease?

A-8 Increase. The frequency increases with the square root of the force constant.

Q-9 Which system will vibrate with the greater frequency, spring **A** (force constant = $4.0 \times 10^{-5}$ N/cm) attached to a 1.0 g ball or spring **A** attached to a 4.0 g ball?

A-9 Frequency for spring A attached to 1.0 g ball:

$$\nu = \frac{1}{\pi}\ s^{-1}$$

Frequency for spring A attached to 4.0 g ball:

$$\nu = \frac{1}{2\pi}\sqrt{\frac{\frac{4.0\ g\ cm}{s^2\ cm}}{4.0\ g}}$$

$$\nu = \frac{1}{2\pi}\ \text{cycle/s}$$

The frequency of the system with a 1.0 g ball is greater.

Q-10 A spring ($k = 4.0 \times 10^{-5}$ N/cm) with a 1.0 g ball attached is set into vibration so that the maximum displacement of the ball from the rest position is ± 1.0 cm. A second spring ($k = 4.0 \times 10^{-5}$ N/cm) with a 1.0 ball attached is set into vibration so that the maximum displacement of the ball from the rest position is ± 2.0 cm. Which vibrating system has the greater frequency? (Hint: What are the factors which determine the frequency?)

A-10 Both systems have the same frequency. The factors affecting frequency are the force constant and the mass of the ball attached.

Q-11 Which of the two spring systems described in Q-10 has the greater *potential* energy?

A-11 The potential energy of the system with the 2.0 cm displacement is greater.

$$E_p = 1/2\, k\, (d)^2$$

The potential energy is dependent upon the displacement and the force constant.

Q-12 As more energy is added to a vibrating spring system, will the frequency of the vibrations change? Why?

A-12 No. The frequency remains unchanged because frequency is dependent only upon the force constant and the mass of the system.

Q-13 As energy is added to a vibrating spring system, the frequency remains unchanged. What characteristic of the vibrating system does change with increasing potential energy?

A-13 The displacement or *amplitude* of the vibration.

**R** Frequency of a harmonic oscillator (one ball on a spring)

$$\nu = \frac{1}{2\pi}\sqrt{\frac{k}{m}}$$

S-3 For a harmonic oscillator consisting of two balls of mass $m_1$ and $m_2$ connected by a spring, the equation for the potential energy is a function of the force constant, $k$, and the total displacement of the two balls from the equilibrium position $(d_1 + d_2)$. The frequency of vibration for the system is a function of the force constant and the mass of each ball.

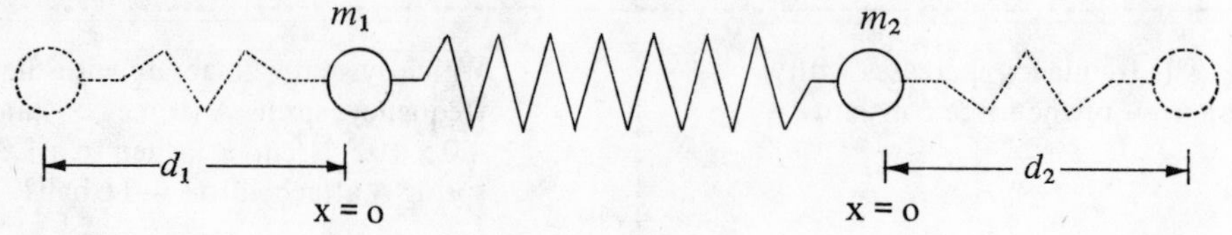

$m_1$ $m_2$

$d_1$ $d_2$

x = o x = o

$$E_p = \frac{1}{2} k\, (d_1 + d_2)^2$$

$$\nu = \frac{1}{2\pi}\sqrt{\frac{k}{\frac{m_1 m_2}{m_1 + m_2}}}$$

Q-14 A spring ($k = 5 \times 10^{-5}$ N/cm) is connected to two balls of mass 1.0 and 2.0 g, respectively. If the 1.0 g mass is displaced 2 cm from equilibrium and the 2.0 g mass is displaced 1.0 cm from equilibrium, what will be the potential energy of the system?
(Hint: 1 N cm = $1 \times 10^{-2}$ J)

A-14 $E_p = 1/2\,(5.0 \times 10^{-5}\text{ N/cm})\,(1.0 + 2.0)^2\text{ cm}^2$
$= 23 \times 10^{-5}$ N cm
$= 2.3 \times 10^{-6}$ J

Q-15 A spring system is stretched so that the potential energy is $3.6 \times 10^{-6}$ J. If the force constant of the spring is $9 \times 10^{-5}$ N/cm and the two balls attached to the spring each have a mass of 1.0 g, what must be the total displacement of the balls from their equilibrium position?

A-15 $3.6 \times 10^{-6}\text{ J} = 1/2\,(9.0 \times 10^{-5}\text{ N/cm})\,(d_1 + d_2)^2$

$8\text{ cm}^2 = (d_1 + d_2)^2$

Total displacement $= \sqrt{8\text{cm}^2}$

$(d_1 + d_2) = 2.8$ cm

Q-16 What is the *frequency* of vibration for the following spring system?

$k = 5 \times 10^{-5}$ N/cm

$m = 1.0$ g $m = 1.0$ g

A-16

$$\nu = \frac{1}{2\pi}\sqrt{\frac{5.0 \times 10^{-5}\text{ N/cm}}{\frac{(1.0)(1.0)\text{g}^2}{(1.0 + 1.0)\text{g}}}}$$

$$= \frac{1}{2\pi}\sqrt{5.0 \times 10^{-5}\text{ N/cm}/0.5\text{g}}$$

$$= \frac{1}{2\pi}\sqrt{10}\ \text{s}^{-1}$$

$\nu = 0.5$ cycle/s

Q-17 What is the frequency of vibration for the following spring system?

$k = 5 \times 10^{-5}$ N

$m = 1.0$ g $m = 2.0$ g

A-17

$$\nu = \frac{1}{2\pi}\sqrt{5.0 \times 10^{-5}\text{ N/cm}/\frac{(1.0)(2.0)\text{g}^2}{(1.0 + 2.0)\text{g}}}$$

$$\nu = \frac{1}{2\pi}\sqrt{7.5}\ \text{s}^{-1} = 0.43\text{ cycle/s}$$

Q-18 As the mass of the balls on a vibrating system increases, how does the frequency change?

A-18 As the mass of the balls increases, the frequency decreases.

Q-19 Each ball of the spring system shown below can be displaced to any value of $d_1$ or $d_2$. What are the possible values for the potential energy of this system?

$k = 5.0 \times 10^{-5}$ N/cm

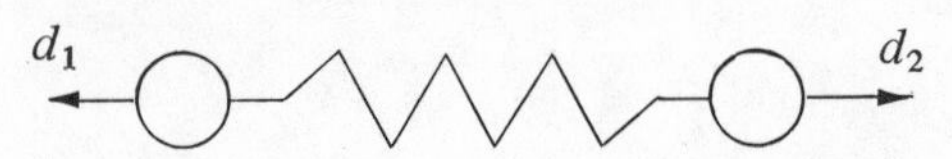


A-19 $E_p = 1/2\,k\,(d_1 + d_2)^2$

Because $d_1$ and $d_2$ can have any value from $0 \rightarrow \infty$, the potential energy, $E_p$, can have any value from $0 \rightarrow \infty$.

Q-20 If two balls are connected by a spring with a force constant $k$, what is the potential energy of the system when $d_1 + d_2 = 0$?

A-20 $E_p = 1/2\,k\,(0)^2 = 0$

The potential energy = 0.

Q-21 If two balls are connected by a spring with a force constant $k$ N/cm, what is the potential energy of the system (in terms of $k$) for each of the following values of $(d_1 + d_2)$?

| $\pm (d_1 + d_2)$ | $E_p$ |
|---|---|
| 0 | 0 $k$ |
| 1 | |
| 2 | |
| 3 | |
| 4 | |
| 5 | |

A-21

| $\pm (d_1 + d_2)$ | $E_p$ |
|---|---|
| 0 | 0 $k$ |
| 1 | 0.5 $k$ |
| 2 | 2 $k$ |
| 3 | 4.5 $k$ |
| 4 | 8 $k$ |
| 5 | 12.5 $k$ |

Q-22 On the graph below, plot the $E_p$ for the $\pm (d_1 + d_2)$ values found in A-21.

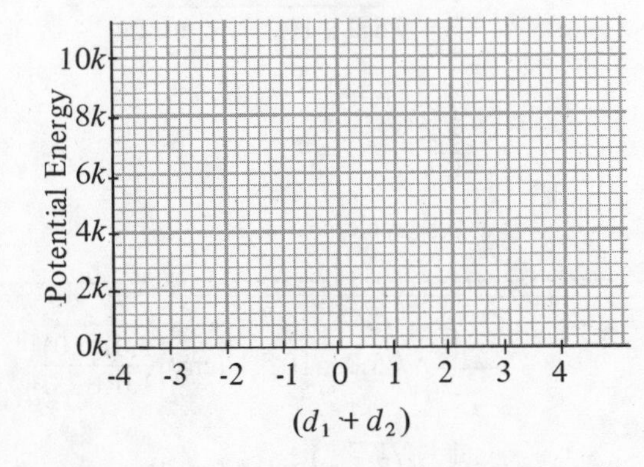

A-22

Potential Energy

10$k$
8$k$
6$k$
4$k$
2$k$
0$k$

-4 -3 -2 -1 0 1 2 3 4

$(d_1 + d_2)$

Q-23 The potential energy diagram below is for a spring system where the force constant is equal to $k$. Sketch on the graph the potential energy diagram for a spring system where the force constant is $2k$.

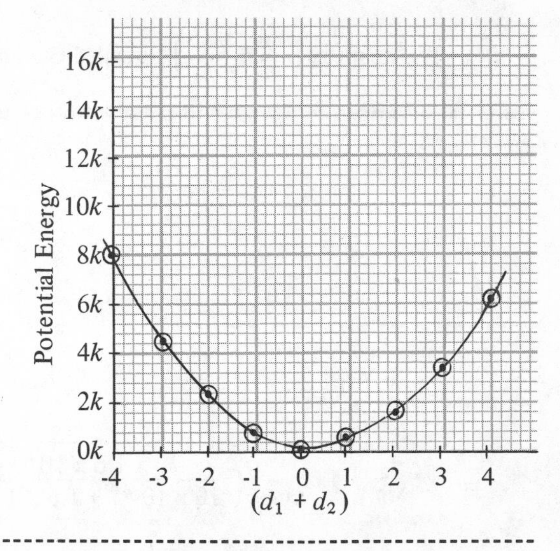


A-23

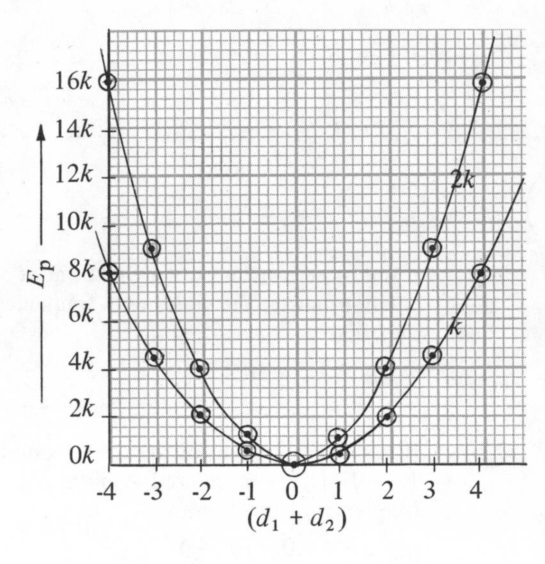


Q-24 The potential energy diagrams below are for four spring systems, each with a different force constant. Which spring system has the largest force constant? Which system has the smallest force constant?

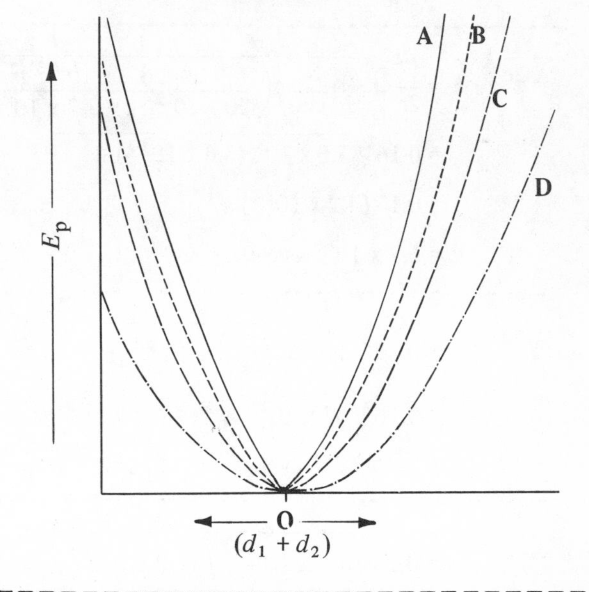


A-24 Spring system **A** has the largest force constant, and system D has the smallest force constant.

**R** For a harmonic oscillator with two balls on a spring

$E_p = 1/2\, k\, (d_1 + d_2)^2$

$E_k = 1/2\, mv^2$

$$\nu = \frac{1}{2\pi}\sqrt{k \Big/ \left(\frac{m_1 m_2}{m_1 + m_2}\right)}$$

S-4 In molecular systems the frequency of vibration, $\nu$, is a function of the mass of atoms, $m_1$ and $m_2$, and the strength of the bond (force constant), $k$.

$$\nu = \frac{1}{2\pi}\sqrt{k \Big/ \left(\frac{m_1 m_2}{m_1 + m_2}\right)}$$

The quantity $\left(\frac{m_1 m_2}{m_1 + m_2}\right)$ is called the reduced mass, $\mu_{m_1\text{-}m_2}$.

Q-25 If the force constant of a C-H bond is 5.0 N/cm, and the mass of carbon and hydrogen atoms are 20 x $10^{-24}$ g and 1.6 x $10^{-24}$ g, respectively, what is the vibrational frequency of the C-H bond?

(1 N = 1 kg m/s$^2$ = 1 x $10^5$ g cm/s$^2$)

A-25

$$\nu = \frac{1}{2\pi}\sqrt{k/\mu_{\text{C-H}}}$$

$$\nu = \frac{1}{2\pi}\sqrt{5.0\,\frac{\text{N}}{\text{cm}} \Big/ \left(\frac{20 \times 1.6 \times 10^{-48}\,\text{g}^2}{20 \times 10^{-24}\,\text{g} + 1.6 \times 10^{-24}\,\text{g}}\right)}$$

$$= 0.16\sqrt{5.0\,\frac{\text{N}}{\text{cm}}\,(6.7 \times 10^{23}\,\text{g}^{1})}$$

$$= 0.16\sqrt{34 \times 10^{28}\ \text{s}^{-2}}$$

$$= 9.3 \times 10^{13}\ \text{cycle/s}$$

Q-26 If the force constant of a C-D bond is 5.0 N/cm, what is the vibrational frequency of the bond? Compare the vibrational frequency of the C-D and the C-H bond (see A-25).

A-26

$$\nu = \frac{1}{2\pi}\sqrt{5.0\,\frac{\text{N}}{\text{cm}} \Big/ \left(\frac{20 \times 3.2 \times 10^{-48}\,\text{g}^2}{20 \times 10^{-24}\,\text{g} + 3.2 \times 10^{-24}\,\text{g}}\right)}$$

$$= 0.16\sqrt{5.0 \times 10^{5}\ (3.6 \times 10^{23})\ \text{sec}^{-2}}$$

$$= 0.16\,(4.2 \times 10^{14})\ \text{s}^{-1}$$

$$= 6.7 \times 10^{13}\ \text{cycle/s}$$

Q-27 If the force constant of an O-H bond is 5.0 N/cm, calculate the vibrational frequency of the bond.

($\mu_{\text{OH}}$ = 1.5 x $10^{-24}$ g)

A-27

$$\nu = 0.16\sqrt{5.0\,\frac{\text{N}}{\text{cm}} \Big/ (1.5 \times 10^{-24}\,\text{g})}$$

$$= 0.16\sqrt{33 \times 10^{28}\ \text{s}^{-2}}$$

$$= 9.2 \times 10^{13}\ \text{cycle/s}$$

Q-28 If the force constant of the C = C bond is 1.0 x 10 N/cm, calculate the vibrational frequency of the bond.

($\mu_{\text{C=C}}$ = 1.0 x $10^{-23}$ g)

A-28

$$\nu = 0.16\sqrt{1.0 \times 10\,\frac{\text{N}}{\text{cm}} \Big/ (1.0 \times 10^{-23}\,\text{g})}$$

$$= 0.16\sqrt{1.0 \times 10^{29}\ \text{s}^{-2}}$$

$$= 5.1 \times 10^{13}\ \text{cycle/s}$$

**R** The vibrational frequency of a chemical bond is a function of the bond strength and the reduced mass.

$$\nu = \frac{1}{2\pi}\sqrt{k/\mu_{m_1 - m_2}}$$

$$\mu_{m_1 - m_2} = \frac{m_1 m_2}{m_1 + m_2}$$

S-5 The chemical bond *differs* from the "two balls on a spring" system in that for molecules only certain vibrational energy levels are allowed; that is, the vibrational energy is quantized.

The energy levels possible are

$E_{vibrational} = (v + 1/2)h\nu$

where v is a vibrational quantum number (0, 1, 2, . . . ), $h$ is Planck's constant ($6.63 \times 10^{-34}$ J s), and $\nu$ is the vibrational frequency of the bond.

---

Q-29 The vibrational frequency of a bond is given by

$$\nu = \frac{1}{2\pi}\sqrt{k/\mu_{m_1 - m_2}}$$

Derive an expression for $E_{vib}$ in terms of the reduced mass of the atoms and the force constant. What are the units for $E_{vib}$?

---

A-29 $E_{vib} = (n + 1/2)h\nu$

$$\nu = \frac{1}{2\pi}\sqrt{k/\mu_{m_1 - m_2}}$$

$$E_{vib} = \frac{(v + 1/2)h}{2\pi}\sqrt{k/\mu_{m_1 - m_2}}$$

Units: v and $\pi$ are unitless, $h$ is in J s, $\mu$ is in kg, and $k$ is in N/m.

Therefore, the unit of $E_{vib}$ is J.

Q-30 What is the value of $E_{vib}$ when v = 0?

---

A-30 $E_{vib} = (v + 1/2)h\nu$

When v = 0

$E_{vib} = 1/2h\nu$

Q-31 What is the difference in energy ($\Delta E_{vib}$) between the lowest possible energy level of a bond and the next highest energy level?

---

A-31 $\Delta E_{vib} = E_{vib\ v\ =\ 1} - E_{vib\ v\ =\ 0}$

$\Delta E_{vib} = 3/2h\nu - 1/2h\nu$

$= h\nu$

Q-32 What is the difference in energy ($\Delta E_{vib}$) between the energy states v = 1 and v = 2?

---

A-32 $\Delta E_{vib} = E_{vib\ v\ =\ 2} - E_{vib\ v\ =\ 1}$

$\Delta E = 5/2h\nu - 3/2h\nu$

$= h\nu$

Q-33 What is the difference between any two successive vibrational energy levels?

---

A-33 $\Delta E_{vib}$ between any two successive levels = $h\nu$

Q-34 Derive an equation which relates $\Delta E_{vib}$ to the force constant and reduced mass of the atoms.

A-34 $\Delta E_{vib}$ is directly proportional to the frequency of vibration. The frequency is dependent upon the reduced mass of the atoms and the force constant.

$$\Delta E_{vib} = h\nu = \frac{h}{2\pi}\sqrt{k/\mu_{m_1 - m_2}}$$

Q-35 How does the separation between energy levels, $\Delta E_{vib}$, change as the force constant increases?

A-35 $\Delta E_{vib} = \frac{h}{2\pi}\sqrt{k/\mu_{m_1 - m_2}}$ As $k$ increases $\Delta E_{vib}$ increases. The separation increases.

Q-36 Assuming the harmonic oscillator model (ball and spring) is valid for chemical bonds, sketch the potential energy and vibrational energy levels for the two bonds, C – C and C = C. The force constant for the C = C bond is greater than the force constant for the C – C bond.

A-36

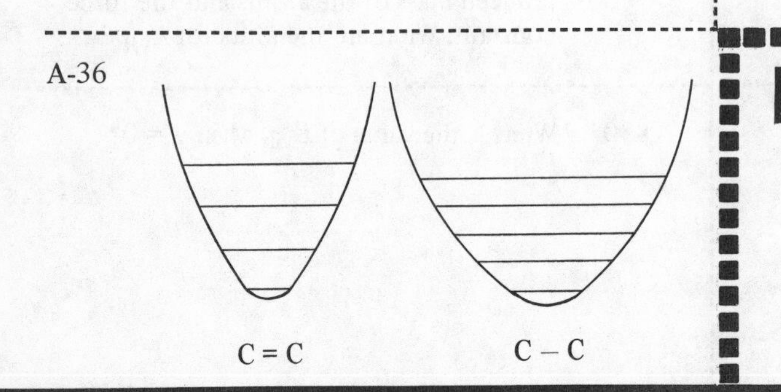


**R** Vibrational energy of a chemical bond is quantized and can have the value

$E_{vib} = (v + 1/2)h\nu$.

The spacing between two successive energy levels ($\Delta E_{vib}$) is

$\Delta E_{vib} = h\nu$

S-6 The potential energy curves for chemical bonds are *not* symmetrical like those for harmonic oscillators. For molecular vibrations the potential energy curves are distorted as shown below.

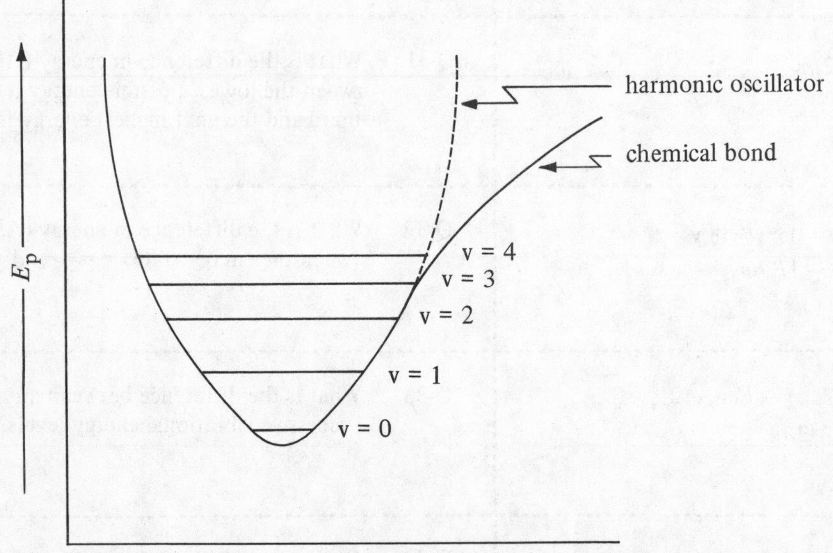


At ordinary temperatures, where molecules are in their lowest vibrational levels, the harmonic oscillator model is a good approximation of the chemical bond.

Absorption of light with energy equal to the difference in energy between two vibrational energy levels ($\Delta E_{vib}$) will cause a vibrational transition to occur. Light with this energy is found in the infrared region of the electromagnetic spectrum.

Transitions from the ground state (v = 0) to the first excited state (v = 1) absorb light strongly giving rise to intense fundamental spectral bands. Transitions from the ground state to the second excited state (v = 2) absorb light and give rise to weak *overtone* bands.

Q-37 The molecule A-B has the potential energy curve shown in S-6. Are the energy levels equally spaced in the molecule?

A-37 No. As the quantum number increases, the energy levels become more closely spaced.

Q-38 At what energy level does the potential energy diagram of the molecule A-B (S-6) deviate appreciably from the harmonic oscillator curve?

A-38 v = 3 or 4

Q-39 Derive an equation which relates the *wavelength* of light absorbed in causing a transition between two energy levels in molecule **A-B**, the force constant of **A-B**, $k_{A\text{-}B}$, and the reduced mass of atoms **A** and **B**, $\mu_{A\text{-}B}$.

A-39 $\lambda = c/\nu$

$$\nu_{AB} = \frac{1}{2\pi}\sqrt{k_{AB}/\mu_{AB}}$$

$$\lambda = \frac{c}{\frac{1}{2\pi}\sqrt{k_{AB}/\mu_{AB}}}$$

Q-40 Derive an equation relating the *wave number* of light required to cause a transition between two adjacent energy levels in the molecule **AB**, $k_{AB}$, and the reduced mass, $\mu_{AB}$.

A-40

$$\lambda = \frac{c}{\frac{1}{2\pi}\sqrt{k_{AB}/\mu_{AB}}}$$

$$\bar{\nu} = 1/\lambda$$

$$\bar{\nu} = \frac{1}{2\pi c}\sqrt{k_{AB}/\mu_{AB}}$$

Q-41 The frequency for the C – H bond was calculated from the force constant and mass of the atoms and found to be $9.3 \times 10^{13}\ s^{-1}$. What is the wavelength (in nm) and wave number (in $cm^{-1}$) of this fundamental (v = 0 to v = 1 transition) absorption band for the C – H bond?

A-41 $\nu = 9.3 \times 10^{13}\ s^{-1}$

$$\lambda = c/\nu = \frac{3.0 \times 10^{10}\ cm\ s^{-1}}{9.3 \times 10^{13}\ s^{-1}}$$

$\lambda = 3.2 \times 10^{-4}$ cm or 3 200 nm

$\bar{\nu} = 1/\lambda = 3.1 \times 10^{3}\ cm^{-1}$ or

3 100 $cm^{-1}$.

Q-42 Assuming that the v = 0, v = 1, and v = 2 energy levels are equally spaced, what would the wavelength (in nm) and wave number (in $cm^{-1}$) be for the first overtone band (v = 0 to v = 2) of the C – H bond?

A-42 $h\nu_{light} = 2h\nu_{vib}$

$\nu_{light} = 2\nu_{vib} = 2(9.3 \times 10^{13})\ s^{-1}$

$$\lambda = \frac{c}{\nu_{light}} = \frac{3.0 \times 10^{10}\ cm\ s^{-1}}{2(9.3 \times 10^{13}\ s^{-1})}$$

$\lambda = 1\ 600$ nm $\quad \bar{\nu} = 6\ 200\ cm^{-1}$

Q-43 Calculate the wavelength and wave number of the fundamental absorption band for the C = O vibration. Assume a force constant of $1.0 \times 10$ N/cm ($\mu_{C=O} = 1.2 \times 10^{-23}$ g).

A-43 $k = 1.0 \times 10$ N/cm

$$\nu_{light} = \nu_{vib} = \frac{1}{2\pi}\sqrt{1.0 \times 10 \frac{N}{cm} / 1.2 \times 10^{-23} g}$$

$$= 0.16 \sqrt{8.3 \times 10^{28} \ s^{-2}}$$

$$= 4.6 \times 10^{13} \ s^{-1}$$

$$\lambda_{light} = c/\nu = \frac{3.0 \times 10^{10} \ cm \ s^{-1}}{4.6 \times 10^{13} \ s^{-1}} = 6.5 \times 10^{-4} \ cm$$

$$= 6\ 500 \ nm$$

$$\bar{\nu} = 1/\lambda = 1/6.5 \times 10^{-4} = 1\ 500 \ cm^{-1}$$

Q-44 Cyclopentanone has an intense absorption band at $\bar{\nu} = 1750$ cm$^{-1}$ due to the $\rangle C = O$. Assuming that this is the fundamental band, calculate the force constant for the $\rangle C = O$ in cyclopentanone ($\mu_{C=O} = 1.2 \times 10^{-23}$ g).

---

A-44 $\bar{\nu} = 1750$ cm$^{-1}$

$\lambda = 1/\bar{\nu}$ $\quad \lambda = c/\nu$

$$\bar{\nu} = \nu/c = \frac{1}{2\pi c}\sqrt{k/\mu_{C=O}}$$

$$1750 \ cm^{-1} = 5.3 \times 10^{-12} \frac{s}{cm}\sqrt{k/1.2 \times 10^{-23} \ g}$$

$$(3.3 \times 10^{14})^2 \ s^{-2} = 8.3 \times 10^{22} \ g^{-1} \ k$$

$$\frac{11 \times 10^{28}}{8.3 \times 10^{22}} \ s^{-2} \ g = k = 1.3 \times 10^{6} \frac{g}{s^2}$$

$$= 1.3 \times 10 \frac{N}{cm}$$

Q-45 Carbon to carbon bond stretch vibrations give rise to absorptions in the following wavelength regions.

| | |
|---|---|
| $\geq C - C \leq$ | 7.0$\mu$m |
| $\rangle C = C \langle$ | 6.0$\mu$m |
| $- C \equiv C -$ | 4.5$\mu$m |

Arrange the three bonds in order of increasing force constant.

(micro = $10^{-6}$ = $\mu$)

---

A-45 Wavelengths are inversely proportional to the square root of force constants.

$$\lambda = \frac{c}{\frac{1}{2\pi}\sqrt{k\mu_{cc}}}$$

smallest → largest force constant

$\geq C - C \leq \ < \ \rangle C = C \langle \ < \ - C \equiv C -$

Q-46 What is the relationship between force constant and the strength of a chemical bond?

---

A-46 The force constant is directly proportional to the strength of a bond.

Q-47 Fundamental stretching vibrations of $\geq C - C \leq$, $\geq C - N -$, and $\geq C - O -$ give rise to absorptions in the following regions.

| | $\lambda$, $\mu$m | $\bar{\nu}$, cm$^{-1}$ |
|---|---|---|
| C – C | 7.0 | 1,430 |
| C – N | 7.5 | 1,330 |
| C – O | 7.8 | 1,280 |

Compare the relative differences between the v = 0 and v = 1 vibrational energy levels for the C – C, C – N, and C – O bonds.

A-47

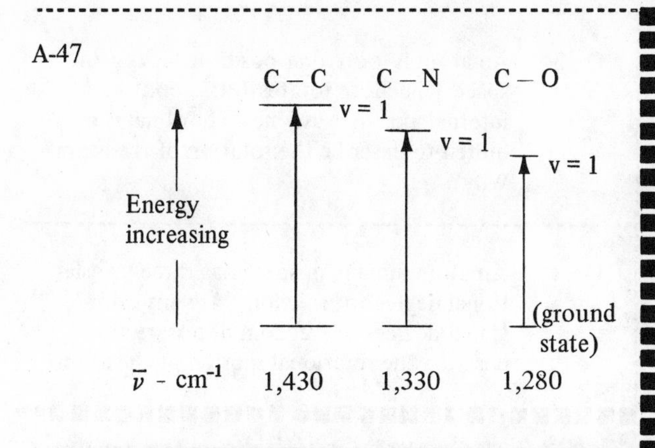


**R** At ordinary temperatures, the potential energy diagram of a molecule approximates that of a harmonic oscillator. For fundamental absorptions (v = 0 to v = 1)

$h\nu_{light} = h\nu_{vibration}$

S-7 Polyatomic molecules may exhibit more than one fundamental vibrational absorption band. The number of these fundamental bands is related to the *degrees of freedom* in a molecule. The number of *degrees of freedom* is equal to the sum of the coordinates necessary to locate all the atoms of the molecule in space.

Q-48 Consider a single atom at position $P_1$ in *space.*

How many *coordinates* are required to describe the position of the atom $P_1$?

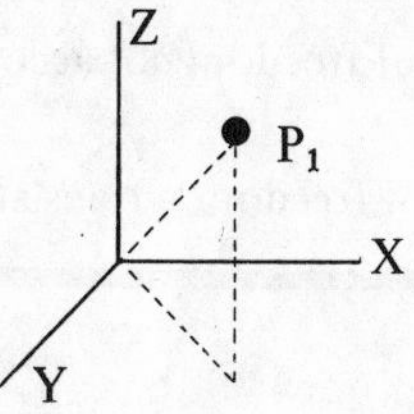


A-48 Three coordinates are required, the X, Y, and Z.

Q-49 How many degrees of freedom does the single atom P have?

A-49 Three. Because three coordinates are required to locate the atom in space.

Q-50 Consider a single atom restricted to movement in a plane. How many *translational* degrees of freedom would such an atom have? Why?

A-50 Two translational degrees of freedom, because only two coordinates (X, Y) would be required to fix the position of the atom in the plane.

Q-51 If there were two atoms in space, how many *total* translational degrees of freedom would the two atoms have? Why?

A-51 Six. Each atom would have three translational degrees of freedom because three coordinates would be required to fix the position of *each* atom.

Q-52 If there were *n* atoms in *space,* how many total translational degrees of freedom would the *n* atoms have? Why?

A-52 There would be $3n$ translational degrees of freedom. Three coordinates would be required to fix the position of *each* atom.

Q-53 An atom is located at position X, Y, Z in space. The atom rotates 180° about its internal axis. Are any new coordinates required to describe the rotation of the atom? Why?

---

A-53 No. The new orientation of the atom is identical to the original orientation because of the symmetry of the atom.

Q-54 An atom moving in space has three translational degrees of freedom. Are any additional degrees of freedom necessary to describe the rotational motion of the atom?

---

A-54 No, because no coordinates are necessary to specify the rotational orientation.

**R** The number of degrees of freedom equals the number of coordinates necessary to fix position in space.

---

S-8 Atoms have three degrees of freedom, all of which are translational. When atoms combine to form molecules, no degrees of freedom are lost. That is, the total number of degrees of freedom of the molecule will be $3n$ where $n$ is the number of atoms in the molecule. The $3n$ degrees of freedom of the molecule will be made up of *rotational, vibrational,* and *translational* degrees of freedom.

*Rotational* degrees of freedom result from the rotation of a molecule about an axis through the center of gravity. These rotations result in a degree of freedom *only if* the positions of the atoms in space change during the rotation.

All degrees of freedom not accounted for by translational and rotational are vibrational degrees of freedom.

$3n$ degrees of freedom = translational + rotational + vibrational.

---

Q-55 How many coordinates are necessary to locate the diatomic *molecule* in space? How many translational degrees of freedom does a diatomic molecule have?

---

A-55 Three coordinates. A diatomic molecule has three translational degrees of freedom.

Q-56 Rotation of a diatomic molecule around the X, Y, and Z axes is illustrated below. Which of these rotations will result in rotational degrees of freedom?

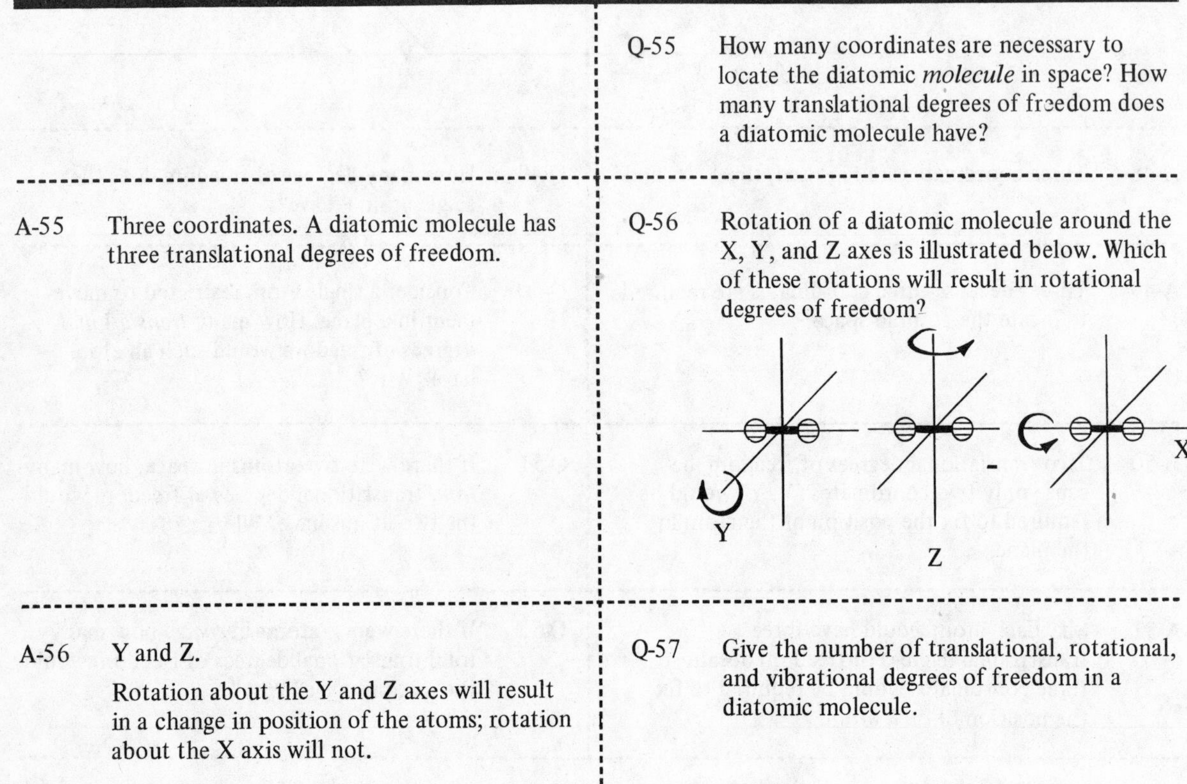


---

A-56 Y and Z.

Rotation about the Y and Z axes will result in a change in position of the atoms; rotation about the X axis will not.

Q-57 Give the number of translational, rotational, and vibrational degrees of freedom in a diatomic molecule.

A-57 Total degrees of freedom = 3 ($n$) = 6

Translational = 3 (A-55)

Rotational = 2 (A-56)

Vibrational = $3n$-5 = 1

Q-58 What motion is associated with the one vibrational degree of freedom in the diatomic molecule?

A-58

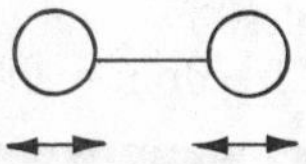

Q-59 Give the number of translational, rotational, and vibrational degrees of freedom for $CO_2$.

O = C = O

A-59 Total degrees of freedom = $3n$ = 9

Translational = 3

Rotational = 2

Vibrational = $3n$-5 = 4

Q-60 Give the number of translational, rotational, and vibrational degrees of freedom for acetylene.

H – C ≡ C – H

A-60 Total degrees of freedom = $3n$ = 12

Translational = 3

Rotational = 2

Vibrational = $3n$-5 = 7

Q-61 Rotations of a non-linear molecule ($H_2O$) around the X, Y, and Z axes are illustrated below. Which of these rotations will result in rotational degrees of freedom?

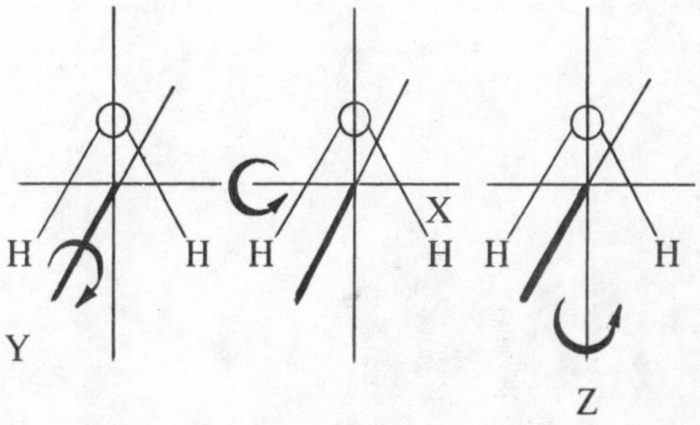

A-61 X, Y, and Z.

Rotation about all axes will result in a change in the position of the atoms.

Q-62 Give the number of translational, rotational, and vibrational degrees of freedom for the non-linear molecule, water.

A-62 Total degrees of freedom = $3n$ = 9

Translational = 3

Rotational = 3

Vibrational = $3n$-6 = 3

Q-63 Give the number of translational, rotational, and vibrational degrees of freedom for the non-linear molecule, $CH_4$.

A-63 Total degrees of freedom = $3n$ = 15

Translational = 3

Rotational = 3

Vibrational = $3n$-6 = 9

Q-64 Give the number of translational, rotational, and vibrational degrees of freedom for benzene, $C_6H_6$.

A-64 Total degrees of freedom = $3n = 36$

Translational = 3

Rotational = 3

Vibrational = $3n$-6 = 30

**R** A molecule with $n$ atoms has $3n$ degrees of freedom.

$3n$ degrees of freedom = translational + rotational + vibrational.

For linear molecules:
Vibrational degrees of freedom = $3n$-5.

For non-linear molecules:
Vibrational degrees of freedom = $3n$-6.

S-9 An infrared absorption band will be observed for each vibrational degree of freedom provided:

1. A change in the dipole moment of the molecule occurs during the vibration.
2. The band does not coincide in frequency with some other fundamental vibration (the energy levels are degenerate).
3. The absorption falls within the infrared region.
4. The intensity of the absorption is strong enough to be detected.

Additional bands may appear in the spectrum due to overtones and *combination* tones (combinations of two or more fundamental vibrations).

Q-65 In the following diagram, the arrows represent the direction and magnitude of the dipole moment of each bond during various stages of a vibration. The dotted arrow represents the resultant dipole of the molecule. Does the dipole moment of the molecule change during the vibration illustrated? Will the vibration be infrared active?

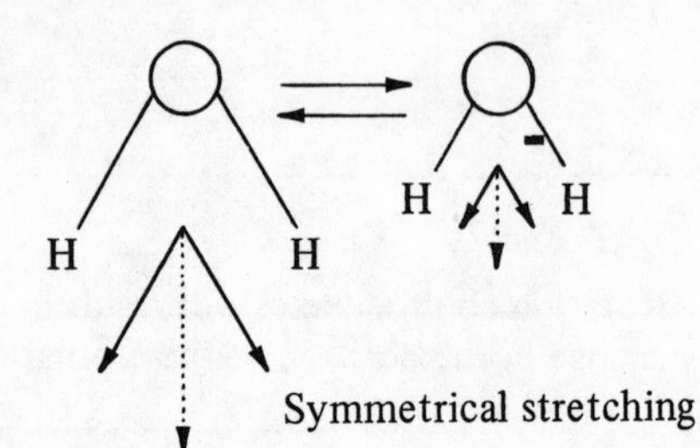


Symmetrical stretching

A-65 Yes, the dipole changes. The vibration will be infrared active.

Q-66 The two other fundamental vibrational modes of water are shown below. The arrows represent the motion of the atoms. Does the dipole moment of the molecule change during the vibrations illustrated? Will the vibrations be infrared active?

unsymmetrical stretching

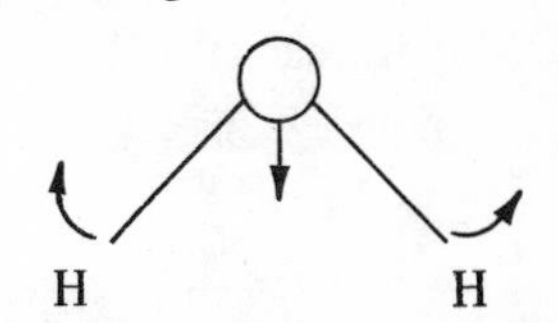


bending vibration

A-66 Both unsymmetrical stretching and bending vibrations will result in a change in dipole of the molecule. Both vibrations are infrared active.

Q-67 How many bands would be expected in the infrared spectrum of water?

A-67 Three. The infrared spectrum of water is shown below.

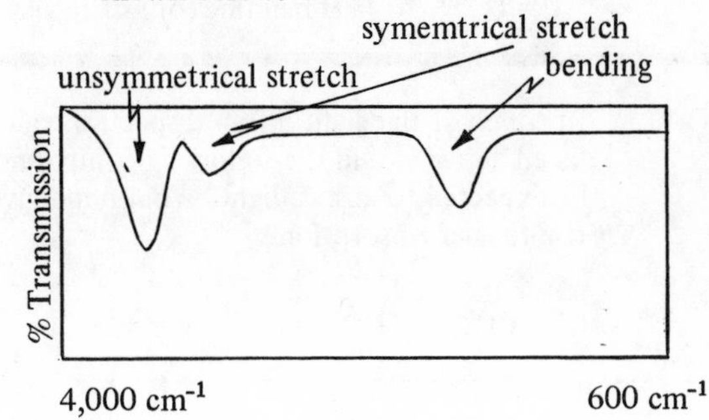


Q-68 The fundamental vibrational modes for $CO_2$ are given below. Which of the vibrations illustrated will be infrared active?

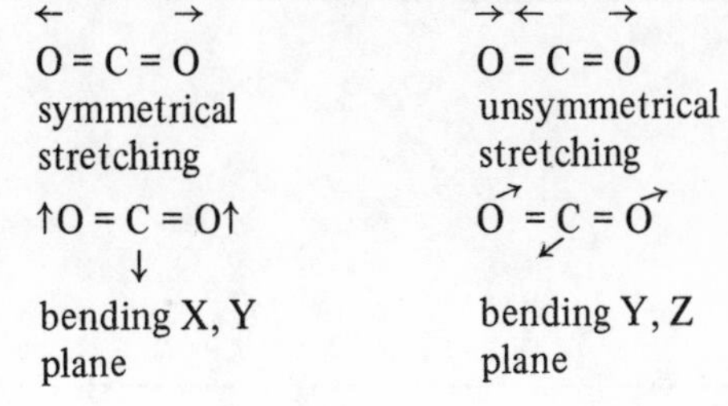


A-68

| | |
|---|---|
| Symmetrical stretch | inactive |
| Unsymmetrical stretch | active |
| Bending X, Y plane | active |
| Bending Y, Z plane | active |

Q-69 The two fundamental bending modes for $CO_2$ are degenerate. How many bands would be expected in the infrared spectrum of $CO_2$?

A-69 Two bands would be expected.

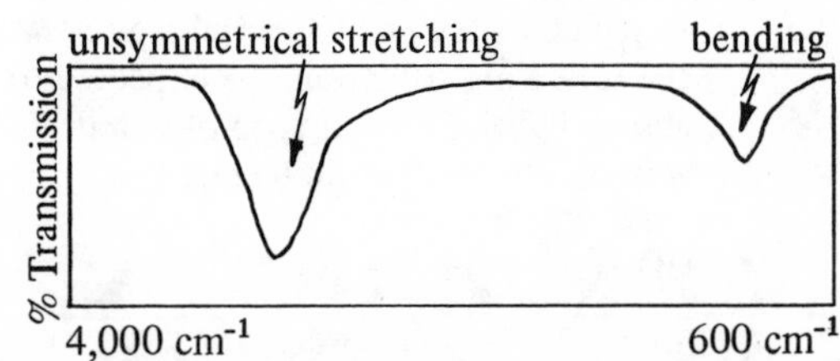


**R** Vibrational modes are infrared active if a change in dipole moment occurs during the vibration.

# INFRARED SPECTROSCOPY

## PART II: Interpretation of Infrared Spectra

S-1 Identification of characteristic absorption bands caused by different functional groups is the basis for the interpretation of infrared spectra. Thus, O – H stretching frequencies give rise to strong absorption bands in the 3350 $cm^{-1}$ region. The presence of a strong band in the 3350 $cm^{-1}$ region of the infrared spectrum of a compound is strongly suggestive that the molecules contain the O – H functional group. The following section will demonstrate the technique used in interpreting infrared spectra.

The eight most important and well-defined areas used in a preliminary examination of the spectra are given below.

| Spectral region: Wavelength Micrometers | Spectral region: Wave number $\bar{\nu}$ - $cm^{-1}$ | Bond causing absorption |
|---|---|---|
| 2.7 – 3.3 | 3750 – 3000 | O – H, N – H stretching |
| 3.0 – 3.4 | 3300 – 2900 | – C ≡ C – H, >C = C<H, Ar – H, (C – H stretch) |
| 3.3 – 3.7 | 3000 – 2700 | $CH_3$–, –$CH_2$–, ≡C – H, C(=O) – H (C – H stretch) |
| 4.2 – 4.9 | 2400 – 2100 | C ≡ C, C ≡ N stretching |
| 5.3 – 6.1 | 1900 – 1650 | C = O (acids, aldehydes, ketones, amides, esters, anhydride) stretching |
| 5.9 – 6.2 | 1675 – 1500 | >C = C< (aliphatic and aromatic) >C = N– stretching |
| 6.8 – 7.7 | 1475 – 1300 | –C – H bending |
| 10.0 – 15.4 | 1000 – 650 | >C = C<H, Ar – H bending (out of plane) |

Q-1 In which of the eight infrared spectral regions listed above would the following compound be expected to absorb light? What bond gives rise to each absorption?

$CH_3$ $CH_2$ C(=O) – H

A-1 Region $\bar{\nu}$ – $cm^{-1}$

| | |
|---|---|
| 3000 – 2700 | C – H stretch of $CH_3$, $CH_2$ and aldehyde proton |
| 1900 – 1650 | C = O stretch |
| 1475 – 1300 | C – H bending of $CH_3$–, –$CH_2$– |

Q-2 In which of the eight spectral regions would the following compound be expected to absorb light?

[Structure: six-membered ring with =O and N–H]

A-2 Region $\bar{\nu}$ – $cm^{-1}$

| | |
|---|---|
| 3750 – 3000 | N – H stretch |
| 3000 – 2700 | C – H stretch |
| 1900 – 1650 | C = O stretch |
| 1475 – 1300 | C – H bending |

Q-3 In which of the eight spectral regions would the following compound be expected to absorb light? To what type of vibration would each absorption be due?

HO–(benzene ring)–C(=O) – H

A-3

| Region $\bar{\nu}$ – $cm^{-1}$ | |
|---|---|
| 3750 – 3000 | O – H stretch |
| 3300 – 3000 | C – H stretch of Ar – H |
| 3000 – 2700 | aldehyde hydrogen |
| 1900 – 1650 | C = O stretch |
| 1675 – 1500 | C = C stretch (Ar) |
| 1000 – 650 | Ar – H bending (out of plane) |

Q-4 In which of the eight spectral regions would the following compound be expected to absorb light? To what type of vibration would the absorptions be due?

$CH_3\overset{O}{\overset{\|}{C}}OCH_2C\equiv CH$

A-4

| Region $\bar{\nu}$ – $cm^{-1}$ | |
|---|---|
| 3300 – 2900 | C≡C – H stretch |
| 3000 – 2700 | C – H stretch of $CH_3$– and $-CH_2-$ |
| 2400 – 2100 | C ≡ C stretch |
| 1900 – 1650 | C = O stretch |

Q-5 In which of the eight spectral regions would the following compound be expected to absorb light? To what type of vibration would each absorption be due?

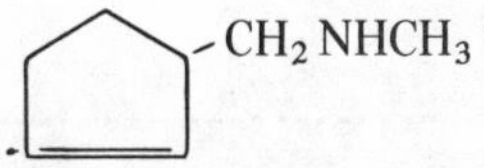


A-5

| Region $\bar{\nu}$ – $cm^{-1}$ | |
|---|---|
| 3750 – 3000 | N – H stretch |
| 3300 – 3000 | >C = C<H (C – H stretch) |
| 3000 – 2700 | $CH_3$–, $-CH_2-$, ≥C – H (C – H stretch) |
| 1475 – 1300 | ≥C – H bending |
| 1675 – 1625 | C = C stretch |
| 1000 – 650 | >C = C<H bending |

Q-6 Below is a portion of the infrared spectrum of a compound. What types of groups are possible? What types of groups are known to be absent?

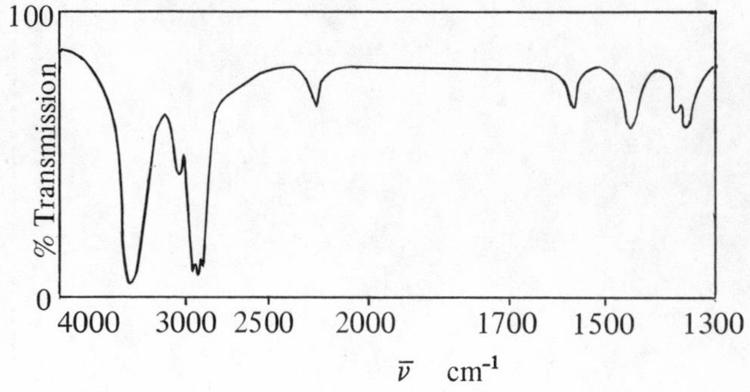

A-6

| Region $\bar{\nu}$ – cm$^{-1}$ | |
|---|---|
| 3750 – 3000 | NH or OH stretch |
| 3300 – 2700 | C – H stretch of $CH_3$–, –$CH_2$–, $\geq$C – H, Ar – H, $>$C = C$<$H, – C ≡ C – H or aldehyde hydrogen |
| 2400 – 2100 | C ≡ C or C ≡ N stretch |
| 1475 – 1300 | C – H bending |
| **Absent** | |
| 1900 – 1650 | C = O stretch |
| 1675 – 1625 | C = C stretch |

Q-7 Below is a partial spectrum of a compound which is known to have either structure I or II. Which structure is *not* consistent with the spectrum? Why?

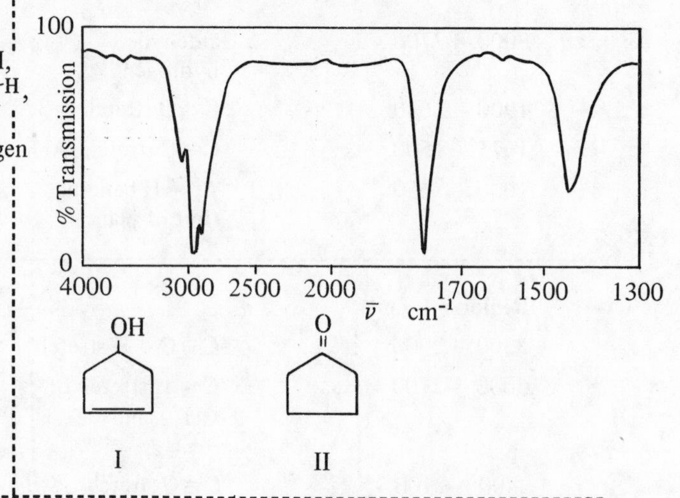


A-7 Structure I is not consistent with the spectrum because of the lack of bands in the 3750 – 3300 cm$^{-1}$ (O – H stretch) region and the 1675 – 1625 cm$^{-1}$ (C = C stretch) region. Absorption in the 1900 – 1650 cm$^{-1}$ region (C = O stretch) is consistent with structure II.

Q-8 Below is a partial spectrum of a compound which is known to have either structure I or II. Which structure is *not* consistent with the spectrum? Why?

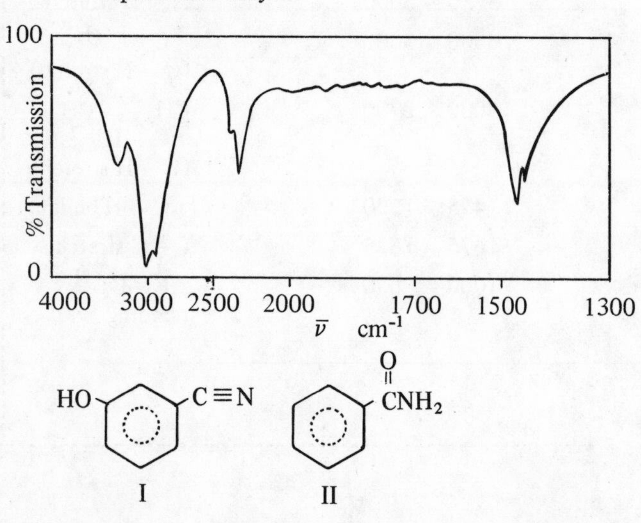

A-8 Structure II is *not* consistent with the spectrum because of the lack of C = O stretch (1900 – 1650 $cm^{-1}$) and the presence of absorption in the 2400 – 2100 $cm^{-1}$ region (C ≡ N stretch).

Q-9 A portion of the infrared spectrum of a molecule known to have either structure I or II is given below. Which structure *is consistent* with the spectrum? Why?

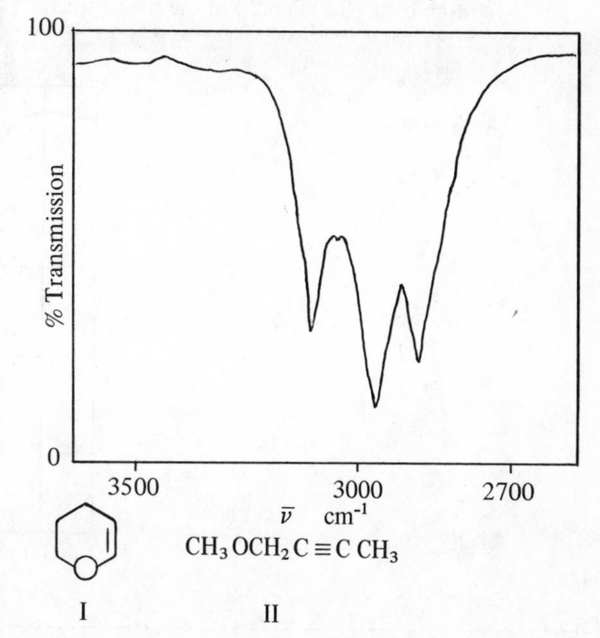


A-9 Structure I. The $\geq C = C\diagdown H$ absorption at 3080 $cm^{-1}$ is consistent with structure I but not structure II.

Q-10 A portion of the infrared spectrum of a molecule known to have structure I, II, or III is given below. Which structure is consistent with the spectrum? Why?

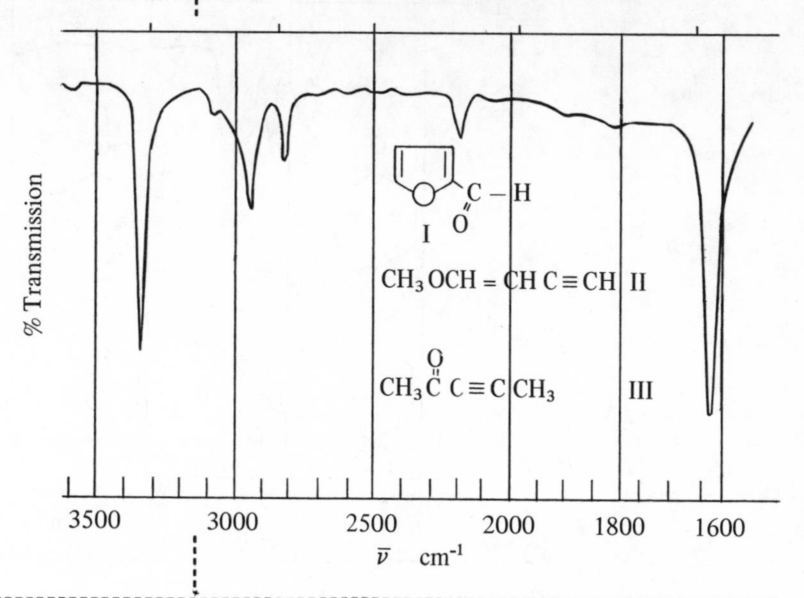

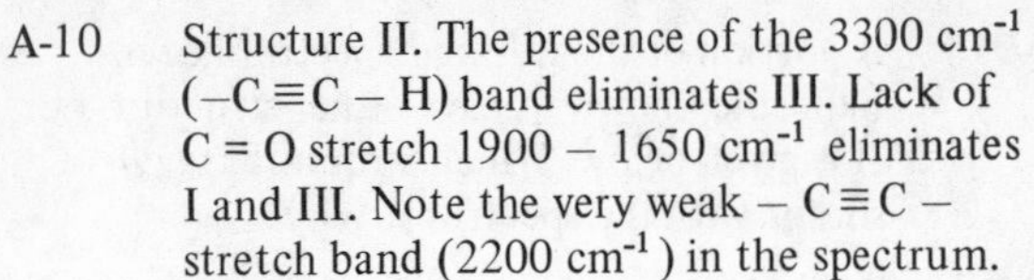

A-10 Structure II. The presence of the 3300 $cm^{-1}$ ($-C\equiv C-H$) band eliminates III. Lack of $C=O$ stretch 1900 – 1650 $cm^{-1}$ eliminates I and III. Note the very weak $-C\equiv C-$ stretch band (2200 $cm^{-1}$) in the spectrum.

Q-11 A portion of the infrared spectrum of a molecule known to have structure I, II, or III is given below. Which structure is consistent with the spectrum? Why?

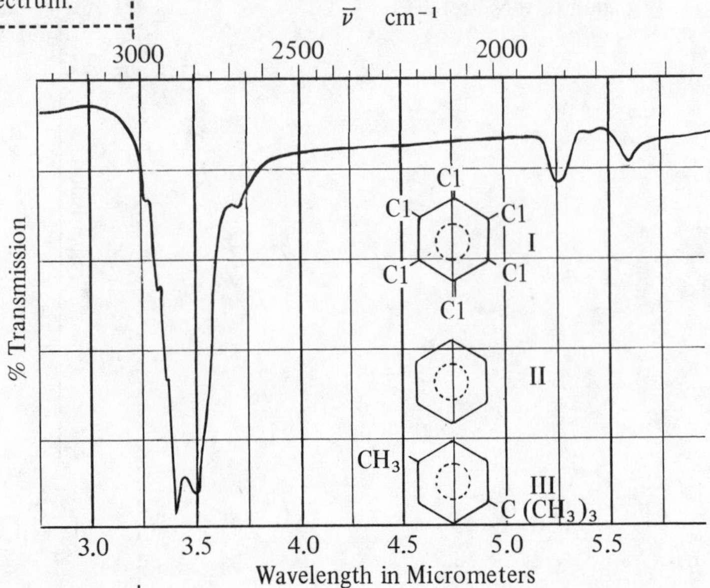


A-11 Structure III. The presence of both aromatic $C-H$ stretch (above 3000 $cm^{-1}$) and aliphatic $C-H$ stretch (3000 – 2700 $cm^{-1}$) is consistent with only structure III.

Q-12 A compound whose spectrum appears below was thought to have structure I, II, III, or IV. Which structure is most nearly consistent with the spectrum?

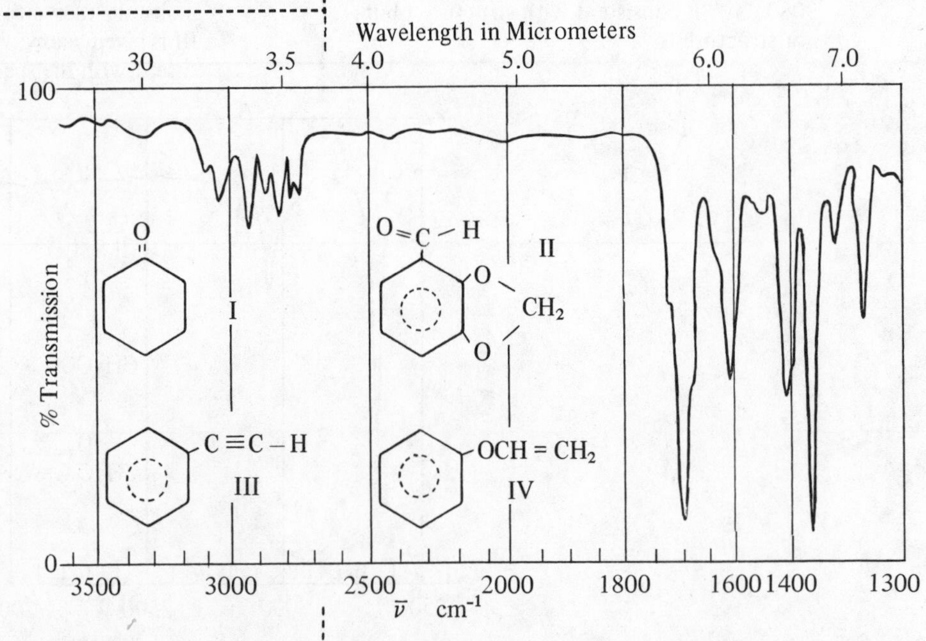


A-12 Structure II. The presence of $C-H$ stretch above 3000 $cm^{-1}$ eliminates I. The presence of $C=O$ stretch at 1700 $cm^{-1}$ eliminates structures III and IV.

S-2 Certain important distinctions can be made in the O – H, N – H stretching region of the spectrum (3750 – 3000 $cm^{-1}$).

**Free (non-hydrogen bonded) O – H stretching vibrations** are in the 3700 – 3500 $cm^{-1}$ range. Free phenol OH vibrations will tend to have absorption at the low (3500) end of this range. The free OH bands have lower intensity than bonded OH bands and will be evident in only dilute solutions (or in gas phase).

**A hydrogen bonded OH absorption** appears in the range 3450 – 3200 $cm^{-1}$ as a rather broad and intense band.

**Non-bonded amines** exhibit bands in the 3500 – 3300 $cm^{-1}$ region while bonded amines give bands in the 3500 – 3100 region; these bands are weaker than the bonded OH bands but much sharper. Primary amines show two bands, 2° amines and imines show only one band, and 3° amines show no bands. Amides and lactams also show N – H absorption in the 3500 – 3300 region.

**Carboxylic acids** in the solid state and even in relatively dilute solutions exist as dimers and show no OH absorption in the expected region. Instead, strong broad absorptions in the 3000 – 2500 $cm^{-1}$ region occur.

---

Q-13 Indicate the absorptions expected in the spectrum of the following compound. The spectrum was obtained of a *neat* sample (i.e., no solvent was present).

$CH_3-CH_2-OH$

---

A-13 A strong band in the bonded OH stretching region (3450 – 3200 $cm^{-1}$) and bands in the C – H stretch region (3000 – 2700 $cm^{-1}$) characteristic of $CH_3-$ and $-CH_2-$.

Q-14 Indicate the absorption expected for the following compound.

$CH_3-\overset{\overset{\displaystyle O}{\|}}{C}-\overset{\overset{\displaystyle H}{|}}{N}-CH_2-CH_3$

---

A-14 A weak but sharp N – H absorption in the 3500 – 3300 $cm^{-1}$ region. Absorption in the C – H region (3000 – 2700 $cm^{-1}$) due to $CH_3-$ and $-CH_2-$ stretch and a $>C=O$ absorption in the carbonyl region (1900 – 1650 $cm^{-1}$).

Q-15 A very dilute solution of *cis*-cyclopentane-1, 2-idol in $CCl_4$ shows bands at 3620 and 3455 $cm^{-1}$. Explain.

---

A-15 The absorption at 3620 $cm^{-1}$ is due to free OH stretch. Absorption at 3485 $cm^{-1}$ is due to bonded OH stretch. The bonded stretch in a dilute solution is indicative of intramolecular hydrogen bonding, i.e.,

O
H··· O – H

Q-16 How would the infrared spectra of the following compounds differ?

$C_6H_5CH_2NH_2$

$CH_3\overset{\overset{\displaystyle O}{\|}}{C}N(CH_3)_2$

---

A-16 The benzylamine would have two peaks in the 3500 – 3100 $cm^{-1}$ region and no C = O stretch, while N,N-dimethylacetamide would have no H–N stretch at 3500 – 3100 $cm^{-1}$ but would have a C = O stretch near 1650 $cm^{-1}$.

Q-17 How would the infrared spectra of the following compounds differ?

$(t-C_4H_9)_2N-H$ $(C_2H_5)_3N$

**A-17** Di-*t*-butylamine would have an H–N stretch in the 3500 – 3100 $cm^{-1}$ region while trimethylamine would have no N–H stretch.

**Q-18** How would the infrared spectra of the following compounds differ (neat samples)?

$CH_3CH(OH)CH_3$

$(CH_3CH_2CH_2CH_2)_2NH$

---

**A-18** *i*-propyl alcohol would have a strong broad OH stretch (3400 – 3200 $cm^{-1}$) and C–H stretch bands characteristic of $CH_3$– and $\geq$C–H, while *n*-dibutylamine would have a weak, sharp single band in the N–H stretch region (3500 – 3100 $cm^{-1}$) and bands characteristic of $CH_3$– and –$CH_2$–.

**Q-19** How would the infrared spectra of the following compounds differ (neat samples)?

$CH_3(CH_2)_6COOH$ (**A**) $(CH_3)_3CCHOHCH_2CH_3$ (**B**)

$(CH_3)_3CCH(CH_2CH_3)N(CH_3)_2$ (**C**)

---

**A-19** **A** would show no OH absorption in the 3500 – 3100 $cm^{-1}$ region but would absorb in the 3000 –2500 $cm^{-1}$ region as well as in the C = O region (1900 – 1650 $cm^{-1}$). **B** would give a strong OH band in the 3450 – 3200 $cm^{-1}$ region and no C = O absorption while **C** would give neither an absorption in the 3400 – 3100 $cm^{-1}$ region nor in the C = O region (1900 – 1650 $cm^{-1}$).

**Q-20** How would the infrared spectra of the following compounds differ (neat samples)?

Phenol (OH on benzene ring) Cyclohexanol (OH on saturated ring, S)

---

**A-20** Both would exhibit OH stretching in the 3450 – 3200 $cm^{-1}$ region (the phenol at lower wave numbers). However, the phenol would show Ar–H absorption and cyclohexanol would give –$CH_2$– and $\geq$C–H absorption. Also, phenol would give some aromatic C–H bending in the 1000 – 650 $cm^{-1}$ region while cyclohexanol would not.

**R**

| | | |
|---|---|---|
| Non-bonded | OH | 3700 – 3500 $cm^{-1}$ |
| Bonded | OH | 3450 – 3300 $cm^{-1}$ |
| Non-bonded | NH | 3500 – 3300 $cm^{-1}$ |
| Bonded | NH | 3500 – 3100 $cm^{-1}$ |
| Carboxylic Acids | | 3000 – 2500 $cm^{-1}$ |

S-3 Different types of C–H bonds show absorption within well-defined areas of the C–H stretching region (3300 – 2700 $cm^{-1}$). The *approximate* positions of the C–H bands for different types of groups are shown below.

| Type H | $\bar{\nu}$ $cm^{-1}$ | Band intensity |
|---|---|---|
| Ar–H | 3030 | moderate intensity |
| C≡C–H | 3300 | high intensity |
| C = C–H | 3040 – 3010 | moderate intensity |
| $-CH_3$ | 2960 & 2870 | high intensity |
| $-CH_2-$ | 2930 & 2850 | high intensity |
| >CH | 2890 | low intensity |
| –C(=O)–H | 2720 | low intensity |

Note that C≡C–H, C = C–H and Ar–H absorb **above** 3000 $cm^{-1}$ while aliphatic and aldehydic C–H absorb below 3000 $cm^{-1}$.

Also note that $-CH_3$ and $-CH_2-$ give rise to **two bands.**

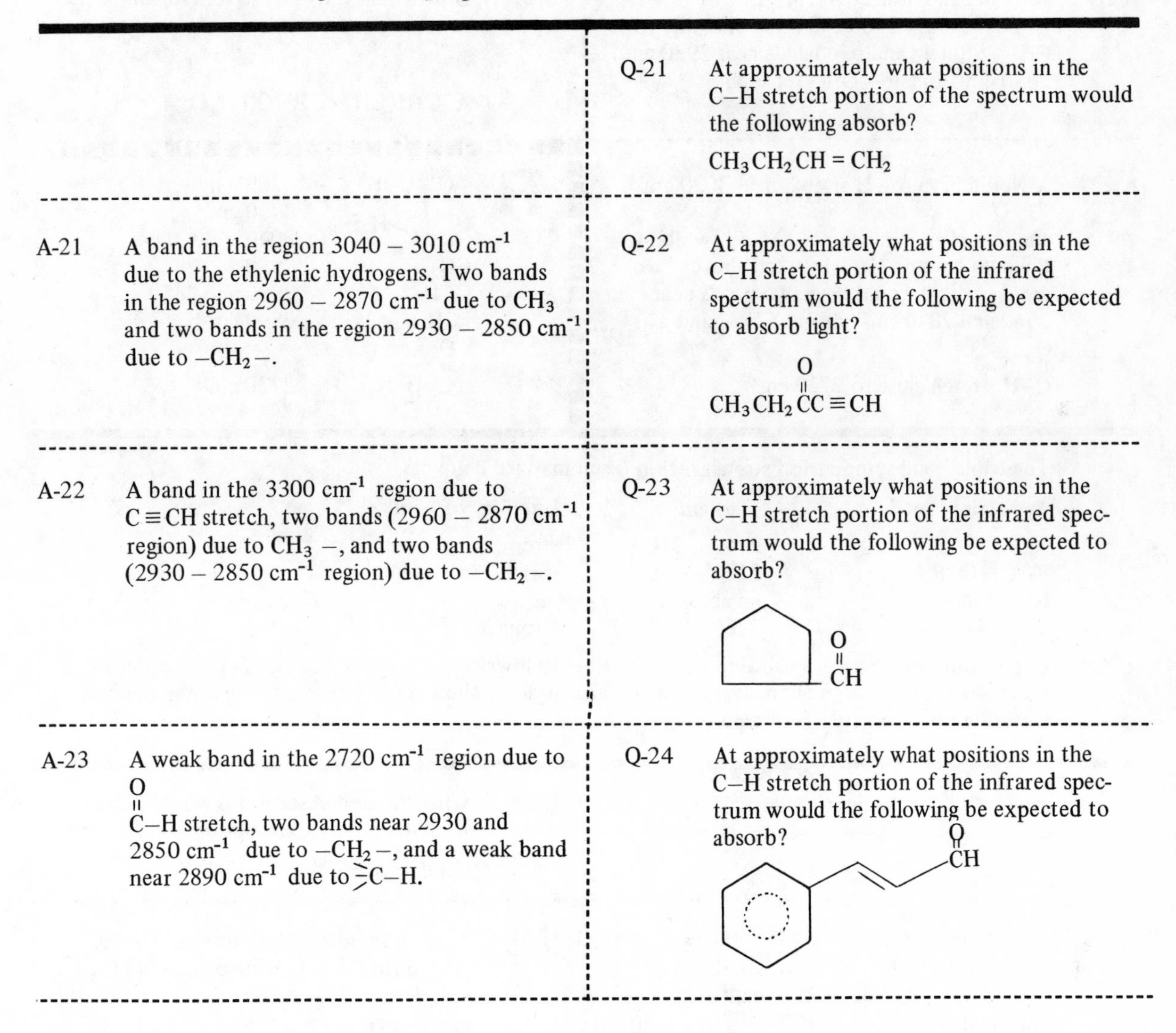

Q-21 At approximately what positions in the C–H stretch portion of the spectrum would the following absorb?

$CH_3CH_2CH = CH_2$

A-21 A band in the region 3040 – 3010 $cm^{-1}$ due to the ethylenic hydrogens. Two bands in the region 2960 – 2870 $cm^{-1}$ due to $CH_3$ and two bands in the region 2930 – 2850 $cm^{-1}$ due to $-CH_2-$.

Q-22 At approximately what positions in the C–H stretch portion of the infrared spectrum would the following be expected to absorb light?

$CH_3CH_2C(=O)C \equiv CH$

A-22 A band in the 3300 $cm^{-1}$ region due to C≡CH stretch, two bands (2960 – 2870 $cm^{-1}$ region) due to $CH_3-$, and two bands (2930 – 2850 $cm^{-1}$ region) due to $-CH_2-$.

Q-23 At approximately what positions in the C–H stretch portion of the infrared spectrum would the following be expected to absorb?

A-23 A weak band in the 2720 $cm^{-1}$ region due to –C(=O)–H stretch, two bands near 2930 and 2850 $cm^{-1}$ due to $-CH_2-$, and a weak band near 2890 $cm^{-1}$ due to >C–H.

Q-24 At approximately what positions in the C–H stretch portion of the infrared spectrum would the following be expected to absorb?

A-24 A band near 3030 $cm^{-1}$ due to aromatic C–H stretch along with a band in the 3040 – 3010 $cm^{-1}$ region due to H–C = C. (These bands may overlap and not be resolved as separate peaks.) A weak band near 2720 $cm^{-1}$ due to –C(=O)–H stretch.

Q-25 At approximately what positions in the C–H stretch portion of the infrared spectrum would the following be expected to absorb?

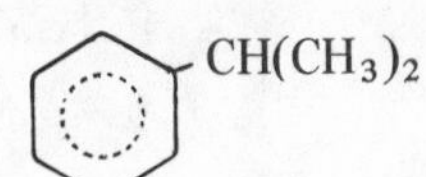


A-25 A band near 3030 $cm^{-1}$ due to Ar–H stretch, a weak band near 2890 $cm^{-1}$ due to >C–H, and two bands near 2960 and 2870 $cm^{-1}$ due to $CH_3$– stretch.

Q-26 At approximately what positions in the C–H stretch portion of the infrared spectrum would the following be expected to absorb?

$CH_3CH_2CH(CH_3)_2$

A-26 A weak band near 2890 $cm^{-1}$ due to >C–H, two bands near 2960 and 2870 $cm^{-1}$ due to $CH_3$– stretch, and two bands near 2930 and 2850 $cm^{-1}$ due to $-CH_2-$ stretch.

Q-27 How would the C–H stretch region of the following compounds differ?

**A** $C_6H_5CH$(=O) **B** $(CH_3)_3CCH$(=O)

A-27 **A** would have Ar–H stretch near 3030 $cm^{-1}$ and C(=O) – H stretch near 2720 $cm^{-1}$, while **B** would have no C – H stretch above 3000 $cm^{-1}$. **B** would have two bands near 2960 and 2870 $cm^{-1}$ due to $CH_3$– and a C(=O)–H stretch close to 2720 $cm^{-1}$.

**R**

| | |
|---|---|
| –C≡C–H | 3300 $cm^{-1}$ |
| Ar–H | 3030 $cm^{-1}$ |
| >C=C<H | 3040 – 3010 $cm^{-1}$ |
| $CH_3$– | 2960 and 2870 $cm^{-1}$ |
| $-CH_2-$ | 2930 and 2850 $cm^{-1}$ |
| >CH | 2890 $cm^{-1}$ |
| –C(=O)–H | 2720 $cm^{-1}$ |

S-4 The triple bond symmetrical stretch region is summarized below:

| Type triple bond | $\bar{\nu}$–$cm^{-1}$ | Intensity of bond |
|---|---|---|
| H–C≡C–R | 2140 – 2100 | strong |
| R–C≡C–R′ | 2260 – 2190 | variable intensity |
| RC≡CR | no absorption | |
| RC≡N | 2260 – 2240 | strong |

Conjugation will cause a small shift of these values to **lower** wave number, e.g., aryl cyanides absorb in the 2240 – 2190 $cm^{-1}$ region. The symmetrical acetylenes show no absorption because **symmetrical vibration causes no change in dipole.**

Q-28 What infrared absorptions would be expected for the following compound?

$CH_3CH_2C \equiv CH$

A-28 A band near 3300 $cm^{-1}$ (C≡C–H), two bands near 2960 – 2870 $cm^{-1}$ ($CH_3$), two bands near 2930 – 2850 $cm^{-1}$ ($CH_2$), one band near 2140 – 2100 $cm^{-1}$ (symmetrical – C≡C – stretch) and aliphatic C – H bending, 1475 – 1300 $cm^{-1}$.

Q-29 What infrared absorptions would be expected for the following compound (neat sample)?

$HC \equiv CCH_2OH$

A-29 A strong broad band in the 3450 – 3300 $cm^{-1}$ region due to OH, which might cover up the 3300 $cm^{-1}$ band of the C ≡CH stretch. Two bands in the 2930 – 2850 $cm^{-1}$ regions due to – $CH_2$ – stretch, a band near 2140 – 2100 $cm^{-1}$ due to symmetrical C ≡C stretch and aliphatic C – H bending near 1475 – 1300 $cm^{-1}$.

Q-30 What infrared absorptions would be expected for the following compound?

$C_6H_5$–C≡C–$C_6H_5$

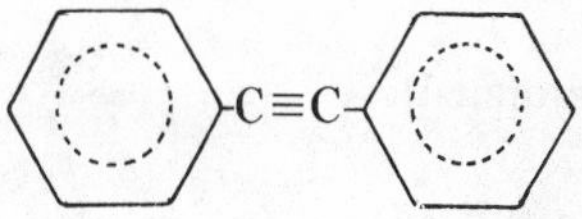

A-30 A band near 3030 $cm^{-1}$ due to Ar–H stretch but **no** band near 2260 – 2190 $cm^{-1}$ because the symmetrical C ≡C stretch does not give rise to a change in dipole. Aromatic C = C stretch in the 1675 – 1500 $cm^{-1}$ region and Ar–H bending in the 1000 – 650 $cm^{-1}$ region.

Q-31 What infrared absorptions would be expected for the following compound (neat sample)?

$CH_3C \equiv CCH_2CH_2NH_2$

A-31 Two sharp but not very strong bands in the 3500 – 3100 $cm^{-1}$ region due to H–N stretch, two bands in the 2960 – 2870 $cm^{-1}$ region ($CH_3$), at least two bands in the 2930 – 2850 $cm^{-1}$ region ($CH_2$), a band (probably low intensity) in the 2260 – 2190 $cm^{-1}$ region due to C ≡C stretch, and aliphatic C–H bending, 1475 – 1300 $cm^{-1}$.

Q-32 What infrared absorptions would be expected for the following compound (saturated $CCl_4$ solution)?

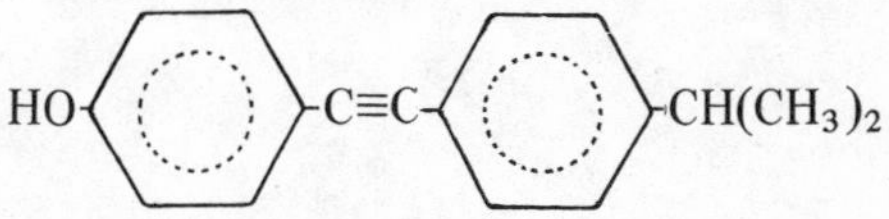

HO–$C_6H_4$–C≡C–$C_6H_4$–$CH(CH_3)_2$

A-32 A band (strong, wide) in the 3450 – 3300 $cm^{-1}$ region due to OH (bonded) stretch. A band near 3030 $cm^{-1}$ due to Ar–H stretch, two bands in the 2960 – 2870 $cm^{-1}$ region ($CH_3$), one band (weak) near 2890 $cm^{-1}$ (⋺C–H), a band in the 2260 – 2190 $cm^{-1}$ region (although it may be very weak) due to symmetrical C ≡ C stretch, aromatic C = C stretch 1675 – 1500 $cm^{-1}$ region, aliphatic C–H bending 1475 – 1300 $cm^{-1}$ region, and aromatic C–H bending 1000 – 650 $cm^{-1}$ region.

Q-33 What infrared absorptions would be expected for the following compound?

COOH, CN (substituents on benzene ring)

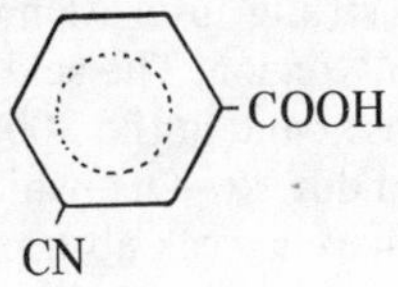

A-33 A band in the 3030 $cm^{-1}$ region due to Ar–H stretch. This may be masked by a rather wide diffuse band in the 3000 – 2500 $cm^{-1}$ region due to COOH. The C ≡ N would give a band in the 2260 – 2240 $cm^{-1}$ region. There would be aromatic C = C stretch, 1675 – 1500 $cm^{-1}$, and bending, 1000 – 650 $cm^{-1}$.

**R**

| | |
|---|---|
| HC ≡ CR | 2140 – 2100 $cm^{-1}$ |
| RC ≡ CR′ | 2260 – 2190 $cm^{-1}$ |
| RC ≡ CR | no absorption |
| RC ≡ N | 2260 – 2240 $cm^{-1}$ |

S-5 Many important bands appear in the carbonyl region of the spectra.

| Type carbonyl | $\bar{\nu}$ cm$^{-1}$ | Intensity of band |
|---|---|---|
| $RCH(=O)$ (saturated) | 1740 – 1720 | strong |
| COOH (saturated) | 1705 – 1725 | strong |
| $R{-}C(=O){-}R$ (saturated) | 1705 – 1725 | strong |
| $R{-}C(=O){-}OR$ (6 and 7 membered lactones) | 1750 – 1730 | strong |
| 5 membered lactones | 1780 – 1760 | strong |
| Esters (non-cyclic) | 1740 – 1710 | strong |
| Acid halides | 1815 – 1720 | strong |
| Anhydrides | Two bands separated by approximately 60 cm$^{-1}$ at 1850 – 1800 cm$^{-1}$ and 1780 – 1740 cm$^{-1}$. | |
| Amides | 1700 – 1640 cm$^{-1}$ | strong |

Conjugation will shift all the absorption bands to smaller wave numbers (note that strain on lactone rings shifts absorptions to higher frequencies).

Q-34 How would the absorption spectra of the following differ in the region 4000 – 1650 cm$^{-1}$?

$CH_3CH_2COOH$    $CH_3CH_2CH(=O)$

$CH_3C(=O)CH_3$

A-34 All would have a strong absorption in the 1740 – 1700 cm$^{-1}$ region. The acid would have a wide diffuse band in the 3000 – 2500 cm$^{-1}$ region due to –OH, while the aldehyde would have a weak absorption in the 2720 cm$^{-1}$ region due to CH stretch.

Q-35 How would the absorption spectra of the following differ in the region 4000 – 1650 cm$^{-1}$?

A: $C_6H_5C(=O)NH_2$    B: $C_6H_5NHC_6H_5$

C: $C_6H_5COOH$

A-35 All three compounds would have Ar–H stretch in the 3030 cm$^{-1}$ region. **A** would have two N–H stretch peaks in the 3500 – 3100 cm$^{-1}$ region, while **B** would have only one N–H stretch peak in that region, and **C** would have none. **A** would have a C = O stretch peak in the 1700 – 1640 cm$^{-1}$ region while **B** would have none. **C** would have the C = O stretch at approximately 1700 cm$^{-1}$. **C** would have a wide diffuse band in the 3000 – 2500 cm$^{-1}$ region due to OH.

Q-36 How would the absorption spectra of the following differ in the region 4000 – 1650 cm$^{-1}$?

$CH_3C(=O)OC(=O)CH_3$    **A**

$CH_3C(=O)OCH_3$    **B**

$CH_3C(=O)N(CH_3)_2$    **C**

A-36 All three compounds would have at least two bands (2960 – 2870 $cm^{-1}$ region) due to $CH_3$– stretch. **A** would have two C = O stretch bands separated by approximately 60 $cm^{-1}$ in the 1850 – 1740 $cm^{-1}$ region, while **B** would have a single C = O band in the 1740 – 1710 $cm^{-1}$ region, and **C** would have a single C = O stretch in the 1700 – 1640 $cm^{-1}$ region.

Q-37 How would the absorption spectra of the following (concentrated solutions) differ in the 4000 – 1650 $cm^{-1}$ region?

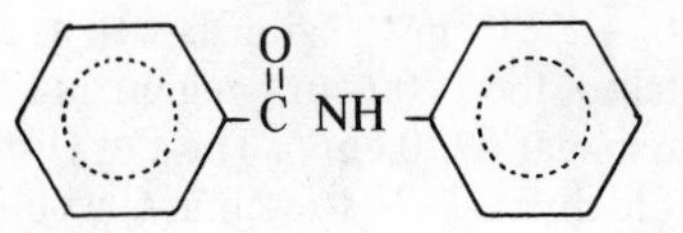


**A**

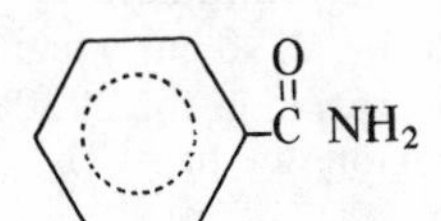


**B**

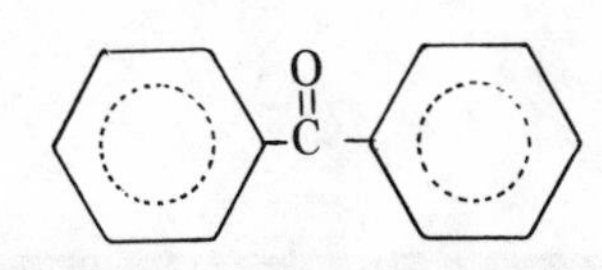


**C**

A-37 Compound **A** would have a single HN stretch peak in the 3500 – 3100 $cm^{-1}$ region, **B** would have two NH stretch peaks, and **C** would have none. **A** and **B** would have absorption in the 1640 $cm^{-1}$ region (low because of conjugation). Certainly the surest way of distinguishing between these compounds is the inspection of the NH stretch region.

Q-38 How would the absorption spectra of the following (concentrated solutions) differ in the 4000 – 1650 $cm^{-1}$ region?

O
C

**A**

COH
3

**B**

O
CCH
3

**C**

A-38 **B** would show an OH stretch absorption in the 3450 – 3300 $cm^{-1}$ range while **A** and **C** would show none. **A** would have a carbonyl stretch absorption in the 1740 – 1720 $cm^{-1}$ region. **C** would also have a weak absorption at 2720 $cm^{-1}$ due to the C–H stretch of the aldehyde.

Q-39 How would the absorption spectra of the following (concentrated solutions) differ in the 4000 – 1650 $cm^{-1}$ region?

C ≡ C

**A**

C ≡ C
O
CH

**B**

$CH_2CH_2$
O
CH

**C**

A-39 **A** would show *no* C ≡ C stretch (no change in dipole) and no C = O stretch. **B** would have a weak C ≡ C absorption in the 2260 – 2190 $cm^{-1}$ region as well as a C = O stretch in the 1700 $cm^{-1}$ region and a C–H stretch near 2720 $cm^{-1}$. The C = O stretch and aldehyde C–H stretch in **C** would be like those in **B**, but **C** would have *no* C ≡ C stretch in the 2260 – 2190 $cm^{-1}$ region. **C** would have absorption in the 2930 and 2850 $cm^{-1}$ region due to $-CH_2-$.

**R**

| | |
|---|---|
| COOH (saturated) | 1705 – 1725 $cm^{-1}$ |
| RCOR (saturated) | 1705 – 1725 $cm^{-1}$ |
| RCHO (saturated) | 1720 – 1740 $cm^{-1}$ |
| Lactones (6 and 7 membered) | 1730 – 1750 $cm^{-1}$ |
| Lactones (5 membered) | 1760 – 1780 $cm^{-1}$ |
| R COOR | 1710 – 1740 $cm^{-1}$ |
| RCOX | 1720 – 1815 $cm^{-1}$ |
| RCOOCOR | 1800 – 1850 and 1740 – 1780 $cm^{-1}$ |
| $RCONH_2$ | 1680 – 1700 $cm^{-1}$ (free)<br>1640 – 1660 $cm^{-1}$ (bonded) |

Conjugation shifts bands to lower wave numbers.

S-6 The symmetrical double bond stretch region from 1680 – 1600 $cm^{-1}$ contains bands due to the following groups.

| Type Double Bond | $\bar{\nu}$ $cm^{-1}$ | Intensity of Band |
|---|---|---|
| >C = C< | 1680 – 1620 | variable |
| >C = N– | 1690 – 1640 | variable |
| –N = N– | 1630 – 1575 | variable |

The >C = C< band will be quite weak for a relatively symmetrical molecule. Aromatic systems have one or more strong bands in the 1400 – 1500 $cm^{-1}$ region.

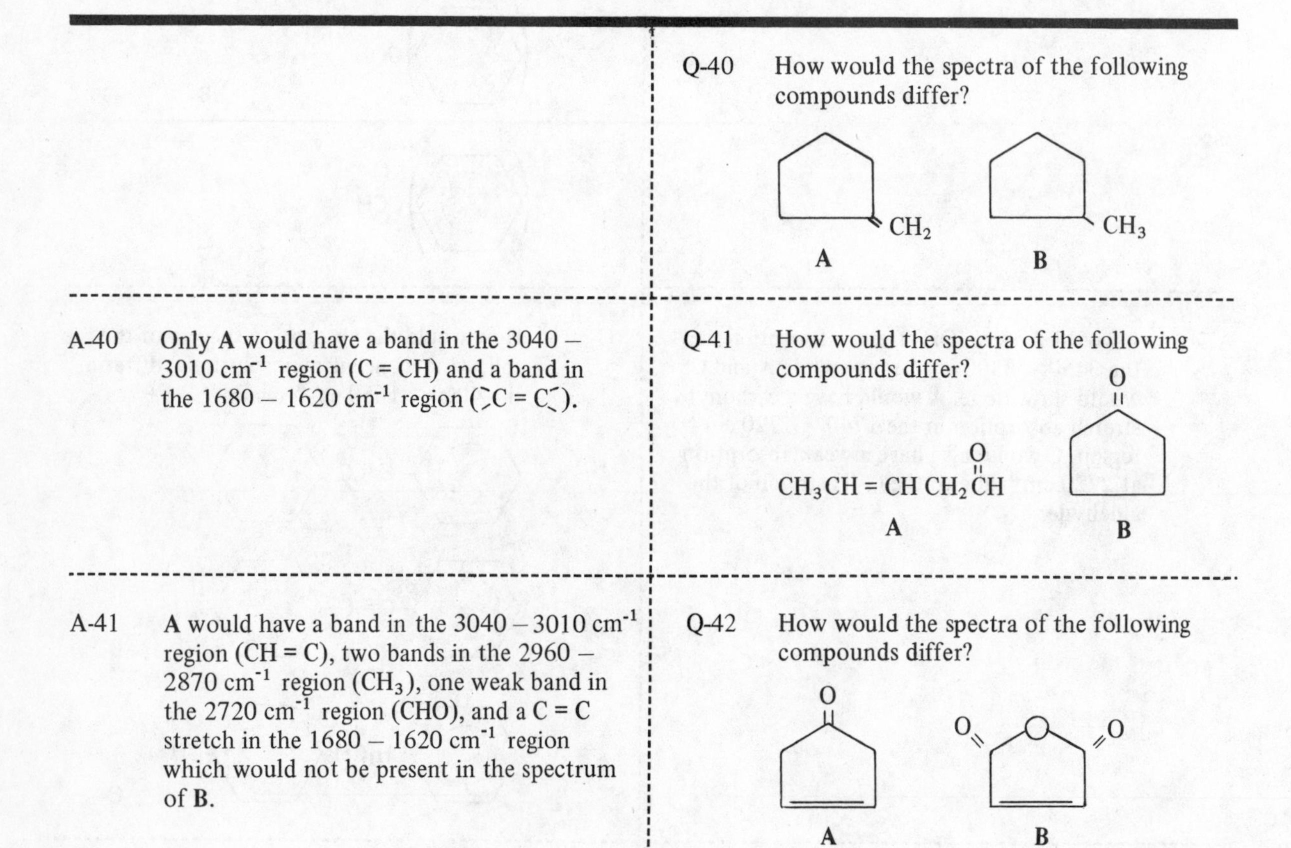

Q-40 How would the spectra of the following compounds differ?



A-40 Only **A** would have a band in the 3040 – 3010 $cm^{-1}$ region (C = CH) and a band in the 1680 – 1620 $cm^{-1}$ region (>C = C<).

Q-41 How would the spectra of the following compounds differ?



A-41 **A** would have a band in the 3040 – 3010 $cm^{-1}$ region (CH = C), two bands in the 2960 – 2870 $cm^{-1}$ region ($CH_3$), one weak band in the 2720 $cm^{-1}$ region (CHO), and a C = C stretch in the 1680 – 1620 $cm^{-1}$ region which would not be present in the spectrum of **B**.

Q-42 How would the spectra of the following compounds differ?

A-42 Both **A** and **B** would have C = CH stretch (3040 – 3010 $cm^{-1}$). **A** would have $-CH_2-$ stretch (2930 – 2850 $cm^{-1}$) which **B** would not have. **A** would have a single C = O stretch in the 1725 – 1705 $cm^{-1}$ region while **B** would have two C = O stretch bands about 60 $cm^{-1}$ apart in the 1850 – 1740 $cm^{-1}$ region. Both **A** and **B** would have **very** weak C = C stretch in the 1690 – 1620 $cm^{-1}$ region.

**R**

| | |
|---|---|
| $>C=C<$ | 1620 – 1680 $cm^{-1}$ |
| $>C=N-$ | 1640 – 1690 $cm^{-1}$ |
| $-N=N-$ | 1575 – 1630 $cm^{-1}$ |

S-7 The C–H bending region (1000 – 650 $cm^{-1}$) yields information useful in characterizing olefins and substitution positions on aromatic rings. A list of useful characteristic absorptions in this region is given below:

Ethylenic C–H Bending:

| **Type Olefin** | $\bar{\nu}$ **cm**$^{-1}$ | **Intensity of Band** |
|---|---|---|
| $RCH=CH_2$ | 990 and 910 | strong |
| $RCH=CRH$ (*cis*) | 690 | moderate to strong |
| $RCH=CRH$ (*trans*) | 970 | moderate to strong |
| $R_2C=CH_2$ | 890 | moderate to strong |
| $R_2C=CHR$ | 840 – 790 | moderate to strong |

Substituted Benzene:

| **Type Substitution** | $\bar{\nu}$ **cm**$^{-1}$ | **Intensity of Band** |
|---|---|---|
| monosubstituted aromatic (5 adjacent H) | 750 and 700 | usually moderate to strong |
| *ortho* aromatic (4 adjacent H) | 750 | usually moderate to strong |
| *meta* aromatic (3 adjacent H) | 780 – 810 | usually moderate to strong |
| *para* aromatic (2 adjacent H) | 850 – 800 | usually moderate to strong |

Q-43 In what regions of the infrared spectrum would the following compounds exhibit different characteristic absorptions?

**A** $CH_3CH=CHCH_3$ **B** $CH_3CH=CH_2$

A-43 **A** would have a weaker C = C stretch in the 1680 – 1620 $cm^{-1}$ region than would **B. B** would have bands at 910 and 990 $cm^{-1}$ characteristic of the $>C=CH_2$, while **A** would have a band at 970 $cm^{-1}$ (if *trans*) or 690 $cm^{-1}$ if *cis.*

Q-44 In what regions of the infrared spectrum would the following compounds exhibit different characteristic absorptions?

$(CH_3)_2C=CHCH_3$ **A**

$CH_3(CH_2)_2CH=CH_2$ **B**

A-44 While both compounds would show some absorption in the 1680 – 1620 $cm^{-1}$ region, **B** would have two bands, 990 and 910 $cm^{-1}$ (characteristic of $RCH = CH_2$) while **A** would have a single band in the 840 – 790 $cm^{-1}$ region.

Q-45 In what regions of the infrared spectrum would the following compounds exhibit different characteristic absorptions?

(m-xylene: benzene ring with $CH_3$ and $CH_3$) (p-xylene: benzene ring with $CH_3$ and $CH_3$)

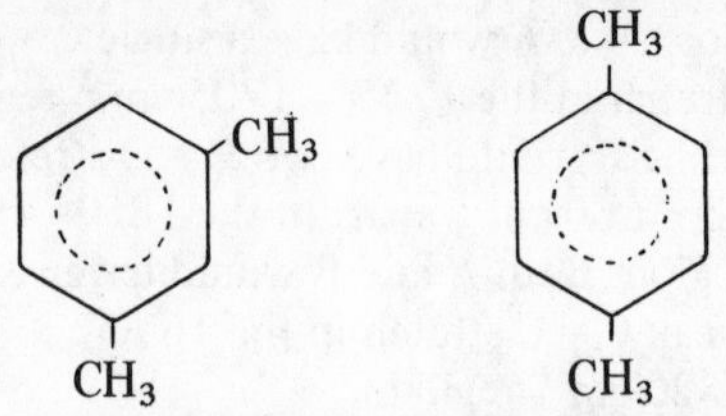

A-45 *m*-xylene has three adjacent H and would show a single band in the 780 – 810 $cm^{-1}$ region. *p*-xylene has only two adjacent H and would show a single peak in the 800 – 850 $cm^{-1}$ region.

Q-46 In what regions of the infrared spectrum would the following compounds exhibit different characteristic absorptions?

**A** *p*-tolunitrile

**B** *o*-tolunitrile

A-46 While both A and B would have a C≡N stretch band in the 2260 – 2240 $cm^{-1}$ region, B has four adjacent H and would have a band in the 750 $cm^{-1}$ region. A has only two adjacent H and would have a band in the 850 – 800 $cm^{-1}$ region.

Q-47 In what regions of the infrared spectrum would the following compounds exhibit different characteristic absorptions?

ethylbenzene *o*-xylene

A-47 Ethylbenzene has five adjacent H and would have two bands in the 750 and 700 $cm^{-1}$ region. *o*-xylene has only four adjacent H and would have but one band near 750 $cm^{-1}$ Ethylbenzene would have two peaks in the 2930 – 2850 $cm^{-1}$ region ($-CH_2-$) which would not be present in *o*-xylene.

Q-48 In what regions of the infrared spectrum would the following compounds exhibit different characteristic absorptions?

*cis*-2-butene *trans*-2-butene

A-48 The *cis*-compound would have a strong absorption in the 690 $cm^{-1}$ region, while the *trans*-isomer would have an absorption in the 970 $cm^{-1}$ region.

Q-49 In what regions of the infrared spectrum would the following compounds exhibit different characteristic absorptions?

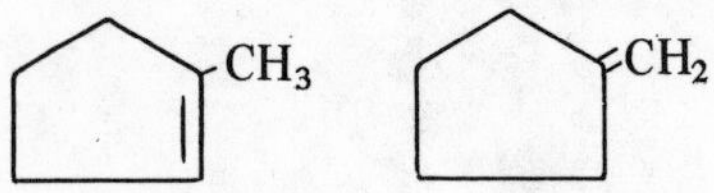

**A**: 1-methylcyclohexene ($CH_3$) **B**: methylenecyclohexane ($=CH_2$)

A-49 **A** would have a single band in the 840 – 790 $cm^{-1}$ region and a rather weak double bond stretch band (1680 – 1620 $cm^{-1}$ region). **B** would have two bands near 910 and 990 $cm^{-1}$ and a strong double bond stretch (1680 – 1620 $cm^{-1}$ region).

Q-50 In what regions of the infrared spectrum would the following compounds exhibit different characteristic absorptions?

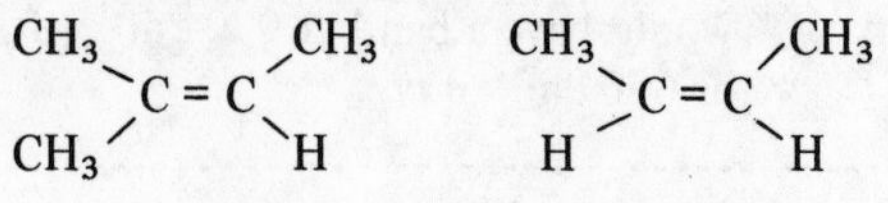

**A**: $(CH_3)_2C=CH(CH_3)$ **B**: $CH_3CH=CHCH_3$ (trans)

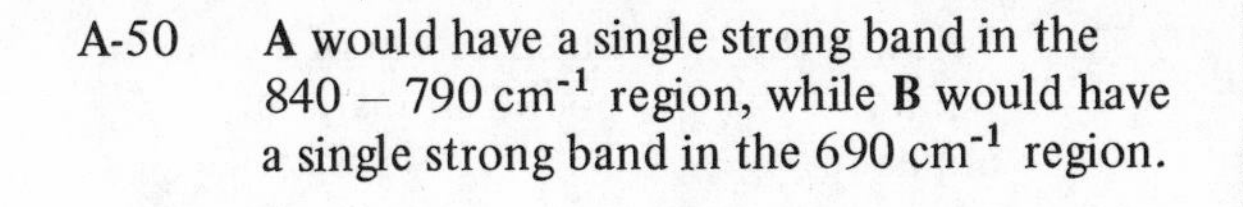

A-50 **A** would have a single strong band in the 840 – 790 $cm^{-1}$ region, while **B** would have a single strong band in the 690 $cm^{-1}$ region.

Q-51 Below is the spectrum of an organic compound containing only C, H, and O. Is the compound aromatic or aliphatic? Is the compound an alcohol?

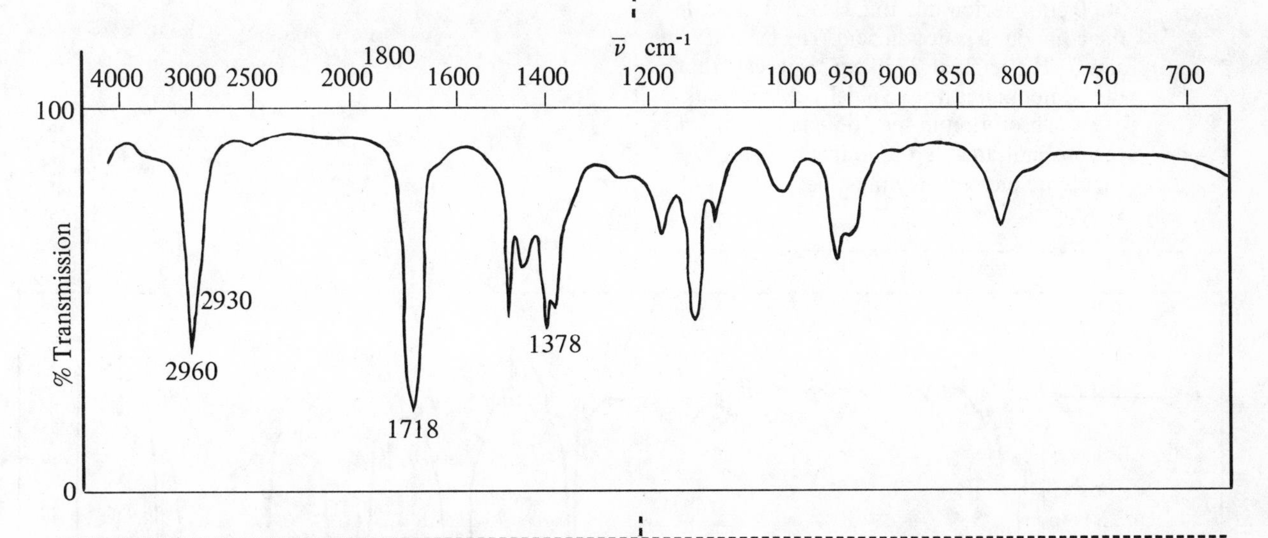


A-51 The lack of any C–H stretch bands above 3000 $cm^{-1}$ would indicate the compound is not aromatic. The C–H stretch at 2960 and 2930 $cm^{-1}$ indicate aliphatic CH. The lack of any strong bands in the 3500 – 3300 $cm^{-1}$ region indicates the compound is not an alcohol.

Q-52 Is the compound whose spectrum is shown above an aldehyde, ketone, ester, anhydride, or acid? Does the compound contain a double or triple bond?

A-52 The compound has absorption in the carbonyl region (1718 $cm^{-1}$) at the frequency typical of aldehydes, ketones, or acids. The lack of any strong, broad bands in the 3000 $cm^{-1}$ region due to OH would indicate the compound is not an acid. The lack of any C–H stretch at 2720 $cm^{-1}$ suggests that the compound is not an aldehyde. The lack of any absorption in the 1650 and 2200 $cm^{-1}$ regions indicates that if double or triple bonds are present they must be symmetrical.

Q-53 Below is the infrared spectrum of a compound. Is the compound aromatic or aliphatic?

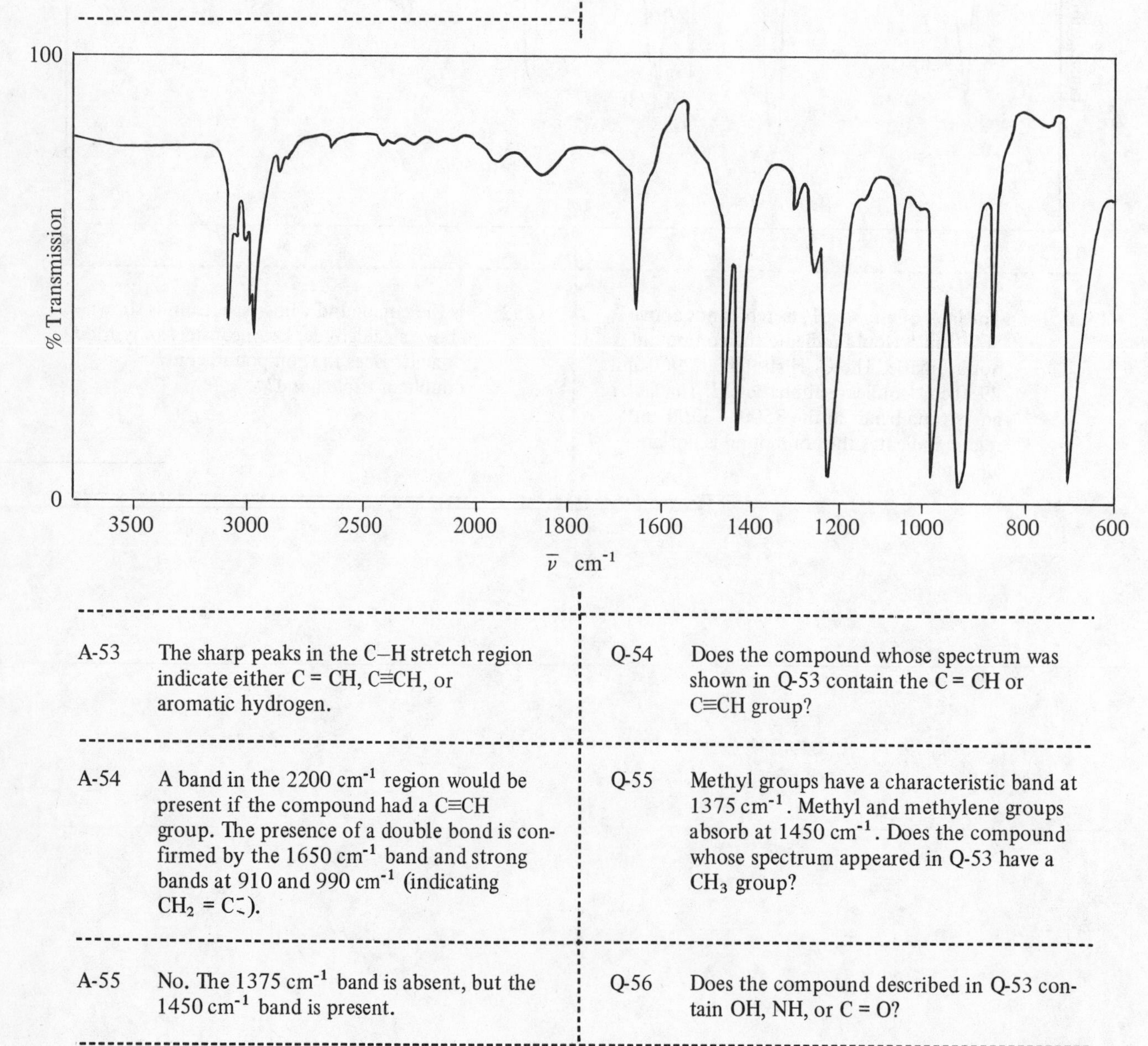


A-53 The sharp peaks in the C–H stretch region indicate either C = CH, C≡CH, or aromatic hydrogen.

Q-54 Does the compound whose spectrum was shown in Q-53 contain the C = CH or C≡CH group?

A-54 A band in the 2200 $cm^{-1}$ region would be present if the compound had a C≡CH group. The presence of a double bond is confirmed by the 1650 $cm^{-1}$ band and strong bands at 910 and 990 $cm^{-1}$ (indicating $CH_2$ = C$\lessgtr$).

Q-55 Methyl groups have a characteristic band at 1375 $cm^{-1}$. Methyl and methylene groups absorb at 1450 $cm^{-1}$. Does the compound whose spectrum appeared in Q-53 have a $CH_3$ group?

A-55 No. The 1375 $cm^{-1}$ band is absent, but the 1450 $cm^{-1}$ band is present.

Q-56 Does the compound described in Q-53 contain OH, NH, or C = O?

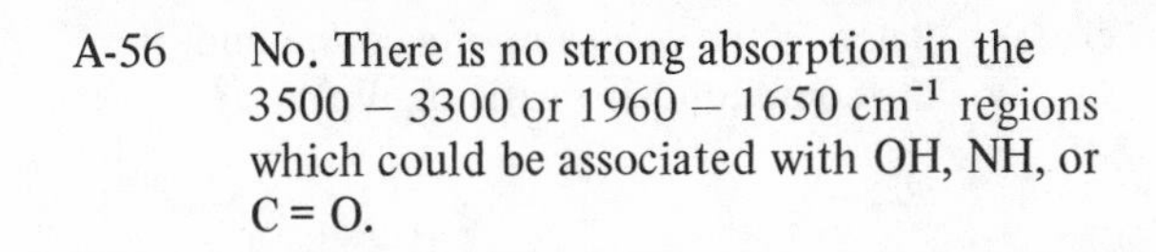

A-56 No. There is no strong absorption in the 3500 – 3300 or 1960 – 1650 $cm^{-1}$ regions which could be associated with OH, NH, or C = O.

Q-57 Below is the infrared spectrum of a compound. What indications are there that the compound is *not* aromatic?

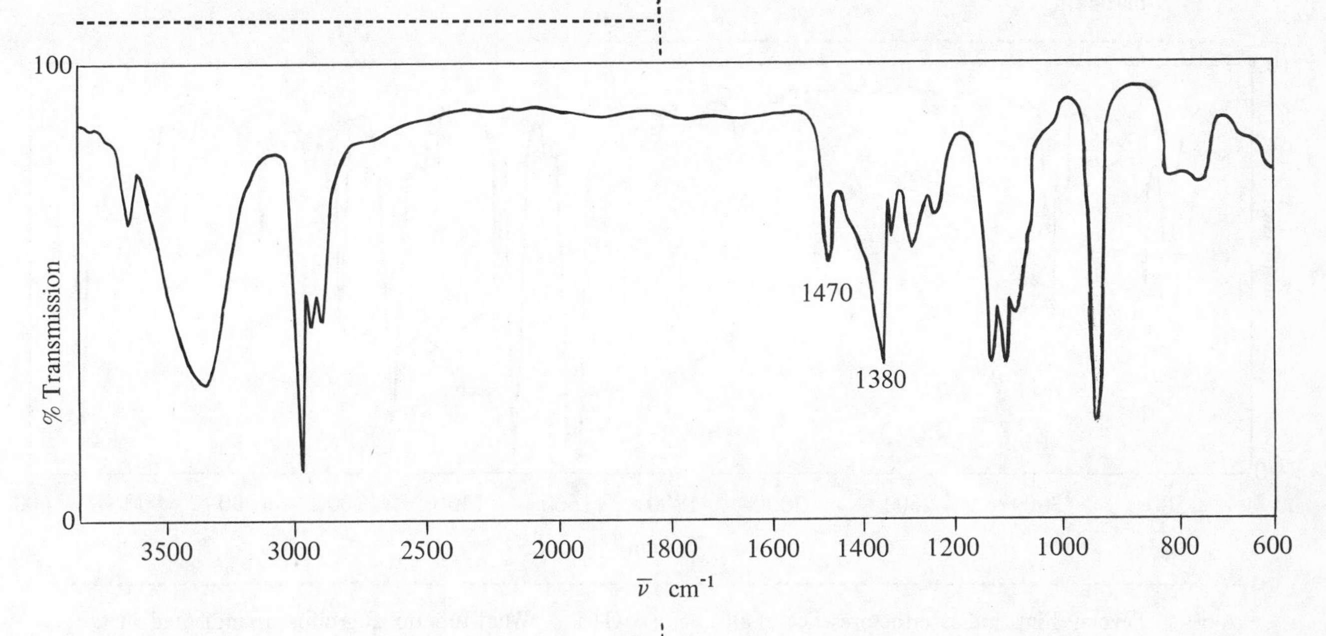


A-57 There is no C–H stretch above 3000 $cm^{-1}$, no aromatic absorption between 1650 and 1450 $cm^{-1}$, and no strong absorption in the 850 – 700 $cm^{-1}$ region.

Q-58 For the compound described in Q-57, what groups are indicated in the spectral region 4000 – 2000 $cm^{-1}$? What groups are known to be absent from the trace shown in this region?

A-58 The strong, wide band at 3350 $cm^{-1}$ is indicative of OH. The absorption in the 2850 – 3000 $cm^{-1}$ region indicates $-CH_2-$, and/or $-CH_3$, and/or $\geqslant$CH. At least two of these C–H type stretchings are indicated by the three bands. The sharp 2970 $cm^{-1}$ band is very suggestive of $-CH_3$. The C=CH, C≡CH, Ar–H, C≡N, and unsymmetrical C≡C groups are *not* indicated.

Q-59 From the spectrum of the compound described in Q-57, what groups are indicated in the 2000 – 1200 $cm^{-1}$ region? What groups are known to be absent?

A-59 The bands at 1380 and 1470 $cm^{-1}$ are indications that $CH_3-$ and perhaps $-CH_2-$ are present. The groups obviously absent are carbonyl, unsymmetrical C = C, and aromatic rings.

Q-60 Below is the spectrum of a compound. Is the compound aromatic or aliphatic?

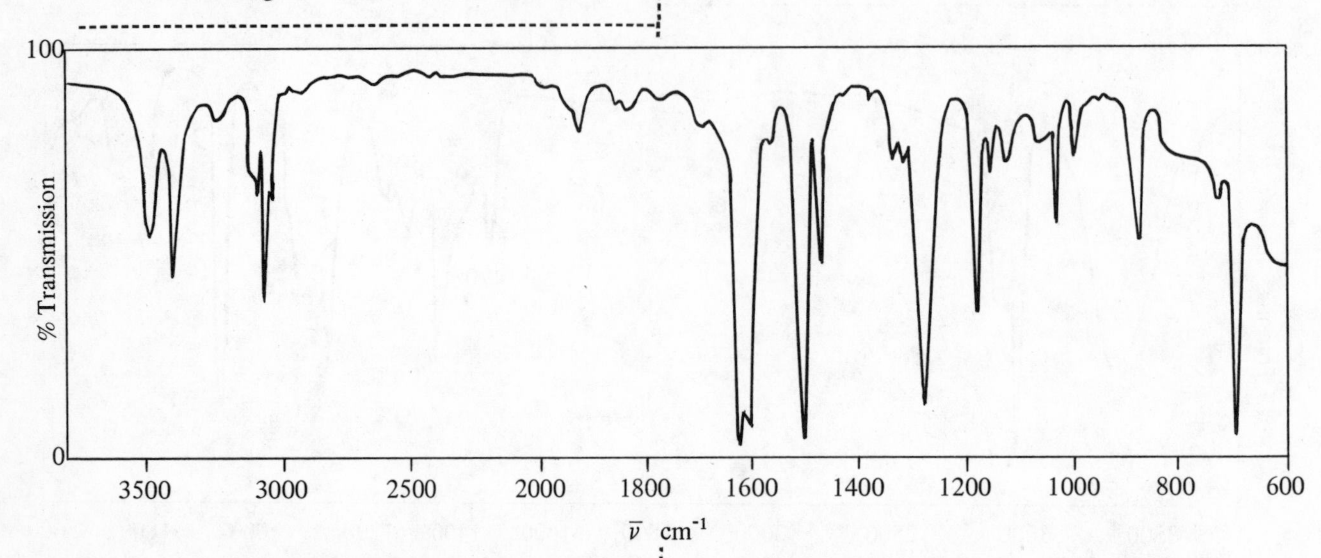


A-60 Several things indicate the presence of an aromatic ring. All C–H stretch absorption is above 3000 $cm^{-1}$ indicating the possible presence of AR–H, C = C–H or C ≡ C–H, and the absence of $-CH_3$, $-CH_2-$, and $\geq$CH. The strong bands at 1620, 1500, and 700 are all characteristic of aromatic compounds.

Q-61 What functional groups are indicated in the spectrum between 4000 and 1000 $cm^{-1}$? What groups are known to be absent?

A-61 The two sharp bands of medium strength in the 3500 region are suggestive of $NH_2$. The groups known to be absent are OH, C = O, C ≡ N, and unsymmetrical C ≡ C.

Q-62 Following is the spectrum of a compound whose formula is $C_8H_8O$. Indicate the type functional group with which the oxygen is associated.

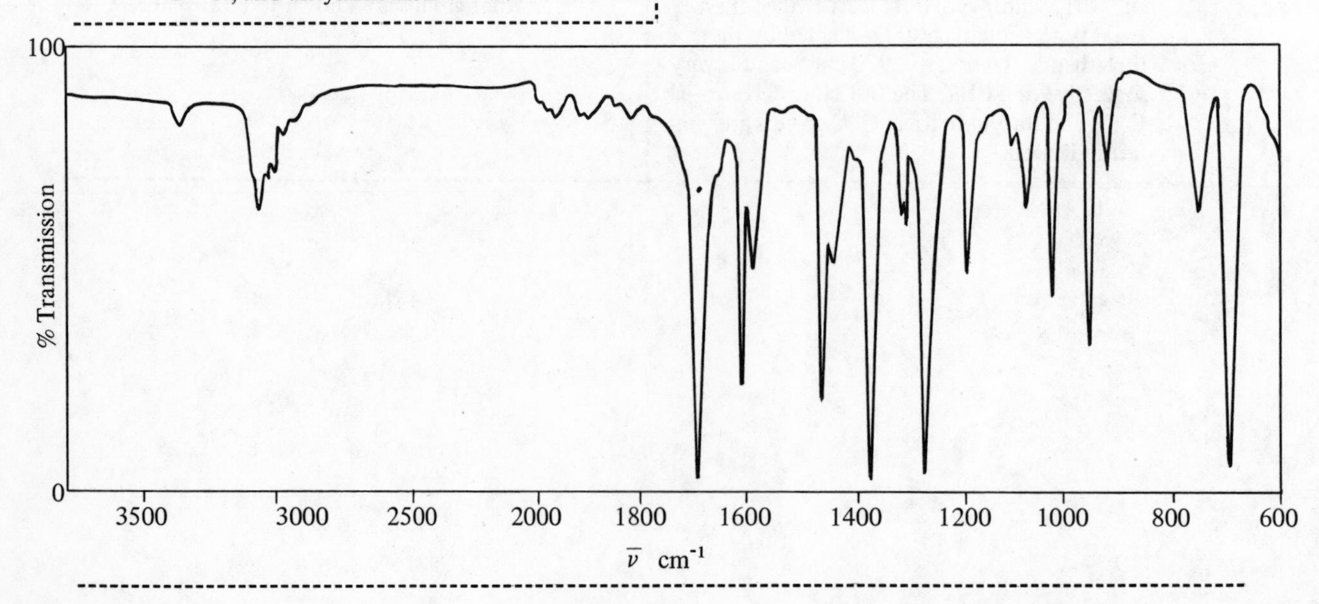

A-62 The lack of any strong band at 3500 – 3300 $cm^{-1}$ indicates the lack of OH. The carbonyl band at 1690 is suggestive of aldehyde, ketone, or amide. Because there is no N in the compound, amide is eliminated. The oxygen must be in a carbonyl of an aldehyde or ketone. The lack of absorption near 2720 $cm^{-1}$ (–COH) suggest the compound is ketone.

Q-63 Is the compound described in Q-62 aromatic? Are any aliphatic carbons indicated?

A-63 The absorption in the C–H stretch region above 3000 $cm^{-1}$ is very suggestive of an aromatic ring. The strong bands at 1600, 1580, and 680 $cm^{-1}$ are also suggestive of an aromatic ring. The absorptions at 2920, 2960 and 1360 $cm^{-1}$ are indicative of $CH_3$.

Q-64 What type of aromatic substitution does the compound described in Q-62 most likely have?

A-64 The two bands near 700 and 750 $cm^{-1}$ are indicative of a monosubstituted aromatic ring.

**R** A complete list of characteristic infrared absorption bands is given in Table V of the Appendix.

## REFERENCES

**Introductory**:

Barrow, Gordon M. 1964. *The Structure of Molecules,* Chapter 3. New York: W. A. Benjamin.

Dyer, John R. 1965. *Application of Absorption Spectroscopy of Organic Compounds,* Chapter 3. Englewood Cliffs, N.J.: Prentice-Hall.

**Intermediate**:

Barrow, Gordon M. 1962. *Introduction to Molecular Spectroscopy,* Chapter 6. New York: McGraw-Hill.

Bauman, Robert P. 1962. *Absorption Spectroscopy.* New York: John Wiley and Sons.

Bellamy, L. J. 1968. *Advances in Infrared Group Frequencies.* London: Methuen and Co.

Conn, G. K. T., and Avery, D. G. 1960. *Infrared Methods.* New York: Academic Press.

Szymanski, Herman A. 1963-66. *Infrared Band Handbook, Theory and Practice of Infrared Spectroscopy* and *Interpreted Infrared Spectra.* New York: Plenum Press.

# Chapter 4 NUCLEAR MAGNETIC RESONANCE SPECTROSCOPY

*Nuclear magnetic resonance (NMR)* spectroscopy is a powerful tool in determining the structure of organic molecules. This technique provides information about the different kinds of hydrogen atoms in a molecule. The NMR spectrum yields information about the chemical environment of hydrogen atoms, the number of hydrogens in each different kind of environment, and the structure of groups adjacent to each hydrogen atom.

This chapter will develop the theory of nuclear magnetic resonance spectroscopy and the way it is employed in structure elucidation.

After completing this chapter, you should

a) understand the type of energy transitions giving rise to nuclear magnetic resonance spectroscopy
b) understand the relationship between energy, precession frequency, magnetic field, and oscillator frequency
c) understand what conditions are necessary to produce nuclear resonance and how resonance is detected
d) understand what factors determine nuclear shielding and how these affect chemical shifts
e) understand spin-spin coupling and the interaction of neighboring nuclei
f) understand anisotropy effects
g) understand why the rapid exchange of protons gives rise to an average resonance signal
h) be able to use chemical shifts and coupling constants in determining structural features of molecules
i) be able to recognize and calculate types of splitting patterns produced by spin-spin interaction
j) be able to estimate the value for various chemical shifts and coupling constants.

S-1 Nuclear spin angular momentum results from the spinning of a nucleus about its axis. This angular momentum is quantized and is designated by the nuclear spin quantum number **I**. The nuclear spin quantum number may have a value of 0, 1/2, 1, 1½, . . . depending on the nucleus involved. Only certain nuclei possess a spin angular momentum. If the mass number (**A**) is odd, then the spin quantum number **I** is a half integer (n+1/2, where n is a whole number or zero). If the mass number (**A**) and the atomic number (**Z**) are both **even**, the spin quantum number **I** is zero. If the mass number (**A**) is **even** and the atomic number (**Z**) is **odd**, the spin quantum number **I** is an integer (1, 2, 3, etc.). Only those nuclei with an **I** value greater than zero have a spin angular momentum.

Q-1 Indicate which of the following nuclei will have a non-zero **I** value (**I** ≠ 0). Will the **I** value be a half integer or an integer value?

$F_9^{19}$ $C_6^{12}$

A-1 $F_9^{19}$ has an odd mass number (19) and odd atomic number (9). Therefore, **I** is a half integer. $C_6^{12}$ has an even mass number (12) and an even atomic number (6). Therefore, **I** is zero.

Q-2 Indicate which of the following nuclei will have a non-zero **I** value. Will the **I** value be a half integer or an integer value?

$P_{15}^{31}$ $O_8^{16}$ $H_1^1$ $N_7^{14}$

A-2 $P_{15}^{31}$: **A** = odd, **Z** = odd, **I** is a half integer.
$O_8^{16}$: **A** = even, **Z** = even, **I** is 0.
$H_1^1$: **A** = odd, **Z** = odd, **I** is a half integer.
$N_7^{14}$: **A** = even, **Z** = odd, **I** is an integer.

Q-3 Which of the following nuclei do not have a spin angular momentum?

$Li_3^7$, $He_2^4$, $C_6^{12}$

A-3 Nuclei with an **I** value of zero do not have a spin angular momentum.
$He_2^4$ : **A** = even, **Z** = even, **I** = 0.
$C_6^{12}$ : **I** = 0.

**R** Certain nuclei possess angular momentum which is quantized. Angular momentum is designated by the nuclear spin quantum number **I**.

S-2 When placed in a magnetic field, nuclei with an angular momentum may assume only 2**I** + 1 orientations relative to the applied field. Each of these orientations corresponds to a given energy state for the nucleus.

Q-4 How many different orientations relative to the applied magnetic field will a nucleus with a value of **I** = 0 be able to assume when in a magnetic field?

A-4 Number of orientations = 2**I** + 1, **I** = 0.
Therefore, 2(0) + 1 = 1 orientation.

Q-5 How many different energy states will a nucleus with a value of **I** = 1 have in a magnetic field?

A-5 Number of orientations = 2**I** + 1, **I** = 1.
Therefore, 2(1) + 1 = 3 orientations.

Q-6 How many different orientations relative to the applied magnetic field will a proton ($H_1^1$) with a value of **I** = 1/2 be able to assume when in a magnetic field?

A-6 Number of orientations = 2I + 1, I = 1/2. Therefore, 2(1/2) + 1 = 2 orientations.

**R** The number of orientations of a nucleus in an applied magnetic field = 2I + 1. Each orientation corresponds to a given energy state of the nucleus.

S-3 For a proton ($H_1^1$) the energy of the orientation corresponds to $\pm \mu_H B_0$ where $\mu_H$ is the magnetic moment of the proton and $B_0$ is the strength of the applied magnetic field. The magnetic moment is a constant for a given type of nucleus. The higher energy level for the proton is the $+\mu_H B_0$ state.

Q-7 Draw an energy level diagram to represent the two energy levels of the proton in a magnetic field $B_0$. Label the energy levels.

A-7

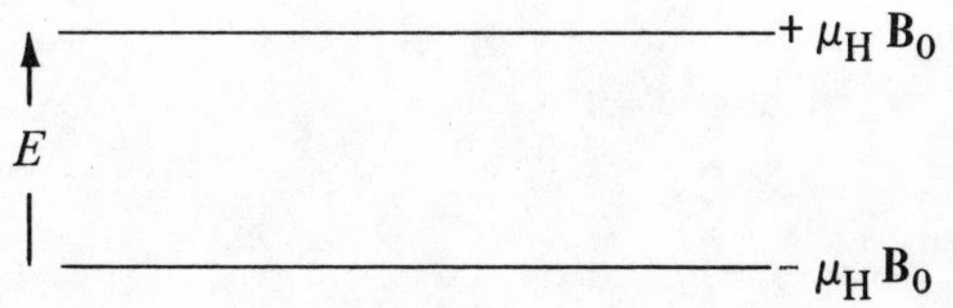

Q-8 A diagram is drawn below which shows the variation in the energy levels as a function of the strength of the applied magnetic field.

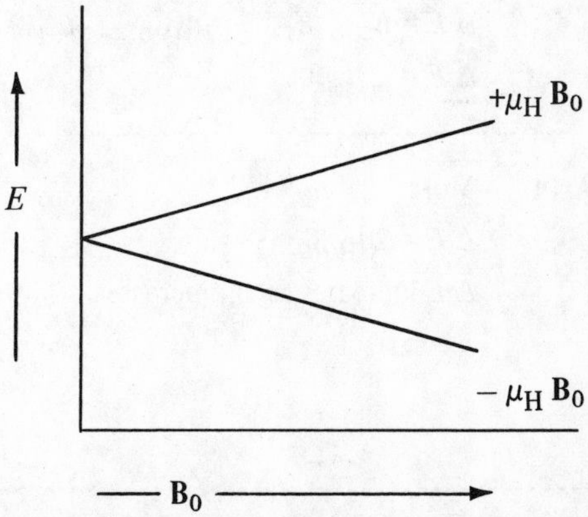

What are the relative values of the two energy levels for a proton in a magnetic field of zero strength?

A-8 At zero magnetic field the two energy levels are equal.

Q-9 Below is an energy diagram for the energy levels of the proton in a magnetic field equal to $B_0$.

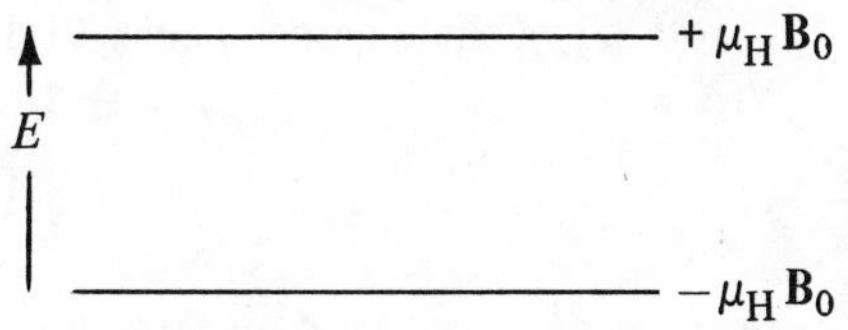

Draw the energy levels for the proton in a magnetic field equal to $B_0'$ and $B_0''$ where $B_0'' > B_0 > B_0'$.

A-9

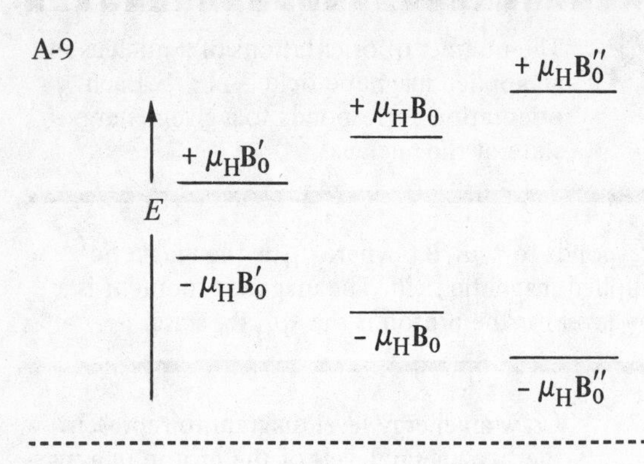


Q-10 The energy transition that is important in nuclear magnetic resonance is the transition of a nucleus from the lower energy level to the higher one. The difference between the two energy levels is $\Delta E$. In terms of the applied magnetic field and magnetic moment, what is the energy necessary to cause a proton to be excited from the lower energy state ($E_1$) to the higher state ($E_2$)? The proton has a magnetic moment of $\mu_H$ and is in a magnetic field of strength $\mathbf{B_0}$.

A-10

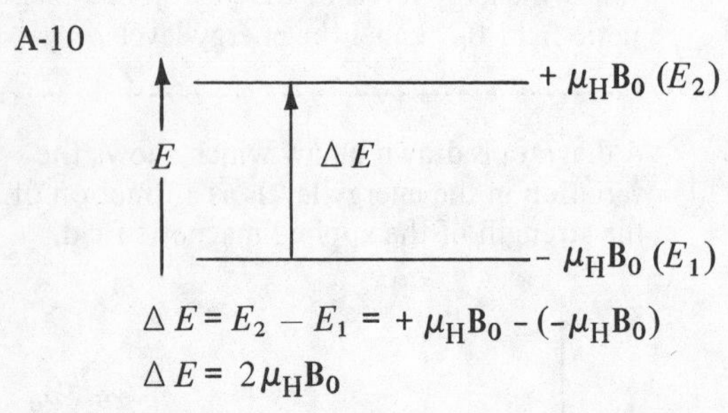


$\Delta E = E_2 - E_1 = +\mu_H \mathbf{B_0} - (-\mu_H \mathbf{B_0})$

$\Delta E = 2\mu_H \mathbf{B_0}$

Q-11 If the strength of the applied magnetic field $\mathbf{B_0}$ is increased, will more or less energy be required to cause a transition from energy level $E_1$ to $E_2$?

A-11 More

$\Delta E = 2\mu_H \mathbf{B_0}$

$\Delta E$ increases as $\mathbf{B_0}$ increases.

Q-12 The magnetic moment for a fluorine nucleus $\mu_F$ is smaller than that for a proton. Fluorine has an **I** value of 1/2.

Compare the energy required for the nuclear transition of the fluorine and hydrogen nuclei in an applied magnetic field of the same strength.

A-12 The hydrogen nucleus requires the greater energy.

$\Delta E = \mu \mathbf{B_0}$

$\Delta E$ increases as $\mu$ increases.

**R** The energy of a given orientation for a nucleus is given by $\pm \mu \mathbf{B_0}$ where $\mu$ is the magnetic moment of the nucleus in question and $\mathbf{B_0}$ is the strength of the applied magnetic field.

S-4 A physical picture will be developed for the energy levels of a nucleus. Consider a spinning top whose spin axis does not correspond to the direction of the gravitational field. The top will precess; that is, rotate in a circle perpendicular to the gravitational field.

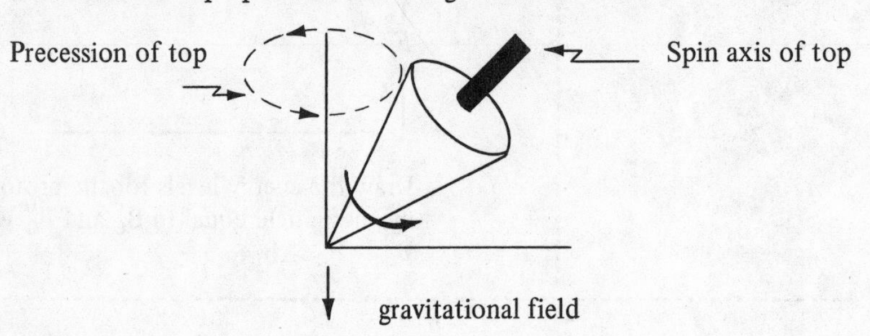

A nucleus with a spin 1/2 is similar to a top precessing about a point. The magnetic moment of the spinning nucleus lines up with or against the applied magnetic field. Because the magnetic moment cannot exactly parallel the applied magnetic field, it will precess about the field. The stronger the applied magnetic field, the faster will be the precession frequency of the magnetic moment.

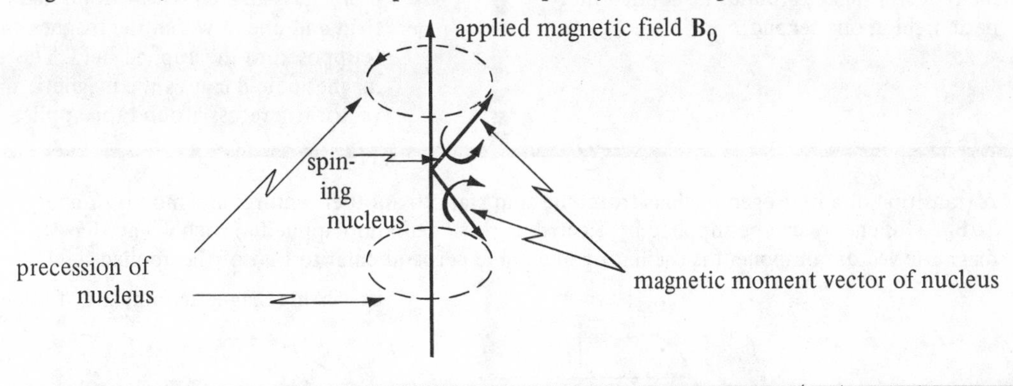


---

**Q-13** The hydrogen nucleus having its magnetic moment aligned with the applied magnetic field is in the lower of the two possible energy states. Draw and give the energy for this orientation (spin state).

**A-13**

$B_0$

magnetic
moment
vector

$$E = -\mu_H B_0$$

applied magnetic field

**Q-14** Draw and give the energy for the high energy spin state of the hydrogen nucleus.

**A-14**

$H_0$
$B_0$

$$E = +\mu_H B_0$$

magnetic moment vector

applied magnetic field

**Q-15** In terms of the nuclear spin states, what happens when the nucleus undergoes a transition from its low energy level to its higher energy level?

**A-15** The nucleus flips from its spin state where the magnetic moment is aligned with the applied field to the spin state where the magnetic moment is opposed to the applied field.

**Q-16** What is meant by *the frequency of precession* of the magnetic moment vector around the applied magnetic field?

A-16 The frequency of precession is the number of complete revolutions the magnetic moment vector makes around the applied magnetic field in one second.

**R** A hydrogen nucleus can exist in two different spin states, one in which the magnetic moment is aligned with the applied magnetic field and one in which the magnetic moment is opposed to the applied field. The applied magnetic field causes the magnetic moment vector to precess around the applied field.

S-5 A transition of a hydrogen nucleus from one spin state to another requires an amount of energy equal to $2\mu \mathbf{B_0}$. This energy can be supplied by electromagnetic radiation applied in such a way that the radiation's magnetic vector component is oscillating in a plane perpendicular to that of the applied field.

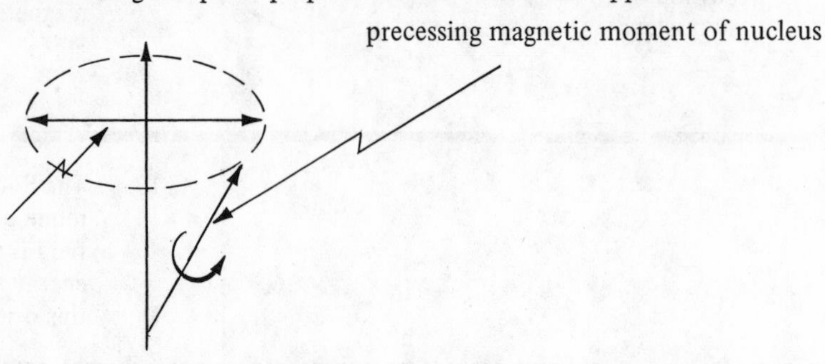


When the frequency of the oscillating magnetic vector component of the electromagnetic radiation is equal to the precessing frequency of the magnetic moment of the nucleus, the two are said to be in *resonance,* and energy can be transferred from the electromagnetic radiation to the hydrogen nucleus. When resonance occurs, the proton is transformed from one spin state to the other.

Q-17 What is the relationship between energy of a transition, $\Delta E$, and frequency of electromagnetic radiation, $\nu$?

A-17 $\Delta E = h\nu$, where $h$ is Planck's constant.

Q-18 Derive a relationship between the frequency of electromagnetic radiation necessary to cause resonance and the strength of the applied magnetic field, $\mathbf{B_0}$.

A-18 $\Delta E = 2\mu \mathbf{B_0}$

$\Delta E = \mathrm{h}\nu$

At resonance the two $\Delta E$'s are equal; therefore, $\mathrm{h}\nu = 2\mu \mathbf{B_0}$

$$\nu = \frac{2\mu}{h} \mathbf{B_0}$$

Q-19 As the strength of the applied field ($\mathbf{B_0}$) is increased, how must the frequency of electromagnetic radiation ($\nu$) be changed in order to cause resonance?

A-19 It must be increased. The frequency is directly proportional to the strength of the applied magnetic field.

$$\nu = \frac{2\mu}{h} \mathbf{B_0}$$

Q-20 Using a constant frequency electromagnetic radiation, will fluorine or hydrogen nuclei require the greater applied magnetic field for resonance to occur? Why?

$\mu_F < \mu_H$

A-20 Fluorine. A frequency of precession equal to that of the electromagnetic radiation is necessary for resonance to occur.

$$\nu_H = \frac{2\mu_H}{h} \mathbf{B_0};\ \nu_F = \frac{2\mu_F}{h} \mathbf{B_0}$$

Since $\mu_F < \mu_H$, $\mathbf{B_0}$ for fluorine systems must be larger than $\mathbf{B_0}$ for hydrogen systems in order to make the two frequencies equal to the electromagnetic radiation.

**R** Resonance occurs when the frequency of the oscillating magnetic vector component of electromagnetic radiation is equal to the precession frequency of the magnetic moment of the nucleus. Resonance allows energy to be transferred from electromagnetic radiation to the nucleus thus causing the nucleus to be transferred from one spin state to another. The resonance condition for a proton is

$$\nu = \frac{2\mu}{h} \mathbf{B_0}$$

S-6 To cause the transition of a proton from one spin state to another the sample can be placed in a fixed applied magnetic field, $\mathbf{B_0}$, and the frequency of the oscillating magnetic vector component of the electromagnetic radiation, $\nu$, varied until the resonance situation is reached. Alternatively, a fixed electromagnetic radiation frequency may be imposed and the magnetic field varied until the resonance situation is reached. The latter procedure is experimentally more satisfactory. A fixed oscillator frequency is utilized and the magnetic field is slowly varied until the resonance situation is reached.

With a fixed oscillator frequency of 60 megahertz (megacycles per second) a magnetic field of approximately 14,000 gauss is required for proton resonance. Very large magnetic fields ($\mathbf{B_0}$) are used to assure a measurable separation of energy levels.

Q-21 What is the expression for the resonance frequency of a proton in an applied magnetic field of $\mathbf{B_0}$?

A-21 $\nu = \frac{2\mu}{h} \mathbf{B_0}$

Q-22 The magnetic moment for hydrogen, fluorine, and phosphorus decreases as $\mu_H > \mu_F > \mu_p$. Using a constant oscillator frequency, which nuclei would require the largest applied magnetic field for resonance to occur?

A-22 $\mu_P$ is smaller than $\mu_F$ and $\mu_H$. Therefore, a larger value of $\mathbf{B_0}$ would be necessary for phosphorus to obtain the precession frequency required for resonance.

Q-23 How does the electron density about the hydrogen nuclei in the following species differ?

a) $CH_4$

b) $H^+$

A-23 In methane each hydrogen nucleus will be surrounded by an electron cloud. The hydrogen ion has no electron cloud.

Q-24 Electrons surrounding a nucleus produce a secondary magnetic field which opposes the applied field. What effect would the electrons surrounding a nucleus be expected to have on the magnetic field felt by that nucleus?

A-24 The electrons act as a shield for the nucleus from the applied magnetic field so that the magnetic field felt by the nucleus would be slightly less.

Q-25 For a nucleus surrounded by electrons, the *shielding effect* of the electrons is directly proportional to the applied magnetic field. If $\sigma$ is the proportionality constant between the shielding effect and the applied magnetic field, write an equation for the shielding effect.

---

A-25 *Shielding effect* = $\sigma \mathbf{B}_0$, where $\sigma$ is called the shielding constant.

Q-26 The constant, $\sigma$, is directly proportional to the density of the electron cloud around the nucleus. As the density of an electron cloud increases, will the magnitude of the magnetic field felt by the nucleus increase or decrease? Why?

---

A-26 Decrease. As the magnitude of $\sigma$ increases, the shielding effect increases, reducing the magnetic field felt by the nucleus.

Q-27 The shielding effects on the protons indicated below are different. Explain.

```
     H
     |
   H-C-O-H <--
     |
     H <--
```

---

A-27 The electron withdrawing effect of oxygen is greater than that for carbon. Therefore, the electron density about the C–H protons will be greater than about the O–H proton; the C–H protons will be shielded to a greater extent than the O–H proton.

Q-28 Would the value for $\sigma$ be expected to be the same for every proton in an organic compound such as

$CH_3-CH_2-Cl$ ?

---

A-28 No. The value of $\sigma$ is dependent upon the electron density about a proton. In ethyl chloride there are two types of protons (methyl and methylene). The electron density would be different because the methylene group has a chloro and a methyl group attached to it while the methyl group has a $-CH_2Cl$ group attached.

Q-29 What factor will affect the electron density about a hydrogen nucleus?

---

A-29 The relative electron pulling or donating power (polar or inductive effect) of the atoms or groups attached to the atom bearing the hydrogen nuclei under consideration.

Q-30 If $\mathbf{B}_N$ represents the magnitude of the magnetic field felt by the nucleus, derive an expression for the magnitude of the field felt by the nucleus in terms of the applied field, $\mathbf{B}_0$, and the shielding constant, $\sigma$.

A-30 $\mathbf{B_0}$ = applied magnetic field
$\sigma\mathbf{B_0}$ = shielding effect
The shielding effect acts to reduce the applied magnetic field.
$\mathbf{B_N} = \mathbf{B_0} - \sigma\mathbf{B_0} = \mathbf{B_0}\ (1 - \sigma)$

Q-31 Below is an energy diagram for an unshielded proton in an applied magnetic field $\mathbf{B_0}$. Show the effect of electronic shielding on the nuclear energy levels.

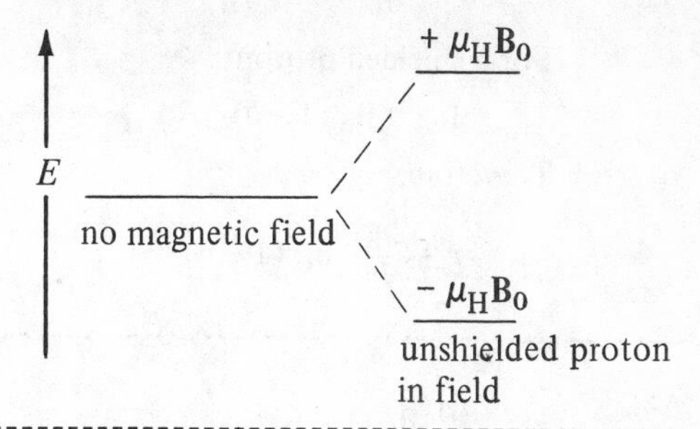


A-31

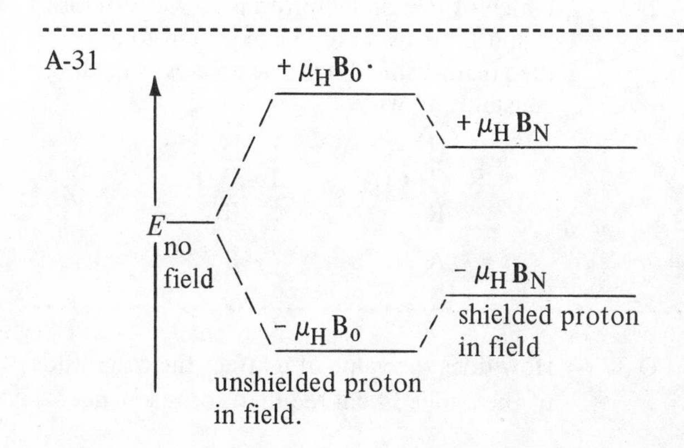


Q-32 The transition energy between nuclear energy levels for an unshielded proton was determined earlier to be $2\mu_H\mathbf{B_0}$. What is the transition energy for the shielded proton?

A-32 $+\mu_H\mathbf{B_N}$

$-\mu_H\mathbf{B_N}$

$$\Delta E = \mu_H\mathbf{B_N} - (-\mu_H\mathbf{B_N})$$
$$= 2\mu_H\mathbf{B_N}$$

Q-33 What is the transition energy for the shielded proton in terms of the applied magnetic field, $\mathbf{B_0}$, and the shielding constant, $\sigma$?

A-33 $\mathbf{B_N} = \mathbf{B_0}\ (1 - \sigma)$
$\Delta E = 2\mu_H\mathbf{B_N} = 2\mu_H\mathbf{B_0}\ (1 - \sigma)$

Q-34 For an unshielded proton ($H^+$), the relationship between the frequency of the oscillating magnetic vector component of electromagnetic radiation and the applied magnetic field when resonance occurs is

$$\nu = \frac{\mu_H}{h}\mathbf{B_0}$$

Rewrite this equation for a shielded proton in terms of $\mathbf{B_0}$ and $\sigma$.

A-34 For an unshielded proton the field felt by the nucleus ($\mathbf{B}_N$) is identical to $\mathbf{B}_0$.

$$\nu = \frac{2\mu_H}{h}\mathbf{B}_0 = \frac{2\mu_H}{h}\mathbf{B}_N$$

For a shielded proton

$$\mathbf{B}_N = \mathbf{B}_0 (1 - \sigma)$$

Therefore,

$$\nu = \frac{2\mu_H}{h}\mathbf{B}_0 (1 - \sigma)$$

Q-35 In nuclear magnetic resonance the frequency is held constant. Rearrange the equation derived in A-34 to show the applied field necessary for any proton to undergo resonance.

A-35

$$\nu = \frac{2\mu_H}{h}\mathbf{B}_0 (1 - \sigma)$$

$$\mathbf{B}_0 = \frac{\nu h}{2\mu_H(1 - \sigma)}$$

Q-36 Which of the circled protons in the systems A and B below would be expected to have the greatest shielding? The greatest shielding constant, $\sigma$? Why?

A: R–C(Ⓗ)(R)–$CH_3$    B: R–C(Ⓗ)(R)–H

A-36 The proton in system A would be surrounded by an electron cloud of higher density than the proton in B (H is a better electron withdrawing group than $-CH_3$). Therefore, the proton in A would be shielded better and would have a higher $\sigma$ value.

Q-37 How does the value of $\sigma$ affect the magnitude of the applied field required for resonance?

A-37

$$\mathbf{B}_0 = \frac{\nu h}{2\mu_H (1 - \sigma)}$$

If $\sigma$ is increased, then the value of $\nu h/[2\mu_H (1 - \sigma)]$ will become larger. Therefore, as $\sigma$ increases, $\mathbf{B}_0$ required for resonance increases.

Q-38 Which of the circled protons in A and B below would be expected to show nuclear resonance with the higher applied field? Why?

A: R–C(Ⓗ)(R)–$CH_3$    B: R–C(Ⓗ)(R)–H

A-38 Since $-CH_3$ is a better electron donating group than H, the shielding constant for the proton in A would be expected to be larger. Therefore, the term $\nu h/2\mu_H(1 - \sigma)$ would be larger for the proton in A than in B. $\mathbf{B}_0$ for system A would be higher.

Q-39 Which of the circled protons in A and B would be expected to show nuclear resonance at the lower applied field? Why?

A: R–C(Ⓗ)(R)–OH

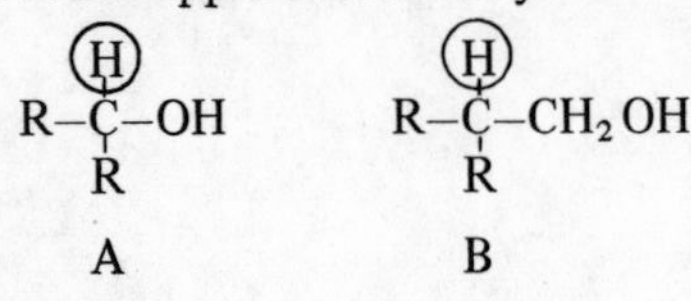

A-39 Because OH is a strong electron withdrawing group, one would expect the shielding constant for the proton in A to be smaller than the shielding constant for the proton in B. Therefore, the applied magnetic field at resonance will be lower for the proton in A.

Shielding effect $= \sigma B_0$, where $\sigma$ is the shielding constant. The magnitude of $\sigma$ depends upon the electron density about a hydrogen nucleus.

$B_N = B_0 (1 - \sigma)$, where $B_N$ is the magnetic field felt by the nucleus. The resonance condition for a shielded proton is

$$\nu = \frac{2\mu_H}{h} B_0 (1 - \sigma)$$

S-7 In practice, the absolute value of $B_0$ cannot be determined very accurately; therefore, it is convenient to compare the $B_0$ required for resonance of a certain chemical system to that required of a standard. The difference is referred to as the *chemical shift.* The chemical shift parameter $\delta$ is defined as

$$\delta = \frac{B_0 - B_0 \text{ (reference)}}{B_0 \text{ (reference)}}$$

where $B_0$ is the magnetic field required for resonance of the proton in question.

Q-40 Derive an expression for the chemical shift between a certain proton and a reference proton in terms of their shielding constants. Remember that the precession frequencies of the sample and the reference are identical at resonance.

A-40
$$\delta = \frac{B_0 - B_0 \text{ (ref)}}{B_0 \text{ (ref)}}$$

$$B_0 = \frac{h\nu}{2\mu_H (1 - \sigma)}$$

$$B_{0(ref)} = \frac{h\nu}{2\mu_H (1 - \sigma \text{ (ref)})}$$

$$\delta = \frac{\frac{1}{1-\sigma} - \frac{1}{1-\sigma \text{ (ref)}}}{\frac{1}{1-\sigma \text{ (ref)}}}$$

$$\delta = \frac{\sigma - \sigma \text{ (ref)}}{1 - \sigma}$$ Since $\sigma << 1$,

$$\delta = \sigma - \sigma_{\text{(ref)}}$$

Q-41 As the difference in the shielding constants of the proton in question and the reference proton increases, what happens to the chemical shift value?

A-41 $\delta = \sigma - \sigma_{\text{(ref)}}$

$\delta$ increases

Q-42 In the measurement of a resonance signal, the magnetic field is varied while the oscillator frequency is held constant. In determining chemical shift values, the magnetic field is most conveniently and accurately measured in frequency units and is reported in Hertz (Hz) or cycles/sec units. From the relationship, $B_0 = \kappa\nu$, where $\kappa$ is a proportionality constant, derive an equation for $\delta$ in frequency units.

A-42 $B_{0(ref)} = \kappa\nu_{(ref)}$

Therefore,

$$\delta = \frac{\kappa\nu - \kappa\nu_{(ref)}}{\kappa\nu_{(ref)}}$$

$$\delta = \frac{\nu - \nu_{(ref)}}{\nu_{(ref)}}$$

**R**

$$\delta = \frac{B_0 - B_{0\,(ref)}}{B_{0\,(ref)}}$$

$$\delta = \sigma - \sigma_{(ref)}$$

S-8 The quantities $\nu$ and $\nu$ (reference), are very large values which are only slightly different from the fixed oscillator frequency, $\nu_0$, used in the nuclear magnetic resonance spectrometer (usually 40, 60, or 100 megahertz for proton resonance). For convenience, therefore, the equation for the chemical shift is re-written as

$$\delta = \frac{\nu - \nu_{(reference)}}{\nu_0} \times 10^6$$

where $\nu_0$ is substituted for $\nu$ (reference) in the denominator. Chemical shift defined in this manner is dimensionless and is expressed in parts per million (ppm).

The compound usually used as a reference standard is tetramethylsilane, $(CH_3)_4Si$, which is abbreviated TMS. TMS makes a convenient reference substance because its protons reach resonance at a higher field than all common types of organic protons. The frequency of TMS, $\nu_{TMS}$, is arbitrarily assigned a value of zero. By convention, any proton which is in resonance at a lower field is assigned a positive chemical shift relative to that of TMS.

Q-43 Using a 60.0 MHz (megahertz = $10^6$ Hertz) instrument, the difference in frequency between TMS absorption and a certain proton in a compound was found to be 120 Hz. What is the chemical shift value for this absorption in parts per million?

A-43

$$\delta = \frac{\Delta\nu \times 10^6}{\text{oscillator frequency (Hz)}}$$

$$\delta = \frac{120 \times 10^6}{60.0 \times 10^6} = 2.00 \text{ ppm}$$

Q-44 Using a 60.0 MHz instrument, the difference in frequency between TMS absorption and a certain proton in a compound was found to be 430 Hz. What is the chemical shift value for this absorption in ppm?

A-44

$$\delta = \frac{\Delta\nu \times 10^6}{\text{oscillator frequency (Hz)}}$$

$$\delta = \frac{430 \times 10^6}{60.0 \times 10^6} = 7.17 \text{ ppm}$$

Q-45 Using a 60.0 MHz instrument, the difference in frequency between TMS absorption and a certain proton in a compound was found to be 180 Hz. What would be the frequency difference between these same protons if a 40 MHz instrument were used?

A-45

$$\delta = \frac{180 \times 10^6}{60.0 \times 10^6} = 3.00 \text{ ppm}$$

$$\Delta\nu = \frac{\delta\ (\text{oscillator frequency})}{10^6}$$

$$\Delta\nu = \frac{3.0\ (40 \times 10^6)}{10^6}$$

$$\Delta\nu = 120 \text{ Hz}$$

Q-46 Three different protons, A, B, and C, have shielding constants in the order

$$\sigma_B > \sigma_A > \sigma_C.$$

Arrange the three protons in order of applied field necessary for resonance.

A-46 $B_{0(B)} > B_{0(A)} > B_{0(C)}$

Q-47 The applied field necessary for resonance for three protons decreases in the following order

$$B_{0(D)} > B_{0(E)} > B_{0(F)}$$

Arrange the protons in order of decreasing chemical shift relative to TMS.

(Hint: Any proton which has resonance at a field below that of TMS has a positive chemical shift.)

A-47 $\delta_F > \delta_E > \delta_D$

Q-48 As $\sigma$ increases, does $\delta$ increase or decrease?

A-48 $\delta$ will decrease.

Q-49 Two absorption peaks are shown on the graph below. Which proton has the greater shielding constant?

Proton A, Proton B, TMS

4 3 2 1 0

← $\delta$ ppm

A-49 Proton B has the larger shielding constant. As $\sigma$ becomes larger, $\delta$ (relative to TMS) becomes smaller.

Q-50 Which of the protons described in Q-49 resonates at the higher field?

A-50 Proton B resonates at the higher field. The larger the shielding constant, the higher the field necessary for resonance.

**R**

$$\delta = \frac{\nu_{(sample)} - \nu_{(ref)}}{\nu_0} \times 10^6$$

$\nu_{(ref)}$ is usually tetramethylsilane (TMS). By convention, any proton which reaches resonance at a lower field than TMS is assigned a positive chemical shift.

S-9 When TMS is used as a reference, the chemical shifts are often reported as $\tau$ values where

$$\tau = 10.00 - \delta$$

and $\delta$ is the chemical shift relative to TMS in parts per million.

Q-51 What is $\tau$ for tetramethylsilane (TMS)?

A-51 $\delta_{TMS} = 0$

$\tau = 10.00 - \delta = 10.00$

Q-52 What are the $\tau$ values for the following chemical shifts relative to TMS?

2.00 ppm

7.17 ppm

A-52 $\tau = 10 - \delta$

$\delta = 2.0$

$\tau = 10 - 2.0 = 8.0$

$\delta = 7.17$

$\tau = 10 - 7.17 = 2.83$

Q-53 As $\delta$ increases, will $\tau$ increase or decrease?

---

A-53 Decrease.

Q-54 As $\sigma$ decreases, will $\tau$ increase or decrease?

---

A-54 Decrease. As $\sigma$ decreases, $\delta$ increases.

Q-55 As $\mathbf{B_0}$ required for proton resonance increases, will $\tau$ increase or decrease?

---

A-55 Increase. As $\mathbf{B_0}$ increases, $\delta$ decreases.

Q-56 For the compound $CH_3CH_2Br$, which protons would be expected to have the lowest $\sigma$ values? Why?

---

A-56 Because of the strong electron pulling power of the bromine, the shielding effect ($\sigma$) on the $-CH_2-$ should be lower than the shielding effect of the $CH_3-$.

Q-57 For the compound $CH_3CH_2Br$, which protons would be expected to have the lowest $\tau$ values? Why?

---

A-57 Since the shielding effect on the $-CH_2-$ is lower than that of its $CH_3-$, the chemical shift relative to TMS of the $-CH_2-$ will be larger than that of $CH_3-$. Therefore, $\tau$ ($\tau = 10 - \delta$) will be smaller for $-CH_2-$ and larger for $CH_3-$. The actual values are:

$$\tau_{CH_2} = 6.60 \text{ and } \tau_{CH_3} = 8.35$$

Q-58 Qualitatively, what generalization can be made about the relationship between $\sigma$, $\delta$, and $\tau$?

---

A-58 Large shielding effects result in small chemical shifts and large $\tau$ values. $\sigma \approx 1/\delta \approx \tau$.

Q-59 As protons become more acidic, what happens to the $\tau$ values?

---

A-59 As the electron density about a proton decreases, it becomes more acidic. Consequently, the chemical shift becomes greater and the $\tau$ value becomes less. (This generalization is only valid when comparing the same general type of protons, i.e., O–H protons.)

Q-60 In the following compound, predict which proton will have the largest $\tau$ values. Explain.

```
     CH3   CH3 (b)
     |     |
   H-C--O--C-H (a)
     |     |
     CH3   CH3
```

---

A-60 Since the $H_a$ proton is just one atom removed from the strongly electronegative oxygen, it should have a smaller shielding constant than the $H_b$ proton which is two atoms removed from the oxygen. Consequently, $\tau$ for $H_a$ should be smaller than $\tau$ for $H_b$. Actual values:

$\tau(a) = 6.44$

$\tau(b) = 8.95$

Q-61 In the following compound, which proton would be expected to have the largest $\tau$ value? Why?

```
       H   H
       |   |
  Cl - C - C - Br
       |   |
       H   H
      (a) (b)
```

A-61 Chlorine is more electronegative than bromine; therefore, the shielding for the $H_a$ proton would be smaller than the shielding for the $H_b$ proton. Consequently, $\tau$ for $H_a$ should be less than $\tau$ for $H_b$. Actual values:

$\tau_{(a)} = 6.23$

$\tau_{(b)} = 6.43$

Q-62 Given below is the low resolution spectrum for

$CH_3CH_2Cl$
(a) (b)

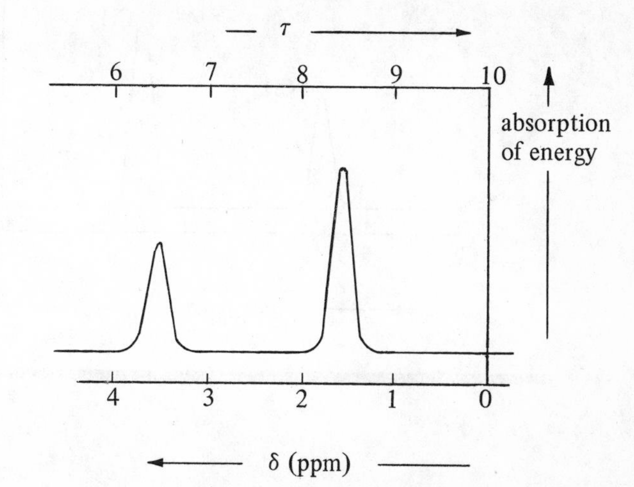


Which absorption peak corresponds to the $H_a$ protons and which to the $H_b$ protons? Explain.

A-62 The chlorine atom should decrease the shielding effect on $H_b$ more than on $H_a$. Consequently, $H_b$ would correspond to the absorption peak with the larger value of $\delta$, while $H_a$ would correspond to the absorption peak with the smaller value of $\delta$.

Q-63 Given below is the low resolution spectrum for

$CH_3CH_2CH_2I$
(a) (b) (c)

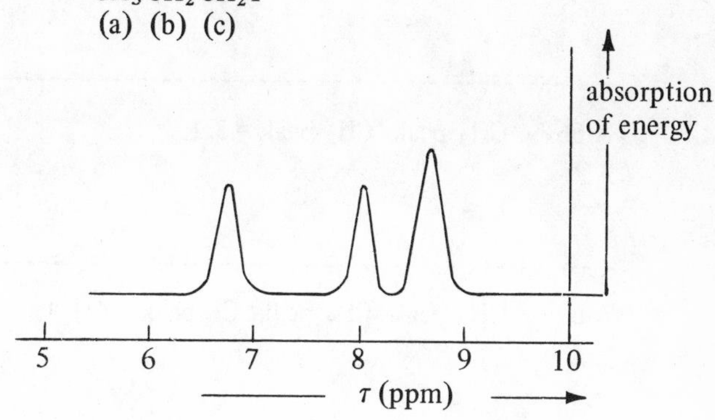


Identify each absorption peak and justify the assignment made.

A-63 The iodine atom should decrease the shielding effect in the order, $H_c$ more than $H_b$ more than $H_a$. Therefore, $H_c$ should have the smallest $\tau$ value.

Assignments:

$H_a = 8.8\ \tau$

$H_b = 8.0\ \tau$

$H_c = 6.9\ \tau$

**R** $\tau = 10 - \delta$ when TMS is used as reference.

Large $\sigma$ values result in small values for $\delta$ and large values for $\tau$.

**S-10** The ratio of the areas under absorption bands is related directly to the number of equivalent protons causing the absorption. A low resolution spectrum of ethyl chloride is given below.

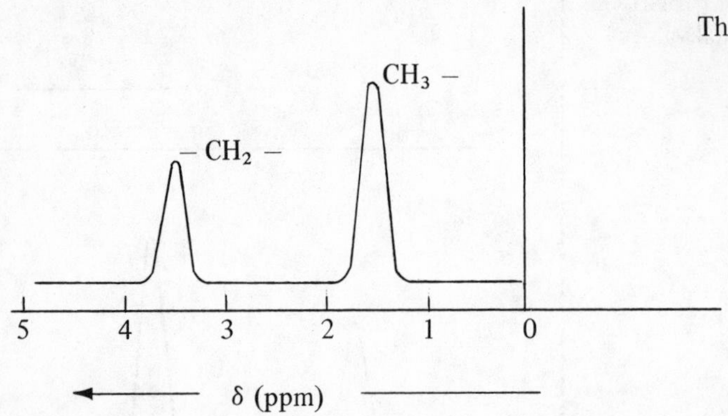


The ratio of areas is 3:2.

---

**Q-64** What should the ratio of peak areas be for the spectrum of ethanol ($CH_3CH_2OH$)?

---

**A-64** $CH_3$ peak: $CH_2$ peak: OH peak = 3:2:1.

**Q-65** How many different kinds of protons are there in the compound

$CH_3CH_2CH_3$?

---

**A-65** Two. The $CH_3$ groups are equivalent.

**Q-66** What should the ratio of peak areas be for the compound

$CH_3CH_2CH_3$?

---

**A-66** $CH_3$ peak: $CH_2$ peak = 3:1.

**Q-67** What should the ratio of peak areas be for the compound

$(CH_3)_2CHCH_2CH(CH_3)_2$?

---

**A-67** $CH_3$ peak: $CH_2$ peak: CH peak = 6:1:1.

**Q-68** What should the ratio of peak areas be for the compound

$HOCH_2CHOHCH_2OH$?

---

**A-68** $CH_2$ peak: CH peak: $CH_2$ OH peak: CHOH peak = 4:1:2:1.

**Q-69** What is the ratio of peak areas for the compound:

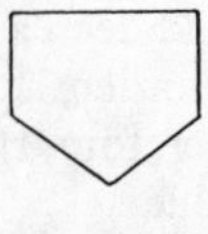

?

---

**A-69** Because all protons are equivalent, there is only one absorption peak in the spectrum. Consequently, no ratio can be determined.

**R** The areas under the peaks in an NMR spectrum are directly proportional to the number of equivalent protons.

S-11 In the *high* resolution spectrum of ethyl chloride, the peaks corresponding to the methyl and methylene protons are "split."

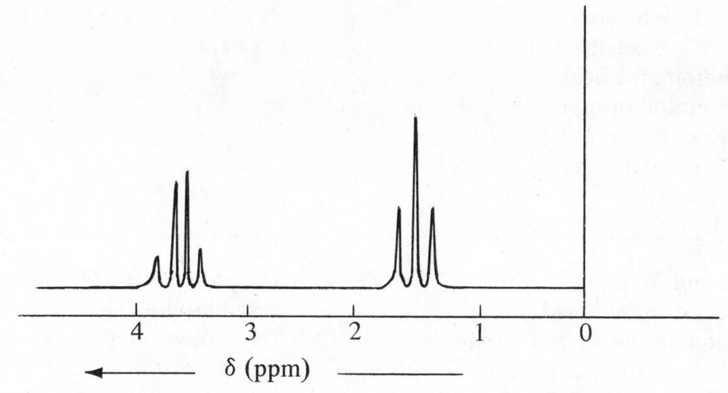


The fine structure in the high resolution spectrum arises from the phenomenon called *spin-spin splitting*. *Spin-spin splitting* is due to the various spin orientations of one nucleus with respect to the spin orientations of a neighboring proton. The magnetic field felt by a proton is affected by the various spin orientations of the neighboring protons.

---

Q-70 In the molecule HF, both H and F have magnetic moments. The spin quantum number, **I**, for fluorine is ½. How many different orientations relative to the applied magnetic field will a $F_9^{19}$ nucleus be able to assume?

A-70 Number of orientations = 2**I** + 1, **I** = ½

Number of orientations = 2

Q-71 The orientation of the fluorine nucleus with spin = –½ is aligned with the applied magnetic field and the orientation with spin = +½ is aligned against the applied magnetic field. Will the magnetic moment resulting from the spin = –½ orientation augment or oppose the applied magnetic field? Why?

A-71 Augment.

Spin = –½ is oriented with the applied field. Therefore, the magnetic moment resulting from this orientation will augment the applied magnetic field. The spin = +½ orientation will produce a magnetic field which opposes the applied magnetic field.

Q-72 Depending upon the orientation of the fluorine nucleus in HF, the magnetic moment of fluorine will either augment or oppose the applied magnetic field. Because of the close proximity, the magnetic field felt by the proton will be affected by the orientation of the fluorine nucleus.

If the spin of the fluorine nucleus is –½, will the magnetic field felt by the proton in HF be larger or smaller than the applied field? Will a larger or smaller applied field be necessary to produce proton resonance?

A-72 Field where resonance occurs = applied field + field transmitted from fluorine nucleus. The magnetic moment of fluorine, with spin = –½, augments the applied field. Therefore, the field transmitted to the proton from the fluorine nucleus is positive, causing the proton resonance to occur at a lower applied field.

Q-73 If the spin of the fluorine nucleus is +½, what effect will this have on the applied magnetic field at which the proton resonance occurs?

A-73 The magnetic moment for fluorine with spin = +½ opposes the applied field. Therefore, the field transmitted by fluorine to the proton will be negative, causing the proton resonance to occur at a higher applied field.

Q-74 The probability of fluorine having spin = +½ and spin = –½ is practically the same. The proton resonance spectrum of HF is shown below.

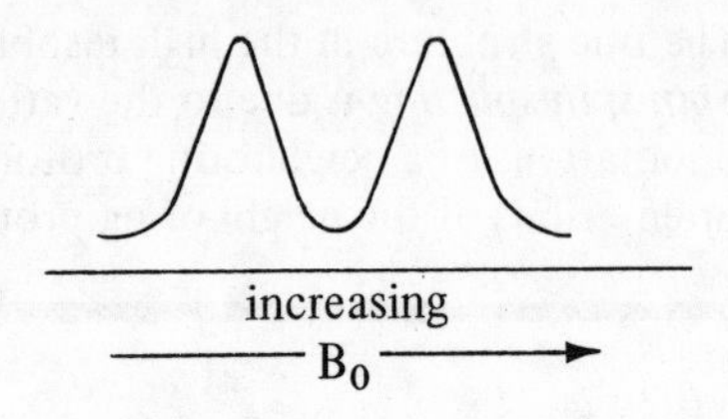


Label the peaks in the spectrum arising from the interaction of the proton with the two different spin states of fluorine.

A-74 Interaction with fluorine spin state

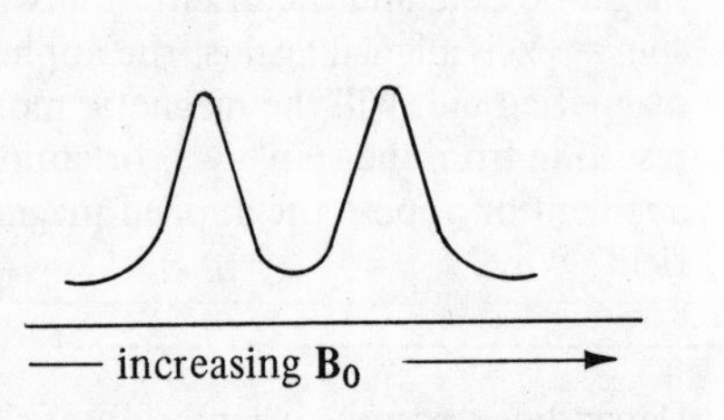


Q-75 Why are the two peaks in the proton resonance spectrum for HF of equal intensity?

A-75 The two peaks are of equal intensity because there is practically equal probability that the fluorine will have –½ and +½ spin states.

Q-76 On the graph below, sketch the position and relative intensity of the proton resonance for HF if there were no interaction between the hydrogen and fluorine nuclei.

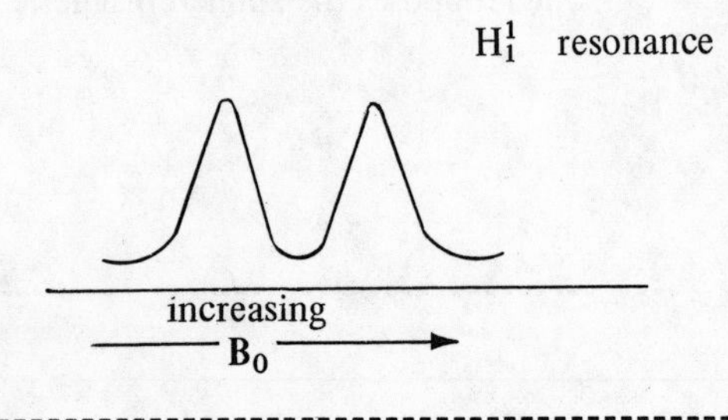

A-76 $H_1^1$ resonance

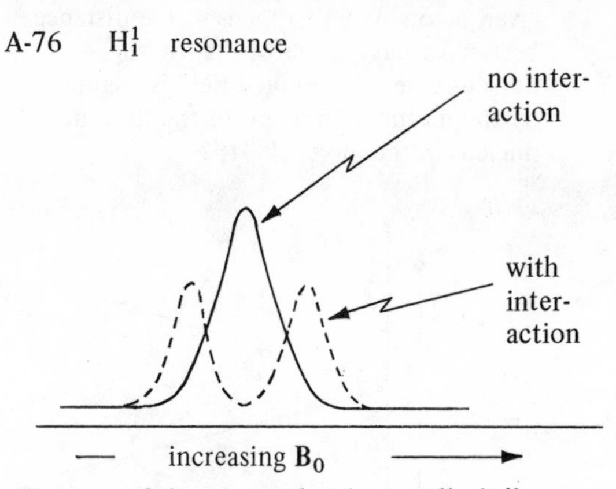


The sum of the area under the two "split" peaks would equal the area under the single peak.

Q-77 Sketch the fluorine resonance spectrum for the molecule HF. The hydrogen nucleus, with $I = ½$, interacts with the fluorine nucleus.

A-77

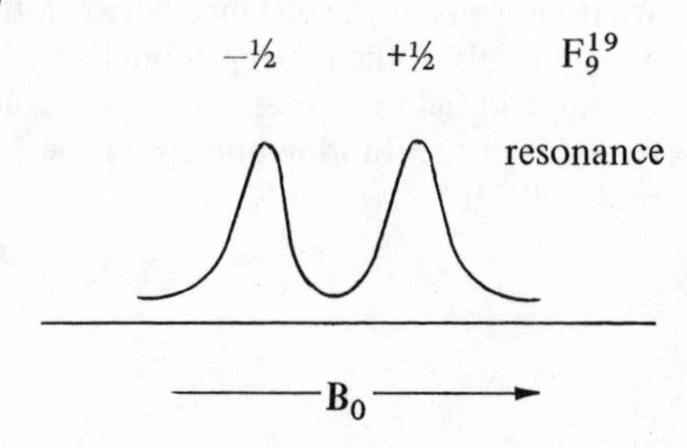


**R** Spin-spin splitting of a nucleus (observed in high resolution spectra) is caused by the nuclear spin interaction of neighboring nuclei. Neighbors with a nuclear spin of −½ augment, while those with spin of +½ oppose the applied magnetic field imposed on a nucleus. The position of observed resonance for a nucleus will depend upon the spin arrangement of neighboring nuclei. A nucleus with $I = ½$ has equal probability of being in the +½ and −½ spin states.

S-12 The distance (in ppm or Hz) between the fine structure of a resonance signal is called the *spin-spin coupling constant, J.* The value of the spin-spin coupling constant is dependent on the nuclei interacting and is independent of the strength of the applied magnetic field. The effect of changing the strength of the applied field for the proton resonance in the molecule HF is illustrated below.

strength of field = 14,000 gauss $J_{HF}$ = 615 Hertz

strength of field = 20,000 gauss $J_{HF}$ = 615 Hertz

Even though the proton and fluorine resonance occurs at different applied fields, the distance between the fine structure in each absorption band will be the same ($J_{HF} = J_{FH}$).

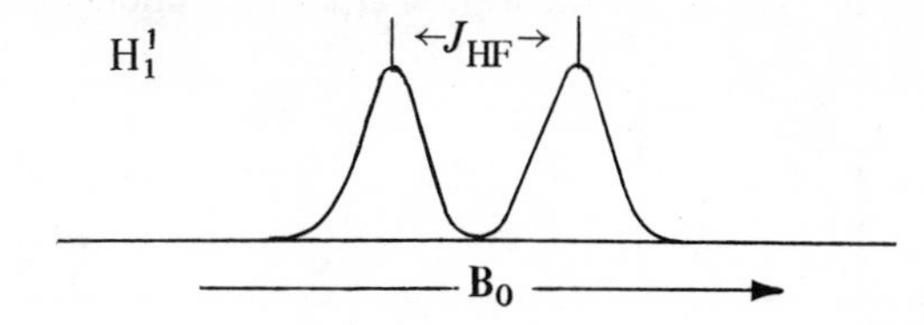


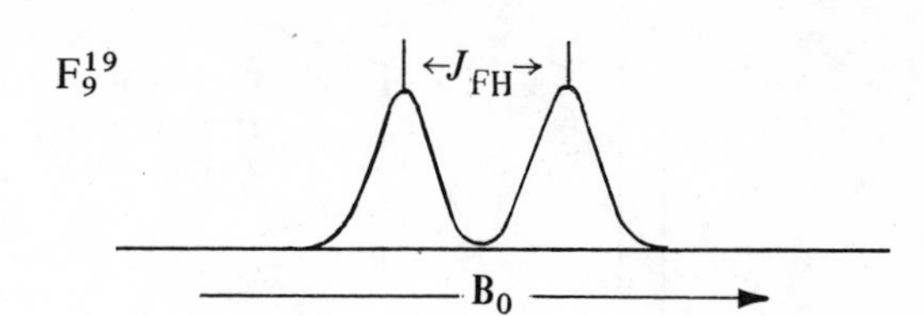


In certain cases, nuclei with spin-spin interaction have such small coupling constants that the interactions may be ignored. Thus, in compounds containing Cl, Br, or I, $J_{HCl}$, $J_{HBr}$, and $J_{HI}$ are so small that the spin-spin interactions are not observed.

Q-78 The energy vs. magnetic field diagram is given below. What happens to the distance between energy levels of the *hydrogen* nucleus when the applied field is augmented by the magnetic moment of the fluorine nucleus in the molecule HF?

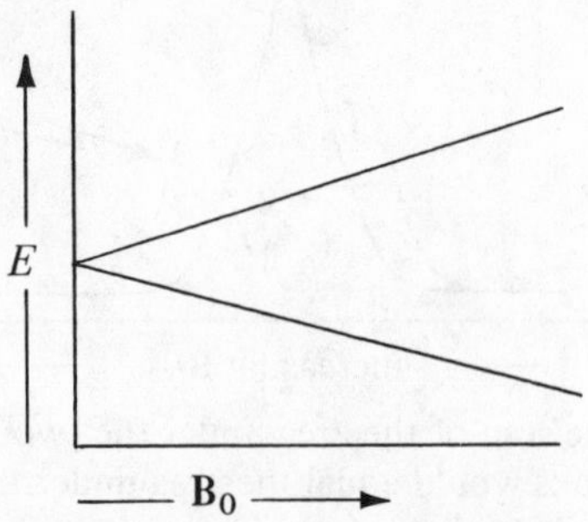

A-78

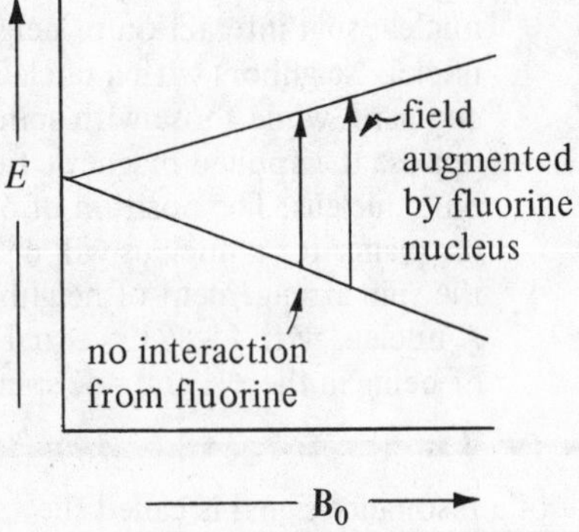

The difference between energy levels increases with an augmented field.

Q-79 What happens to the distance between the energy levels of the hydrogen nucleus when the applied field is opposed by the magnetic moment of the fluorine nucleus in the molecule HF?

A-79

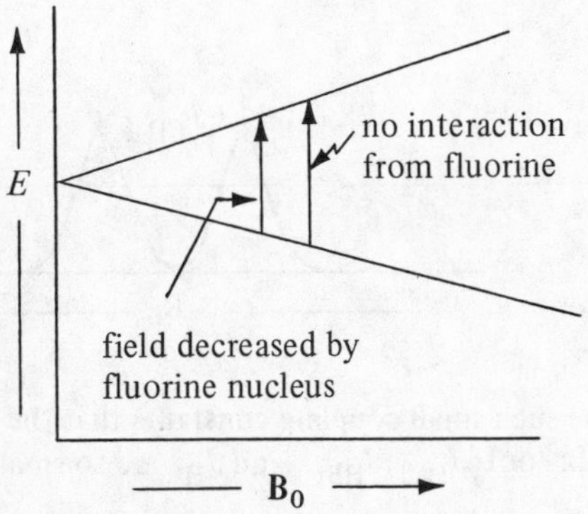

The difference between energy levels decreases.

Q-80 In HF, the interaction by the fluorine nucleus increases or decreases the difference between the proton energy levels by $J/4$. On the diagram below, sketch the energy levels of the proton after interaction with the fluorine nucleus.

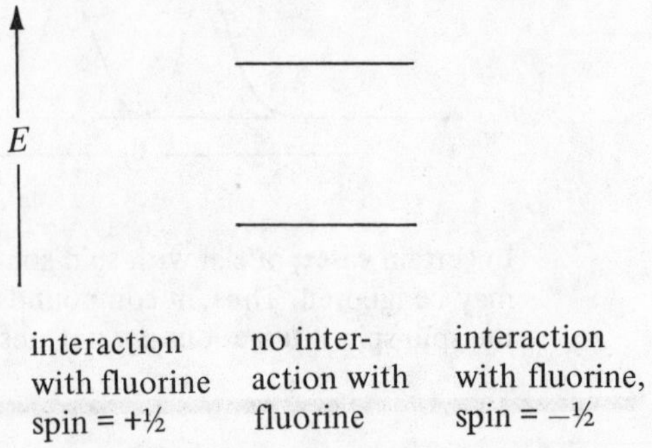

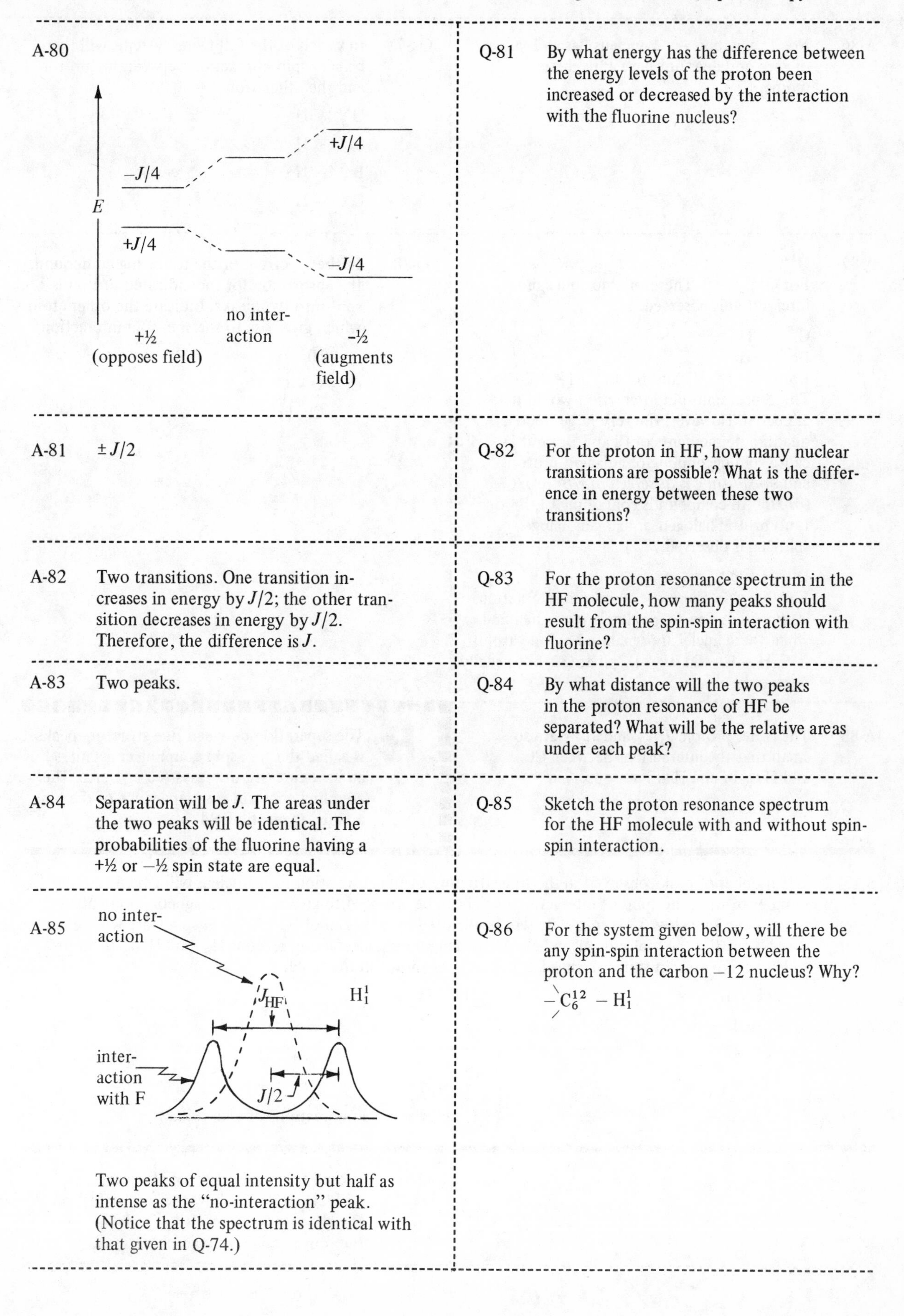

A-80



Q-81 By what energy has the difference between the energy levels of the proton been increased or decreased by the interaction with the fluorine nucleus?

A-81 $\pm J/2$

Q-82 For the proton in HF, how many nuclear transitions are possible? What is the difference in energy between these two transitions?

A-82 Two transitions. One transition increases in energy by $J/2$; the other transition decreases in energy by $J/2$. Therefore, the difference is $J$.

Q-83 For the proton resonance spectrum in the HF molecule, how many peaks should result from the spin-spin interaction with fluorine?

A-83 Two peaks.

Q-84 By what distance will the two peaks in the proton resonance of HF be separated? What will be the relative areas under each peak?

A-84 Separation will be $J$. The areas under the two peaks will be identical. The probabilities of the fluorine having a $+½$ or $-½$ spin state are equal.

Q-85 Sketch the proton resonance spectrum for the HF molecule with and without spin-spin interaction.

A-85



Two peaks of equal intensity but half as intense as the "no-interaction" peak. (Notice that the spectrum is identical with that given in Q-74.)

Q-86 For the system given below, will there be any spin-spin interaction between the proton and the carbon $-12$ nucleus? Why?

$$-C_6^{12} - H_1^1$$

A-86 No. The carbon – 12 nucleus has an **I** value of zero and does not have a magnetic moment.

Q-87 In which of the following systems will there be spin-spin interaction between the proton and the other atom?

$O^{16}$ – H

$Cl^{35}$ – H

$Br^{79}$ – H

D – H

---

A-87 $O^{16}$ – H:
For $O^{16}$, **I** = 0. Therefore, no spin-spin interaction is observed.

$Cl^{35}$ – H
$Br^{79}$ – H:
For $Cl^{35}$, **I** = 3/2 and for $Br^{79}$, **I** = 3/2. Therefore, spin-spin interaction would be expected. However, the very large electric quadrapole moments of Cl and Br (and I), cause spin decoupling of adjacent protons and *no splitting is observed in HBr or HCl (or HI).* In compounds containing Cl, Br, or I, no proton-halogen or halogen-halogen splitting is observed.

D – H:
For D, **I** = 1. Therefore, spin-spin interaction is expected. The $J_{HD}$ value is very small and when these nuclei are separated by one atom (i.e., H – C – D) H – D splitting is not usually observed.

Q-88 In the spectrum of the following compound, the absorption for the indicated proton is split into two peaks. Indicate the other atom which gives rise to the spin-spin interaction.

```
     (H)  H
      |   |
 Cl – C – C – Br
      |   |
      Cl  Br
```

---

A-88 The other proton. $J_{HBr}$ and $J_{HCl}$ are so small that the interactions between Cl and H or Br and H are not observed.

R The separation between fine structure peaks is called the spin-spin coupling constant, *J*, and is expressed in cycles/sec or Hertz. The coupling constant is independent of the applied magnetic field.

---

S-13 The number of peaks observed in the fine structure of the absorption for a nucleus will depend on the number of ways the spins of interacting nuclei can be arranged to give different magnetic moments. In the illustration below a +½ spin state will be denoted by the symbol ↑ while a –½ spin state will be indicated by ↓. For the molecule given below, the neighboring interacting protons $H_1$ and $H_2$ (neighbors to $H_A$) can have spin states arranged in the manner shown in the table.

```
     HA  H1
      |   |
 Cl – C – C – H2
      |   |
      Cl  Cl
```

| $H_1$ | $H_2$ |
|---|---|
| ↑ | ↑ |
| ↑ | ↓ |
| ↓ | ↓ |
| ↓ | ↑ |

(four possible combinations of spin states)

---

Q-89 If each spin interaction of $H_1$ and $H_2$ with $H_A$ is ± $J/2$, show the total effect on the $H_A$ resonance resulting from each of the four possible spin combinations.

A-89

| Spin Combination $H_1$ | $H_2$ | Total Effect |
|---|---|---|
| ↑ | ↑ | $J/2 + J/2 = J$ |
| ↓ | ↓ | $-J/2 - J/2 = -J$ |
| ↑ | ↓ | $J/2 - J/2 = 0$ |
| ↓ | ↑ | $-J/2 + J/2 = 0$ |

Q-90 For the molecule given in S-13, how many *different* magnetic moments result from the various spin combinations of $H_1$ and $H_2$?

A-90 Three. The ↑↓ and ↓↑ combinations give the same total effect.

Q-91 How many peaks will be present in the fine structure for the proton resonance of $H_A$ in the molecule shown in S-13?

(Hint: Remember that $J_{HCl}$ will not be observed.)

A-91 Three. There are three different spin arrangements for the neighboring protons $H_1$ and $H_2$.

Q-92 The probability of +½ and −½ spin states is equal. Predict the area under the three peaks in the fine structure of $H_A$.

A-92 1 : 2 : 1.
The area under the peaks resulting from the total spin interaction of zero (↑↓ and ↓↑) will be twice as probable as the other two interactions (↓↓ and ↑↑).

Q-93 Sketch on the graph below the positions of the three peaks in the fine structure of $H_A$. Label the distance between peaks and their relative areas. The arrow on the graph indicates the position of the $H_A$ resonance if no spin-spin interactions occur.

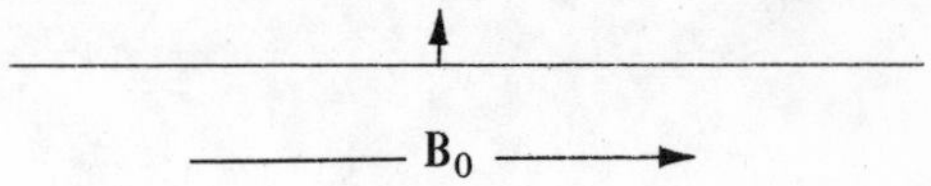

$B_0$ →

A-93 Ratio of peak areas 1 : 2 : 1.

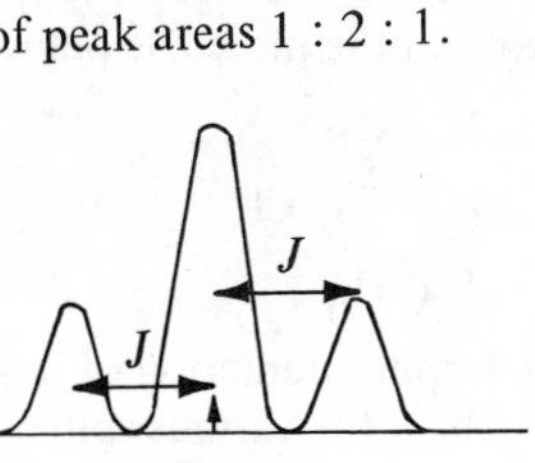

J
J
$B_0$ →

Q-94 Now consider the fine structure of the identical protons $H_1$ and $H_2$ (S-13) due to the spin-spin interaction with their neighbor, $H_A$. If each spin interaction of $H_A$ with $H_1$ and $H_2$ is $\pm J/2$, show the various spin combinations of $H_A$ and the total effect on the $H_1$ and $H_2$ resonance.

**A-94**

| **Spin Combination** | **Total Effect** |
|---|---|
| $H_A$ | |
| ↑ | $+J/2$ |
| ↓ | $-J/2$ |

Identical protons (such as $H_1$ and $H_2$ in the molecule shown in S-13) do not **appear** to interact with one another.

**Q-95** Sketch on the graph below the positions of the fine structure of the $H_1$ and $H_2$ protons. Label the distance between peaks and their relative areas. The arrow on the graph indicates the position of the $H_1$ and $H_2$ resonance if no spin-spin interaction occurs. (Hint: Remember that $J_{HCl}$ will not be observed.)

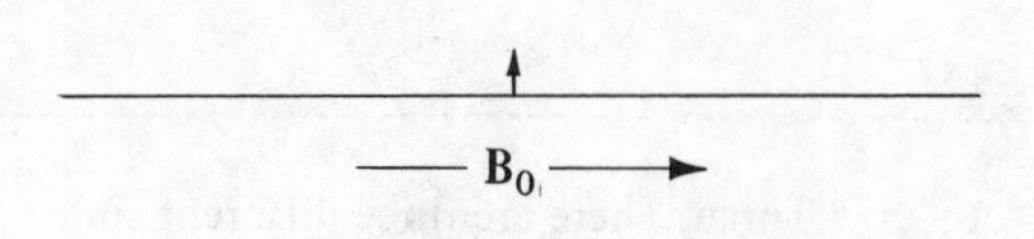


**A-95** Ratio of peak areas 1 : 1.

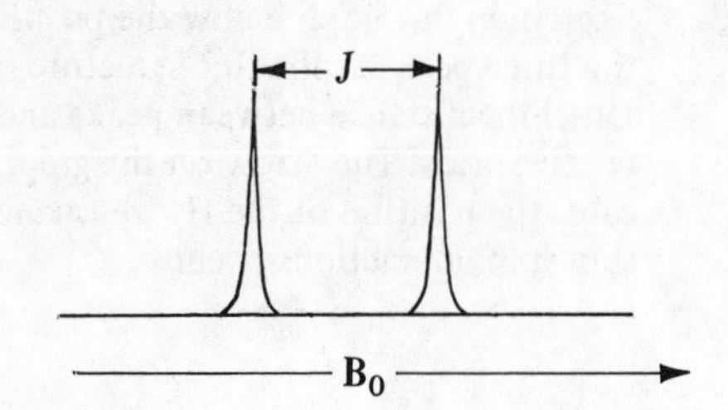


**Q-96** The chemical shift for the $H_A$ protons is 4.23 $\tau$ while the chemical shift for $H_1 = H_2$ = 6.05 $\tau$. The coupling constant is equal to 0.10 ppm. Sketch the proton spectrum for the molecule given in S-13 on the graph below.

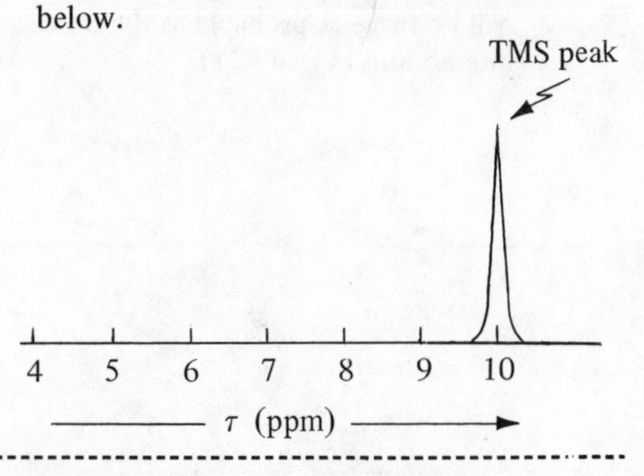


**A-96**

$H_1$ and $H_2$

$H_A$

TMS

4 5 6 7 8 9 10

$\tau$ (ppm)

Chemical shifts are measured from the center of the fine structure. The relative areas of $H_A : H_1 + H_2$ proton resonance are 1 : 2. The relative areas of the fine structure peaks are:

| | |
|---|---|
| $H_A$ | 1 : 2 : 1 |
| $H_1$ and $H_2$ | 1 : 1 |

**Q-97** Consider the fine structure for the $H_A$ protons in the molecule below.

$$\begin{array}{ccccc} & H_A & & H_B & \\ & | & & | & \\ H_A - & C & - & C & - Cl \\ & | & & | & \\ & H_A & & H_B & \end{array}$$

If each spin interaction of $H_B$ with $H_A$ is $\pm J/2$, show the various spin combinations of the $H_B$ protons and the total effect on the $H_A$ proton resonance.

A-97

| Spin Combination | | Total Effect |
|---|---|---|
| $H_B$ | $H_B$ | |
| ↑ | ↑ | $J/2 + J/2 = +J$ |
| ↓ | ↓ | $-J/2 - J/2 = -J$ |
| ↓ | ↑ | $-J/2 + J/2 = 0$ |
| ↑ | ↓ | $J/2 - J/2 = 0$ |

Q-98 Consider the fine structure for the $H_B$ protons in the molecule given in Q-97. If each spin interaction of $H_A$ with $H_B$ is $\pm J/2$, show the various spin combinations of the $H_A$ protons and the total effect on the $H_B$ resonance.

A-98

| Spin Combination | | | Total Effect |
|---|---|---|---|
| $H_A$ | $H_A$ | $H_A$ | |
| ↑ | ↑ | ↑ | $J/2 + J/2 + J/2 = 3/2J$ |
| ↓ | ↓ | ↓ | $-J/2 - J/2 - J/2 = -3/2J$ |
| ↑ | ↑ | ↓ | $+J/2 + J/2 - J/2 = J/2$ |
| ↑ | ↓ | ↓ | $+J/2 - J/2 - J/2 = -J/2$ |
| ↑ | ↓ | ↑ | $+J/2 - J/2 + J/2 = J/2$ |
| ↓ | ↑ | ↑ | $-J/2 + J/2 + J/2 = J/2$ |
| ↓ | ↑ | ↓ | $-J/2 + J/2 - J/2 = -J/2$ |
| ↓ | ↓ | ↑ | $-J/2 - J/2 + J/2 = -J/2$ |

Q-99 How many peaks should the $H_B$ fine structure have? What should the relative areas under these peaks be?

A-99 Four peaks ($3/2J$, $+J/2$, $-J/2$, $-3/2J$). Relative areas 1 : 3 : 3 : 1.

Q-100 For the molecule described in Q-97, the chemical shifts for the protons are

$\delta\ H_A = 1.48$ ppm

$\delta\ H_B = 3.57$ ppm

$J_{AB} = 0.15$ ppm

Sketch the proton spectrum for $CH_3CH_2Cl$ on the graph below. Label the distance between fine structure peaks and give relative areas.

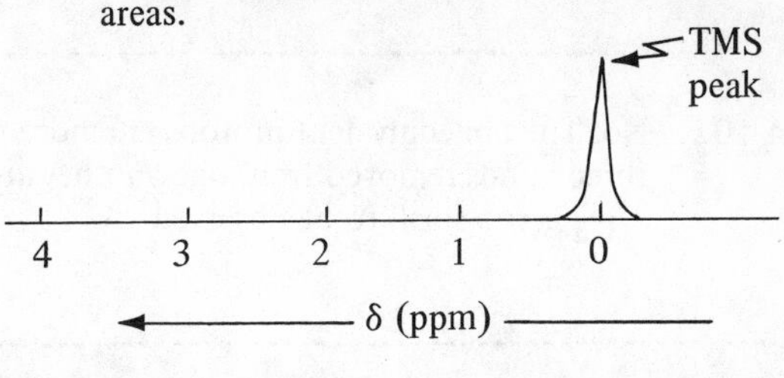

A-100

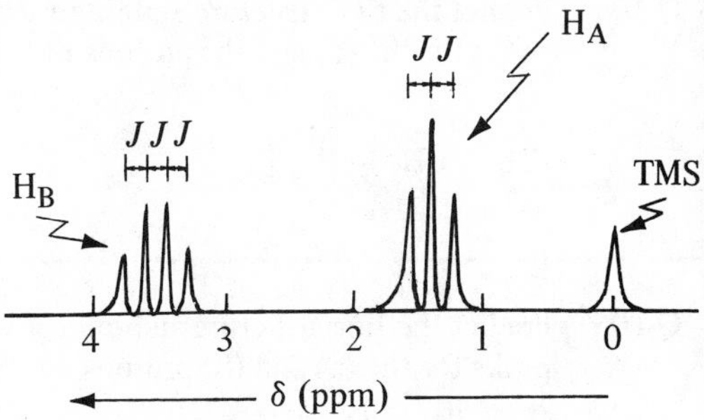

Relative areas of $H_A : H_B = 3 : 2$. The relative areas of fine structure peaks are: $H_A$ 1:2:1; $H_B$ 1:3:3:1.

**R** The number of peaks in the fine structure depends on the number of ways the spin of neighboring nuclei can be arranged to give different magnetic moments.

The area under the peaks for fine structure is relative to the number of ways the spin of neighboring nuclei can be arranged to give the same magnetic moments.

The chemical shift between two different types of hydrogens is measured from their centers of fine structure.

**S-14** Spin-spin splitting occurs through the bonds separating the interacting nuclei. Usually this interaction will occur only between nuclei that are separated by three or fewer bonds. In general the number of peaks (#) in the fine structure of an absorption band will be given by the formula

$$\# = 2n\mathbf{I} + 1$$

where **I** is the nucleus spin quantum number of the nuclei causing the splitting and $n$ is the number of nuclei causing the splitting.

The relative areas under peaks in the fine structure of an absorption band are given by the coefficients preceding each term in the expansion formula of:

$$(x + 1)^m$$

where $m$ equals the number of peaks minus one (# – 1). These rules can be illustrated by using

$CH_3 - CH_2 - Cl$
(a) (b)

$\#_{(a)} = 2n\mathbf{I} + 1 = 2\,(2)\,(1/2) + 1 = 3$

$\#_{(b)} = 2n\mathbf{I} + 1 = 2\,(3)\,(1/2) + 1 = 4$

Area Ratio for (a):
$(x + 1)^2 = x^2 + 2x + 1$
Peak area ratios 1 : 2 : 1
Area Ratio for (b):
$(x + 1)^3 = x^3 + 3x^2 + 3x + 1$
Peak area ratios 1 : 3 : 3 : 1

---

**Q-101** Will there be any spin-spin splitting in the spectrum of the molecule
$Cl\,CH_2 - CH_2\,Cl$?
(a) (b)
Why?

---

**A-101** No. All the protons are equivalent.

**Q-102** Will there be any spin-spin interaction observed in the molecule
$CH_3 - CCl_2 - CH_2\,Cl$?
(a) (b)
Why?

---

**A-102** No. The nonequivalent protons are more than three bonds removed from one another and $J_{HCl}$ is too small to be observed.

**Q-103** Predict the fine structure and intensity of peaks for the (a) and (b) protons in
$CHCl_2 - CH_2\,Cl$.
(a) (b)

---

**A-103**

| Protons being split | Interacting neighbors | $n$ | $\# = 2n\,\mathbf{I} + 1$ | $(x+1)^m$ |
|---|---|---|---|---|
| $CHCl_2 -$ | $-CH_2\,Cl$ | 2 | 3 | 1:2:1 |
| $-CH_2\,Cl$ | $-CHCl_2$ | 1 | 2 | 1:1 |

**Q-104** Predict the fine structure and intensity of peaks for the (a) and (b) protons in
$CH_3 - CH_2 - \overset{\overset{O}{\|}}{C} - Cl$.
(a) (b)

---

**A-104**

| Protons being split | Interacting neighbors | $n$ | $\# = 2n\,\mathbf{I} + 1$ | $(x+1)^m$ |
|---|---|---|---|---|
| $CH_3 -$ | $-CH_2 -$ | 2 | 3 | 1:2:1 |
| $-CH_2 -$ | $-CH_3Cl_2$ | 3 | 4 | 1:3:3:1 |

**Q-105** Predict the fine structure and intensity of peaks for the (a) and (b) protons in
$Cl - CH_2 - O - CH_3$.
(a) (b)

A-105 No splitting. The protons are more than three bonds apart. Each proton would give a single peak (singlet). The ratio of a:b peaks is 2:3.

Q-106 Predict the fine structure and intensity of peaks for the (a), (b), and (c) protons in

$CH_3 - O - CH_2 - CH_3$.
(a) (b) (c)

A-106

| Protons being split | Interacting neighbors | $n$ | $\# = 2n\,I + 1$ | $(x+1)^m$ |
|---|---|---|---|---|
| $CH_{3\,(a)}$ | No splitting | | | |
| $-CH_2-_{(b)}$ | $-CH_3$ | 3 | 4 | 1:3:3:1 |
| $-CH_{3\,(c)}$ | $-CH_2-$ | 2 | 3 | 1:2:1 |

Q-107 Predict the fine structure and intensity of peaks for the equivalent protons of ammonia, $NH_3$.

(Hint: $I_{N^{14}} = 1$)

A-107

| Protons being split | Interacting neighbor | $n$ | $\# = 2nI+1$ | Intensity |
|---|---|---|---|---|
| $H_3N$ | N | 1 | 3 | 1:2:1 |

Q-108 Predict the fine structure and intensity of peaks for the protons of difluoromethane,

```
    H
    |
H - C - F
    |
    F
```

(Hint: $I_{F^{19}}=1/2$)

A-108

| Protons being split | Interacting neighbor | $n$ | $\# = 2nI+1$ | Intensity |
|---|---|---|---|---|
| $-CH_2-$ | $-CF_2-$ | 2 | 3 | 1:2:1 |

Q-109 Predict the fine structure and intensity of peaks for the methylene protons of propane, $CH_3 - CH_2 - CH_3$ .

A-109

| Protons being split | Interacting neighbor | $n$ | $\# = 2nI+1$ | Intensity |
|---|---|---|---|---|
| $-CH_2-$ | $(CH_3-)_2$ | 6 | 7 | |
| | | | (1:6:15:20:15:6:1) | |

Q-110 For a hydrocarbon, the absorption spectrum of a proton ($H_A$) is split into three peaks with relative areas of 1:2:1. How many equivalent neighboring protons interact with $H_A$?

A-110 There must be only two equivalent protons no more than three bonds removed from the given proton.

Q-111 Give possible structures for the system which gives rise to the absorption spectrum described in Q-110. Label proton $H_A$ and the two interacting protons $H_X$.

A-111

```
  \    H_X
   \  /
 - C - C -
   |   |
  H_A H_X
```

or

```
  H_X   |   H_X
   |    |    |
 - C -  C -  C -
   |    |    |
       H_A
```

Q-112 The fine structure for proton $H_A$ in a molecule is shown below. How many equivalent neighboring protons interact with $H_A$? Give possible structures for systems which could give rise to the absorption spectrum below. Label proton $H_A$ and the interacting proton $H_X$.

Area ratio 1:3:3:1

Proton $H_A$

A-112 $H_A$ must have only three equivalent protons, no more than three bonds removed. Possible structure is:

```
    HX HA
    |  |
HX - C - C -
    |  |
    HX      I
```

Structure II below is not a possibility since the shielding around $H_X$ & $H'_X$ would not be identical.

```
 H'X HA HX
  |   |   |
- C - C - C -
  |   |   |
          HX  II
```

Q-113 The partial spectrum of a compound is given below.

Ratio $H_X : H_M : H_A$ = 1:1:3

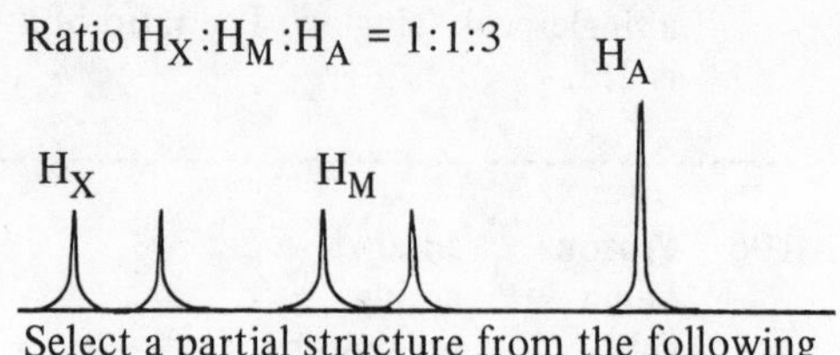


Select a partial structure from the following which is consistent with the above spectrum.

a. $CH_3 - \overset{O}{\overset{\|}{C}} - CH_2 - CH<$

b. $CH_3 - CH_2 - O -$

c. $CH_3 - O - \overset{/}{C}H - CH<$

d. $CH_3 - \overset{H}{\overset{|}{C}} - O -$

A-113 Structure c.

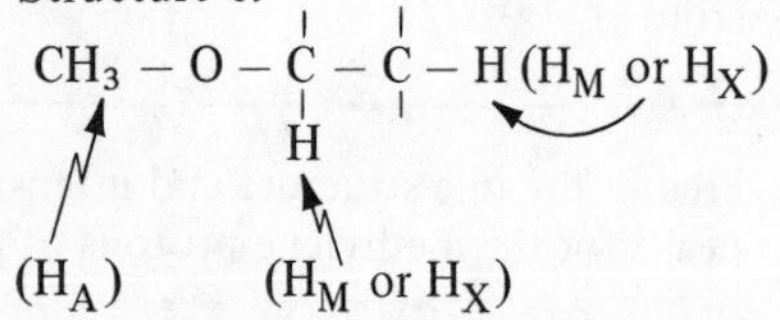


**R** Spin-spin coupling usually will occur only between atoms that are three or fewer bonds removed from each other. The number of peaks in fine structure of an absorption band = $2nI + 1$. The relative intensities of fine structure peaks are obtained from coefficients resulting from the expansion of $(x+1)^m$ where $m$ = number of peaks in fine structure – 1.

S-15 Another way of analyzing a spectrum is by use of spin-spin coupling diagrams. The technique used in this method is illustrated for the molecule 1,1,2-trichloroethane.

```
    HA HX
    |  |
Cl - C - C - HX
    |  |
    Cl Cl
```

The two different protons, A and X, interact and have a coupling constant of $J_{AX}$. The splitting of the X protons into a doublet by the A proton is diagramed below.

The height of the line representing the absorption peak is proportional to the area of the peak.

$-CH_2-$ (X)

no spin-spin interaction (no coupling)

$J_{AX}$ Coupling with A proton

1 : 1 ratio of areas

The splitting of the A proton by each of the equivalent X protons can be diagramed in two ways.

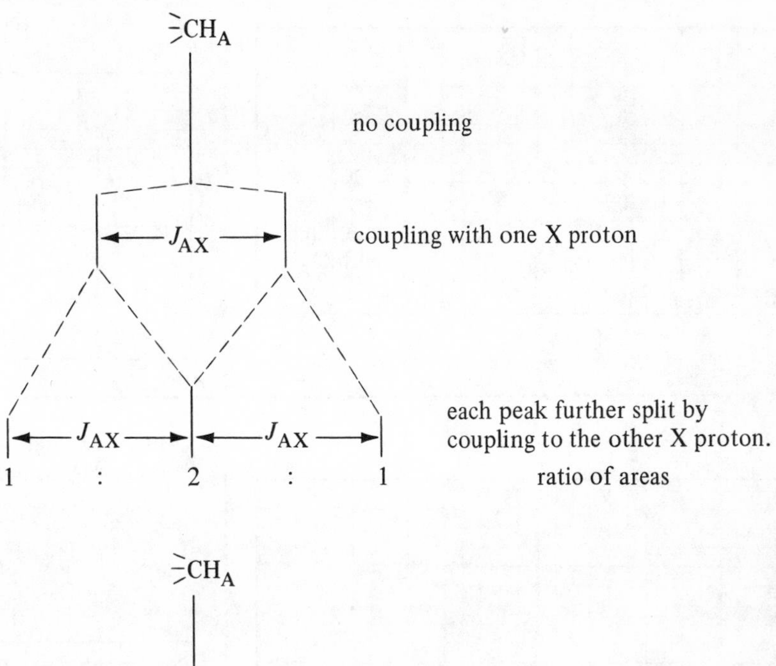


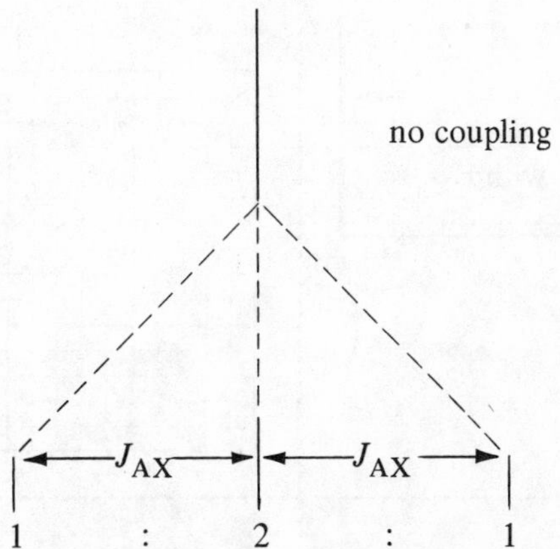


Q-114 Diagram and label the splitting pattern for $H_X$ in the system shown below.

$J_{AX} = 10$ Hz

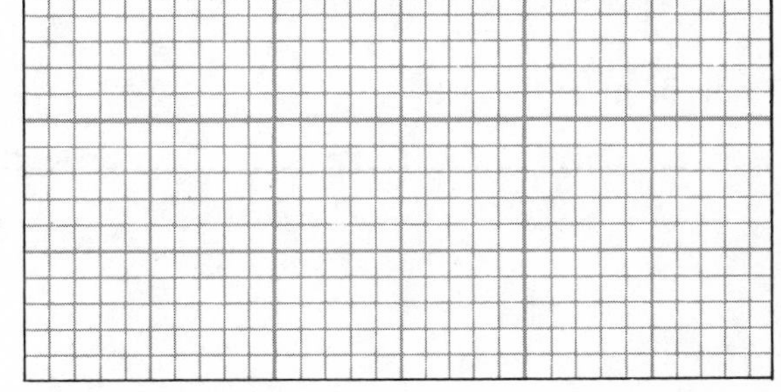

A-114

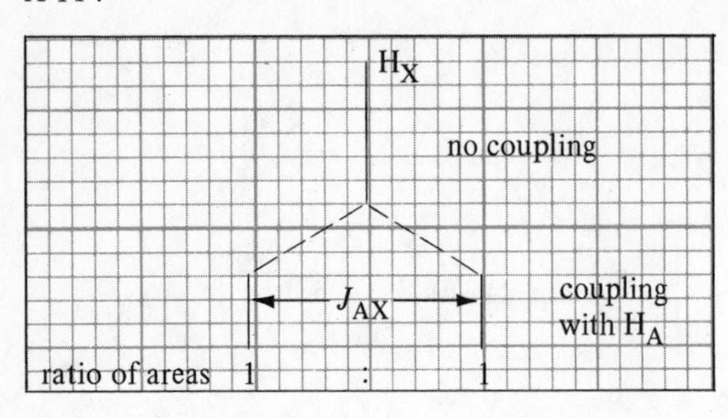


Q-115 Diagram and label the splitting pattern for $H_A$ in the system described in Q-114.

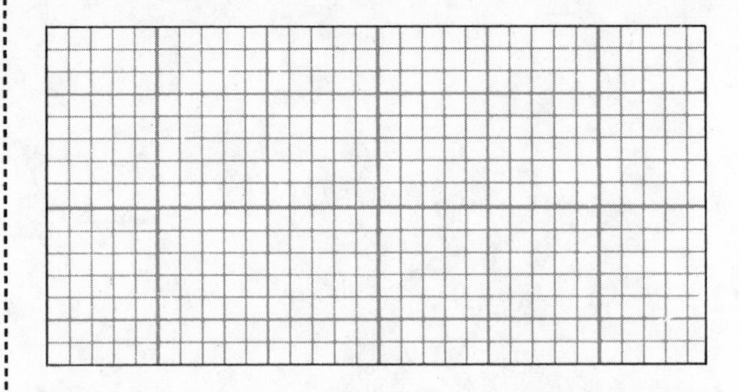

A-115

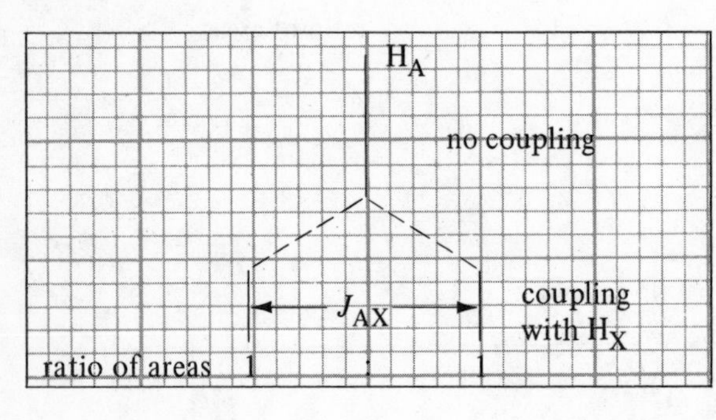


Q-116 Diagram and label the splitting pattern for $H_X$ in the system shown below.

$J_{AX}$ = 15 Hz

$$
\begin{array}{ccc}
H_X & H_A & \\
| & | & \\
-C- & C- & H_A \\
| & | &
\end{array}
$$

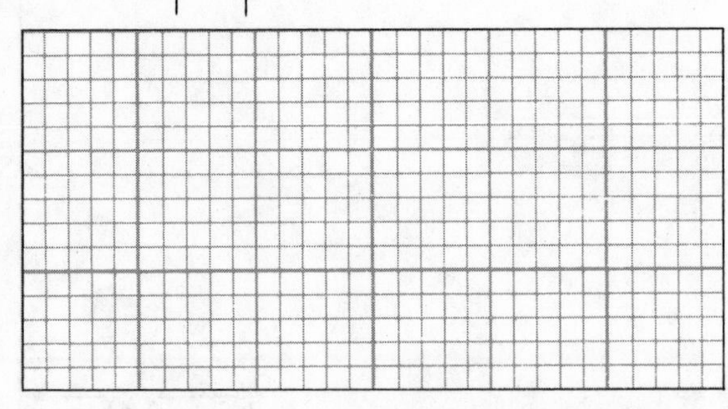

A-116

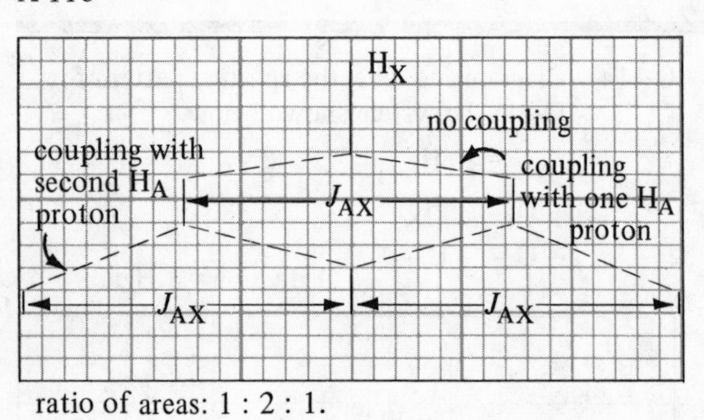


ratio of areas: 1 : 2 : 1.

Q-117 Diagram and label the splitting pattern for $H_A$ in the system described in Q-116.

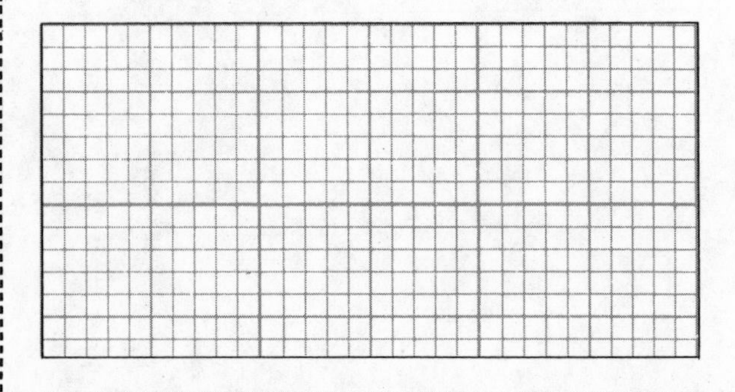

A-117

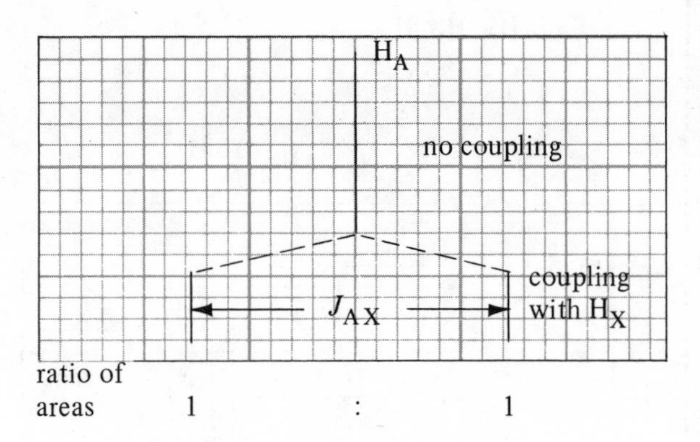


ratio of areas 1 : 1

Q-118 Diagram and label the splitting pattern for $H_X$ in the system shown below.

$J_{AX}$ = 7 Hz

$$H_A - \underset{H_A}{\overset{H_A}{C}} - \underset{H_X}{\overset{H_X}{C}} - O - R$$

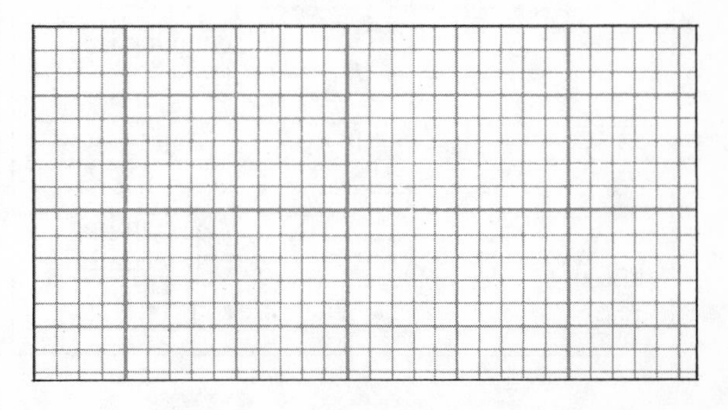

A-118

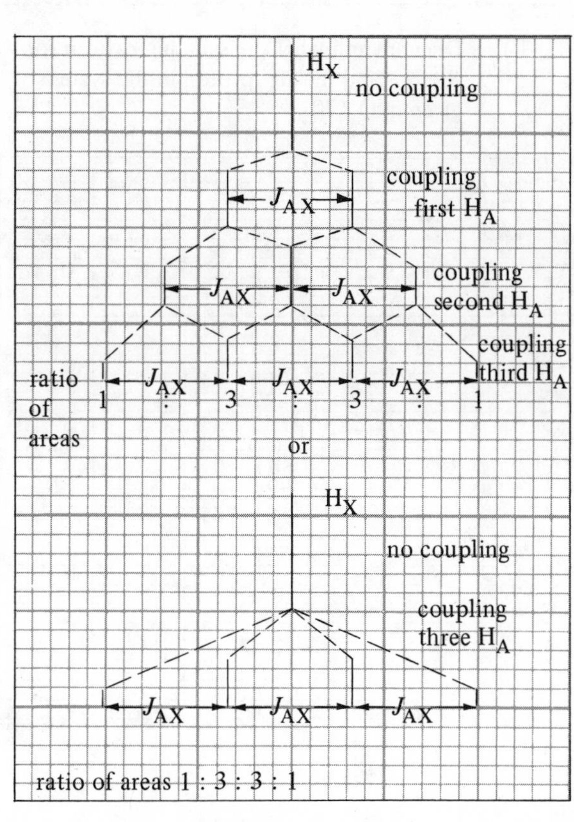


Q-119 For the system described in Q-118, diagram and label the splitting pattern for the $H_A$ proton.

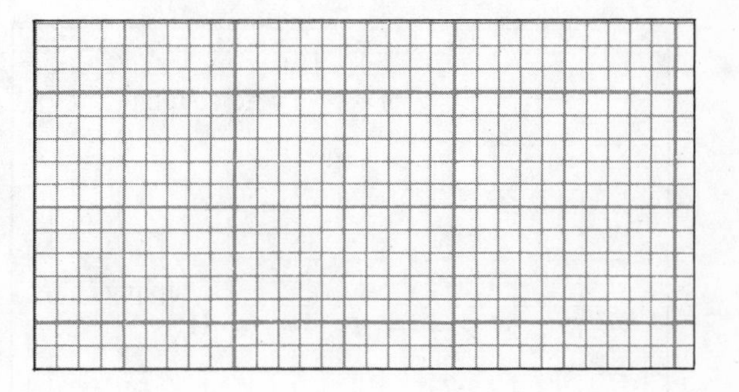

A-119

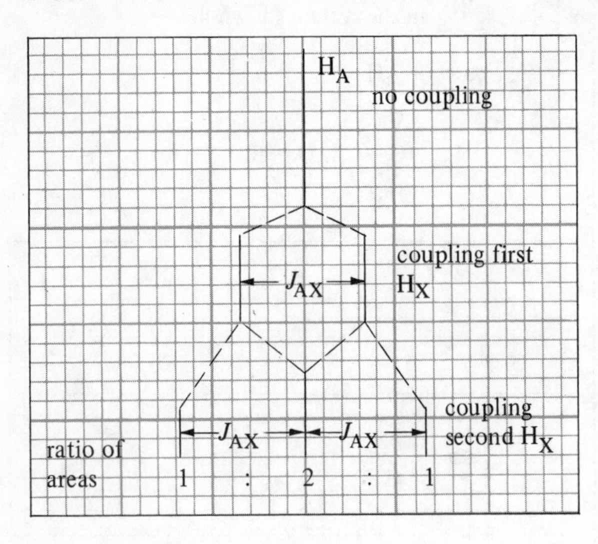


Q-120 Diagram and label the splitting pattern for the $H_M$ proton in the system shown below.

$$\begin{array}{ccc} H_A & H_M & H_X \\ -C- & C- & C- \end{array}$$

$J_{MX} = 5$ Hz $\quad J_{AM} = 10$ Hz

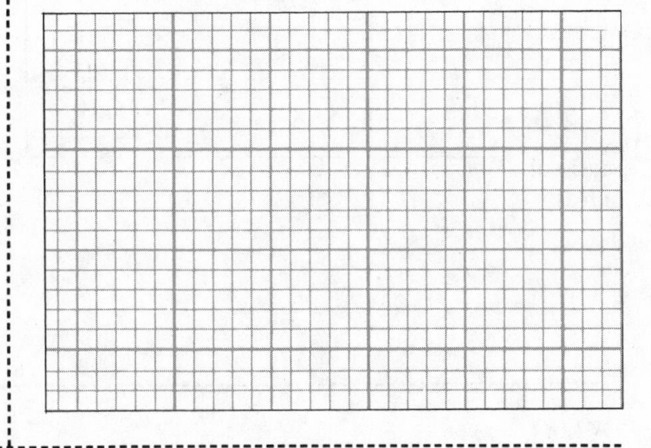

A-120

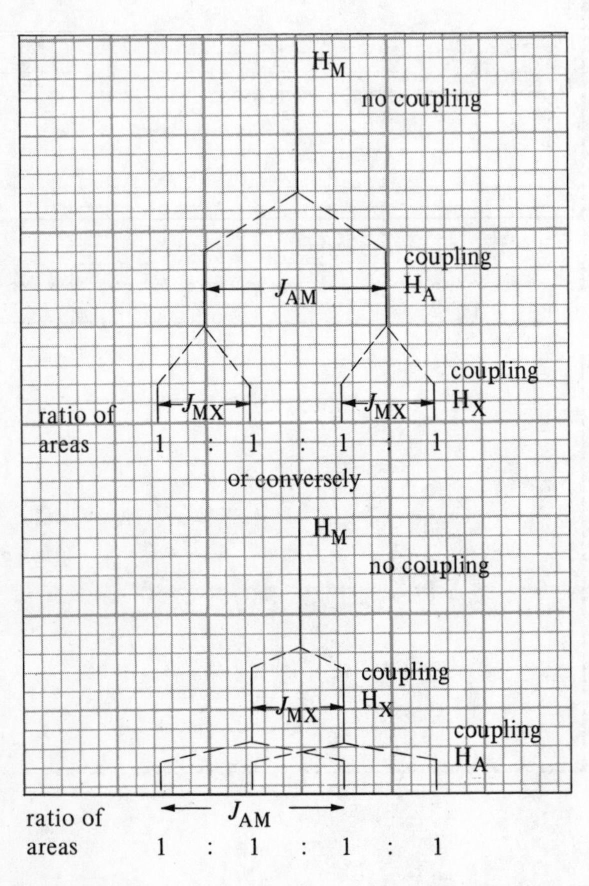


ratio of areas $\quad$ $J_{AM}$

1 : 1 : 1 : 1

Q-121 Diagram and label the splitting pattern for the $H_A$ and the $H_X$ proton in the system given in Q-120.

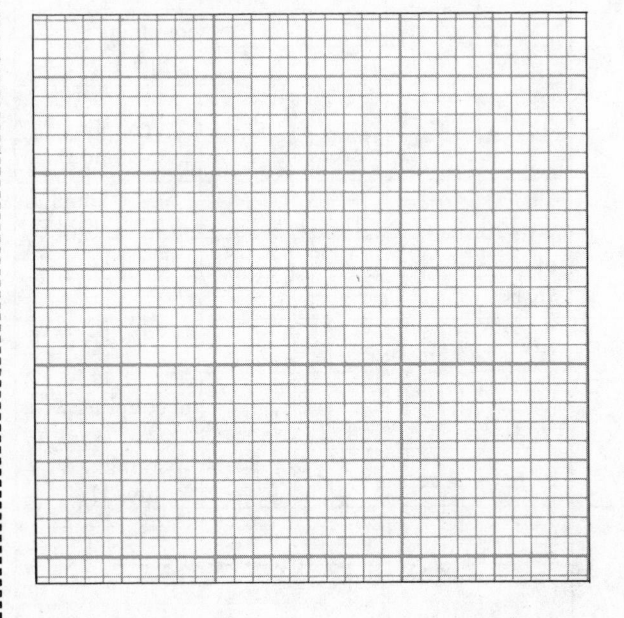

A-121

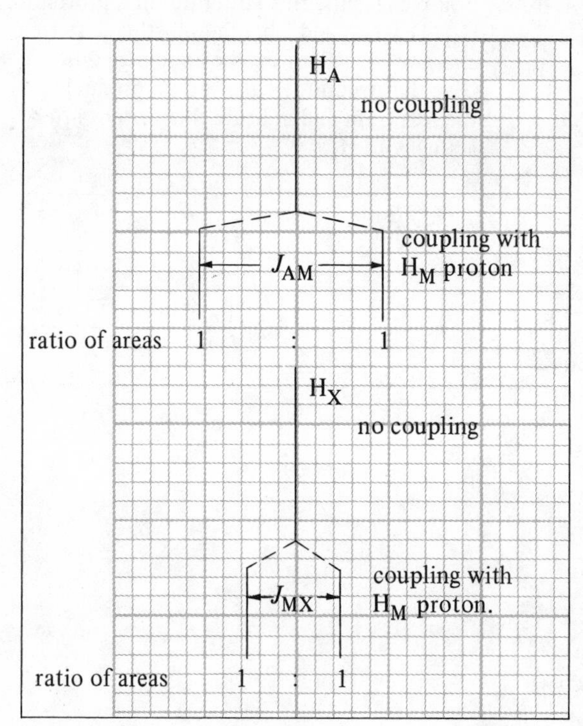


Q-122 Diagram and label the splitting pattern for the $H_M$ proton in the system shown below.

$J_{AM}$ = 4 Hz

and

$J_{MX}$ = 10 Hz

$$
\begin{array}{cccccc}
 & H_A & & & & \\
 & | & & | & & | \\
H_A - & C & - & C & - & C - \\
 & | & & | & & | \\
 & H_A & & H_M & & H_X
\end{array}
$$

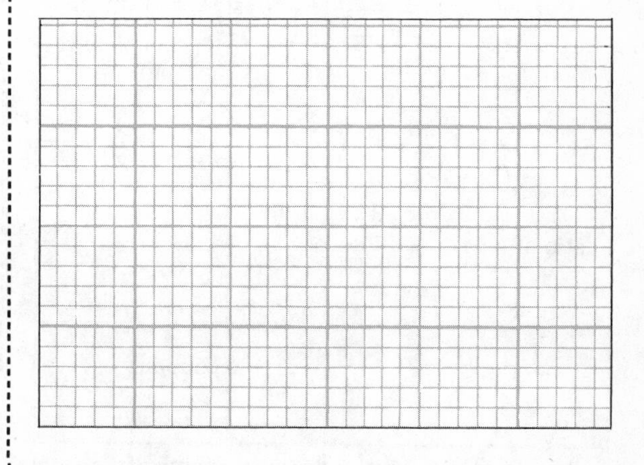

A-122

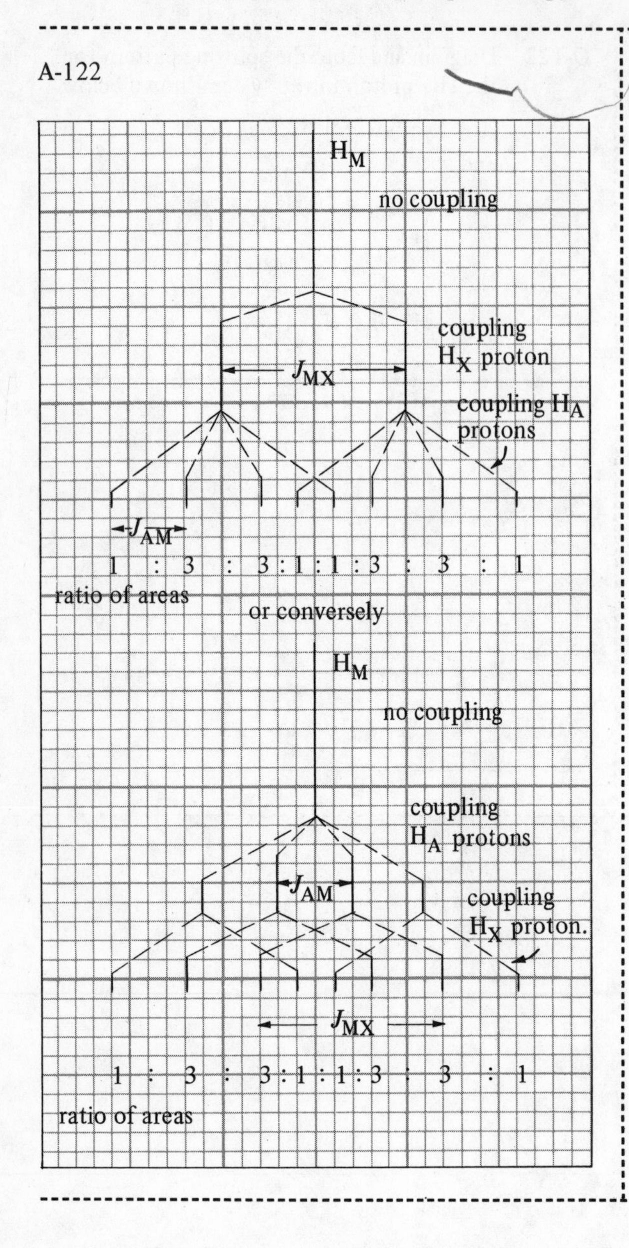


Q-123 Following is the fine structure for a proton $H_M$. Diagram and label the splitting pattern and give the values for the coupling constants. Give the structure of the system producing the spectrum and indicate the protons interacting with $H_M$.

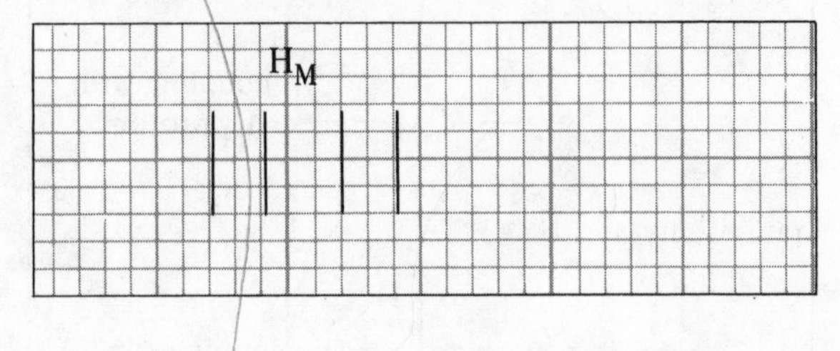

A-123

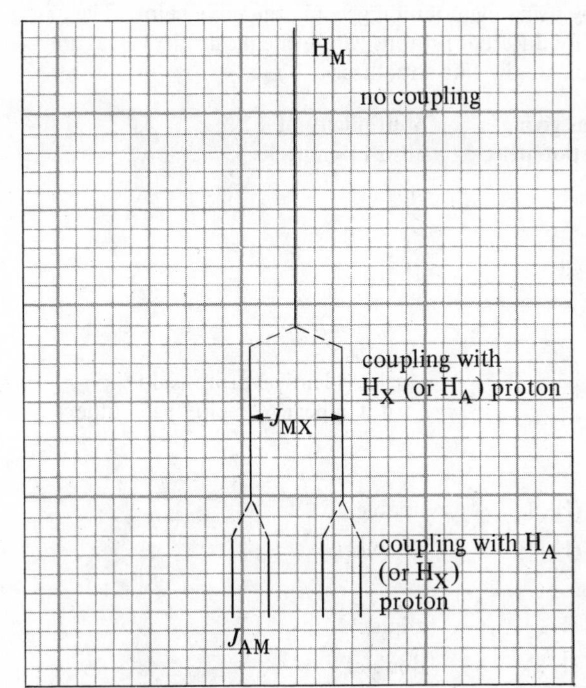


```
 HX HM HA
  |  |  |
– C – C – C –
  |  |  |
```

$J_{MX}$ = 5 Hz

$J_{AX}$ = 2 Hz

Q-124 Following is the fine spectrum for a proton $H_M$. Diagram and label the splitting pattern and give the value for the coupling constants. Give the structure of the system producing the spectrum and indicate the protons interacting with $H_M$.

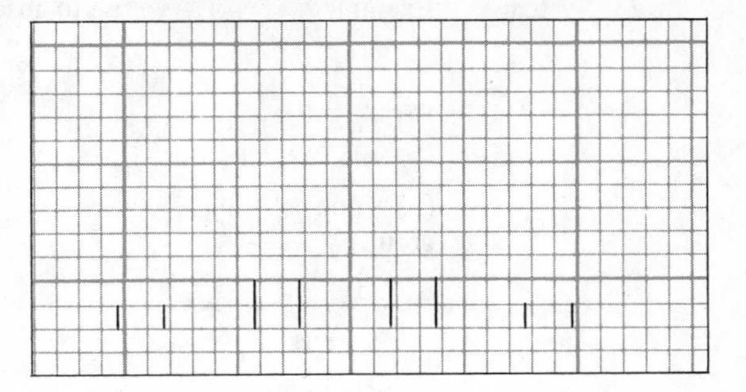

A-124

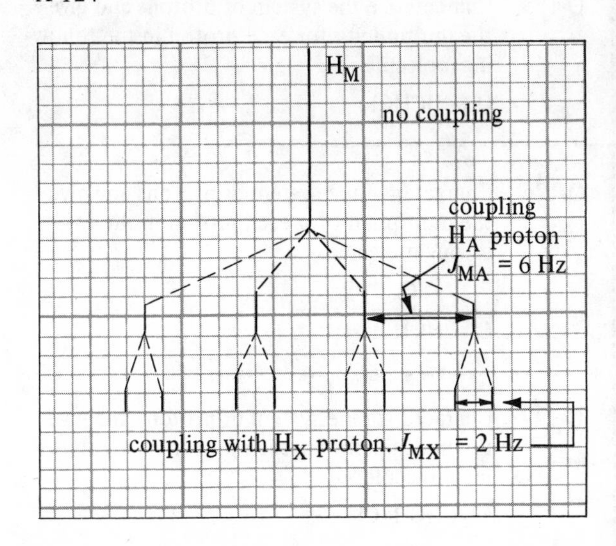


**R** The splitting diagram for the $H_M$ proton in the system

```
 HA HM HX
  |  |  |
– C – C – C –
  |  |  |
```

is shown as

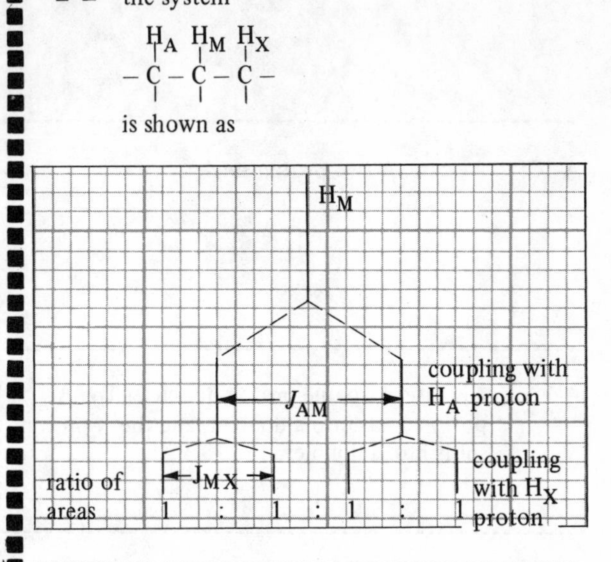

S-16 Interacting systems of protons are characterized by using the notation given in the table below. The letters A, M, and X are used to designate chemically different protons whose chemical shift values are very different. The term multiplicity is used to indicate the fine structure.

| System | Example | No. of protons giving rise to absorption peak | Multiplicity of absorption peak |
|---|---|---|---|
| A | $H_A$–C(ns)(ns)–ns, i.e. ns – C($H_A$) – ns with ns below (ns = non-splitting group) | 1 | singlet |
| $A_2$ | ns – C($H_A$)($H_A$) – ns | 2 | singlet (identical protons do not appear to "split" each other) |
| AX | ns – C($H_A$)(ns) – C($H_X$)(ns) – ns | 1 ($H_A$)<br>1 ($H_X$) | doublet<br>doublet |
| $AX_2$ | ns – C($H_A$)(ns) – C($H_X$)(ns) – $H_X$ | 1 ($H_A$)<br>2 ($H_X$) | triplet<br>doublet |
| AMX | ns – C($H_A$)(ns) – C($H_M$)(ns) – C($H_X$)(ns) – ns | 1 ($H_A$)<br>1 ($H_M$)<br>1 ($H_X$) | doublet<br>four peaks (two doublets)<br>doublet |

Q-125 Characterize the system of protons and give the multiplicity for each proton in the following molecule.

$Cl_2CHCHCl_2$

---

A-125 $A_2$ case. A **two** proton singlet.

Q-126 Characterize the system of protons and give the multiplicity for each proton in the following molecule.

$Cl_2CHCH{=}O$ (i.e. $Cl_2CHCHO$)

---

A-126 AX case. Two **one** proton doublets. The A proton splits into a doublet, and the X proton splits into a doublet.

Q-127 Characterize the system of protons and give the multiplicity for each proton in the following molecule.

$Cl_2CHCH_2Cl$

---

A-127 $AX_2$ case. A **one** proton triplet (A proton) and a **two** proton doublet (X protons).

Q-128 Characterize the system of protons and give the multiplicity for each proton in the following molecule.

$Cl_2CHCHClCH{=}O$ (i.e. $Cl_2CHCHClCHO$)

---

A-128 AMX case. A **one** proton doublet (A proton), a **one** proton doublet (X proton), and **one** proton (M proton) with four peaks (two doublets).

Q-129 Characterize the system of protons and give the multiplicity for each proton in the following molecule.

$CH_3OCHClCH_2Cl$

A-129 The $CH_3-$ protons do not couple with any others. Therefore, they are an $A_3$ case. The other protons couple together to form an $AX_2$ case. The spectrum would consist of a **three** proton singlet ($A_3$ case), a **one** proton triplet ( A proton), and a **two** proton doublet ($X_2$ protons).

Q-130 Characterize the system of protons and give the multiplicity for each proton in the following molecule.

$CHCl_2CH_2\overset{O}{\overset{\|}{C}}H$

A-130 An $AM_2X$ case. A **one** proton triplet (A proton), a **one** proton triplet (X proton), and a **two** proton pattern consisting of four peaks (two doublets).

**R** Systems of protons are characterized by using the notations given in S-16. Each particular system has a characteristic multiplicity (fine structure) for the absorption peaks.

S-17 The chemical shift parameter, $\delta$, has been shown to be a function of the electron density around the proton. The electron density about a proton is affected by the inductive effect of various groups operating through the chemical bonds attached to the proton. In addition to inductive effects, chemical shift parameters are also influenced by magnetic fields operating through space. Chemical bonds produce small magnetic fields which can operate through space. The magnetic fields produced by chemical bonds are anisotropic, that is, unsymmetrical about the bond. This anisotropic effect may result in a magnetic field in the area of the proton which either opposes or augments the applied field. An anisotropic effect which opposes the applied field will result in the proton resonance shifting to a higher field (a shielding effect). An anisotropic effect which augments the applied field results in the proton resonance shifting to lower fields (deshielding effect). The anisotropic effect becomes very important in molecules containing $\pi$ bonds, such as aromatic systems and carbonyl groups.

The magnetic anisotropic effect produced by the $\pi$ electron system of benzene in a uniform magnetic field, $B_0$, is diagramed below. The arrows indicate that the anisotropic effect opposes the applied field in the region above and below the ring while it augments the applied field in the region outside the perimeter of the ring.

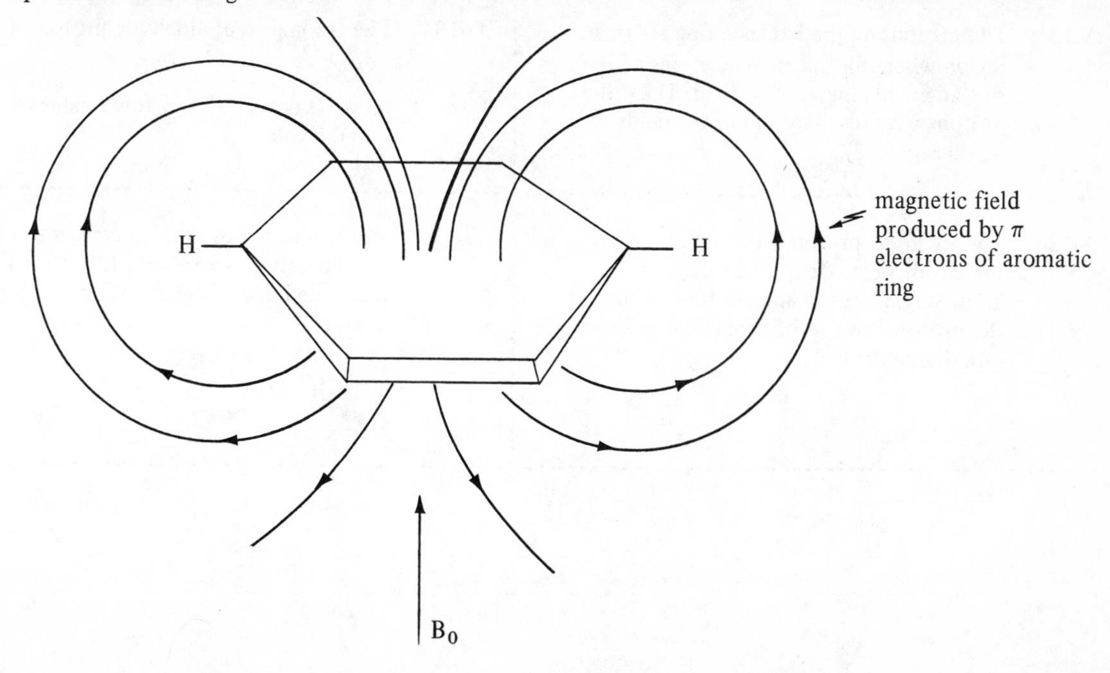

The magnetic anisotropic effect produced by the $\pi$ – electrons of the carbonyl group in a uniform magnetic field, $\mathbf{B_0}$, is diagramed below.

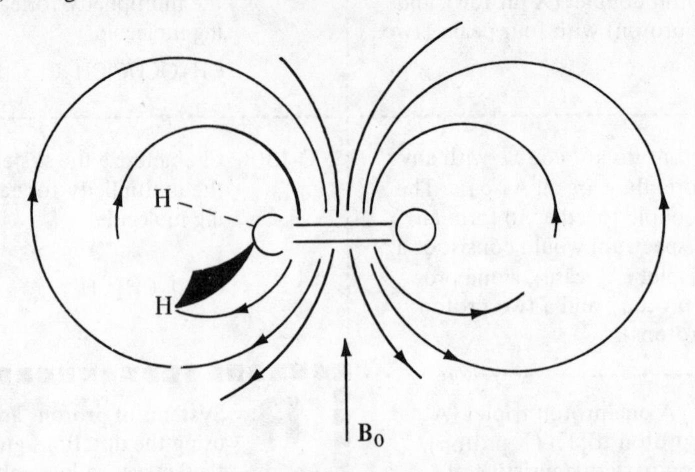


Q-131 In an area where a secondary magnetic field augments the applied field, will proton resonance be shifted to a higher or a lower field?

A-131 When the magnetic field is augmented in the vicinity of a proton, the proton resonance will be shifted to a lower field (deshielding effect).

Q-132 In the vicinity where the applied magnetic field is opposed by a secondary field, will shielding or deshielding of the proton occur? Which way will the resonance peak shift?

A-132 The proton will be more shielded, which causes the proton resonance peak to shift to higher magnetic fields.

Q-133 Explain the fact that aromatic proton resonance occurs at low $\tau$ values (1.0 – 4.0).

A-133 The protons on the benzene ring are in a region where the anisotropic effect of the ring augments the applied field. Thus, the protons have resonance at lower fields (are deshielded).

Q-134 The resonance of aldehyde protons

O
//
(C – H) occurs at very low $\tau$ values (0 – 0.6$\tau$). Explain.

A-134 The aldehyde proton is in a region where the anisotropic effect of the carbonyl group augments the applied field. Thus, the protons have resonance at lower fields (are deshielded).

Q-135 The resonance of ethylenic protons occurs at relatively low $\tau$ values (4.0 – 5.5$\tau$). Explain the large deshielding shown by these protons.

H       R
  C = C
H       R

A-135 The ethylenic protons are in a region where the anisotropic effect of the $\pi$ bond augments the applied field. Thus, the protons have resonance at lower fields (are deshielded).

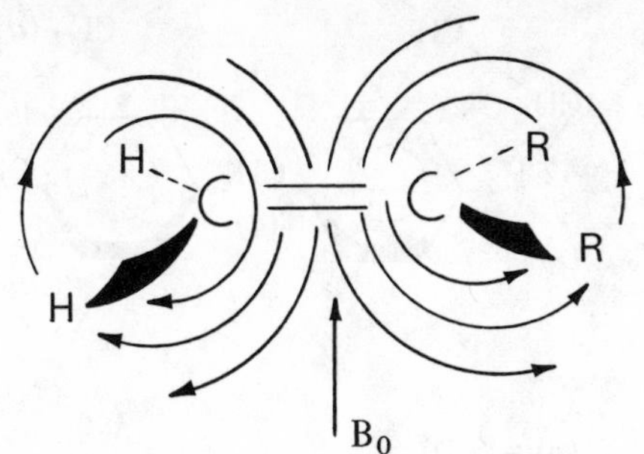


Q-136 The $\tau$ values for the methyl protons in $\alpha$-pinene are given below.

$CH_3$ (*a*)

$CH_3$

$CH_3$ (*c*)

(*b*)

(*a*) = 8.37 $\tau$

(*b*) = 8.73 $\tau$

(*c*) = 9.15 $\tau$

Explain the differences in chemical shift of the (*a*), (*b*), and (*c*) protons.

A-136 The (*a*) protons would be expected to have less shielding than the other two because of the inductive effect of the ethenyl group; thus, (*a*) protons would be expected further "down field" (lower $\tau$) than the other two. The (*c*) protons would be above the $\pi$-electron cloud of the double bond, and thus be *shielded* by the anisotropic effect. The (*c*) protons would be expected to be found further "up field" (higher $\tau$) than the (*b*) protons which are not in the proximity of the double bond.

Q-137 Account for the $\tau$ values for the protons indicated in the compounds below.

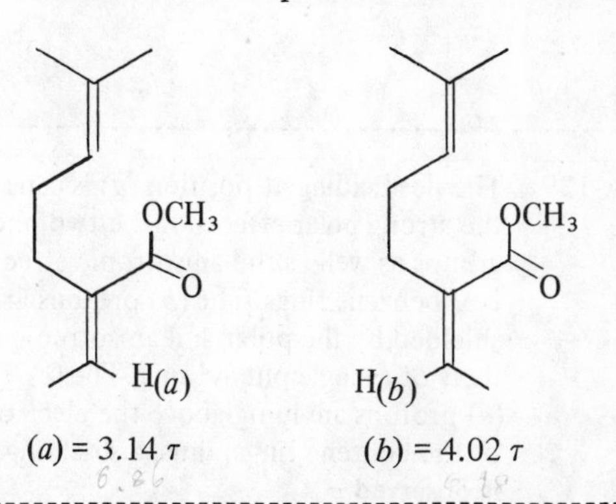


(*a*) = 3.14 $\tau$ (*b*) = 4.02 $\tau$

A-137 The (*a*) proton lies close to the C = O and is in a position in which the applied field is augmented by the anisotropic effect of the carbonyl. Thus, the (*a*) proton would have a lower $\tau$ value than the (*b*) proton which is remote from the carbonyl.

Q-138 Explain the differences in $\tau$ values for the indicated protons in the following compound.

H–C=O

H(*a*)

H

H(*b*)

(*a*) = 2.28 $\tau$

(*b*) = 2.60 $\tau$

A-138 All of the aromatic protons are deshielded by the anisotropic effect of the benzene ring. Because of its position, proton (*a*) is also deshielded by the inductive and the anisotropic effect of the carbonyl group.

Q-139 The $\tau$ values for various methylene groups in the following compound are given. Explain the observed order.

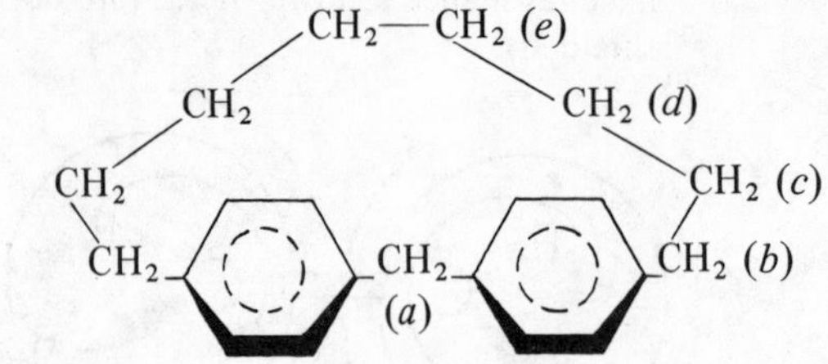

$(a) = 6.19\ \tau$

$(b) = 7.50\ \tau$

$(c) = 8.70\ \tau$

$(d) = 9.10\ \tau$

$(e) = 9.70\ \tau$

A-139 The deshielding at position (*a*) is caused by the strong polar effects of the two phenyl groups as well as the anisotropic effect of two benzene rings. The (*b*) protons are deshielded by the polar and anisotropic effects of a single phenyl ring. The (*c*), (*d*), and (*e*) protons are lying above the electron cloud of the benzene rings; thus, a shielding effect is observed.

**R** The magnetically anisotropic character of a $\pi$ bond system produces an unsymmetrical magnetic field operating through space which either opposes or augments the applied field in the vicinity of neighboring protons. Protons in a region where the anisotropic effect augments the applied field will be deshielded (lower $\tau$ values) while protons in a region where the anisotropic effect opposes the applied field are shielded (higher $\tau$ values).

S-18 The positions of certain proton absorption peaks are also affected by exchange reactions. Acidic hydrogens (i.e., those attached to oxygen, nitrogen, and sulfur) undergo rapid exchange reactions such as the one shown below.

$$ROH_{(a)} + R'OH_{(b)} \rightleftharpoons ROH_{(b)} + R'OH_{(a)}$$

When such a rapid exchange occurs, separate absorption peaks for the (*a*) and (*b*) protons are not observed. Rather, a single peak is observed whose chemical shift is a weighted average of the $\delta$ values of the two types of protons. The observed chemical shift, $\delta_{obs}$, is given by the expression

$$\delta_{obs} = N_a\delta_a + N_b\delta_b$$

where $N_a$ and $N_b$ are the mole fractions of the (*a*) and (*b*) protons, respectively.

Q-140 In a solution of acetic acid in $H_2O$, very rapid exchange occurs between the OH proton of acetic acid and the protons of water. How many peaks would be expected for the spectrum of this solution?

A-140 Two. One for the $CH_3$ protons of acetic acid and one for the weighted average of the OH proton of acetic acid and the $H_2O$ protons of water.

Q-141 In the solution described in Q-140, the mole fraction of acetic acid is 0.5 and that of water is 0.5. The chemical shifts of $H_2O$ protons in pure water and the OH proton of acetic acid in pure acetic acid are 4.8 $\tau$ and $-1.6\tau$, respectively. What is the observed chemical shift for these protons in the solution? (Hint: Both protons in water take part in the exchange reaction.)

A-141 $\delta_{obs} = N_a\delta_a + N_b\delta_b$

Mole fraction of $H_2O$ protons

$$= \frac{2(0.5)}{1.5}$$

Mole fraction of OH protons

$$= \frac{0.5}{1.5}$$

$$\delta_{obs} = \frac{2(0.5)(4.8\tau)}{1.5} + \frac{0.5(-1.6\tau)}{1.5}$$

$$\delta_{obs} = 2.7\tau$$

Q-142 What is the observed chemical shift for the OH protons in a solution containing 4.0 moles of acetic acid and 2.0 moles of $H_2O$?

A-142 Mole fraction of $H_2O$ protons $= \frac{2(2.0)}{8.0}$

Mole fraction of OH protons of acetic acid = 4.0/8.0

$$\delta_{obs} = \frac{2(2.0)(4.8\tau)}{8.0} + \frac{4.0(-1.6\tau)}{8.0}$$

$$\delta_{obs} = 1.6\tau$$

**R** Observed chemical shifts, $\delta_{obs}$, for rapidly exchanging protons are given by

$$\delta_{obs} = N_a\delta_a + N_b\delta_b$$

S-19 Protons involved in rapid exchange reactions do not show fine structure (are unsplit) and do not cause splitting of neighboring protons. Coupling does not occur because only the *average* effect of the spin states is felt by the neighboring nuclei. Occasions may be encountered when the time required for the exchange reaction to occur is approximately the same time necessary for the nuclear transition. In these cases of slower exchange rates a broad resonance peak is observed.

Q-143 In a very pure sample of methyl alcohol, the rate of exchange of OH protons is nearly zero. What would be expected for the spectrum of pure methyl alcohol?

A-143 A three proton doublet for the $CH_3$ proton and a one proton quartet for the OH proton.

Q-144 Acids and bases catalyze the rate of exchange of the OH protons in methyl alcohol so that the exchange rate becomes extremely fast. Will any splitting occur in the methyl and OH resonance peaks in a solution of methyl alcohol and acid?

A-144 No. When rapid exchange occurs, only the average effect of the spin states is felt by the OH and methyl protons.

Q-145 In a very pure ethanol sample, how much splitting of the OH proton peak would be expected?

A-145 In pure ethanol the exchange rate is very slow (nearly zero). Therefore, the OH proton would appear as a triplet. This is actually the case.

Q-146 In an acidic sample of ethanol, how much splitting of the $-CH_2-$ protons would be expected?

A-146 Since fast exchange would be occurring in the acid solution, the $-CH_2-$ protons would be split only by the methyl protons. Therefore, a quartet would result.

Q-147 Besides adding acids or bases to an alcohol solution, how can the rate of exchange between OH protons be increased?

A-147 By increasing the temperature. An increase in temperature speeds up the exchange of protons.

Q-148 At room temperature the OH absorption peak of very pure methanol is a quartet. What would happen to the fine structure as the temperature of the sample is raised?

A-148 At a higher temperature the exchange rate may be rapid enough to result in a single peak for the OH proton.

Q-149 The following spectra are for OH protons undergoing exchange reactions at fast, slow, and intermediate rates. Identify each spectrum.

a)

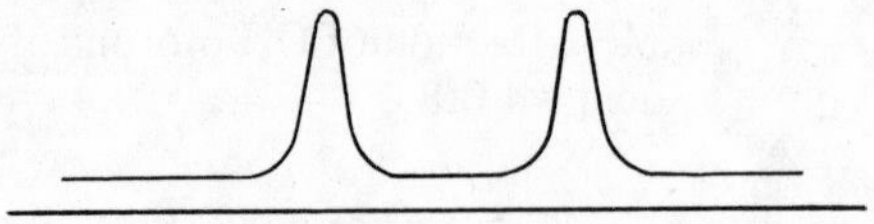

b)

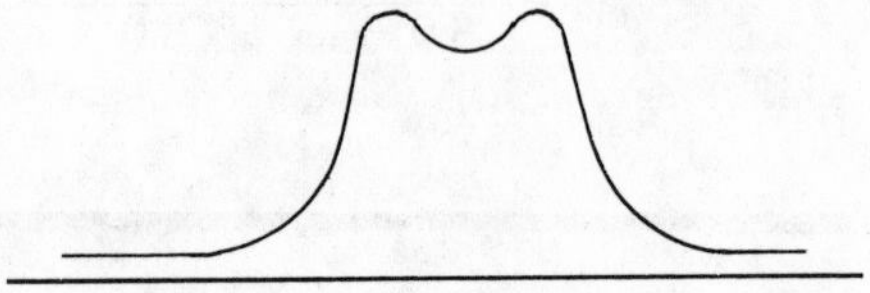

c)

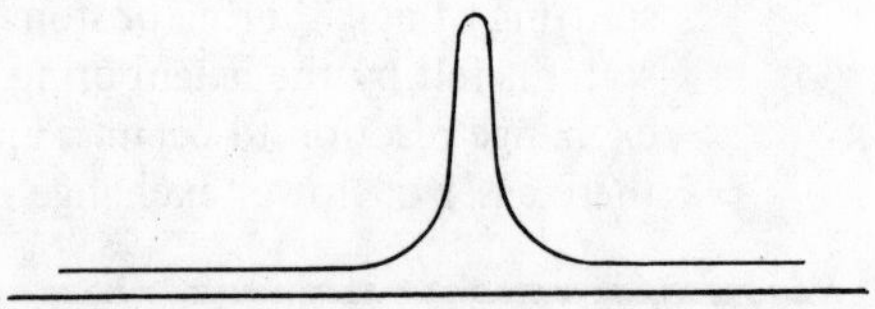

A-149 a) Slow exchange, the signal for both exchanging protons is observed.
c) Fast exchange, the weighted average of the signals for the two exchanging protons is observed.
b) Intermediate, a broad signal is observed.

**R** The spectrum of a molecule where proton exchange is possible is dependent upon the rate at which the protons are exchanging. When rapid exchange occurs, no spin-spin coupling between the exchanging protons and any neighbors is observed. Exchange rates intermediate between fast and slow result in broad resonance signals.

S-20 A hydrogen bonded proton is less shielded and absorbs at lower fields (smaller $\tau$ values) than non-bonded protons. Rapid equilibrium between protons in their hydrogen bonded and non-hydrogen bonded form results in a single resonance peak.

---

Q-150 In methanol the following equilibrium exists.

$$CH_3OH \rightleftharpoons CH_3OH \cdots \underset{}{\overset{H}{\overset{|}{O}}}CH_3$$

non-bonded — bonded

Dilution with an inert solvent will shift the equilibrium toward the non-bonded form. As methanol is diluted with $CCl_4$, will the OH proton resonance shift to a higher or lower field? Why?

---

A-150 Higher field. The hydroxyl protons in pure methanol are largely in the hydrogen bonded form. When an inert solvent is added, the equilibrium is shifted in the direction of the non-hydrogen bonded form. Since the non-hydrogen bonded chemical shift is at a higher field, the observed chemical shift moves in this direction.

Q-151 In acetic acid, an equilibrium exists between the acid proton in hydrogen bonded form and free form. As acetic acid is diluted with an inert solvent, in which direction should the acid proton resonance peak shift?

---

A-151 Higher field.

Q-152 Explain the down field shift of the chloroform proton when diluted with triethylamine.

---

A-152 The proton of chloroform forms a hydrogen bond with the nitrogen of triethylamine. This deshields the chloroform proton, shifting the resonance signal to a lower field.

Q-153 The following reaction is exothermic (gives off heat). In which direction will the equilibrium shift as the temperature is raised?

$$ROH \rightleftharpoons ROH \cdots \overset{H}{\overset{|}{O}} - R$$

non-bonded — bonded

---

A-153 The equilibrium is shifted toward the non-hydrogen bonded form.

Q-154 What happens to the OH resonance signal for phenol as the temperature is increased?

A-154 The OH proton of phenol is hydrogen bonded. The equilibrium is shifted to the unassociated form as the temperature is increased, causing the resonance signal to shift up-field.

Q-155 Chloroform forms a hydrogen bond with the $\pi$ electrons of benzene in the manner shown below.

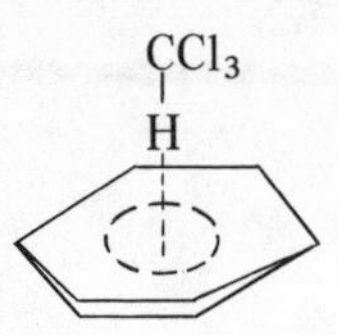


In which direction would the C-H proton of chloroform be expected to shift in forming the hydrogen bond with benzene? Why? (Hint: Remember the anisotropic effect of benzene.)

A-155 Up-field shift (larger $\tau$) is actually observed. A proton in the vicinity above or below a benzene ring will experience a shielding anisotropic effect. The anisotropic effect is larger than the ordinary deshielding effect observed on hydrogen bond formation.

**R** A hydrogen bonded proton is deshielded, causing the resonance to shift down-field. The position of the resonance signals of a proton that can form a hydrogen bond is concentration and temperature dependent.

S-21 The position of proton resonance signals has been shown to be dependent on shielding and deshielding arising from inductive effects, anisotropic effects, hydrogen bonding, and rapid exchange of protons. However, proton resonance in certain regions of the NMR spectrum is indicative of certain *types* of protons. For example:

**Aliphatic** – C–H protons, which do not have any adjacent atoms except hydrogen and $sp^3$ hybridized carbons, absorb in the region **8.2 – 10$\tau$.**

**Aliphatic** – C–H protons, which have adjacent halogen, O or N atoms (CL–C–H, O–C–H, N–C–H) or $sp^2$ or sp hybridized carbons (>C=C–C–H, –C≡C–C–H), absorb in the region **5.0 – 8.5$\tau$.**

**Acetylenic** – C≡C–H protons absorb in the region **7.0 – 8.2$\tau$.**

**Olefinic** – C≡C–H protons absorb in the region **2.5 – 5.5$\tau$.**

**Aromatic** – (Benzenoid and heterocyclic) protons absorb in the region **0.5 – 4.0$\tau$.**

**Aldehydic** – C(=O) – H protons absorb in the region **0.0 – 1.0$\tau$.**

Even though considerable overlap of the regions exists, the generalizations delineated above are very useful.

Q-156 A compound was believed to be either an aliphatic aldehyde or a ketone. The NMR spectrum showed an absorption at 0.10$\tau$. What conclusion could be drawn?

A-156 The compound was an aldehyde. A C(=O)-H proton absorbs in the region 0.0 – 1.0$\tau$.

Q-157 A compound was believed to be either diphenyl ether or diphenylmethane. Could the NMR spectrum be used to distinguish between these two compounds? How?

A-157 Yes. Diphenylmethane would have a two proton peak in the region 5–8.5$\tau$, while diphenyl ether would have absorption in the 0.5 – 4.0$\tau$ region only.

Q-158 A hydrocarbon was known to be either an alkene or alkane. The NMR spectrum showed absorptions at 5.2, 7.3, 8.1 and 9.2 $\tau$. What structure is indicated?

A-158 The absorption at 5.2$\tau$ is indicative of an ethylenic proton. The hydrocarbon must be an alkene.

Q-159 A compound $C_5H_{10}O$ could be an aldehyde or a ketone. Could the NMR spectrum be used to distinguish between the two possible functional groups? How?

A-159 Yes. The aldehyde would have an absorption in the 0.0 – 1.0 $\tau$ region.

Q-160 In what region of the spectrum would the labeled protons of the following compound be expected to absorb?

$CH_3$ (c), N, H (a), S, $CH_3$ (b)

A-160 (a) in the aromatic region (0.5–4 $\tau$)
(b) and (c) in the region 5–8.5 $\tau$

Q-161 In what region of the spectrum would the labeled protons of the following compound be expected to absorb?

$CH_2$ (a)
(c) (b)

A-161 (a) in the 2 – 5.5 $\tau$ region
(b) in the 5 – 8.5 $\tau$ region
(c) in the 8.2 – 10 $\tau$ region.

Q-162 In what region of the spectrum would the labeled protons of the following compound be expected to absorb?

(b)
(a)
O

A-162 (a) in the 5 – 8.5 $\tau$ region
(b) in the 8.2 – 10 $\tau$ region

Q-163 In what region of the spectrum would the labeled protons of the following compound be expected to absorb?

$$\underset{(d)}{CH_3} - \underset{(c)}{CH_2} - \underset{(b)}{CH_2} - \overset{O}{\overset{\|}{C}} - \underset{(a)}{H}$$

A-163 (a) in the 0.0 – 1.0 $\tau$ region
(b) in the 5 – 8.5 $\tau$ region
(c) and (d) in the 8.2 – 10 $\tau$ region

Q-164 In what region of the spectrum would the labeled protons of the following compound be expected to absorb?

$$\underset{(a)}{CH_3} - \overset{O}{\overset{\|}{C}} - \underset{(b)}{CH_2} - \underset{(c)}{CH_3}$$

A-164 (*a*) and (*b*) in the 5 – 8.5 $\tau$ region
(*c*) in the 8.2 – 10 $\tau$ region

Q-165 In what region of the spectrum would the labeled protons of the following compound be expected to absorb?

$CH_3 - CH = CH - \overset{O}{\overset{\|}{C}} - H$
(*a*) (*b*) (*c*) (*d*)

A-165 (*a*) in the 5 – 8.5 $\tau$ region
(*b*) and (*c*) in the 2.5 – 5.5 $\tau$ region
(*d*) in the 0.0 – 1.0 $\tau$ region

Q-166 In what region of the spectrum would the labeled protons of the following compound be expected to absorb?

$Cl_2CH\overset{O}{\overset{\|}{C}}OCH_2CH_3$
(*a*) (*b*) (*c*)

A-166 (*a*) on the low end of the 5 – 8.5 $\tau$ region (actual 4.1 – note that there are three strong electron pulling groups attached to the carbon bearing the proton)
(*b*) in the 5 – 8.5 $\tau$ region
(*c*) in the 8.2 – 10 $\tau$ region

Q-167 In what region of the spectrum would the labeled protons of the following compound be expected to absorb?

$CH_3CH(NO_2)CH_3$
(*a*) (*b*)

A-167 (*a*) in the 8.2 – 10 $\tau$ region
(*b*) in the 5 – 8.5 $\tau$ region and probably on the *low* end of the region (actual 5.2)

Q-168 In what region of the spectrum would the labeled protons of the following compound be expected to absorb?

$H - \overset{O}{\overset{\|}{C}} - N(CH_3)_2$
(*a*) (*b*)

A-168 (*a*) in the 0.0 – 1.0 $\tau$ region (Actual 2.0 $\tau$. This is not an aldehydic hydrogen. The compound is a formamide derivative. Formamide protons absorb near 2.0 $\tau$.)
(*b*) in the 5 – 8.5 $\tau$ region

Q-169 In what region of the spectrum would the labeled protons of the following compound be expected to absorb?

$CH_3 - C \equiv C - H$
(*a*) (*b*)

A-169 (*a*) in the 5 – 8.5 $\tau$ region
(*b*) in the 7.0 – 8.2 (actual 8.2) $\tau$ region

Q-170 In what region of the spectrum would the labeled protons of the following compound be expected to absorb?

$Cl - CH_2 - CH_2 - OH$
(*a*) (*b*)

A-170 (*a*) and (*b*) in the 5 – 8.5 $\tau$ region

Q-171 In what region of the spectrum would the labeled protons of the following compound be expected to absorb?

$C_6H_5CH(CH_3)_2$

(*a*) (*b*) (*c*)

A-171 (*a*) in the 0.5 – 4.0 $\tau$ region

(*b*) in the 5 – 8.5 $\tau$ region

(*c*) in the 8.2 – 10 $\tau$ region

Q-172 In what region of the spectrum would the labeled protons of the following compound be expected to absorb?

p – $CH_3C_6H_4NH\overset{O}{\overset{\|}{C}}CH_3$

(*a*) (*b*) (*c*)

A-172 (*a*) and (*c*) in the 5 – 8.5 $\tau$ region

(*b*) in the 0.5 – 4.0 $\tau$ region

Q-173 In what region of the spectrum would the labeled protons of the following compound be expected to absorb?

O

$NCH_2CH_2Br$

(*b*) (*c*)

H(*a*)

O

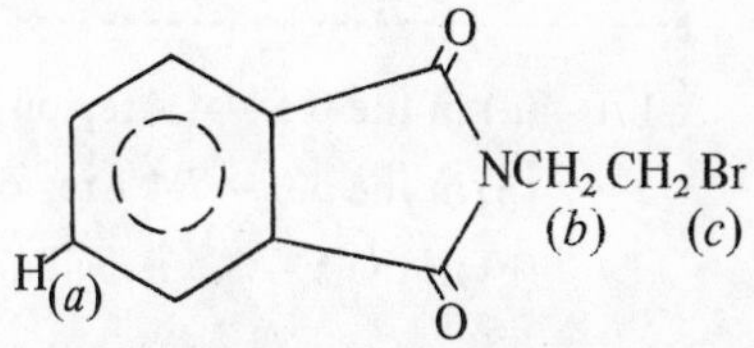

A-173 (*a*) in the 0.5 – 4 $\tau$ region

(*b*) and (*c*) in the 5 – 8.5 $\tau$ region

**R** Chemical shifts expected for various types of protons are given in Table VI in the Appendix.

S-22 Below are listed correlations of proton absorption for hydrogens which are bonded to O, N, and S. Remember that the order of electronegativity is O > N > C.

| Proton | Type of molecule | $\tau$ value | |
|---|---|---|---|
| O–H | Alcohols | 4.5 to 9.5 | On dilution, gradually shifts to high fields. |
| | Phenols | 2.3 to 6.0 | On dilution, gradually shifts to high fields. |
| | Enols | – 6.0 to – 5.0 | Very low fields due to strong intramolecular hydrogen bonds. |
| | Acids | – 2.0 to – 0.5 | Very low fields due to strong intramolecular hydrogen bonds. |
| N–H | Aliphatic amine | 7.8 to 9.7 | |
| | Aromatic amine | 5.0 to 7.4 | |
| | Amide | 1.5 to 5.0 | |
| S–H | Aliphatic sulfhydryl | 8.4 to 8.8 | |
| | Aromatic sulfhydryl | 6.4 | |

Q-174 For the following compound, where would the indicated protons be expected to absorb?

$ClCH_2CH_2OH$

(*a*) (*b*)

A-174 (*a*) in the 5 – 8.5 $\tau$ region
(*b*) in the 4.5 – 9.5 $\tau$ region

Dilute solutions would show OH absorption at the high field end of this region.

Q-175 For the following compound, where would the indicated protons be expected to absorb?

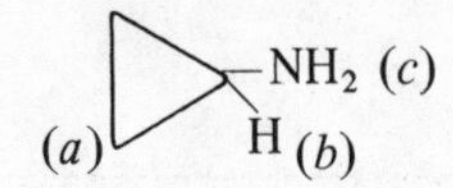

(cyclopropane ring (*a*) bearing $NH_2$ (*c*) and H (*b*))

A-175 (*a*) in the 8.2 – 10 $\tau$ region
(*b*) in the 5 – 8.5 $\tau$ region
(*c*) in the 7.8 – 9.7 $\tau$ region

Q-176 Where would the indicated protons in the following compound be expected to absorb?

(benzene ring with O $CH_3$ (*c*), (*b*) $H_2N$, H (*a*))

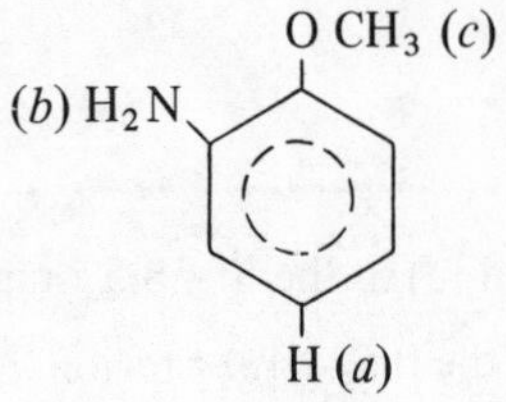

A-176 (*a*) in the 0.5 – 4 $\tau$ region
(*b*) in the 5.0 – 7.4 $\tau$ region
(*c*) in the 5 – 8.5 $\tau$ region

Q-177 Where would the protons indicated in the following compound be expected to absorb?

(pyridine ring with (*a*) H, N, $CH_2$ (*b*)–C(=O)–$NH_2$ (*c*))

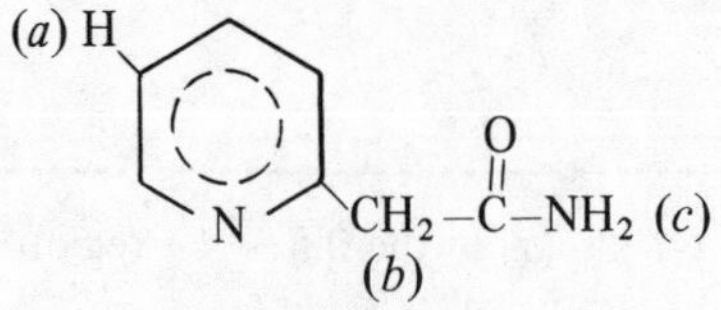

A-177 (*a*) in the 0.5 – 4 $\tau$ region
(*b*) in the 5 – 8.5 $\tau$ region
(*c*) in the 1.5 – 5 $\tau$ region

Q-178 Where would the labeled protons in the compound below be expected to absorb?

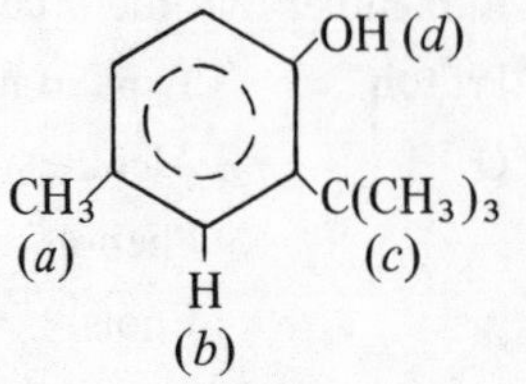

(benzene ring with OH (*d*), $CH_3$ (*a*), H (*b*), $C(CH_3)_3$ (*c*))

A-178 (*a*) in the 5 – 8.5 $\tau$ region
(*b*) in the 0.5 – 4.0 $\tau$ region
(*c*) in the 8.2 – 10 $\tau$ region (low end)
(*d*) in the 2.3 – 6.0 region. Dilute solutions will absorb in the high field end of this region.

Q-179 In the following compound, the OH proton absorption is at –1.0 $\tau$, much lower than expected for phenols. Explain.

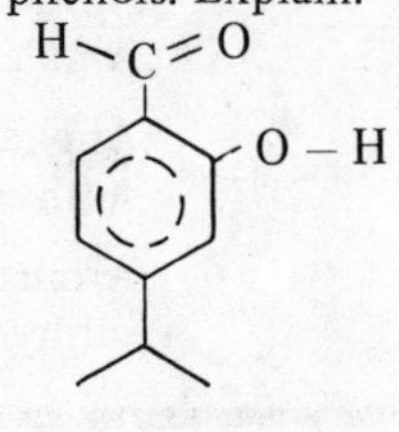

(benzene ring with H–C=O, O – H, and isopropyl group)

A-179 There would be a strong intramolecular hydrogen bond in this compound.

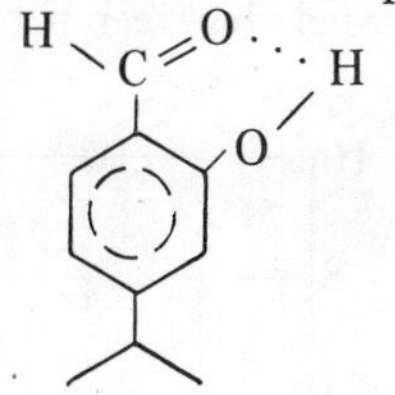

This results in a strong down-field shift of the O–H absorption.

**R** Proton resonance in certain regions of the NMR spectrum indicates specific types of protons.

The chemical shift expected for the various types of protons is given in Table VI in the Appendix.

S-23 Just as chemical shift values can be used to tell what types of protons are present in a molecule, the magnitude of spin-spin coupling constants can be used for structure and confirmation elucidation. Coupling constants are affected by several factors. Coupling constants of vicinal protons (protons on adjacent carbons, i.e., –C(H)–C(H)–) will be considered first. The following generalizations are useful.

1) The magnitude of the vicinal coupling constant, $J$, is a function of the dihedral angle ($\phi$). The relationship is given by the Karpus equation

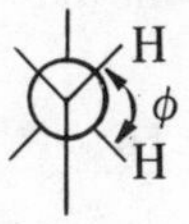

$$J_{\text{(Hertz)}} = 4.2 - 0.5 \cos\phi + 4.5 \cos 2\phi$$

2) The dependence of the vicinal coupling constant ($J$) upon the electronegativity of X ($CH_3 - CH_2 - X$) is given by the equation

$$J_{\text{(Hertz)}} = 7.9 - n0.7\,(\Delta x)$$

where $n$ is equal to the number of hydrogens replaced by an X group and $\Delta x$ is equal to $(x_X - x_H)$ where $x_X$ and $x_H$ are electronegativities of the substituent and the hydrogen, respectively.

3) The vicinal coupling constant ($J$) decreases with increasing C–C bond length.

4) The vicinal coupling constants ($J$) depend upon the carbon-carbon-hydrogen bond angles ($\theta$)

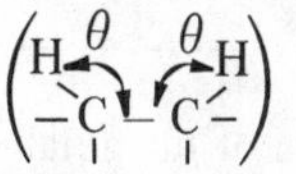

As the angles between the hydrogens are increased, the coupling constant is decreased (bond order remaining the same).

Q-180 The dihedral angle between two protons is zero. What coupling constant would be predicted from the Karpus equation?

A-180 $J = 4.2 - 0.5 \cos\phi + 4.5 \cos 2\phi$

$\cos 0^0 = 1$

$J = 8.2$ Hz

Q-181 What should the coupling constant be when the dihedral angle, $\phi$, between protons is 45°, 60°, 90°, 120°, 135°, and 180°?

A-181

| Angle in degrees | $\cos \phi$ | $J$ (Hz) |
|---|---|---|
| 45 | $1/\sqrt{2}$ | 3.9 |
| 60 | 1/2 | 1.8 |
| 90 | 0 | −0.3 |
| 120 | −1/2 | 2.1 |
| 135 | $-1/\sqrt{2}$ | 4.5 |
| 180 | −1 | 9.2 |

Q-182 Sketch on the following graph the dependence of the vicinal coupling constant on the dihedral angle between protons.

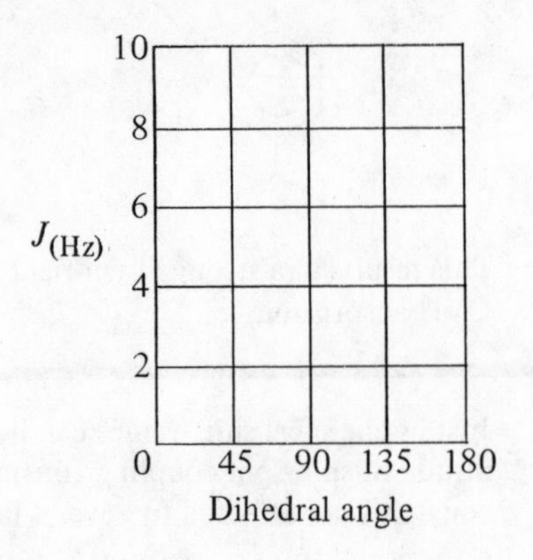


A-182

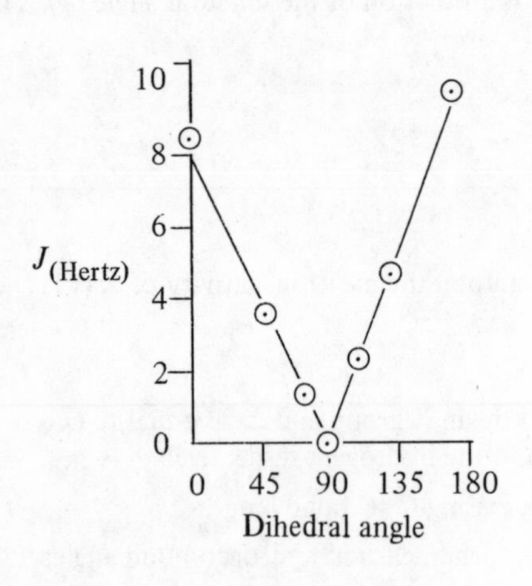


Q-183 Below is a Newman projection (end on view) for the chair form of a cyclohexane ring. Predict the magnitude of the *diaxial* coupling constant.

$H_{(a)}$ $H_{(e)}$ $H_{(e)}$ $H_{(a)}$

A-183 $\phi_{aa} = 180°$ $J_{aa} = 9.2$ Hz
(usually 9–13 Hz)

Q-184 Predict the magnitude of the axial-equatorial coupling constant in cyclohexane.

A-184 $\phi_{ae} = 60°$ $J_{ae} = 1.8$ Hz
(usually 2–5 Hz)

Q-185 Predict the magnitude of the di-equatorial coupling constant in cyclohexane.

A-185 $\phi_{ee} = 60°$ $J_{ae} = 1.8$ Hz
(usually 2–5 Hz)

Q-186 Using the Karpus equation predict $J_{a,b}$ for the following compound.

$H_b(CH_3)C{=}C(H_a)C(=O)H$

A-186 $\phi_{ab} = 180°$ $J_{ab} = 9.2$ Hz
$J_{ab}$ actual = 14 Hz

Q-187 In the last answer, why is the coupling constant larger than predicted by the Karpus equation?

A-187 The Karplus equation is for carbon-carbon single bonds. As the bond order increases, the bond length decreases, causing the coupling constant to increase (rule 3).

Q-188 In which compound would the coupling constant between the indicated protons be larger? Why?

1) 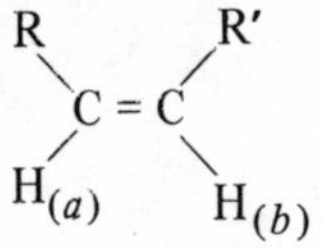 R, R' — C = C — $H_{(a)}$, $H_{(b)}$ Bond order = 2

2) 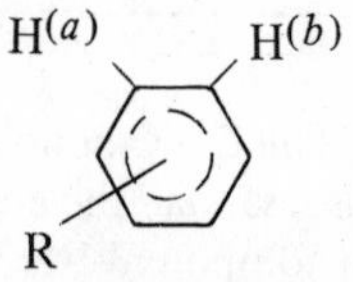 $H^{(a)}$ $H^{(b)}$ (benzene ring with R) Bond order = 1.67

A-188 Compound 1. The bond order in compound 1 is indicative of a shorter bond length. The coupling constant increases with decreasing bond length.

*Actual*

Compound 1 $J_{ab} = 11.5$ Hz

Compound 2 $J_{ab} = 8.0$ Hz

Q-189 Using the equation given in rule 2, predict the $J_{ab}$ coupling constant in the following compound.

$CH_3 - CH_2 Cl$
(*a*) (*b*)

$\chi_H = 2.20$

$\chi_{Cl} = 3.15$

A-189 $J = 7.9 - 0.7\,(\Delta X)$

$J_{ab} = 7.9 - 0.7\,(0.95) = 7.2$ Hz

observed value = 7.5 Hz

(rule 2)

Q-190 Predict the coupling constant, $J_{ab}$, in the following compound.

$CH_2Cl - CHCl_2$
(*a*) (*b*)

A-190 $J_{ab} = 7.9 - (3)\,(0.7)\,(0.95) = 5.9$ Hz

Observed value = 6.0 Hz

Q-191 Which of the following compounds has the larger C = C – H bond angle?

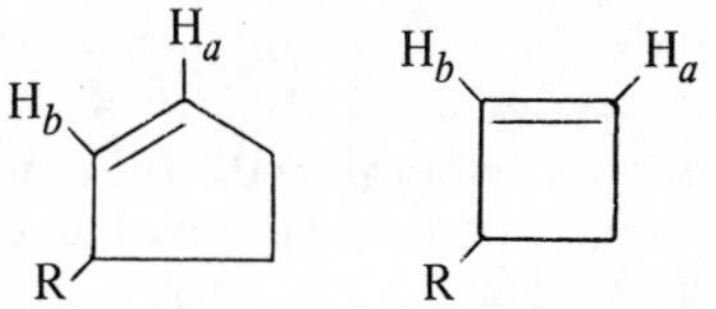

I: cyclopentene with $H_a$, $H_b$, R

II: cyclobutene with $H_b$, $H_a$, R

A-191 Compound II. C = C–H = 135°

Compound I. C = C–H = 120°

Q-192 In which of the *compounds* shown in Q-191 will the $J_{ab}$ coupling constant be larger?

**A-192** Compound I. (rule 4)

Range for compounds of type I:

$J_{ab} = 5 - 7$ Hz.

Range for compounds of type II:

$J_{ab} = 2.5 - 4.0$ Hz.

**Q-193** In which of the compounds given below would the $J_{ab}$ coupling constant be expected to be larger? Why?

I

$H_a$ $H_b$

II

$H_a$ $H_b$

**A-193** Compound I. The C = C(H) bond angle in compound I is less than the corresponding bond angle in compound **II**. (rule 4)

Range for compounds of type **I**:

$J_{ab} = 8.8 - 10.5$ Hz.

Range for compounds of type **II**:

$J_{ab} = 5 - 7$ Hz.

**Q-194** 1, 2-dihalocyclohexane was found to have a $J_{ab}$ coupling constant of 10 Hz. Is the compound the *cis* or *trans* isomer? What is the conformation of the compound?

I

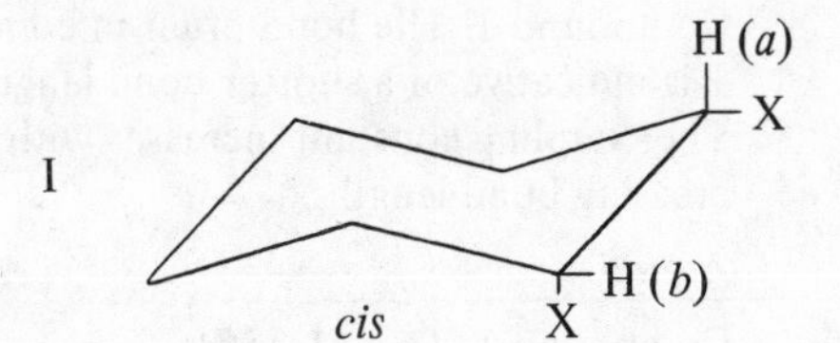


II

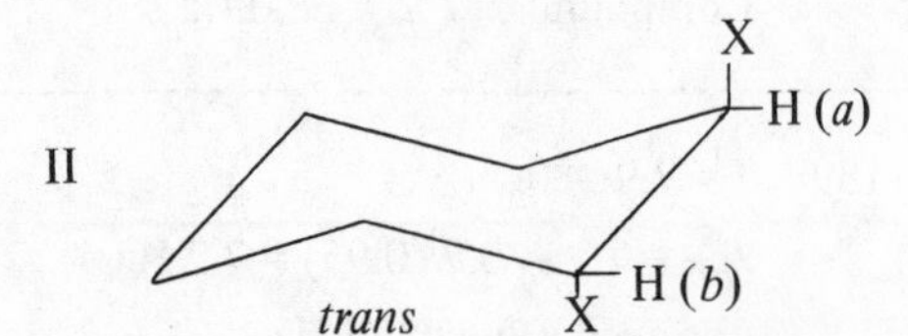


III

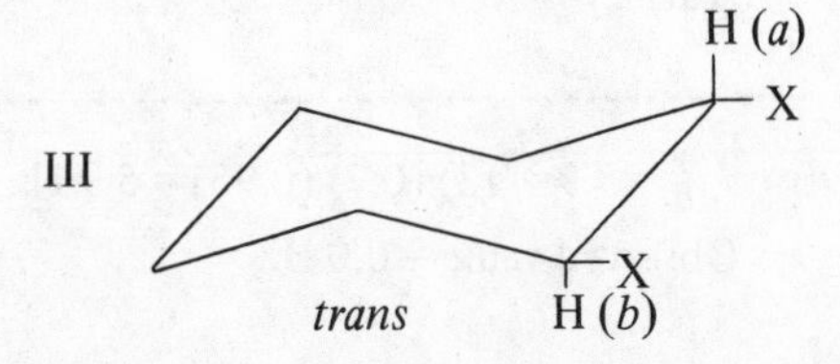


**A-194** *Trans.* Conformation III. The coupling constant is that usually observed for diaxial interactions.

**R** Coupling constants can be used for structure and conformation elucidations. Rules for coupling between vicinal protons are given in S-23. Values for spin-spin coupling constants in some common systems are presented in Table VII of the Appendix.

S-24 Spin-spin coupling is frequently observed between protons (or other nuclei having I > O) separated by more than three bonds. These interactions are called long-range couplings. These coupling constants are on the order of 0 – 3 Hz and are common in $\pi$ – systems. Examples are:

1) *Substituted benzenes.* The ranges of coupling constants observed in substituted benzenes are summarized below:

$J_{(ortho)} = 6 - 10$ Hz

$J_{(meta)} = 1 - 3$ Hz

$J_{(para)} = 0 - 1$ Hz

2) *Allylic coupling.* Long-range splitting between proton (*a*) and proton (*c*) or proton (*b*) and proton (*c*) in the unsaturated system shown below is observed.

$J_{ac\ (trans)} = 1.6 - 3.0$ Hz

$J_{bc\ (cis)} = 0 - 1.5$ Hz

Q-195 Diagram the splitting pattern for hydrogen (*a*) in the following compound:

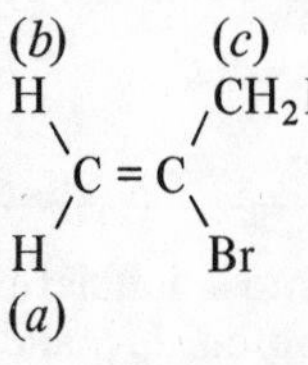


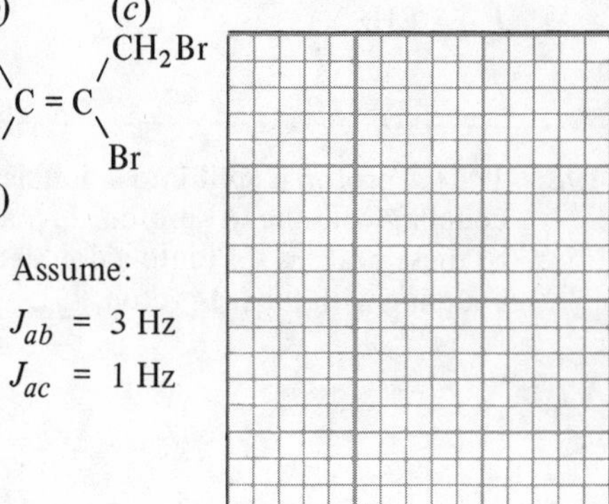

Assume:

$J_{ab} = 3$ Hz

$J_{ac} = 1$ Hz

A-195

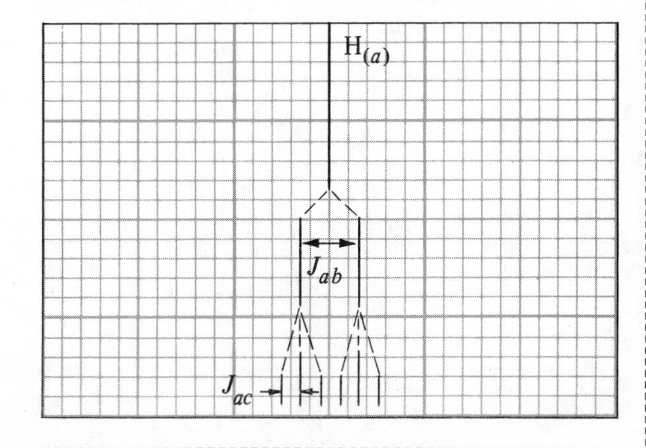


Q-196 Diagram the splitting pattern for the indicated hydrogens in the following compound.

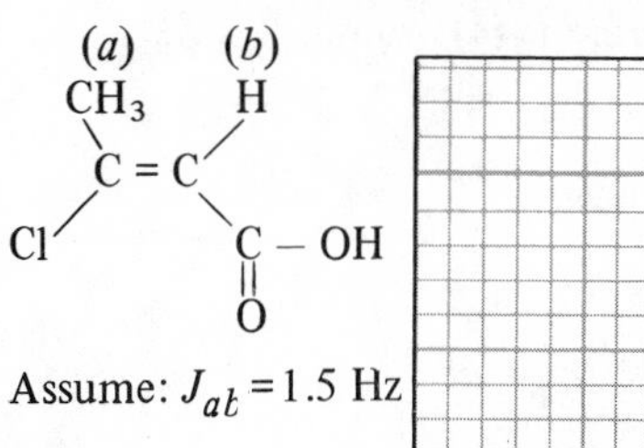


Assume: $J_{ab} = 1.5$ Hz

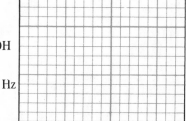

A-196

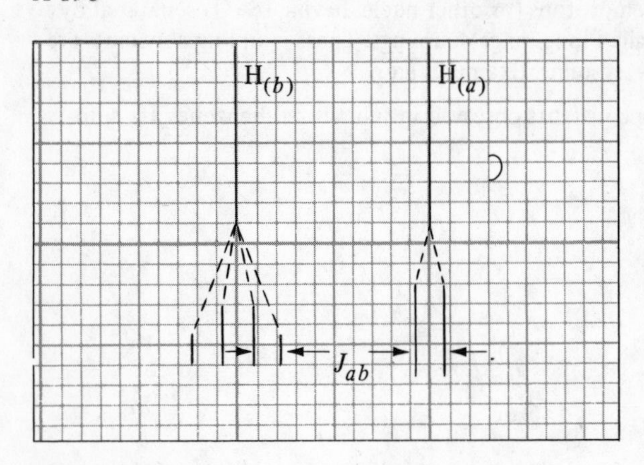


Q-197 Predict the multiplicity for the indicated proton in the following compound.

H H
H S C – $H_{(a)}$
O

A-197 Doublet.

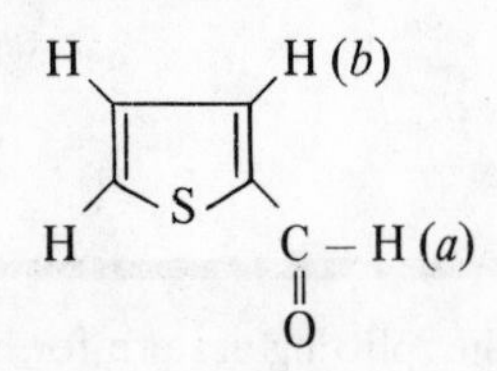


Long-range splitting with the (*b*) proton.

$J_{ab}$ = 1 Hz

Q-198 Account for the splitting pattern for the (*a*) proton in the compound

$CH_3O$
C – H
(*c*)
H – C ≡ C – C
(*a*)
H (*b*)

(*a*) proton

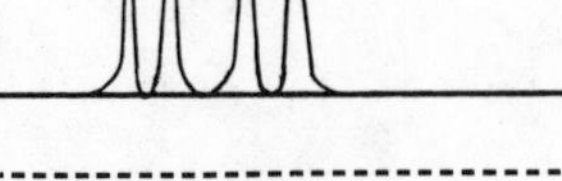

A-198 The (*a*) proton is split into a doublet by coupling with the (*b*) proton, $J_{ab}$, and each of these peaks is split into a doublet by coupling with the (*c*) proton, $J_{ac}$.

Q-199 Predict the splitting pattern for the (*a*) proton in the following compound.

$H_{(c)}$
$NH_2$ $NH_2$
$H_{(b)}$ $H_{(b)}$
$H_{(a)}$

A-199 The (*a*) proton is split into a triplet by coupling with the two (*b*) protons, $J_{ab}$, and each of these peaks is split into a doublet by coupling with the (*c*) proton, $J_{ac}$.

Q-200 In an AMX system containing $\pi$-bonds, long-range spin-spin splitting can occur. In the compound below, what coupling constants are involved?

Br
Br $H_{(A)}$
$CH_3O$ $H_{(M)}$
$H_{(X)}$

A-200 $J_{AM}, J_{AX}, J_{MX}$

Q-201 For the compound given in Q-200, which coupling constant, $J_{AM}$, $J_{MX}$, or $J_{AX}$ would be expected to be smallest? Why?

A-201 $J_{AX}$ will be smallest. *Meta* coupling constants are smaller than *ortho* coupling constants.

Q-202 In the AMX system illustrated below, would long-range spin-spin splitting be expected?

(A) H    H (M)
C = C
C – H (X)

Give all coupling constants for the system.

A-202 Yes.

$J_{AM} \approx 10$ Hz

$J_{MX} \approx 10$ Hz

$J_{AX} \approx 1.6–3.0$ Hz

Q-203 For the molecule given below,

(M) H    H (X)
(A) H    O    C – $CH_3$
‖
O

assume $J_{AM}$ = 2.5 Hz, $J_{AX}$ = 1.2 Hz, $J_{MX}$ = 3.5 Hz. Diagram the splitting pattern for the (A) and (X) proton peaks (1 Hz = 2 divisions).

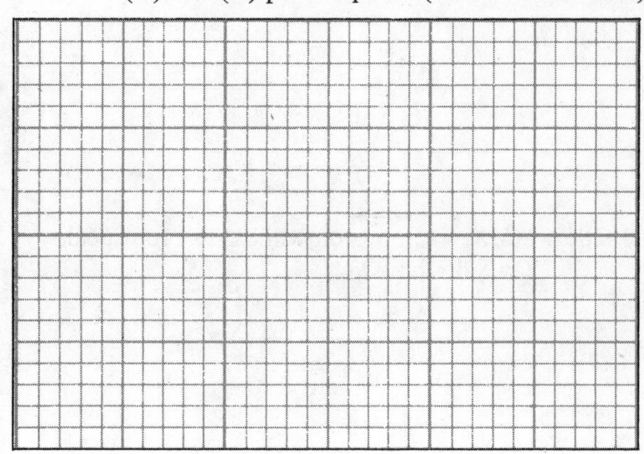

A-203

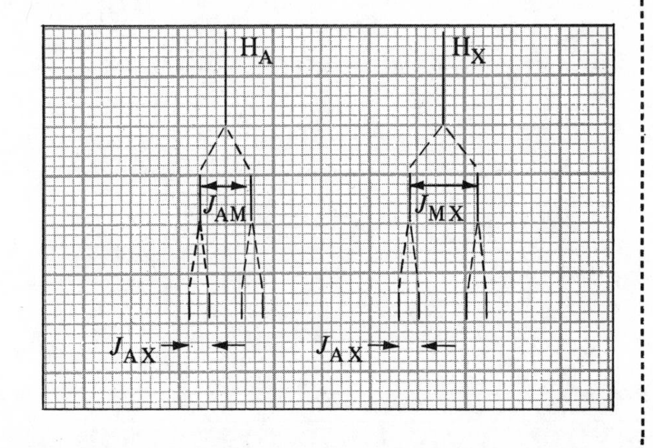


Q-204 Diagram the splitting pattern of the (M) proton in the compound given in Q-203.

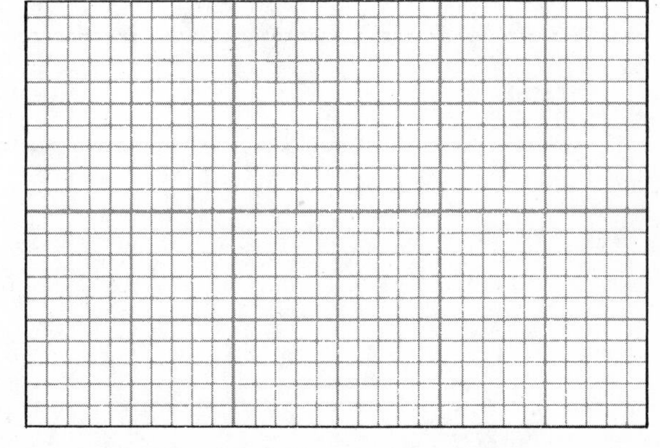

A-204

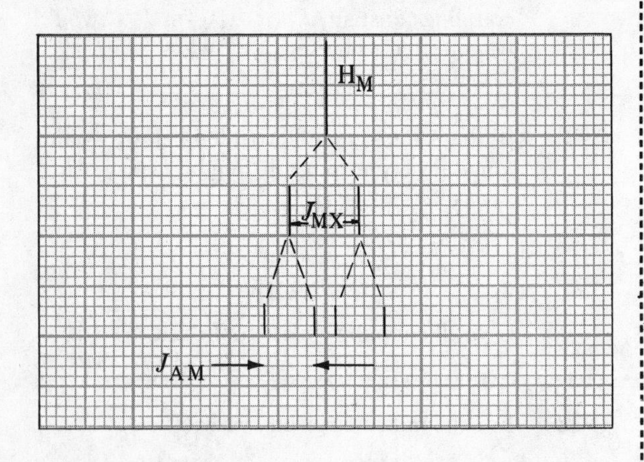


Q-205 What is the multiplicity of the A, the M, and the X proton absorptions in the spectrum of an AMX system in which there is long-range splitting?

A-205 There should be three groups of lines (one for each type of proton) and each group should have four peaks (two doublets).

Q-206 Establish whether the indicated protons in the following compound make up an AX, $AX_2$, AMX, $A_2$, or $A_3$ case and give the expected spectrum for these protons.

$OCH_3$, H, $NO_2$, H, H, $NO_2$

A-206 AMX. Each proton will exhibit four peaks (two doublets).

Q-207 Establish whether the protons in the following compound make up an AX, $AX_2$, AMX, $A_2$, or $A_3$ case and give the expected multiplicity for each proton.

(b)H, $H_{(c)}$, (a)H, C(=O)–O–$CH_3$ (d)

A-207 Protons (*a*), (*b*), and (*c*) constitute an AMX case. Each of these protons will exhibit four peaks (two doublets). The (*d*) protons are an $A_3$ case and yield a three proton singlet.

Q-208 Long-range spin-spin splitting can give rise to complicating features in the spectra of compounds with protons which make up an $A_2X_2$ system. In the spectrum of the compound below, the hydrogens give a characteristic quartet with four extra "inside" lines.

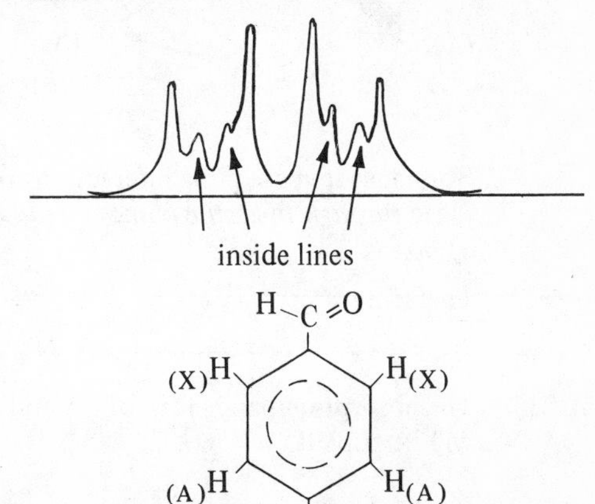


What feature of the above compound gives rise to the complications in its spectrum? (Hint: What coupling constants are present?)

A-208 The following spin-spin interactions are present:

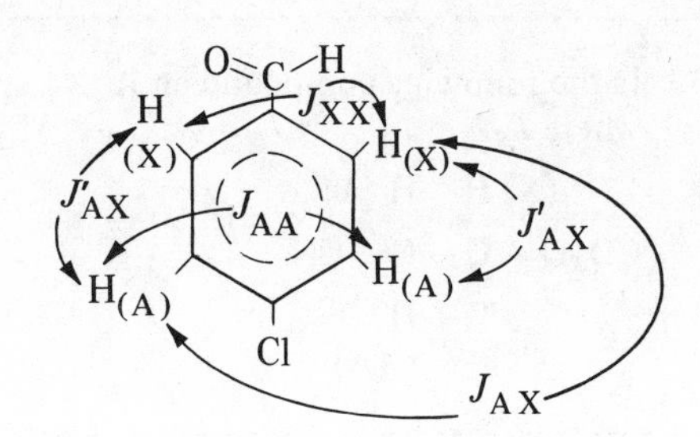


The $J'_{AX}$ and $J_{AX}$ coupling constants are not the same. Thus, a complex spectrum results.

Q-209 Establish whether the protons designated in the following compound make up an $A_2$, $A_4$, AX, $A_2X_2$, or AMX system.

$OCH_3$

H H

H H

$CH_3$

A-209 $A_2X_2$ system where $J_{AX} \neq J'_{AX}$

Q-210 In the $A_2X_2$ system for the compound below, how many different $J_{AX}$ coupling constants would there be? Why?

(A) H F (X)

C = C = C

(A) H F (X)

(Hint: What is the stereochemistry of the compound?)

A-210 One.

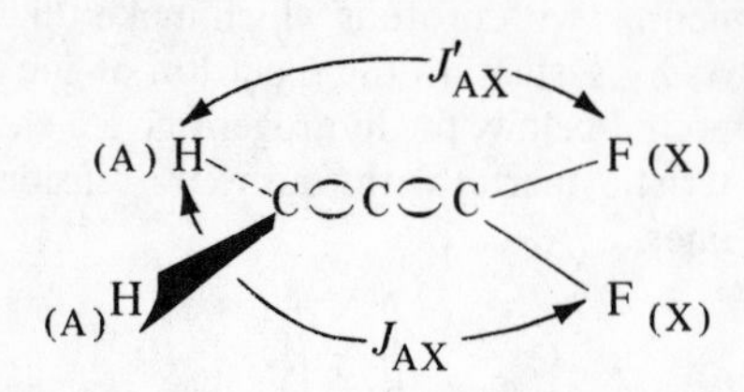


Both spin-spin coupling interactions take place *through the same number and kinds of bonds.*

Therefore, $J_{AX} = J'_{AX}$

Q-211 What would be the multiplicity of each group in an $A_2X_2$ spectrum where $J_{AX} = J'_{AX}$?

---

A-211 The absorption bands for both A and X will be triplets.

Q-212 Is the following compound an $A_2X_2$ system where $J_{AX} = J'_{AX}$ or $J_{AX} \neq J'_{AX}$? Why?

Cl
(X) H H (X)
(A) H H (A)
N

---

A-212 $J_{AX} \neq J'_{AX}$

The spin-spin interactions operate through a different number of bonds.

Q-213 Is the following compound an $A_2X_2$ system where $J_{AX} = J'_{AX}$ or $J_{AX} \neq J'_{AX}$? Why?

(X) H H (A)
$CH_3O - C - C - CN$
H H
(X) (A)

---

A-213 $J_{AX} = J'_{AX}$

There is free rotation about the single bond; therefore, $J_{AX} = J'_{AX}$.

Q-214 Draw the expected spectrum for the ring hydrogens of the compound:

H
H C – H
C
H H
H H
N

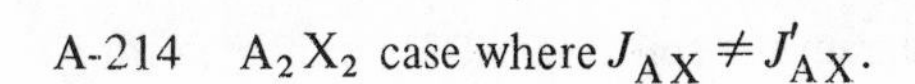

A-214 $A_2X_2$ case where $J_{AX} \neq J'_{AX}$.

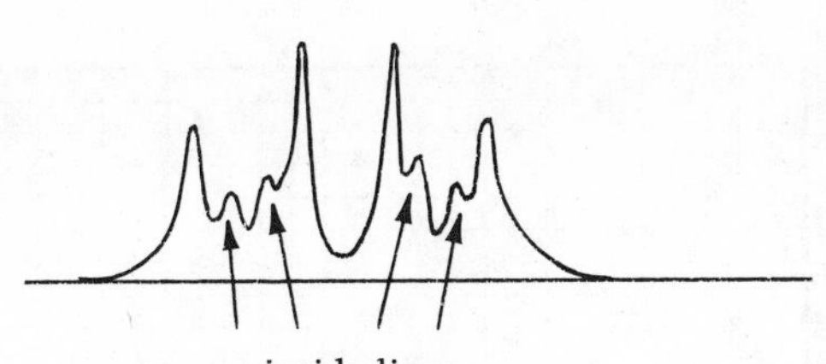


Quartet with 4 inside lines.

**R** Long-range splitting (coupling between nuclei separated by more than three bonds) is observed in $\pi$-bond systems. These interactions give $J$ values of 0 - 3 Hz.

Characteristic spectra are observed for the AMX and $A_2X_2$ cases with long-range interactions.

| System | Spectrum Characteristics |
|---|---|
| AMX (long-range coupling) | Each proton exhibits four peaks (two doublets). |
| $A_2X_2$ (long-range coupling $J_{AX} \neq J'_{AX}$) | **Four** proton quartet with four additional weak "inside" lines. |

S-25 Splitting patterns considered so far have been between interacting nuclei in which $\delta_{AX}$ (the difference between the chemical shift of nucleus A and X) is large compared to the size of the coupling constant, $J_{AX}$. Interacting nuclei for which $\delta_{AX} >> J_{AX}$ are called *first order systems.* First order systems give rise to splitting patterns which are unperturbed by other effects.

Complicating factors arise when the chemical shift value between two interacting nuclei and their coupling constants are of comparable magnitude. Interacting nuclei of this type are designated by the letters A and B. Interacting nuclei for which $\delta_{AB} \cong J_{AB}$ give rise to perturbed peaks in their splitting patterns. In AX cases the ratio $\delta_{AX}/J_{AX}$ is larger than about 25. In AB cases the ratio $\delta_{AB}/J_{AB}$ is approximately 1.

Q-215 Two protons have chemical shift values of 369 Hz and 120 Hz, respectively. The coupling constant between their protons is 5.0 Hz. Are these protons part of an AX or an AB system?

A-215 $\delta = 360\text{ Hz} - 120\text{ Hz} = 240\text{ Hz}$

$$\frac{\delta}{J} = \frac{240}{5.0} = 48$$

This ratio is large; therefore, the protons are a part of an AX system.

Q-216 Two protons have chemical shift values of 6.20 and 6.08 $\tau$. The coupling constant between these protons is 10.0 Hz. Are these protons part of an AX or an AB system? (Assume a 60 MHz oscillator.) (Hint: $\delta$ must be in Hz units for comparison.)

A-216 $\Delta\tau = 6.20 - 6.08 = 0.12$ ppm

$$\frac{(0.12\text{ ppm})\,(60\text{x}10^6\text{ Hz})}{1 \times 10^6\text{ ppm}} = 7.2\text{ Hertz}$$

$$\frac{\delta}{J} = \frac{7.2}{10} = 0.72$$

This ratio is small; therefore, the protons are part of an AB system.

Q-217 Two protons have chemical shifts at 6.76 and 6.25 $\tau$. The coupling constant between these protons is 12.0 Hz. Are these protons part of an AX or an AB system? (Assume a 60 MHz oscillator.)

A-217 $\Delta\tau = 6.76 - 6.25 = 0.51$ ppm

$$\frac{(0.51 \text{ ppm})(60\text{x}10^6 \text{ Hz})}{(1 \text{ x } 10^6 \text{ ppm})} = 31 \text{ Hz}$$

$$\frac{\delta}{J} = \frac{31}{12.0} = 2.5$$

This ratio is small enough so that the protons will make up an AB system.

Q-218 Diagram the splitting pattern expected for first order coupling for two protons, A and B, whose chemical shift differs by only 5 Hz while $J_{AB}$ = 10 Hz.

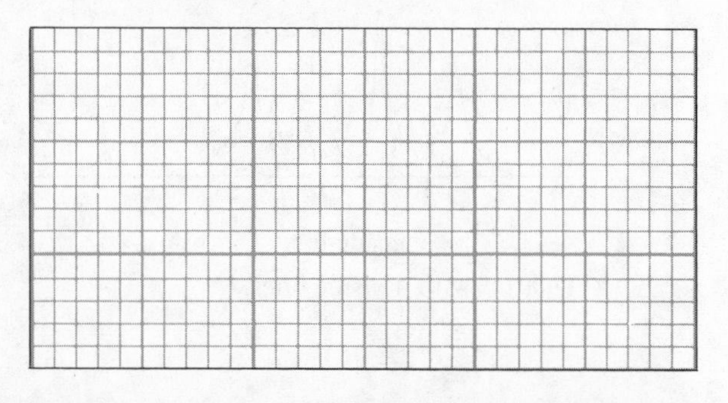

A-218

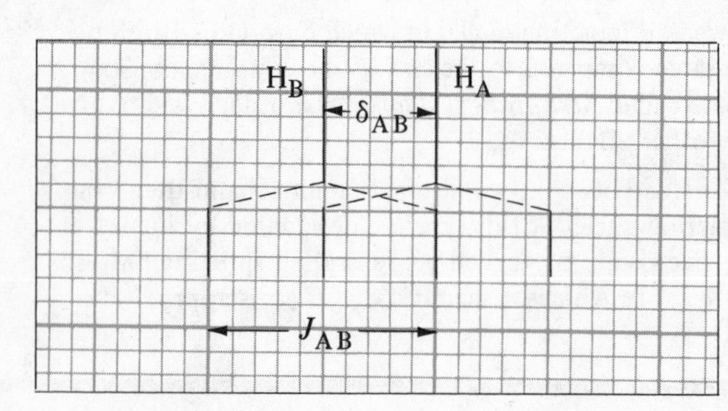


Actually this is not the splitting pattern observed because of perturbation.

**R** A first order unperturbed splitting pattern results from the interaction between nuclei where $\delta_{AX}/J_{AX}$ is larger than about 25.

An AB system, which gives rise to a perturbed splitting pattern, will result from the interaction between nuclei where $\delta_{AB}/J_{AB}$ is approximately 1.

S-26 For a system in which $\delta_{AB}$ approaches the magnitude of $J_{AB}$ (See A-218), the intensity and positions of the split lines do not follow the simple first order rules.

For an AB case the inner split lines become more intense at the expense of the outer lines. Furthermore, the splitting pattern is such that the inner peaks do not cross.

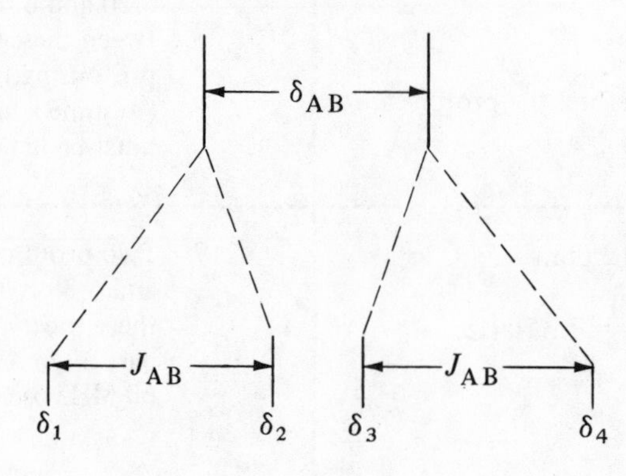

As a result of the perturbation, the difference in chemical shift between protons A and B, $\delta_{AB}$, is given by the equation

$$\delta_{AB} = \sqrt{(\delta_1 - \delta_4)(\delta_2 - \delta_3)}\,.$$

The $J_{AB}$ coupling constant is given by the equation

$$J_{AB} = \delta_1 - \delta_2 = \delta_3 - \delta_4.$$

The relative intensity ($I$) of the lines is given by the equation

$$\frac{I_2}{I_1} = \frac{I_3}{I_4} = \frac{\delta_1 - \delta_4}{\delta_2 - \delta_3}.$$

---

Q-219 A perturbed splitting pattern is given below for an AB system. Will the difference between the midpoints (marked with arrows) of the split lines represent the positions of unsplit lines? Why?

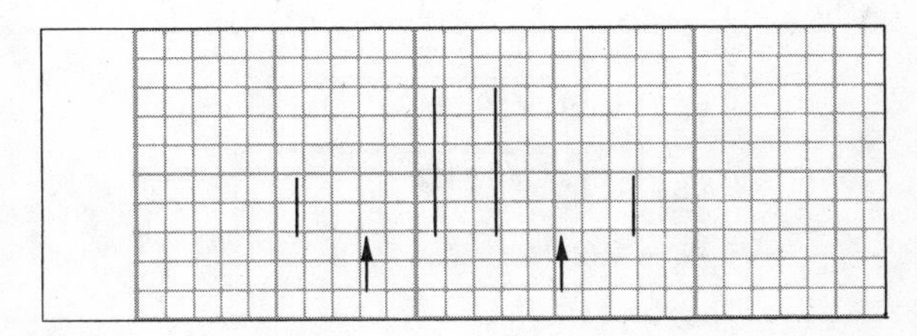

A-219 No. Because the split lines are perturbed and are no longer symmetrical about the original line.

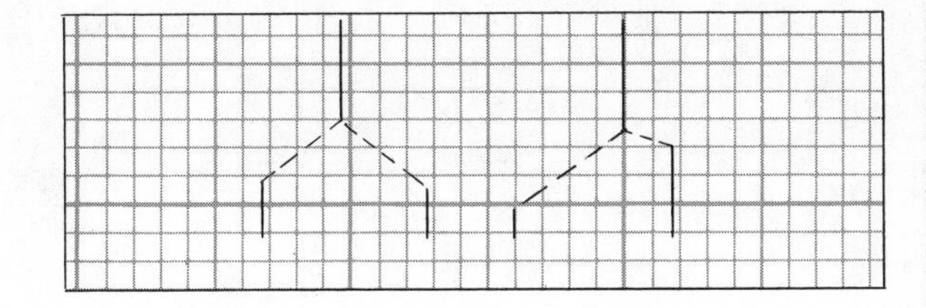


Q-220 On a scale of 1 Hz = 1 division, the splitting pattern for an AB system is diagramed below.

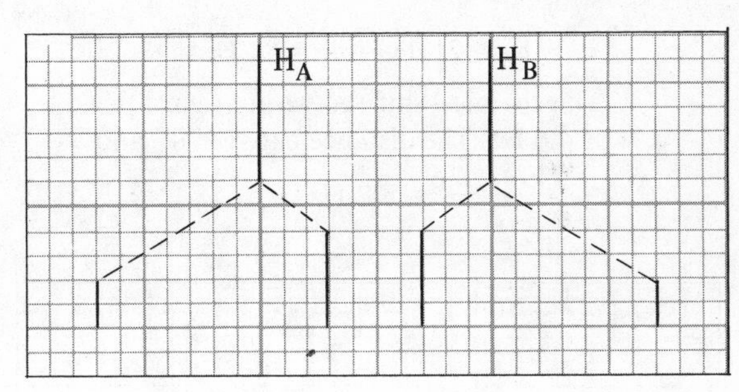


Give the chemical shift difference, $\delta_{AB}$, and the $J_{AB}$ coupling constant for the above splitting pattern.

A-220

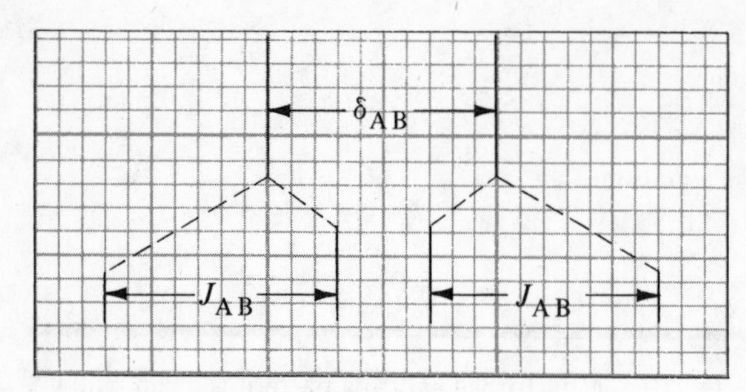


$J_{AB}$ = 10 divisions = 10 Hz

$\delta_{AB}$ = 10 divisions = 10 Hz

Q-221 For the AB pattern of peaks given below (1 Hz = 1 division), determine the chemical shift difference and the coupling constant between the A and B protons.

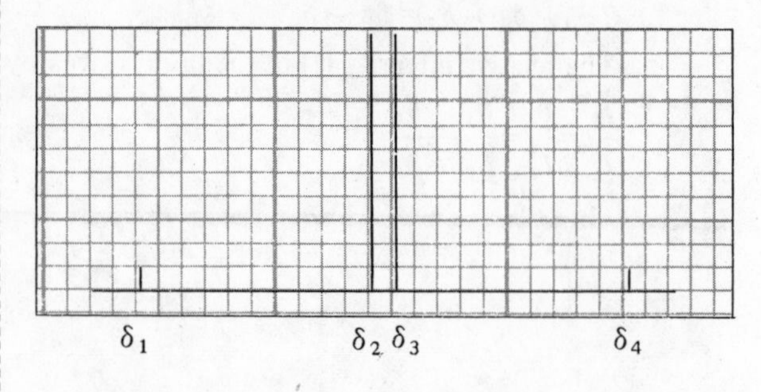


A-221 $J_{AB} = \delta_1 - \delta_2$ = 10 divisions = 10 Hz

$\delta_{AB} = \sqrt{(\delta_1 - \delta_4)(\delta_2 - \delta_3)}$

$\delta_{AB} = \sqrt{(21)(1)\ \text{Hz}^2}$

$\delta_{AB}$ = 4.6 Hz

Q-222 Indicate on the diagram shown in Q-221 the positions of the uncoupled A and B peaks. What is the distance (in Hz) between the $\delta_1$ and the unsplit $\delta_A$ peak?

A-222 The uncoupled lines for the A and B protons, separated by 4.6 Hz, are centered between the two outside peaks of the fine structure.

$\delta_1 - \delta_4$ = 21 divisions = 21 Hz

The chemical shift between original peaks is 4.6 Hz. The distance between $\delta_1$ and $\delta_A$

$= \dfrac{21.0 - 4.6}{2} = 8.2$ Hz

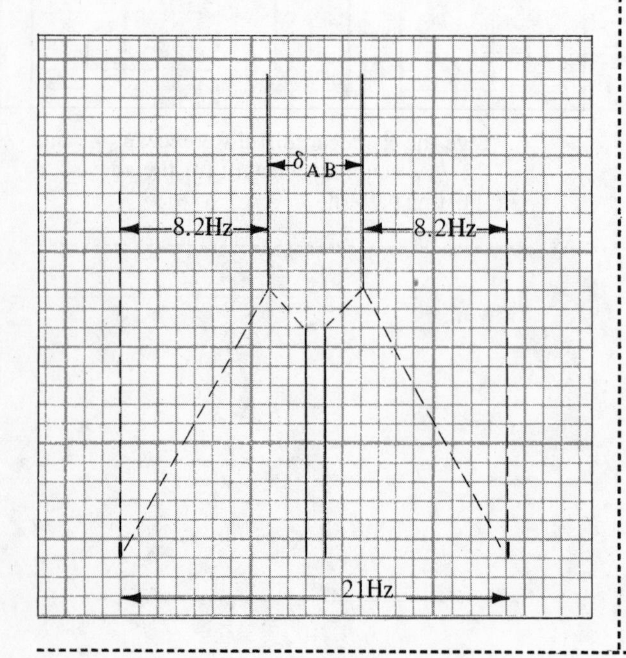


Q-223 Using the equation given in S-26, calculate the relative intensity of the fine structure peaks for the system described in Q-221.

A-223 $\frac{I_2}{I_1} = \frac{21 \text{ Hz}}{1 \text{ Hz}} = \frac{21}{1}$

Q-224 Analyze the AB splitting pattern given below. Find $J_{AB}$, $\delta_{AB}$, relative intensity of lines, and positions of unsplit A and B protons.

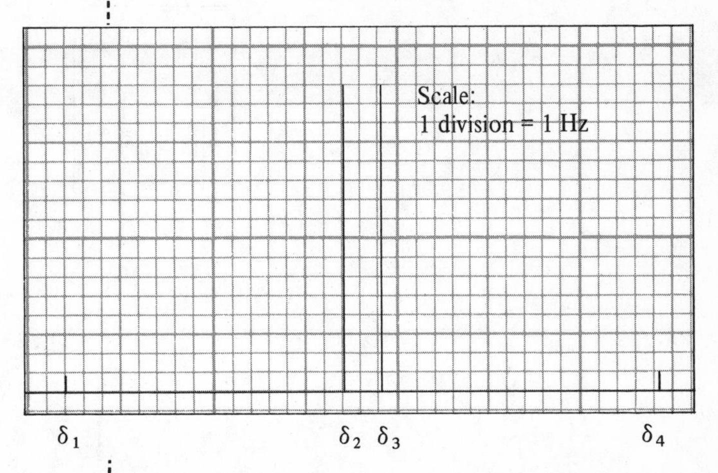


A-224 $J_{AB} = \delta_1 - \delta_2 = 15 \text{ Hz}$

$\delta_{AB} = \sqrt{(\delta_1 - \delta_4)(\delta_2 - \delta_3)} = \sqrt{(32)(2)}$

$= \sqrt{64} = 8 \text{ Hz}$

$\frac{I_2}{I_1} = \frac{32}{2} = \frac{16}{1}$

Distance original peak is from outside peak

$= \frac{32-8}{2} = 12 \text{ Hz}$

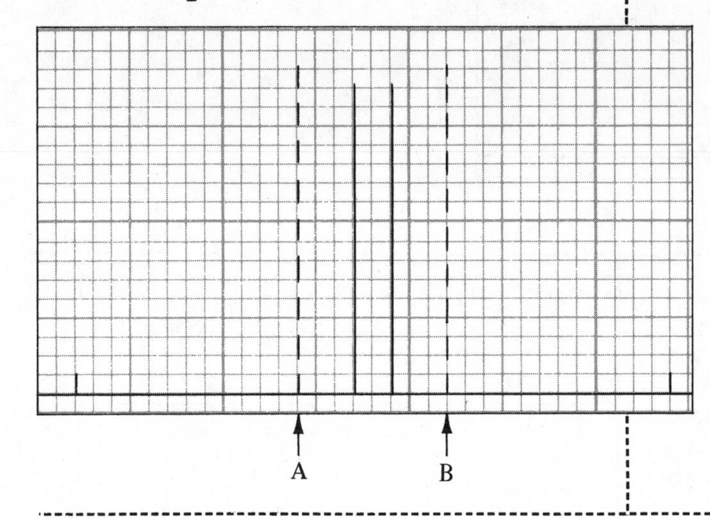

Q-225 Analyze the AB splitting pattern given below. Find $J_{AB}$, $\delta_{AB}$, relative intensity of lines, and position of unsplit A and B protons.

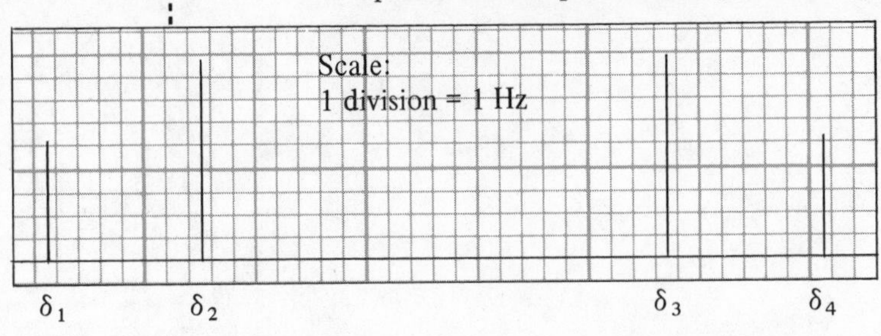


A-225 $J_{AB} = \delta_1 - \delta_2 = 7$ Hz

$\delta_{AB} = \sqrt{(35)(21)} = \sqrt{735} = 27$ Hz

$\frac{I_2}{I_1} = \frac{35}{21} = 1.7$

$\delta_1 - \delta_4 = 35$ Hz

Distance original peak is from outside peak

$= \frac{35 - 27}{2} = 4$ Hz

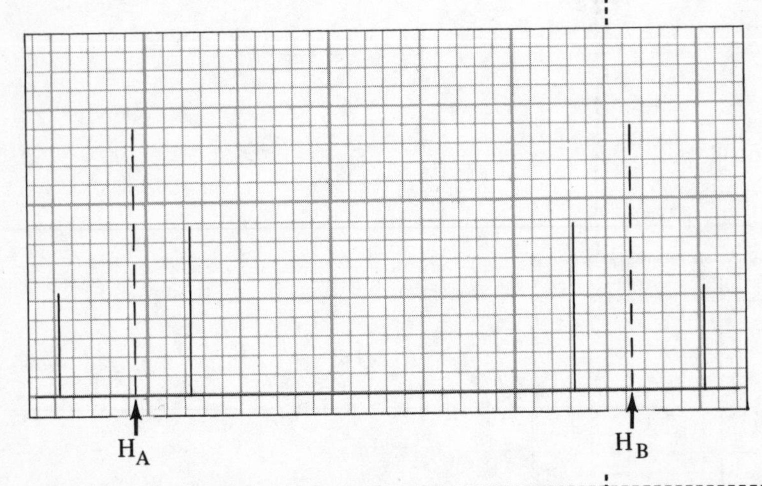


Q-226 A spectrum contains the four lines shown below.

What factor would have to be considered in order to determine if these lines were due to an AB system or the splitting pattern of the X proton in an $A_3X$ system?

A-226 In an $A_3X$ system the four lines are symmetrical with the distance between each line equal to $J_{AX}$. The peak intensities will be in the ratio of 1 : 3 : 3 : 1. In an AB system the distance between the outside pair of lines will be equal to $J_{AB}$, but the distance between the inside lines will rarely be equal to $J_{AB}$. The ratio of intensities will rarely be in the ratio of 1 : 3 : 3 : 1.

Q-227 The following sketch is drawn to scale. Is this an $A_3X$ splitting pattern or an AB splitting pattern?

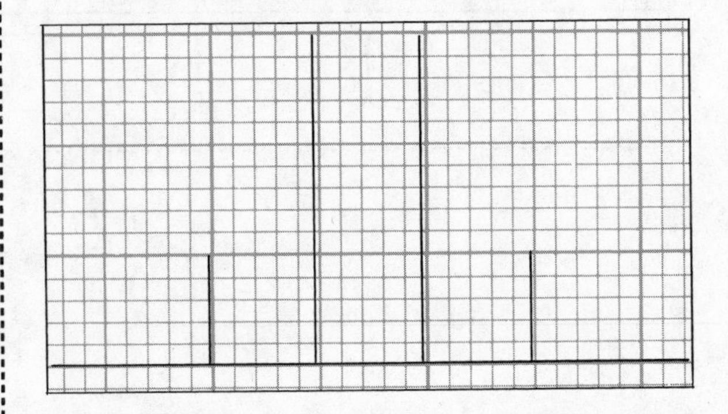

A-227 The distance between each pair of lines is the same and the intensities of the lines are in the ratio of 1 : 3 : 3 : 1. Therefore, this is *most likely* an $A_3X$ system.

Q-228 The following sketch is drawn to scale. Is this an $A_3X$ splitting pattern or an AB splitting pattern?

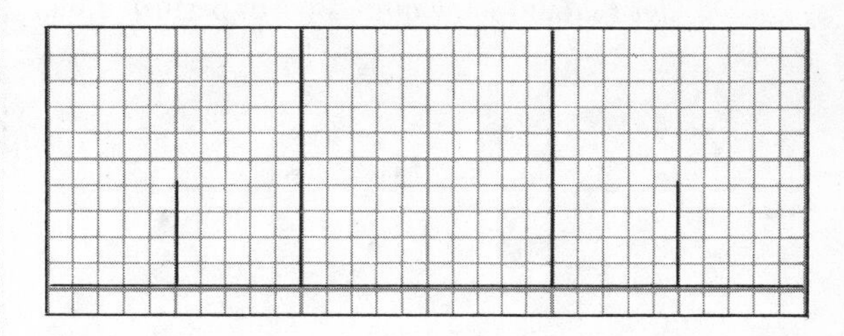

A-228 Because the distance between the pairs of lines are *not* the same and the ratio of intensities are not 1 : 3 : 3 : 1, this set of lines is, in fact, two doublets of an AB system.

Q-229 For the following compound, $\delta_{2,3}$ = 35 Hz and $J_{2,3}$ = 3.0 Hz

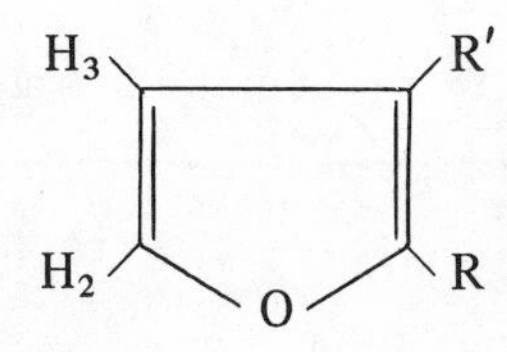


The spectrum for these protons is drawn below. Explain.

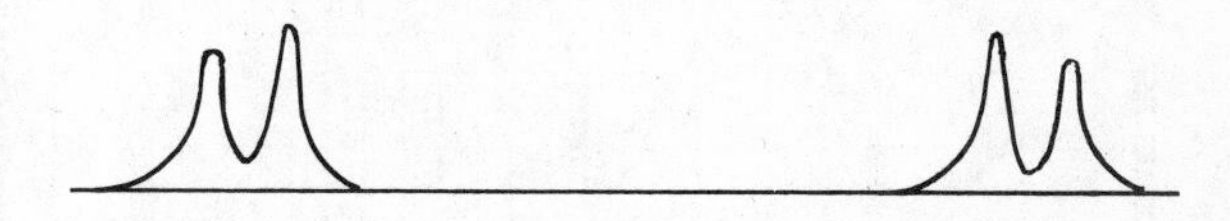

A-229 This spectrum is the result of an AB system where there is very little perturbation, $\delta_{2,3}/J_{2,3}$ = 11.8. This system is nearly an AX system.

Q-230 For the following molecule, $\delta_{AB}$ = 27 Hz and $J_{AB}$ = 8 Hz.

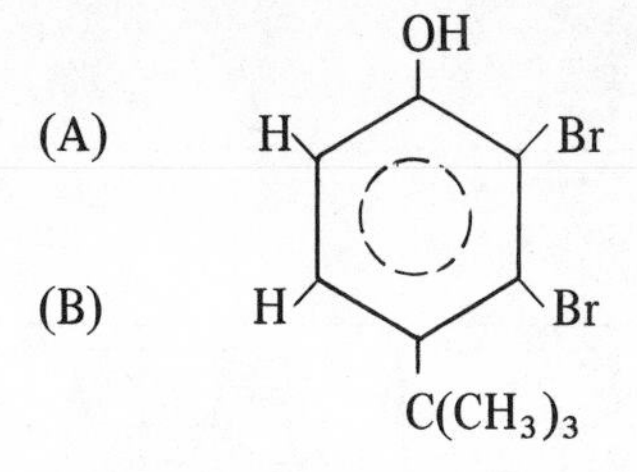


**Sketch** the splitting pattern for the A and B protons.

A-230 $\delta_{AB}/J_{AB} = \frac{27}{8} = 3.3$

The ratio is quite low; therefore, the pattern should show quite a bit of perturbation.

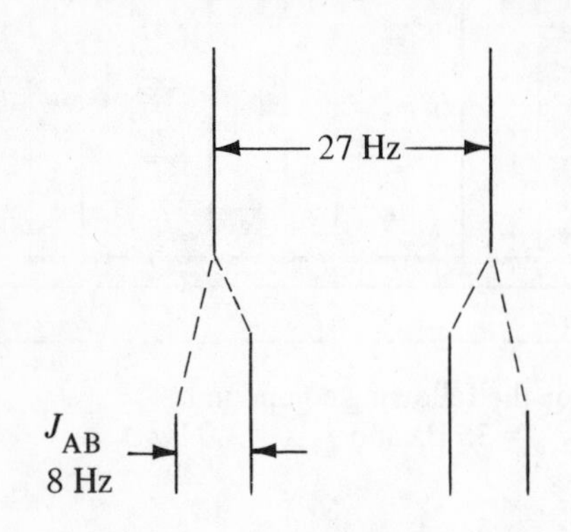


(not drawn to scale)

Q-231 In an AB system, the peaks may further be split by coupling with another proton, X. This type of interaction is called an ABX system. Below is drawn the splitting pattern for an AB system. If protons A and B can interact with another proton, X, with coupling constants $J_{AX}$ = 4 Hz and $J_{BX}$ = 2 Hz, diagram the new pattern of lines which would appear for the A and B protons.

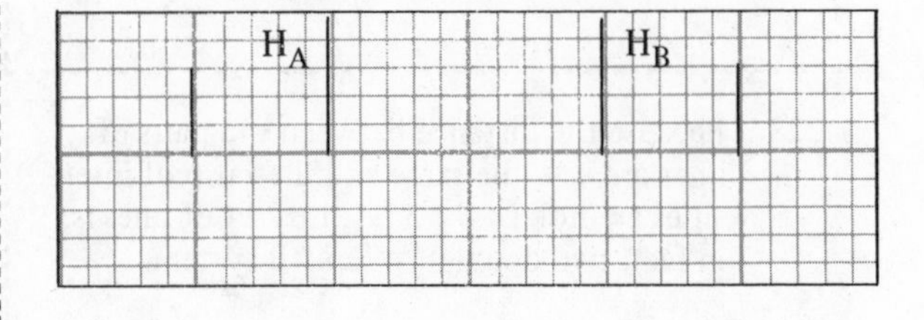


A-231

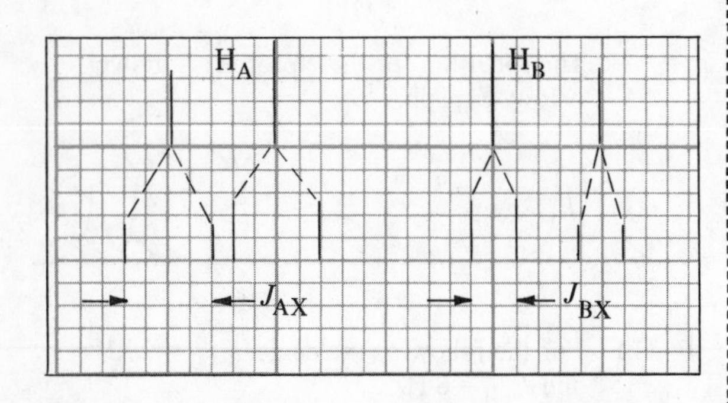


Q-232 Diagram the splitting pattern for the X proton of the system described in Q-231.

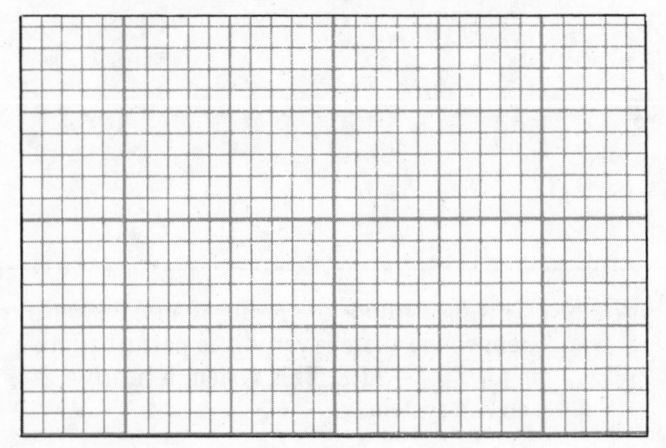

A-232

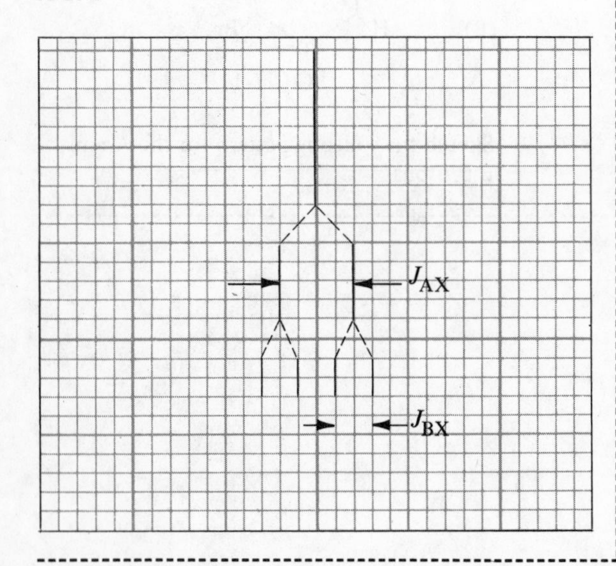


Q-233 An example of an ABX system is 2,3-lutidine.

H, H, H, $CH_3$, $CH_3$, N

Which protons make up the AB part of the spectrum? Which is the X proton?

A-233

(A) H, (B) H, (X) H, $CH_3$, $CH_3$, N

Q-234 The pattern below is for the AB system of 2,3-lutidine (uncoupled to the X proton). If $J_{AX}$ = 1.3 Hz and $J_{BX}$ = 5.0 Hz, diagram the fine structure for the AB protons resulting from coupling with X.

(Scale: 1 division = 1 Hz)

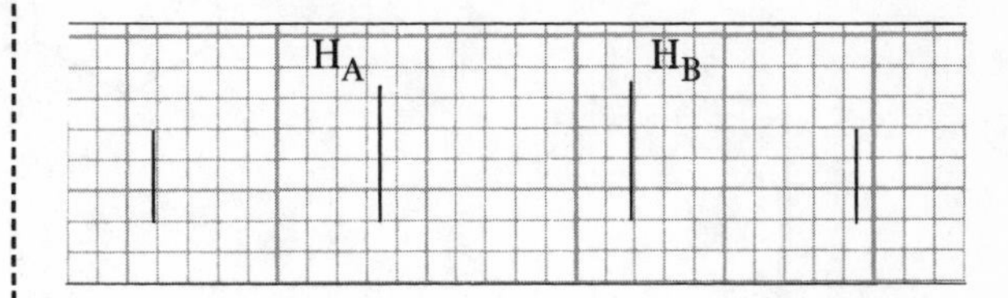


A-234

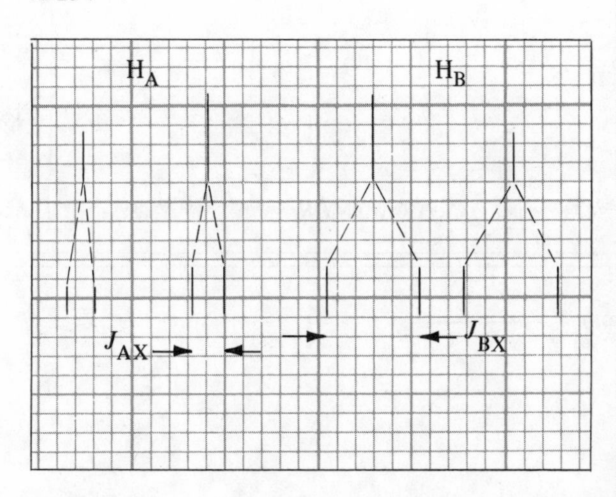


Q-235 What would be the multiplicity of the X proton in 2,3 - lutidine?

A-235 Four lines (two doublets).

**R** Analysis of AB type spectrum:

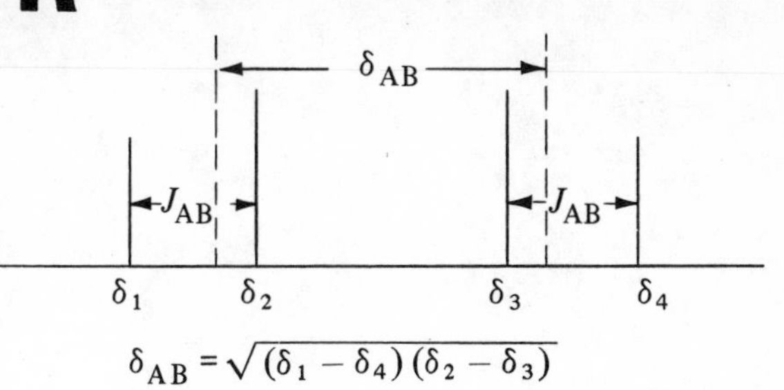


$$\delta_{AB} = \sqrt{(\delta_1 - \delta_4)(\delta_2 - \delta_3)}$$

$$\frac{I_2}{I_1} = \frac{I_3}{I_4} = \frac{\delta_1 - \delta_4}{\delta_2 - \delta_3}$$

In an ABX system the AB splitting pattern is further split by interaction of A and B with X.

S-27 At times, protons that are bonded to a nucleus with a spin quantum number of $I = 1$ or greater will give rise to a broad, unsplit absorption band. This is common for protons bonded to nitrogen. The broad peak in the spectrum below for compound I is due to the two protons on the nitrogen.

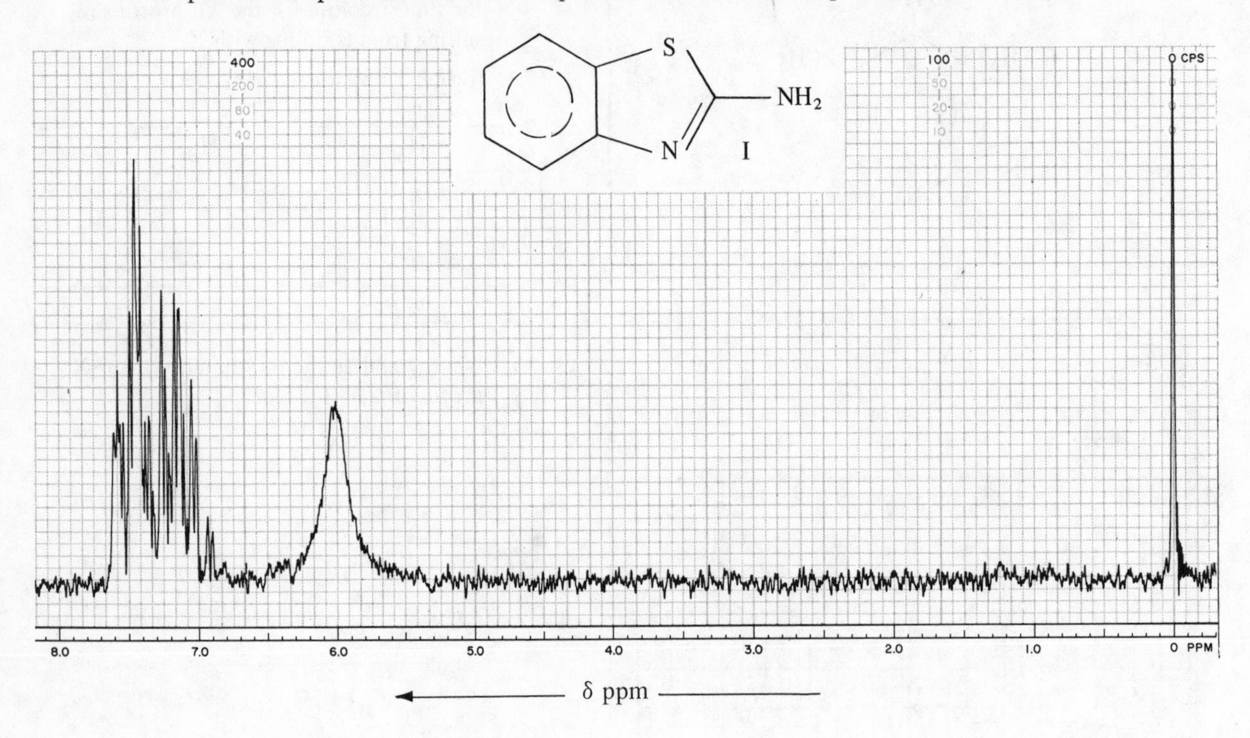


Spectra charts on pages 160-167 reprinted with permission from NMR Spectra Catalog, 1962, Varian Associates, Palo Alto, California.

When several protons with about the same $\tau$ values are present in a molecule, the spectrum may have a large peak with or without fine structure that resembles a haystack. These peaks are very difficult to analyze. The molecule

has the NMR spectrum shown below. The peaks in the 1.6 and 2.2 $\delta$ regions are due to the $-CH_2-$ protons. (The peaks at 5.3 $\delta$ are due to an impurity.)

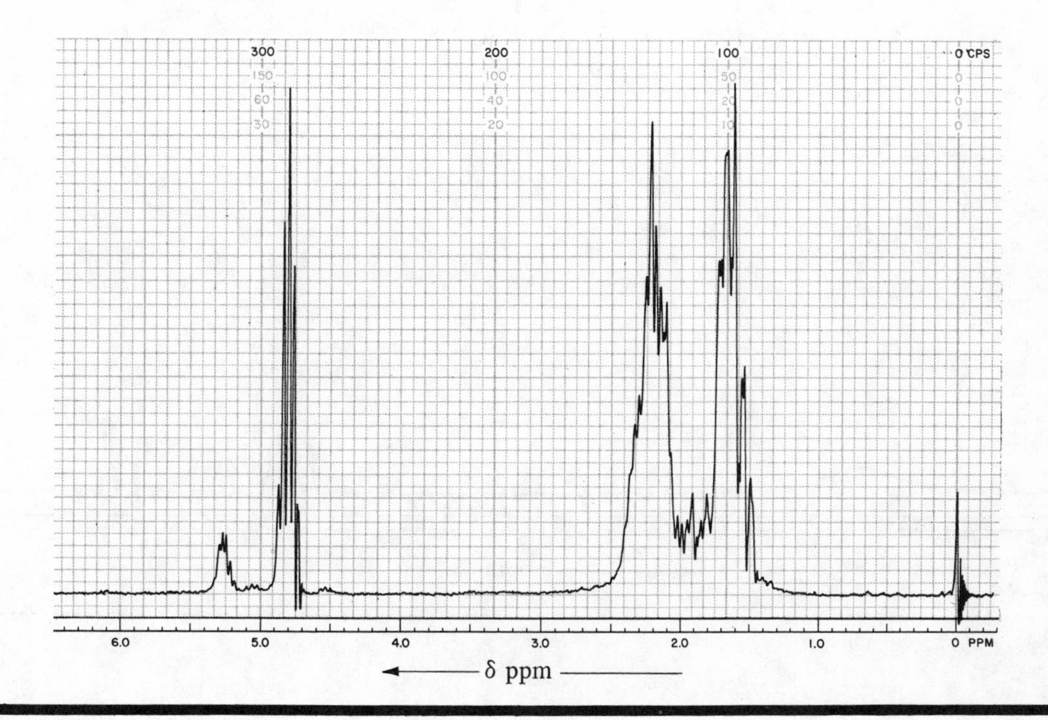

Q-236 The spectrum for propionamide

$\overset{(a)}{CH_3} - \overset{(b)}{CH_2} - C(=O)\underset{(c)}{NH_2}$

is shown below. Give the $\tau$ values for each of the different protons.

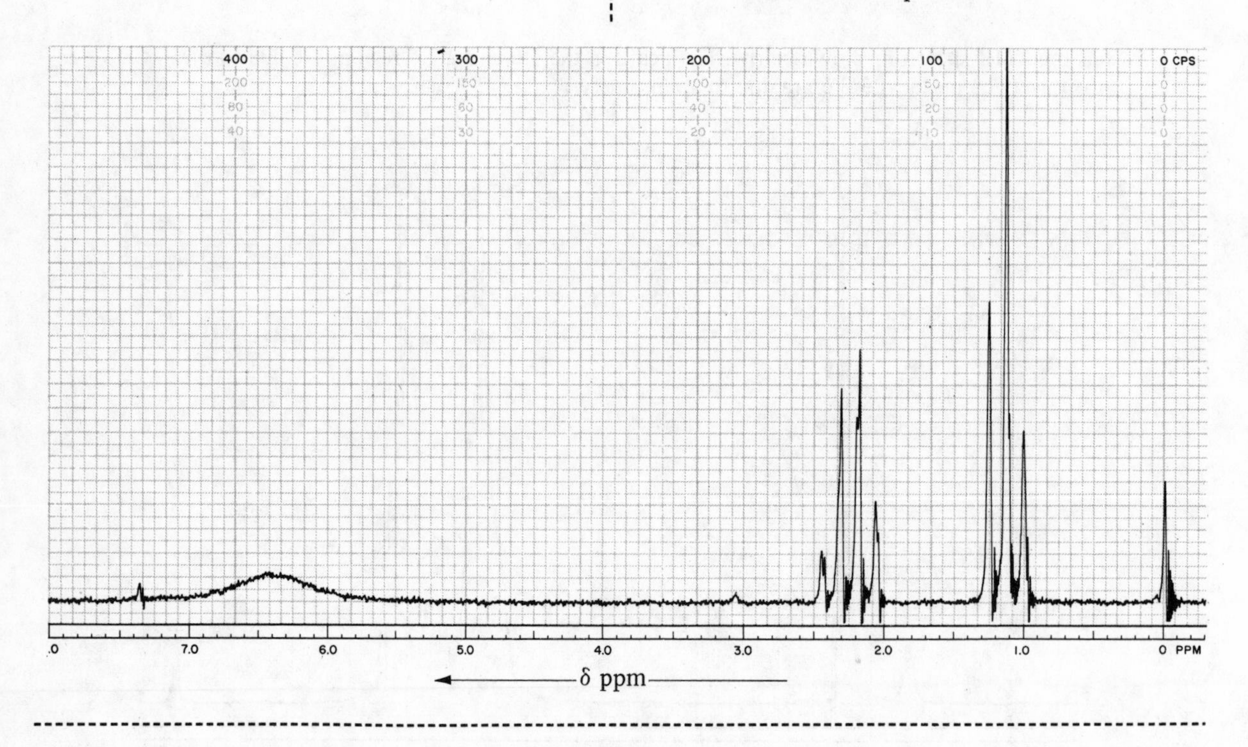

A-236 (*a*) 8.9 $\tau$
(*b*) 7.8 $\tau$
(*c*) 3.6 $\tau$

Q-237 The spectrum for 1-chloro-2-bromoethane

(*a*) (*b*)
$ClCH_2 - CH_2Br$

is shown below. Can the peaks corresponding to the different protons be distinguished in this spectrum?

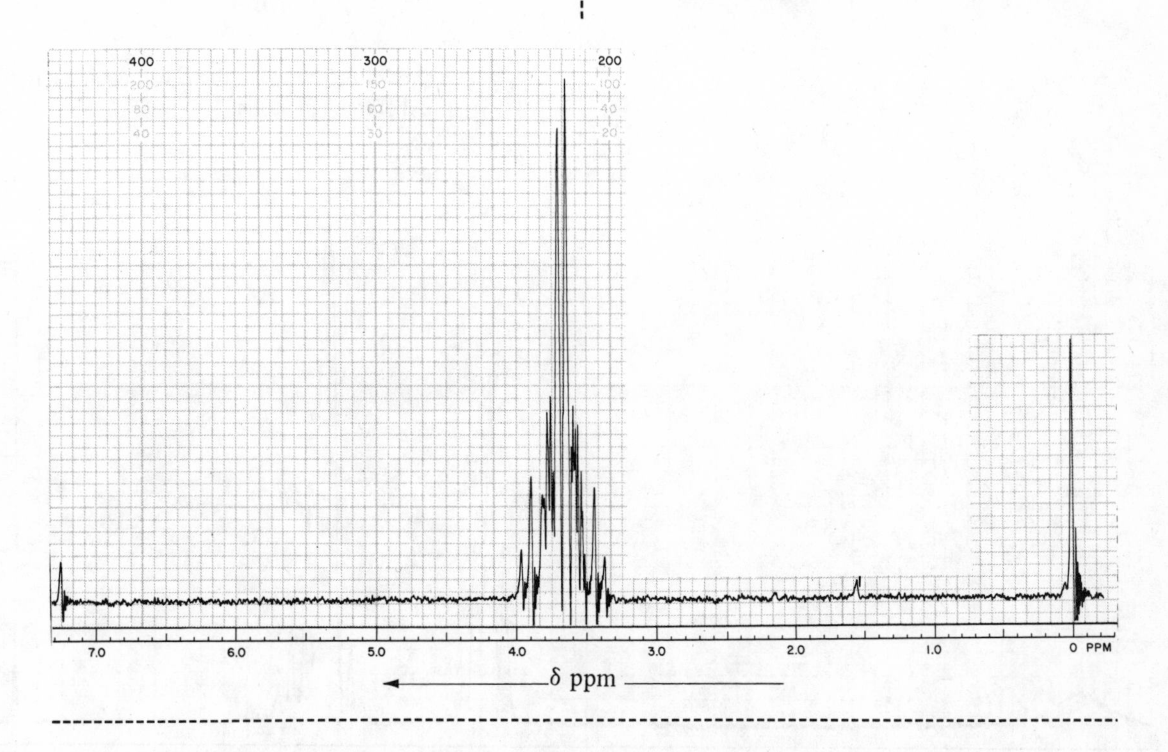

A-237 No. The (*a*) and (*b*) protons absorb in the 3.6 – 3.8 δ region. This pattern is typical of an $A_2B_2$ system.

Q-238 The spectrum for 2-pyrrolidone

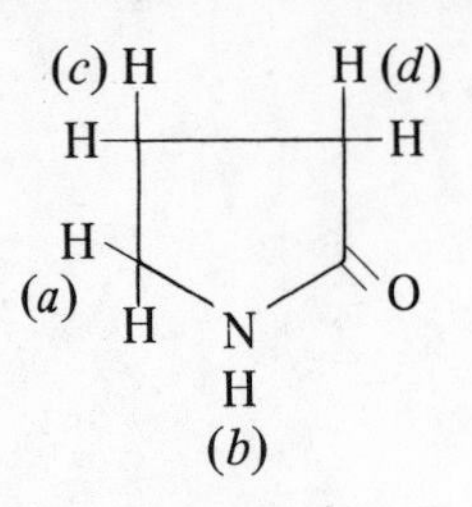


is shown below. Give the δ value for each of the different protons.

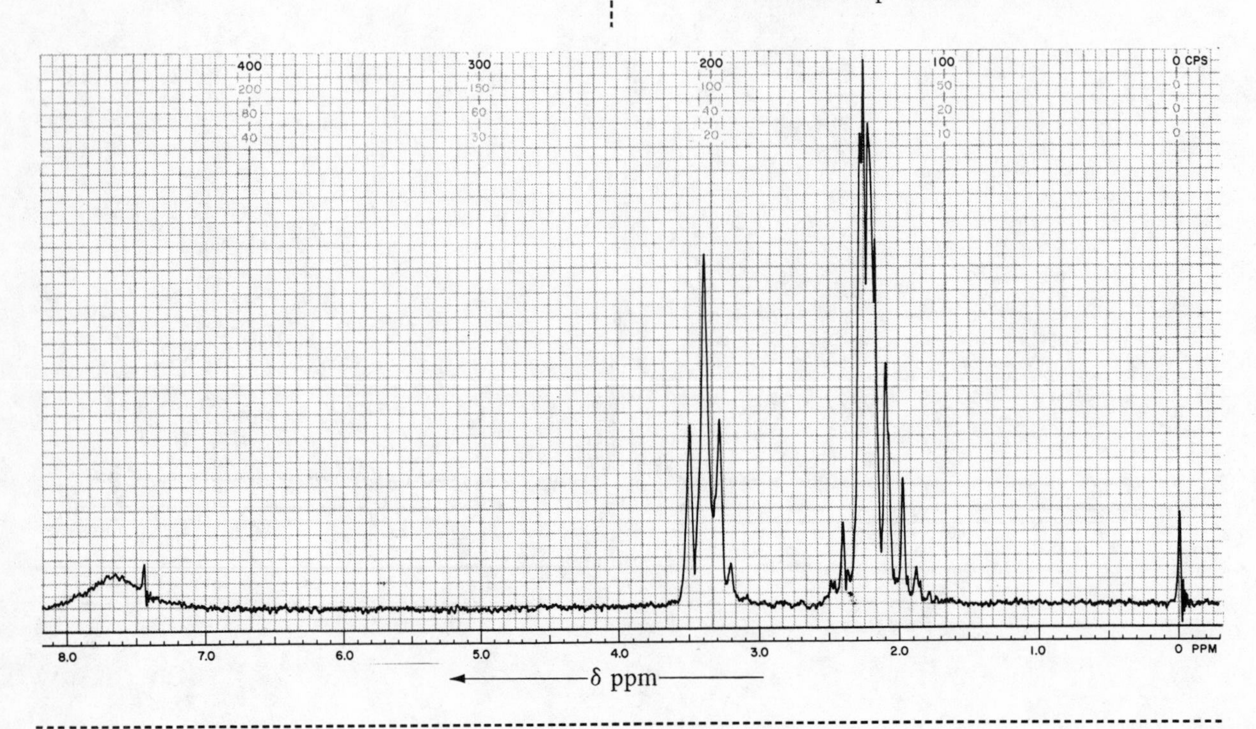

A-238 (*a*) 3.4 δ
(*b*) 7.7 δ
(*c*) (*d*) 2.1 – 2.3 δ

Q-239 The spectrum for morpholine

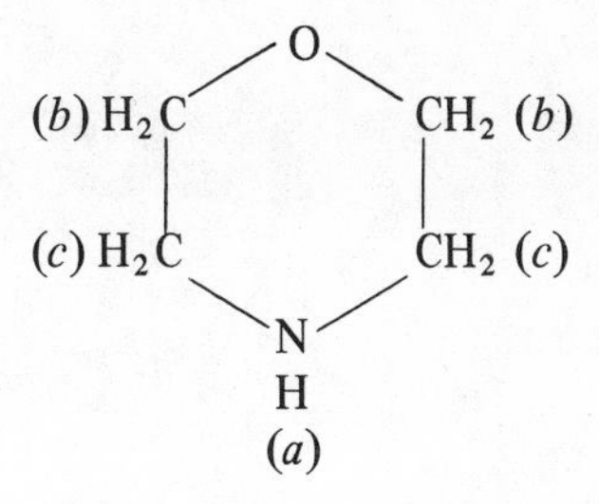


is shown below. Give the τ value for each of the different protons.

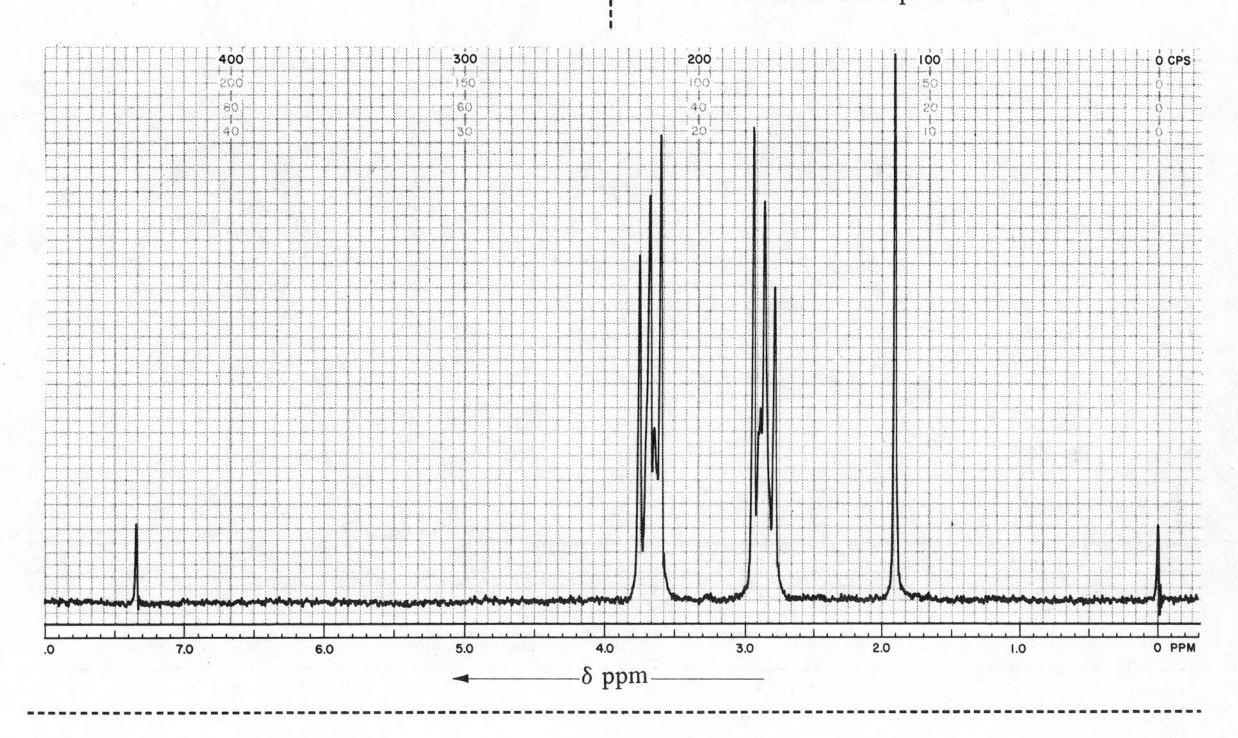

A-239 (*a*) 8.2 $\tau$

(*b*) 6.3 $\tau$

(*c*) 7.1 $\tau$

Q-240 The spectrum for 1-ethynyl-1-cyclohexanol

(*d*) H H H(*c*)
H— —OH (*b*)
—C≡CH (*a*)
H— —H
(*e*) H H (*c*)
H H
(*d*)

is shown below. Give the δ value for each of the different protons.

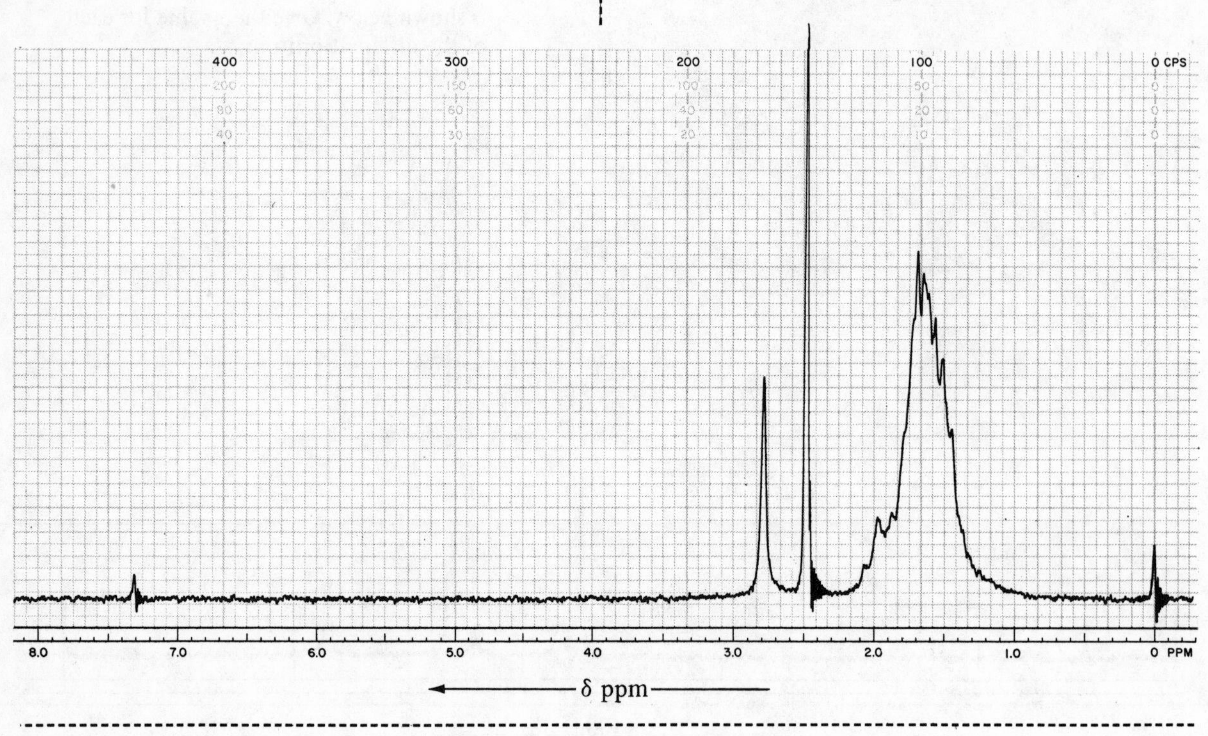

A-240 (*a*) 2.48 δ
(*b*) 2.78 δ
(*c*) (*d*) (*e*) 1.4 – 1.8 δ

Q-241 The spectrum for phenacetin

(c) (d) (e)
H H H
(a) (b)
$CH_3 - CH_2 - O$–C₆H₄–N(H)–C(=O)–$CH_3$ (f)
H H
(c) (d) O

is shown below. Give the δ value for each of the different protons.

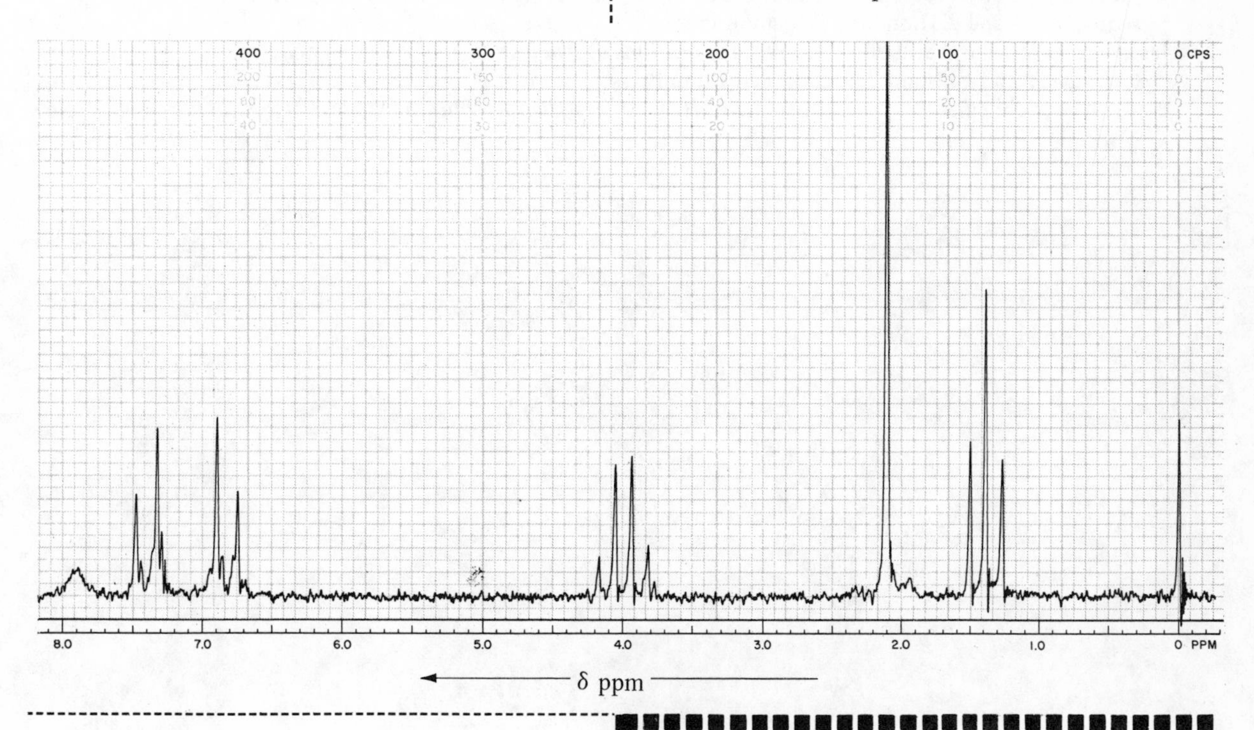


A-241 (*a*) 1.4 δ
(*b*) 4.0 δ
(*c*) 6.8 δ
(*d*) 7.4 δ
(*e*) 7.9 δ
(*f*) 2.1 δ

**R** Protons bonded to nuclei with $I = 1$ or greater can give rise to broad absorption peaks.

Several protons with approximately the same $\tau$ values and complex coupling patterns can give rise to a complicated absorption band referred to as a "haystack".

## REFERENCES

**Introductory**:

Lynden-Bell, Ruth, and Harris, R. K. 1969. *Nuclear Magnetic Resonance Spectroscopy*. New York: Appleton-Century-Crofts.

Roberts, John D. 1959. *Nuclear Magnetic Resonance*. New York: McGraw-Hill.

**Intermediate**:

Bhacca, N. S.; Hollis, D. P.; Johnson, L. F.; and Per, E. A. 1963. *NMR Spectra Catalog*, vol. 1 and 2. Palo Alto, Calif.: Varian Associates.

Bhacca, Norman S., and Williams, Dudley H. 1964. *Applications of NMR Spectroscopy in Organic Chemistry*. San Francisco: Holden-Day.

Emsley, J. W.; Feeney, J.; and Sutcliffe, L. H. 1965. *High Resolution Nuclear Magnetic Resonance Spectroscopy*, vol. 1 and 2. London: Pergamon Press.

Jackman, L. M., and Sternhell, S. 1969. *Applications of Nuclear Magnetic Resonance Spectroscopy in Organic Chemistry*, 2nd edition. London: Pergamon Press.

Pople, J. A.; Schneider, W. G.; and Bernstein, H. J. 1959. *High Resolution Nuclear Magnetic Resonance*. New York: McGraw-Hill.

# Chapter 5 SYNTHESIS OF SPECTRAL DATA

The determination of chemical structure by spectroscopic (or other) methods requires the successful completion of three basic steps, namely:

a) data procurement,
b) interpretation of data, and
c) synthesis of the interpreted data into a chemical structure.

Of the three steps, the synthesis of interpreted data into a chemical structure is by far the most difficult. While every individual will develop his own technique for this synthesizing process, the following chapter demonstrates a systematic method which many chemists employ.

After completing this chapter you should be able to

a) calculate the number of sites of unsaturation in a molecule from the molecular formula of a compound
b) determine the possible molecular formulas of a compound from data provided by molecular weight
c) determine structural and **residual** units present in a molecule
d) combine simple structural units into more complex units
e) arrange simple and complex structural units into molecular structures.

S-1 All compounds containing only C, H, O, or S and having no **units of unsaturation** (multiple bonds or rings) have the general formula $C_nH_{2n+2}O_aS_b$. Each double bond or ring is equivalent to one unit of unsaturation while a triple bond is equal to two units. Each unit of unsaturation reduces the number of hydrogen atoms per molecule by two.

The units of unsaturation ($\Omega$) may be calculated from the formula

$$\Omega = \frac{2n+2 - (\text{number of hydrogen atoms per molecular formula})}{2}$$

where n is the number of carbon atoms per molecule.

| **structural formula** | **molecular formula** | **general formula** | $\Omega$ |
|---|---|---|---|
| $CH_3CH_2CH=CHCH_2OH$ | $C_5H_{10}O$ | $C_nH_{2n}O$ | 1 |
| S = (cyclopentene ring)–$C\equiv C-CH_3$ | $C_8H_{10}S$ | $C_nH_{2n-6}S$ | 4 |

Q-1 Cyclopentanone has the formula $C_5H_8O$. Calculate $\Omega$ for the compound. What is each unit of unsaturation?

A-1 $\Omega = \frac{2n+2-8}{2} = \frac{10+2-8}{2} = 2$

1 double bond + 1 ring

Q-2 Napthalene has the formula $C_{10}H_8$. Calculate $\Omega$ for the compound. What is each unit of unsaturation?

A-2 $\Omega = \frac{2n+2-8}{2} = 7$

5 double bonds + 2 rings

Q-3 Furan has the formula $C_4H_4O$. Calculate $\Omega$ for the compound. What is each unit of unsaturation?

A-3 $\Omega = \frac{2n+2-4}{2} = 3$

2 double bonds + 1 ring

Q-4 A compound has the formula $C_6H_{10}O_2S$. Calculate $\Omega$ and give all the possible combinations of rings and multiple bonds.

A-4 $\Omega = \frac{2n+2-10}{2} = 2$

2 double bonds **or**

1 triple bond **or**

1 double bond and one ring **or**

2 rings.

Q-5 A compound known to be a carboxylic acid has the formula $C_5H_8O_3$. What are the possible combinations of rings and multiple bonds?

A-5 $\Omega = \frac{2n+2-8}{2} = 2$

In addition to the one unit of unsaturation due to the carbonyl group, there could be one double bond **or** one ring.

**R** The units of unsaturation ($\Omega$) in molecules containing only C,H,O,S are given by the expression,

$$\Omega = \frac{2n+2 - (\text{hydrogen atoms per molecule})}{2}$$

S-2 All compounds containing nitrogen and no **units of unsaturation** have the general formula

$C_nH_{2n+2+m}N_m$

Each unit of unsaturation reduces the number of hydrogens by two. The units of unsaturation in a nitrogen-containing compound may be calculated from the formula:

$$\Omega_{\text{N containing compound}} = \frac{2n+2+m - (\text{number of hydrogen atoms per molecule})}{2}$$

where n and m equal the number of carbon and nitrogen atoms in the molecule, respectively.

| structural formula | molecular formula | general formula | $\Omega$ |
|---|---|---|---|
| $(CH_3)_3N$ | $C_3H_9N$ | $C_nH_{2n+2+1}N$ | 0 |
| $CH_2 = CH\,CH_2\,NH_2$ | $C_3H_7N$ | $C_nH_{2n+1}N$ | 1 |
| $(CH_3)_2\,C = N - NH\,CH_3$ | $C_4H_{10}N_2$ | $C_nH_{2n+2}N_2$ | 1 |

All compounds containing halogen and no **units of unsaturation** have the general formula $C_nH_{2n+2-x}X_x$

$$\Omega_{\text{halogen containing compounds}} = \frac{2n+2-x-(\text{number of hydrogen atoms per molecule})}{2}$$

Q-6 A compound has the formula $C_5H_7N$. Calculate $\Omega$ for the compound. What are the possible combinations of rings and double bonds?

A-6 $\Omega = \frac{2n+2+m-7}{2} = \frac{10+2+1-7}{2} = 3$

3 double bonds **or**

1 double and one triple bond **or**

2 double bonds and one ring **or**

1 triple bond and one ring **or**

1 double bond and two rings **or**

3 rings.

Q-7 A compound has the formula $C_6H_{11}N_3$. Calculate $\Omega$.

A-7 $\Omega = \frac{2n+2+3-11}{2} = 3$

Q-8 A compound has the formula $C_8H_7ClO$. Calculate $\Omega$.

A-8 $\Omega = \frac{2n+2-1-7}{2} = \frac{16+2-1-7}{2} = 5$

Q-9 Could a compound with the formula $C_{10}H_{17}Cl_3$ be aromatic (contain a benzene ring)? Why?

A-9 $\Omega = \frac{2n+2-3-17}{2} = 1$

No. To be aromatic there must be at least four units of unsaturation. This compound has only one.

Q-10 What is the general formula of a compound containing C, H, N, and X?

A-10 $C_nH_{2n+2+m-x}N_mX_x$

Q-11 Calculate $\Omega$ for a compound with the formula $C_6H_{11}N_2Cl$.

A-11 $\Omega = \frac{2n+2+2-1-11}{2} = 2$

**R** $\Omega_{N \text{ containing compound}} = \frac{2n+2+m-(\text{H atoms in molecule})}{2}$

$\Omega_{\text{halogen containing compound}} = \frac{2n+2-x-(\text{H atoms in molecules})}{2}$

S-3 From the NMR spectrum, various types of protons present in a molecule may be distinguished. The areas under the NMR peaks are proportional to the number of protons in a particular chemical environment. The total number of protons in the molecule is a multiple of the *sum* of the simplest ratio of these peak areas. Thus, a molecule whose NMR spectrum yields a **three** protron singlet, a **two** proton quartet, and a **three** proton triplet could have 8, 16, 24 . . . protons.

Possible molecular formulas may be determined if the molecular weight, qualitative elemental analysis, and NMR spectrum are available. This information allows the number of carbon atoms per molecule to be calculated by using the formula

$$\#\text{C atoms per molecule} = \frac{\text{molecular weight} - \text{mass of H in molecule} - \text{mass of other atoms}}{12}$$

Q-12 The NMR spectrum of a molecule shows a multiplet (complex splitting pattern), a singlet, a heptet, and a doublet. The peak area ratios are 4 : 3 : 1 : 6. What are the possible number of hydrogens in the molecule?

A-12 The sum of the ratios is 4 + 3 + 1 + 6 = 14. The number of hydrogens could be 14, 28, 42 . . .

Q-13 The compound described in Q-12 contained only C and H and had a molecular weight of 133±2. How many carbon atoms per molecule are present? What is the formula of the compound?

A-13 $\#C = \frac{133\pm2-14}{12} = 10$

Formula = $C_{10}H_{14}$

Higher multiples of 14 for the number of hydrogens are not possible, i.e.,

$\#C = \frac{133\pm2-28}{12} = 9\ (C_9H_{28})$

$C_9H_{28} = C_nH_{2n+10}$ (non-existent)

Q-14 The NMR spectrum of a molecule with a molecular weight of 90±2 showed two singlets with a ratio of 1 : 3. The molecule contained only C,H, and O. What are the possible formulas, assuming **one** oxygen per molecule?

A-14 $C_6H_4O$, $C_5H_{12}O$

Using the possible number of hydrogens per molecule to calculate the number of carbon atoms:

$$\#C = \frac{90\pm2-4-16}{12} = 6\ (C_6H_4O)$$

$$\#C = \frac{90\pm2-8-16}{12} = \frac{66\pm2}{12}\ \text{(not a whole number)}$$

$$\#C = \frac{90\pm2-12-16}{12} = 5\ (C_5H_{12}O)$$

$$\#C = \frac{90\pm2-16-16}{12} = 5\ (C_5H_{16}O\ \text{non-existent})$$

All higher multiples of hydrogen will give non-existent formulas.

Q-15 A molecule contained only C, H and N. The NMR spectrum had three peak areas with ratios of 1 : 2 : 3. How many nitrogen atoms per molecule are possible?
(Hint: Will the number of hydrogens in an N containing compound be odd or even? $C_nH_{2n+2+m}N_m$.)

A-15 Possible number of nitrogen atoms per molecule must be even, i.e., $N_2$, $N_4$, $N_6$ . . . . If a compound contains only C, H, and an odd number of N atoms per molecule, the number of hydrogens will be odd. If the molecule contains an even number of N atoms, the number of hydrogen atoms will be even. The compound in this question can have only an even number of hydrogens (6,12,18 . . .).

Q-16 The molecule described in Q-15 had a molecular weight of 70±2. What are possible formulas for the molecule, assuming only two nitrogen atoms per molecule?

A-16 Only possible formula is $C_3H_6N_2$

$$\#C = \frac{70\pm2-6-28}{12} = 3(C_3H_6N_2)$$

Using other possible multiples for hydrogen will yield only non-existent formulas, i.e.,

$$\#C = \frac{70\pm2-18-28}{12} = 2\ (C_2H_{16}N_2\ \text{non-existent})$$

Q-17 A hydrocarbon has a molecular weight of 140±3 and has two sets of NMR peaks with proton ratios of 3 : 7. What are the possible formulas for the compound?

A-17 $C_{11}H_{10}$ and $C_{10}H_{20}$

$$\#C = \frac{140\pm3-10}{12} = 11\ (C_{11}H_{10})$$

$$\#C = \frac{140\pm3-20}{12} = 10\ (C_{10}H_{20})$$

$$\#C = \frac{140\pm3-30}{12} = 9\ (C_9H_{30}\ \text{non-existent})$$

Q-18 The NMR, IR, and UV spectra for the compound described in Q-17 indicated the presence of an aromatic system. What is the formula of the compound? Give your reasoning.

A-18 $C_{11}H_{10}$

The simplest aromatic system has the general formula $C_nH_{2n-6}$ (three double bonds and a ring). A compound with the formula $C_{10}H_{20}(C_nH_{2n})$ cannot be aromatic. A compound with the formula $C_{11}H_{10}$ $(C_nH_{2n-12})$ could be aromatic. (It would contain other units of unsaturation in addition to one aromatic ring.)

Q-19 A compound has a molecular weight of 110±3 and does not contain halogen, N or S. The NMR spectrum has two sets of peaks with area ratios of 5 : 3. What are the possible formulas for the compound? (Hint: oxygen *may* be present.)

A-19 $C_8H_{16}$, $C_7H_8O$, $C_6H_8O_2$ and $C_3H_8O_4$.

*Formulas with no O present:*

$$\#C = \frac{110\pm3-8}{12} = \frac{102\pm3}{12} \text{ (not a whole number)}$$

$$\#C = \frac{110\pm3-16}{12} = 8 \quad (C_8H_{16})$$

$$\#C = \frac{110\pm3-24}{12} = 7 \text{ } (C_7H_{24} \text{ non-existent})$$

*Formula with one O present:*

$$\#C = \frac{110\pm3-8-16}{12} = 7 \text{ } (C_7H_8O)$$

*Formula with two O present:*

$$\#C = \frac{110\pm3-8-32}{12} = 6 \text{ } (C_6H_8O_2)$$

*Formula with four O present:*

$$\#C = \frac{110\pm3-8-64}{12} = 3 \text{ } (C_3H_8O_4)$$

Q-20 The NMR, IR and UV spectra of the compound described in Q-19 indicate an aromatic system (benzenoid or furanoid). What are the possible formulas for the compound?

A-20 $C_7H_8O$ and $C_6H_8O_2$. If a benzene ring is present, $\Omega$ must be 4 or larger. If a furan ring is present, $\Omega$ must be 3 or larger.

$C_8H_{16}$ $\Omega = 1$

$C_3H_8O_4$ $\Omega = 0$

**R** The total number of protons in a molecule is a whole number multiple of the sum of the simplest ratio of NMR peak areas. The number of carbons in a molecule may be calculated from the formula: #C = (molecular weight – mass of hydrogen – mass of other atoms)/12.

S-4 The NMR, UV, and IR spectra of a compound yield data which allows identification of *structural units* within the molecule (i.e., a carbonyl group, an ethyl group, a *p*-disubstituted benzene ring, etc.). After the identification of these units, two problems still remain: a) the determination of structural units not identified from spectral data, and b) the possible sequences in which the structural units may be attached.

To help determine **residual** structural units not identified from spectral data, the formulas (or formula weights) of all the *unique* known structural units are subtracted from the formula (or molecular weight) of the compound. The number of sites of unsaturation can also be determined for the **residual** units. The **residual** formula (or formula weight) may also give some indication as to the structure(s) possible for the unknown unit(s). Chemical intuition and other data about the compound, such as odor, mp, or origin may be very helpful in elucidating the possible structures for the **residual** units.

**Q-21** The IR spectrum of a compound, $C_9H_{10}O$, indicated the following structural units:

a carbonyl group and

a *p*-disubstituted benzene ring.

Are all units of unsaturation in the molecule accounted for by the two known structural units? What is the formula of the **residual** unit(s)? What are possible structures for the **residual** unit(s)?

---

**A-21** $\Omega = 5$

The carbonyl and benzene ring account for all 5 units of unsaturation.

| | | $\Omega$ |
|---|---|---|
| $C_9H_{10}O$ | | 5 |
| $-\ C_6H_4$ | (aromatic ring) | −4 |
| $-\ C\quad O$ | (carbonyl) | −1 |
| $C_2H_6$ | (residual unit) | 0 |

**Residual** group(s) can have no units of unsaturation and must be two $CH_3$ groups.

**Q-22** The IR spectrum of a compound, $C_{11}H_{14}O_2$, indicated the following structural units:

a carbonyl group,

a hydroxyl group, and

a monosubstituted benzene ring.

Are all units of unsaturation in the molecule accounted for by the three known structural units? What is the formula of the **residual** structural unit(s)? What are some possible structures for the **residual** unit(s)?

---

**A-22** $\Omega = 5$

The carbonyl and benzene ring account for all 5 units of unsaturation.

| | | $\Omega$ |
|---|---|---|
| $C_{11}H_{14}O_2$ | | 5 |
| $-C_6\ H_5$ | (aromatic ring) | −4 |
| $-C\qquad O$ | (carbonyl) | −1 |
| $-\quad H\ O$ | (hydroxyl) | 0 |
| $C_4\ H_8$ | (residual formula) | 0 |

Possible structures are

$-CH_2(CH_2)_2CH_2-$; $-CH_2-CH_3$, $-CH_3$ and

$-\overset{|}{\underset{|}{C}}-$; $CH_3$, $-CH_2-CH_3$, and $-\overset{|}{\underset{|}{C}}-H$; etc.

**Q-23** The IR spectrum of a compound with a molecular weight of 159±3 indicated the following structural units:

a carbon-carbon double bond,

a carboxylic acid group, and

a monosubstituted benzene ring.

What is the formula weight of the **residual** unit(s)? What are some possible structures for the **residual** unit(s)?

---

**A-23**

| | |
|---|---|
| 159±3 | |
| − 45 | (COOH) |
| − 77 | ($C_6H_5$) |
| − 24 | (C=C) |
| 13±3 | (formula weight of residual units) |

Possible structures:

$-\overset{|}{\underset{|}{C}}-$, $-\overset{|}{\underset{|}{C}}-H$, $-CH_2-$, $CH_3-$, $-O-$, $\underset{|}{N}-$

(NH and OH would be detected by IR)

**Q-24** The NMR spectrum of a compound, $C_8H_6O_3$, had an aldehyde peak, a multiple in the aromatic proton region, and a singlet in the 4.0 $\tau$ region. The ratio of peak areas was 1 : 3 : 2, respectively. What is the formula of the **residual** unit(s)? (Hint: The aromatic protons could be due to a benzene or furan ring.)

A-24 If the compound contains a benzene ring, the **residual** is $O_2$ and must contain one unit of unsaturation.

| | | Ω |
|---|---|---|
| $C_8H_6O_3$ | | 6 |
| $-C_6H_3$ | (benzene) | −4 |
| $-CHO$ | (aldehyde) | −1 |
| $-CH_2$ | (4.0 τ singlet) | 0 |
| $O_2$ | (residual unit) | 1 |

If the compound contains a furan ring, the residual unit is $C_2O$ with two sites of unsaturation.

| | | Ω |
|---|---|---|
| $C_8H_6O_3$ | | 6 |
| $-C_4H_3O$ | (furan ring) | −3 |
| $-CHO$ | (aldehyde) | −1 |
| $-CH_2$ | (4.0 τ singlet) | 0 |
| $C_2O$ | (residual unit) | 2 |

Q-25 The NMR spectrum of a compound $C_8H_8O_3$ had an aldehyde peak, a multiple in the aromatic proton region, a broad OH peak and a singlet in the 6 τ region. The ratio of peak areas was 1 : 3 : 1 : 3, respectively. What is the formula of the **residual** unit(s)?

A-25 If the compound contained a benzene ring the **residual** unit would be – O – with no units of unsaturation.

| | | Ω |
|---|---|---|
| $C_8H_8O_3$ | | 5 |
| $-C_6H_3$ | (benzene) | −4 |
| $-CHO$ | (aldehyde) | −1 |
| $-HO$ | (hydroxyl) | 0 |
| $-CH_3$ | (6 τ singlet) | 0 |
| $-O-$ | (residual unit) | 0 |

If the compound contained a furan ring, the **residual** unit would be $-C_2-$ and must have one unit of unsaturation.

| | | Ω |
|---|---|---|
| $C_8H_8O_3$ | | 5 |
| $-C_4H_3O$ | (furan) | −3 |
| $-CHO$ | (aldehyde) | −1 |
| $-HO$ | (hydroxyl) | 0 |
| $-CH_3$ | (6 τ singlet) | 0 |
| $-C_2-$ | (residual unit) | 1 |

**R** The formula (or formula weight) of the **residual** structural units may be found by subtracting the formula (or formula weight) of all known, *unique* structural units from the formula (or molecular weight) of the compound.

S-5 The preceding sections have demonstrated how functional groups and **residual** units in an unknown compound could be determined from spectral data and a knowledge of the units of unsaturation in the molecule. A more detailed consideration of spectral data yields information about the *neighbors* of the various structural units. Thus, larger structural units may be envisioned and the correct sequence of these units in the molecule may be determined.

Q-26 A portion of the NMR spectrum of a molecule is given below. From the splitting pattern and the position of the peaks, what can be determined about the structure and electronegativity of the neighbors of the protons yielding the absorption?

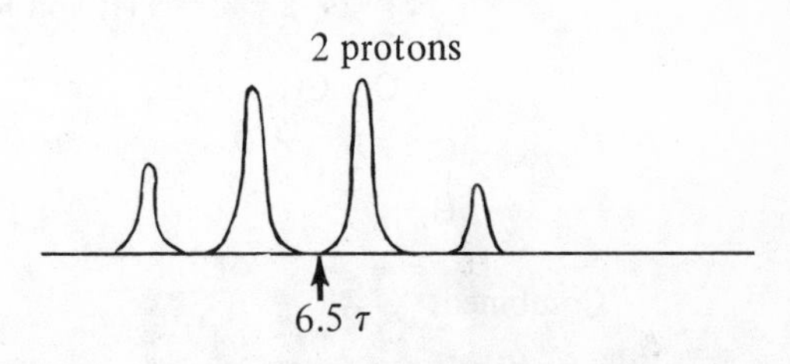


A-26 The two equivalent protons giving the absorption must have three equivalent neighboring protons and at least one neighbor more electronegative than carbon to cause absorption at such a low $\tau$ (i.e., $-CH_2-O$ or $-CH_2-N$).

Q-27 The splitting pattern of a portion of the NMR spectrum for a molecule is given below. The *a* and *b* protons are coupled. Give the structural unit for the *a* protons and the *b* protons separately and then combine them into a single unit. (Hint: The *b* proton is also coupled "long range".)

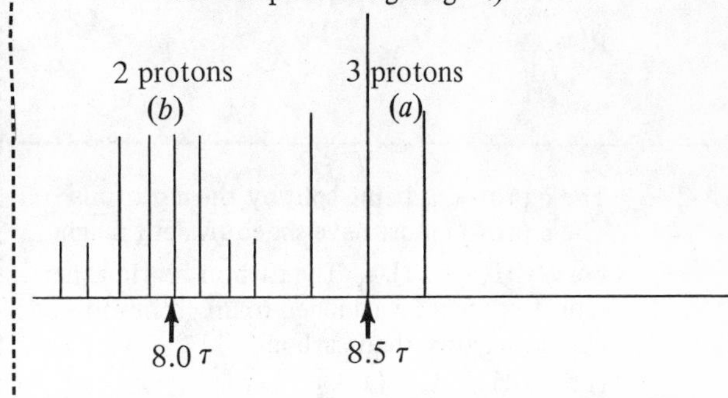

A-27 (*a*) protons (two near neighbors)

$$CH_3\underset{(a)}{-}\overset{H}{\underset{H}{C}}-$$

(*b*) protons (three near neighbors with long-range splitting by one proton)

$$\begin{array}{l} R \\ \quad \backslash C-CH_2-CH_3 \\ \quad\quad\quad\quad (b) \\ H-C// \\ \quad \backslash R' \end{array}$$

or (R and R' $\neq$H)

$$\begin{array}{l} R \\ \quad \backslash C-CH_2-CH_3 \\ \quad\quad\quad\quad (b) \\ R'-C// \\ \quad\ | \\ \quad H \end{array}$$

Combined:

$$\begin{array}{l} R \\ \quad \backslash C-CH_2-CH_3 \\ \quad\quad\quad\quad (b) \quad (a) \\ H-C// \\ \quad | \\ \quad R' \end{array}$$

or

$$\begin{array}{l} R \\ \quad \backslash C-CH_2-CH_3 \\ \quad\quad\quad\quad (b) \quad (a) \\ R'-C// \\ \quad | \\ \quad H \end{array}$$

Q-28 The NMR spectrum of the compound, $C_5H_{10}O_2$, is given below.
What protons are splitting the *c* proton?
Describe the neighbors of the *c* proton.

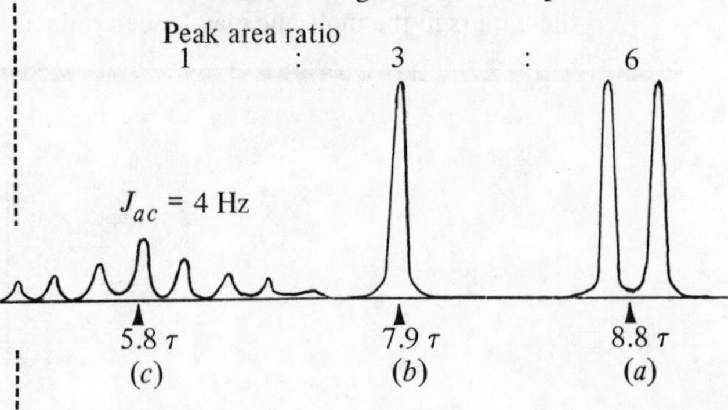


A-28 The *c* proton is being split by the *a* protons. The *c* proton must have six equivalent neighbors $(CH_3)_2\overset{|}{C}-H_{(c)}$. The carbon bearing the *c* proton must be attached to an atom more electronegative than carbon (i.e., $(CH_3)_2\underset{H_{(c)}}{\underset{|}{C}}-O-$).

Q-29 Describe the neighbors of the *a* and *b* protons for the compound given in Q-28.

A-29 The chemical shift of the *b* protons indicates a $CH_3-$ group attached to a C = C or C = O group.

The *a* protons are identical and have one neighbor.

$$\underset{(a)}{CH_3}-\overset{|}{\underset{CH_{3\,(a)}}{\underset{|}{C}}}-H$$

Q-30 How many sites of unsaturation are there in the compound described in Q-28? What is the **residual** group(s)?

A-30 $\Omega = 1$

**Residual** group is oxygen and has one unit of unsaturation.

Q-31 A compound with the formula $C_{10}H_{16}O$ showed a **one** proton multiplet at 4.1 $\tau$, a **three** proton doublet (with very small $J$) at 8.1 $\tau$, a six proton doublet at 9.2 $\tau$ and a **six** proton "haystack" at 7.2 – 8.0 $\tau$. The UV and IR spectra along with the NMR data given above suggested the following structural units.

$-\overset{O}{\overset{\|}{C}}-\overset{H}{\overset{|}{C}}=C\langle$ or $-\overset{O}{\overset{\|}{C}}-\overset{|}{C}=\overset{H}{C}\langle$ (acyclic or six membered ring); $(CH_3)_2-\overset{|}{C}H$; $CH_3-\overset{|}{C}=CH-$.

Which structural units may be combined? Why?

A-31 From the NMR data, only one ethylenic proton can be present in the molecule (peak at 4.1 $\tau$). Therefore, the units,

$\overset{O}{\overset{\backslash\backslash}{}}C-\underset{H}{\underset{|}{C}}=C\langle$ or $\overset{O}{\overset{\|}{C}}-\overset{|}{C}=\overset{H}{C}\langle$ and $CH_3\overset{|}{C}=CH-$

must have some atoms in common and must be combined to give

$\overset{O}{\overset{\|}{C}}-\overset{H}{\overset{|}{C}}=C-CH_3$ or $\overset{O}{\overset{\|}{C}}-\overset{H}{\overset{|}{C}}=\underset{|}{C}-CH_3$

Q-32 Determine the **residual** unit for the molecule described in Q-31. What are possible structures for the **residual** unit(s)?

A-32 **Residual** formula: $C_3H_5$ with one site of unsaturation.

| | | $\Omega$ |
|---|---|---|
| $C_{10}H_{16}O$ | | 3 |
| $-C_4H_4O$ | $(-\overset{O}{\overset{\|}{C}}-CH=\overset{|}{C}CH_3)$ | −2 |
| $-C_3H_7$ | $((CH_3)_2\underset{|}{C}H)$ (9.2 $\tau$ doublet) | 0 |
| $C_3H_5$ | | 1 |

Because there is only one ethylenic proton, the **residual** must be $-CH_2-\overset{H}{\underset{|}{C}}-CH_2-$ or $-CH_2-CH_2-\overset{|}{\underset{|}{C}H}$ or $-\overset{|}{\underset{|}{C}H}-CH_2-$ and $-\overset{|}{\underset{|}{C}H}$, etc., and be part of a saturated ring structure ($-\overset{|}{\underset{|}{C}}-\overset{H}{\underset{H}{C}}-CH_3$ is not probable because of the lack of a three proton triplet in the NMR).

Q-33 A compound, with the formula $C_{10}H_{14}O$, gave the following NMR spectra.

Peak area ratios: 1 : 2 : 5 : 6

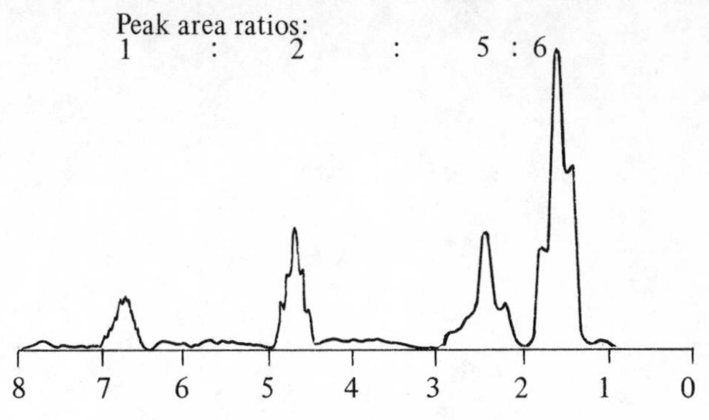


Other spectral evidence indicated the following features:

$\overset{H}{}\!\!\backslash C=\overset{R}{\overset{|}{C}}-\overset{O}{\overset{\|}{C}}-$ (with R on the first C) (acyclic or six membered ring)

and a terminal double bond, $\rangle C=CH_2$. Give structural units and some possible **residual** units.

A-33 $-\overset{O}{\overset{\|}{C}}-\overset{CH_3}{\overset{|}{C}}=C(H)(R)$ and $(CH_3)C=C(H)(H)$

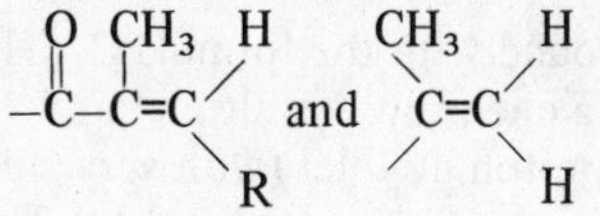

or

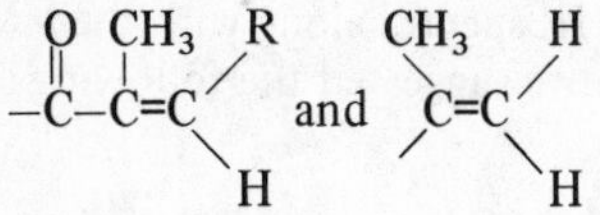

$-\overset{O}{\overset{\|}{C}}-\overset{CH_3}{\overset{|}{C}}=C(R)(H)$ and $(CH_3)C=C(H)(H)$

or

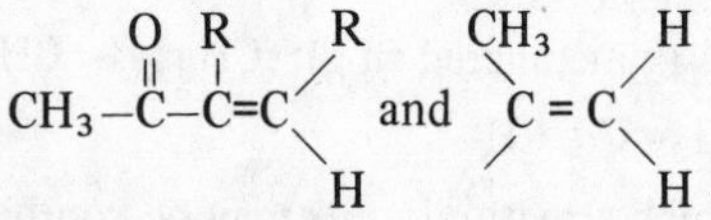

$CH_3-\overset{O}{\overset{\|}{C}}-\overset{R}{\overset{|}{C}}=C(R)(H)$ and $(CH_3)C=C(H)(H)$

or

$CH_3\overset{O}{\overset{\|}{C}}-\overset{CH_3}{\overset{|}{C}}=C(R)(H)$ and $(CH_3)C=C(H)(H)$

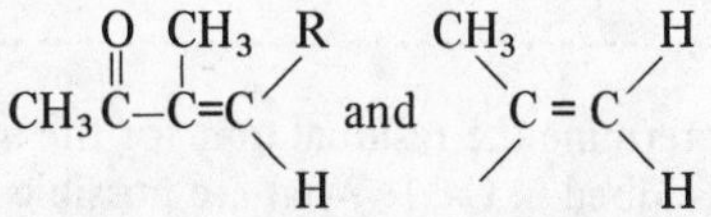

Reasoning: Only three ethylenic protons are present in the molecule. At least one of the $CH_3$ groups (and perhaps both $CH_3$ groups giving the 6 proton peak at 8.2 $\tau$) must have long-range splitting.

**Residual** unit: $C_3H_5$ with one unit of unsaturation. Because none of the hydrogens of the residual unit can be ethylenic, the unsaturation unit must be a ring. Residue could be $-CH_2-\underset{|}{\overset{|}{C}}H-CH_2-$, $-CH_2-\underset{|}{\overset{|}{C}}H-$ and $-CH_2-$, etc.

**R** Detailed analysis of spectra may yield information about neighbors of structural units, allowing larger, more complex structural units to be constructed.

S-6 When constructing structural formulas from structural units, **polyvalent** units are combined first in all possible ways, then **monovalent** groups are attached in all possible ways. Finally, the various possible structures are checked to see if they are consistent with all known information about the compound. An example of the reasoning used in combining structural units follows.

Structural Units: $-C{\equiv}N$; $-\overset{O}{\overset{\|}{C}}-$; $CH_3$–ns; – O – ; ns′– $CH_2$ – $CH_2$ – ns″ (ns = non-splitting group)

Step One: Combine polyvalent units in all possible ways.

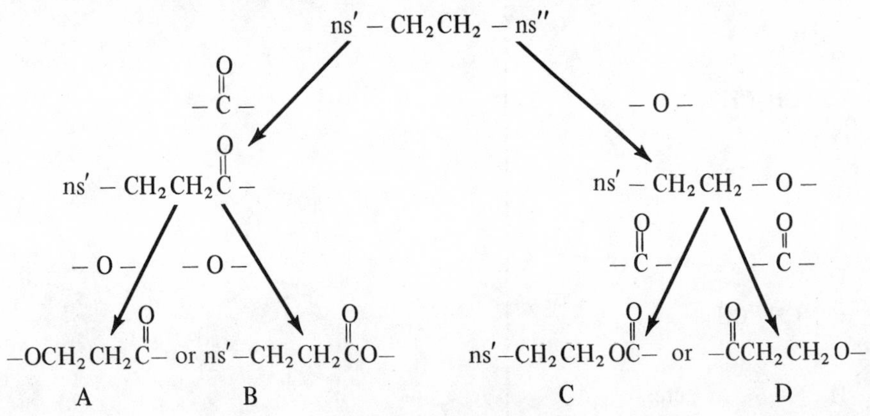


A and D are identical.

Step Two: Combine the units resulting from step one in all possible ways with monovalent units.

A $\xrightarrow[-CN]{CH_3-}$ $CH_3OCH_2CH_2\overset{O}{\overset{\|}{C}}CN$ or $N{\equiv}COCH_2CH_2\overset{O}{\overset{\|}{C}}OCH_3$

$A_1$ $A_2$

B $\xrightarrow[-CN]{CH_3-}$ $N{\equiv}C\,CH_2CH_2\overset{O}{\overset{\|}{C}}{-}OCH_3$ (note that only –CN is a non-splitting group)

$B_1$

C $\xrightarrow[-CN]{CH_3-}$ $N{\equiv}C\,CH_2CH_2O\overset{O}{\overset{\|}{C}}CH_3$

$C_1$

Step Three: Check possible structures with other data. In the preceding case, the compound yielded acetic acid on hydrolysis. The structure $C_1$ is the only structure consistent with this piece of evidence.

Q-34 A compound with the formula $C_{12}H_{14}O_2$ had the following structural units:

$CH_3O-$; $CH_3-CH_2-$ns; $C_6H_5-$ ;

$>C{=}O$; $>C{=}C<^{H}$ .

Give the possible structures for the compound.

A-34

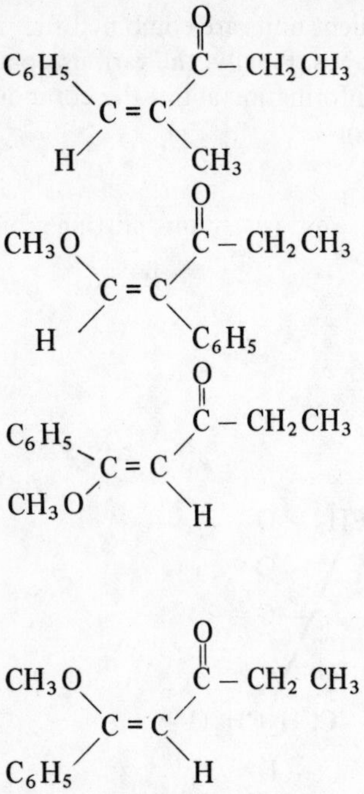

$$C_6H_5(H)C=C(CH_3)-C(=O)-CH_2CH_3$$

$$CH_3O(H)C=C(C_6H_5)-C(=O)-CH_2CH_3$$

$$C_6H_5(CH_3O)C=C(H)-C(=O)-CH_2CH_3$$

$$CH_3O(C_6H_5)C=C(H)-C(=O)-CH_2CH_3$$

, etc.

Reasoning process:

$$>C=C(H)< \; + \; -C(=O)-$$

$$>C=C(H)-C(=O)- \quad \text{or} \quad -C(=O)-C(H)=C<$$

$$>C=C(H)-C(=O)-C_2H_5 \qquad C_2H_5-C(=O)-C(H)=C<$$

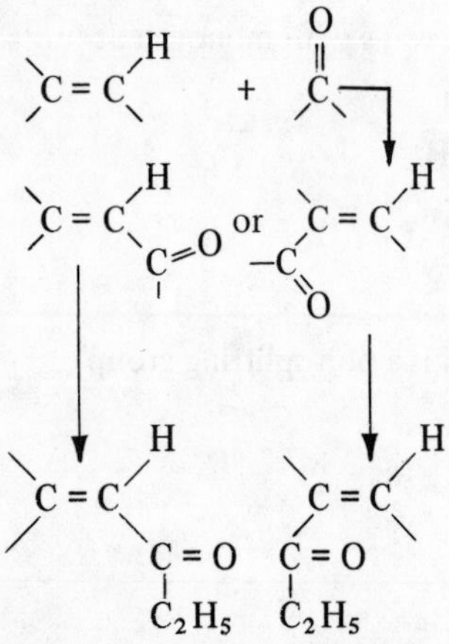

(no long-range splitting of ethyl group)

Further possible combinations of phenyl and methoxyl give structures above.

Q-35 A compound, with the formula $C_6H_{12}O_2$, had the following structural units:

$CH_3$–ns (three identical); $CH_3$–ns; $>C=O$; residual CO with no units of unsaturation. Give all the possible structures.

A-35

$$(CH_3)_3\,C{-}\overset{\overset{\large O}{\|}}{C}{-}O{-}CH_3$$

$$(CH_3)_3\,C{-}O{-}\overset{\overset{\large O}{\|}}{C}{-}CH_3$$

Reasoning process:

Residual unit(s) could be

$-\overset{|}{\underset{|}{C}}-$ and $-O-$ or $-\overset{|}{\underset{|}{C}}-O-$

Combine the multivalent groups:

$-\overset{|}{\underset{|}{C}}-$ , $-O-$ , $-\overset{\overset{O}{\|}}{C}-$

$-\overset{|}{\underset{|}{C}}-O-$ + $-\overset{\overset{O}{\|}}{C}-$ ; $-\overset{\overset{O}{\|}}{C}-\overset{|}{\underset{|}{C}}-$ + $-O-$

$-\overset{\overset{O}{\|}}{C}-\overset{|}{\underset{|}{C}}-O-$ (A) or $-\overset{|}{\underset{|}{C}}-O-\overset{\overset{O}{\|}}{C}-$ (B) or $-O-\overset{\overset{O}{\|}}{C}-\overset{|}{\underset{|}{C}}-$ (C) or $-\overset{\overset{O}{\|}}{C}-\overset{|}{\underset{|}{C}}-O-$ (D)

A and D are identical

Combine with monovalent groups:

A $\xrightarrow[CH_3-]{3\,CH_3-}$ Cannot have three equivalent methyl groups.

B $\xrightarrow[CH_3-]{3\,CH_3-}$ $(CH_3)_3\,CO\overset{\overset{O}{\|}}{C}CH_3$

C $\xrightarrow[CH_3-]{3\,CH_3-}$ $CH_3O\overset{\overset{O}{\|}}{C}C\,(CH_3)_3$

Q-36 A compound, with the formula $C_{10}H_{13}NO$, had the following structural units:

$CH_3CH_2-$ ; $-\overset{|}{N}H$; $-\overset{\overset{O}{\|}}{C}-N\langle$ ; $CH_3$ – ns;

$p$–$C_6H_4\langle$; no **residual** unit.

Hydrolysis of the substance yielded a base soluble, acid insoluble material with a mp of 179-180°. Give all possible structures indicated by the spectral data. Give the structure indicated by spectral and chemical data.

A-36 From spectral data alone, the following structures are indicated:

$CH_3C_6H_4CONHC_2H_5$ A

$C_2H_5C_6H_4CONHCH_3$ B

$CH_3CONHC_6H_4C_2H_5$ C

$C_2H_5CONHC_6H_4CH_3$ D

From spectral and chemical data, the only structure possible is A.

Reasoning process:

multivalent groups combined first

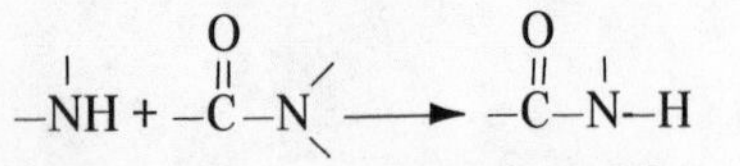

$$-\overset{|}{N}H + -\overset{O}{\overset{\|}{C}}-N\!\!<\; \longrightarrow\; -\overset{O}{\overset{\|}{C}}-\overset{|}{N}-H$$

(only one N atom per molecule)

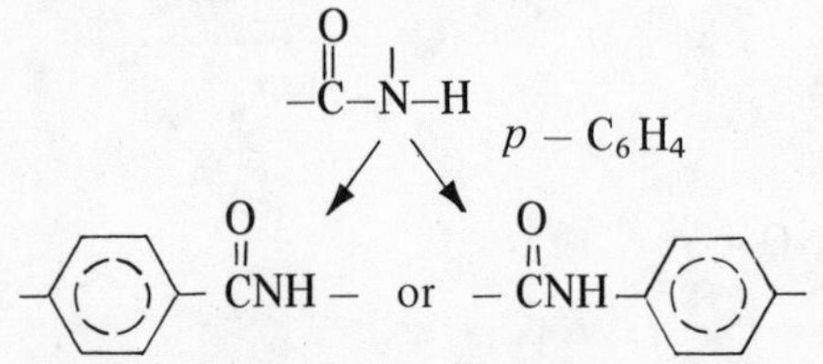

$$-\overset{O}{\overset{\|}{C}}-\overset{|}{N}-H \quad p-C_6H_4$$

$$-C_6H_4-\overset{O}{\overset{\|}{C}}NH- \quad \text{or} \quad -\overset{O}{\overset{\|}{C}}NH-C_6H_4-$$

Further combinations of methyl and ethyl give compounds A–D. On hydrolysis, the substance yielded an acid with a mp of 179-180°. Only A & B would yield solid acids. *p*-Toluic acid melts at 179°. *p*-Ethyl benzoic acid melts at 112°.

Q-37 A compound, with the formula $C_8H_8O$, had the following structural units: $C_6H_5-$;

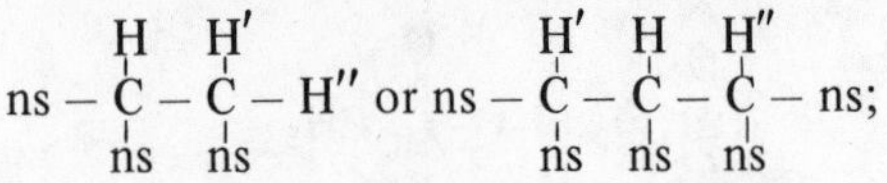

$$ns-\underset{ns}{\overset{H}{\overset{|}{\underset{|}{C}}}}-\underset{ns}{\overset{H'}{\overset{|}{\underset{|}{C}}}}-H'' \text{ or } ns-\underset{ns}{\overset{H'}{\overset{|}{\underset{|}{C}}}}-\underset{ns}{\overset{H}{\overset{|}{\underset{|}{C}}}}-\underset{ns}{\overset{H''}{\overset{|}{\underset{|}{C}}}}-ns;$$

residual unit O with one site of unsaturation. Give all the possible structures. (No aldehydic proton was observed.)

A-37

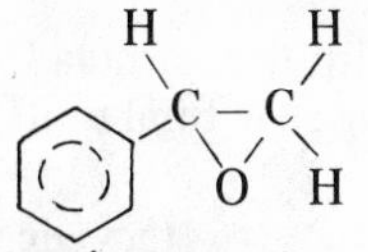

$C_6H_5$–C(H)–C(H)(H), with C–C bridged by O (epoxide)

Reasoning process:

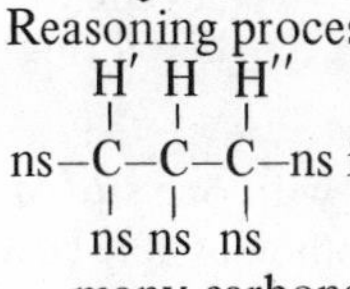

ns–C(H′)(ns)–C(H)(ns)–C(H″)(ns)–ns is not a possible unit (too many carbons).

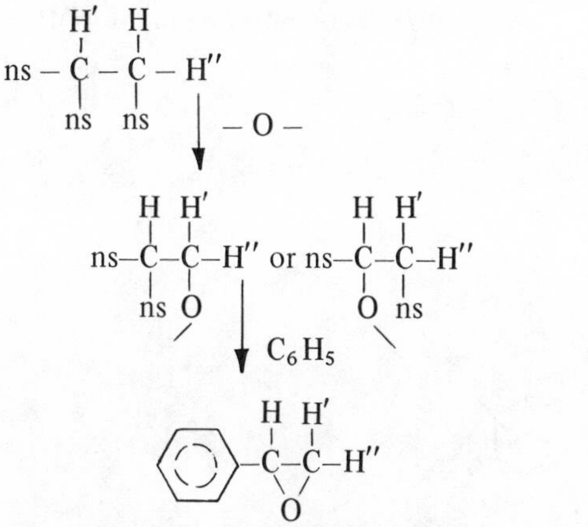

ns – C(H′)(ns) – C(H)(ns) – H″

↓ – O –

ns–C(H)(ns)–C(H′)(O)–H″ or ns–C(H)(O)–C(H′)(ns)–H″

↓ $C_6H_5$

$C_6H_5$–C(H)–C(H′)–H″ (C–C bridged by O)

(H′ and H″ are not identical because $C_6H_5$ – is *cis* to H″ and *trans* to H′.)

Q-38 A compound, with the formula $C_4H_6Cl_2O_2$, had the following structural units:

Cl–; Cl–; – C(=O) –; $CH_3CH_2$–O–; H–C(ns)(ns)–ns

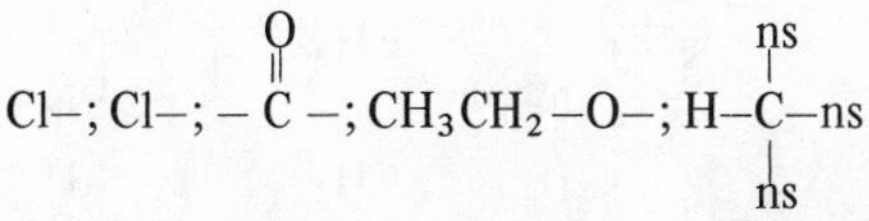

Give possible structures.

A-38

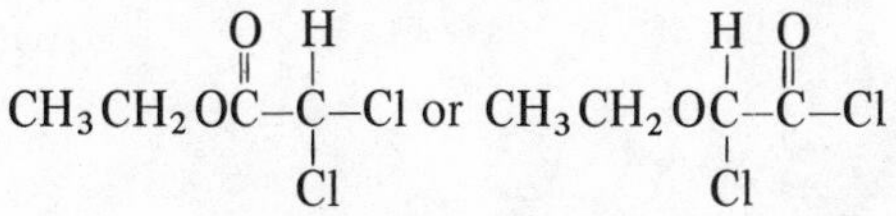

$CH_3CH_2OC(=O)$–C(H)(Cl)–Cl or $CH_3CH_2O$–C(H)(Cl)–C(=O)–Cl

Reasoning process:

H – C(–)(–) –

↓ – C(=O) –

H – C – C(=O) –

↓ 2Cl–, $CH_3CH_2O$ –

H – C(Cl)(Cl) – C(=O) – O – $C_2H_5$ or

H – C(Cl)($OC_2H_5$) – C(=O) – Cl

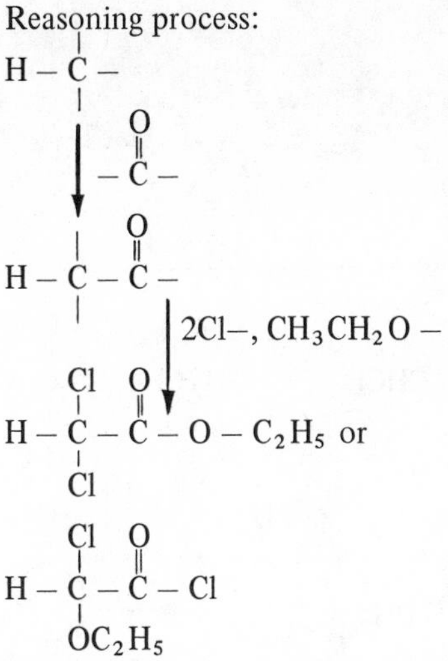

Q-39 A compound, with the formula $C_9H_{10}N_4O_4$, had the following structural units: –$NO_2$; $CH_3$–ns (attached to double bond carbon); $CH_3$–ns (attached to double bond carbon); >C = N–; >N – H; 1, 2, 4- trisubstituted $C_6H_3$; residual unit $NO_2$ with one site of unsaturation. Give some possible structures.

A-39

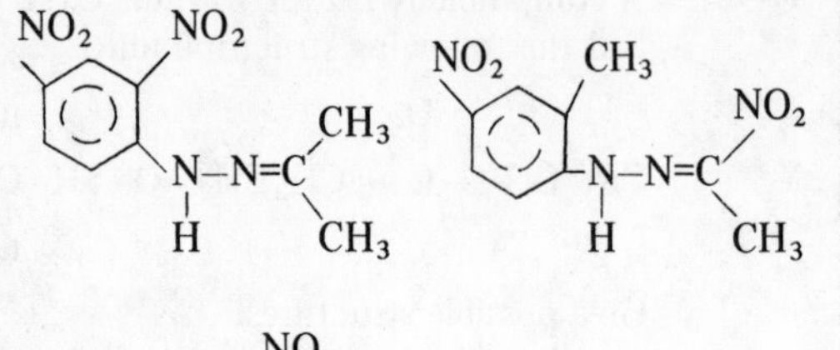

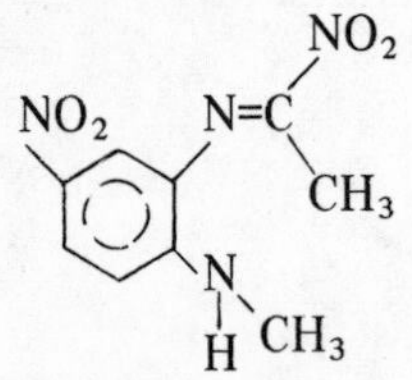

and other combinations.

Reasoning process:

The residue is most likely – $NO_2$, but could also be –N=O, and –O–, etc. Combining the multivalent units in all possible ways gives many structural units. Some are

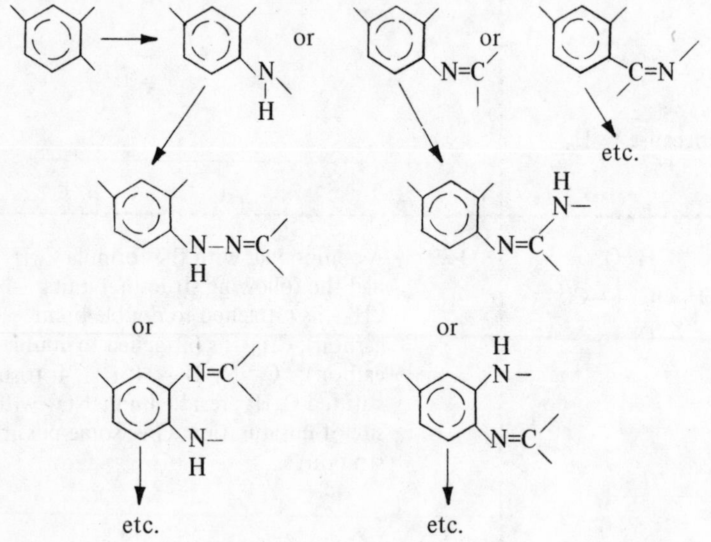

etc.

etc.

Q-40 A compound, with the formula $C_{10}H_{14}O$, had the following structural units:

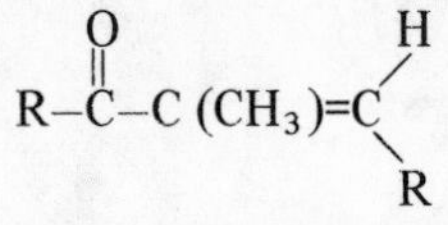

$R{-}\overset{O}{\overset{\|}{C}}{-}C(CH_3){=}C\langle^{H}_{R}$ (acyclic or six membered ring);

$CH_2{=}CCH_3R$; $C_3H_5$ with one site of unsaturation. Give some possible structures for the compound. Only three ethylenic protons were indicated in the NMR spectrum.

A-40

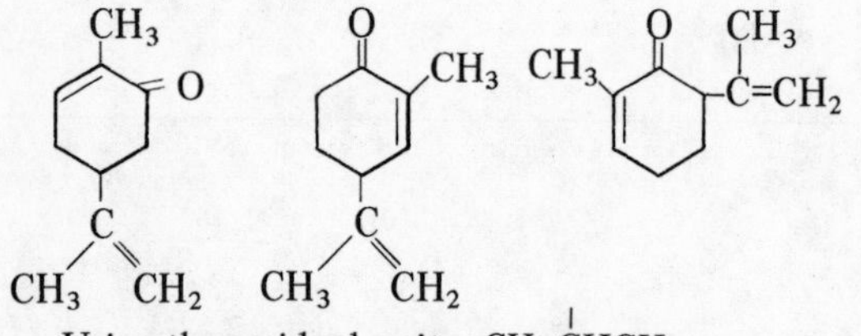

Using the residual unit $-CH_2\overset{|}{C}HCH_2-$ as part of a ring structure the following analysis results:

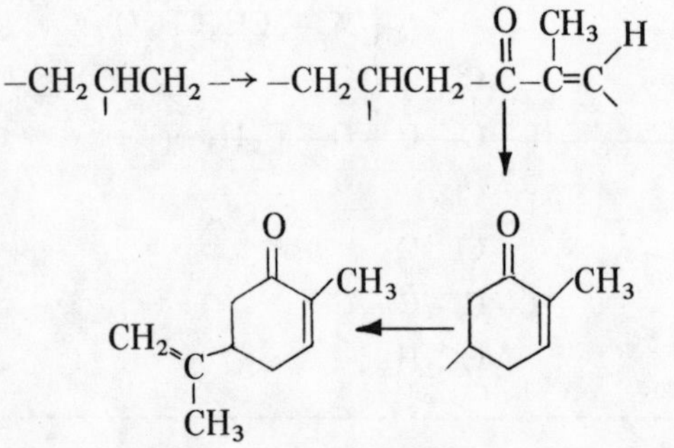

$-CH_2\underset{|}{C}HCH_2- \rightarrow -CH_2\underset{|}{C}HCH_2-\overset{O}{\overset{\|}{C}}-\overset{CH_3}{\overset{|}{C}}{=}C\langle^{H}$

Using the residual unit $\rangle CHCH_2CH_2-$, and a similar analysis to that shown above yields the other two structures.

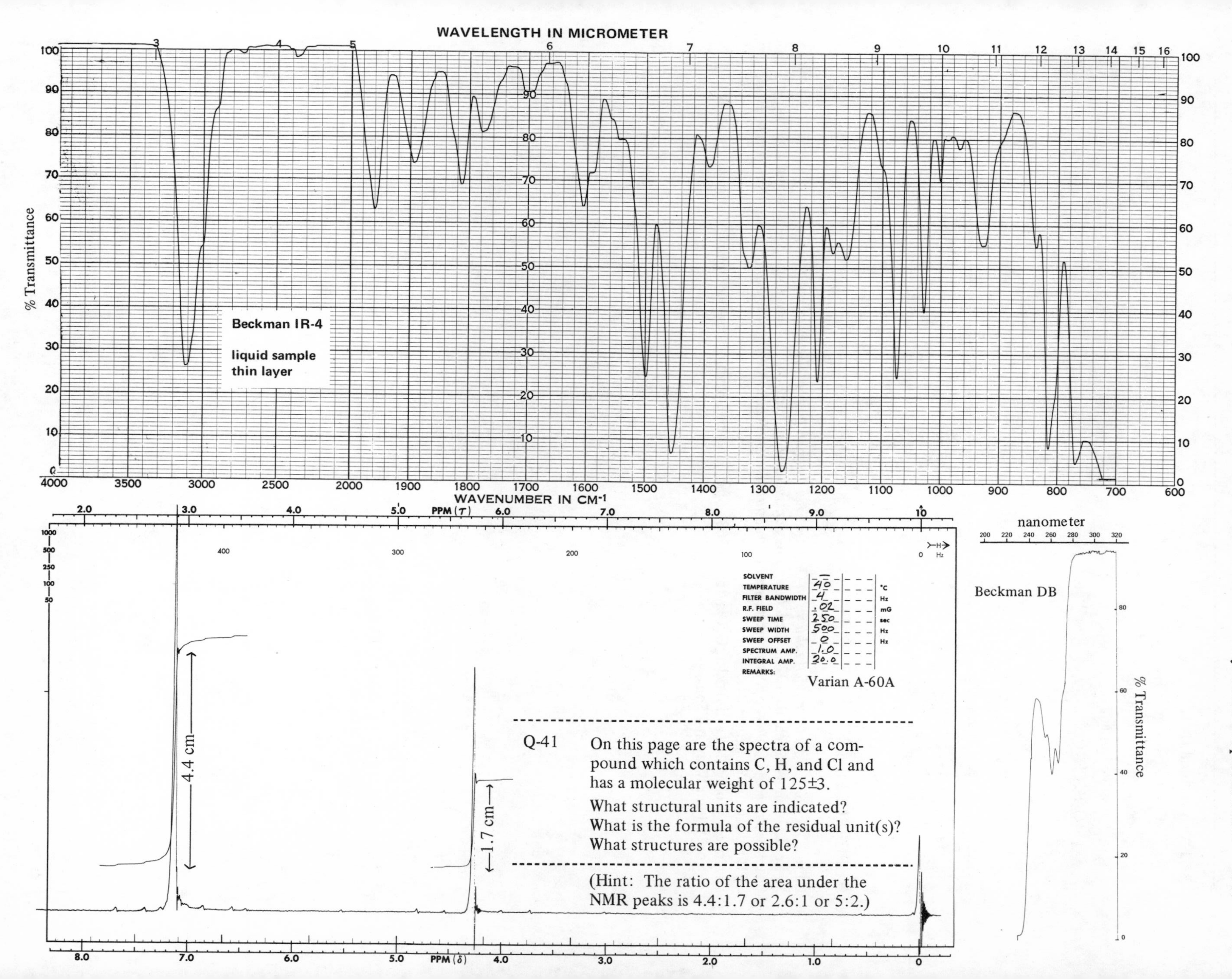


**Q-41** On this page are the spectra of a compound which contains C, H, and Cl and has a molecular weight of 125±3.

What structural units are indicated?
What is the formula of the residual unit(s)?
What structures are possible?

(Hint: The ratio of the area under the NMR peaks is 4.4:1.7 or 2.6:1 or 5:2.)

A-41 **Structural Units:**

Cl–; $C_6H_5$–; ns–$CH_2$–ns

**Residual units:** none.

Possible formula

$C_6H_5CH_2Cl$.

## REFERENCES


Cairns, T. 1964. *Spectroscopic Problems in Organic Chemistry.* London: Heydon and Sons.

Maire, J. C., and Waegell, B. 1971. *Structures, Mechanisms and Spectroscopy.* New York: Gordon and Breach.

Shapiro, Robert H. 1969. *Spectral Exercises in Structural Determination of Organic Compounds.* New York: Holt, Rinehart and Winston.

Silverstein, Robert M., and Bassler, G. Clayton. 1963. *Spectrometric Identification or Organic Compounds.* New York: John Wiley and Sons.

# Chapter 6 PRACTICE PROBLEMS

This chapter contains the IR, UV, NMR spectra and the elemental analysis, molecular weight, or molecular formula of 31 organic compounds. Other chemical data may also be included.

Determine a structure consistent with the data given for each compound.

After the first few problems the answer and a detailed analysis of the data are given. For the latter questions, the answers are given on page 236. The time required to solve successfully each of these problems will vary. Some will require as little as five minutes while others may require thirty or more minutes of diligent work. Do not give up on a problem too soon. It is not necessary to attempt the problems in the order presented.

While each individual is likely to develop his own unique method of determining molecular structure by the interpretation of spectral data, the following step-wise approach is used by many chemists.

Step 1: A cursory examination of all the data to note the presence or absence of various structural features such as, OH, C=O, aromatic protons, sites of unsaturation, etc.

Step 2: A detailed examination of the spectra to determine the kinds of structural features present. (The NMR spectrum is perhaps the best to start with since it allows not only the determination of structural fragments but also the calculation of possible molecular formulas if one is not available.)

Step 3: Prepare a list of all the structural fragments and molecular formulas determined. Elimination or combination of structural fragments by the restrictions of the molecular formula *or*, conversely, elimination of possible molecular formulas through the requirements of the structural fragments may be possible.

Step 4: Determine the residual formula and then combine fragments in all possible ways.

Step 5: Examine the possible structures derived in Step 4 to see if any can be eliminated on the basis of other data that might be available.

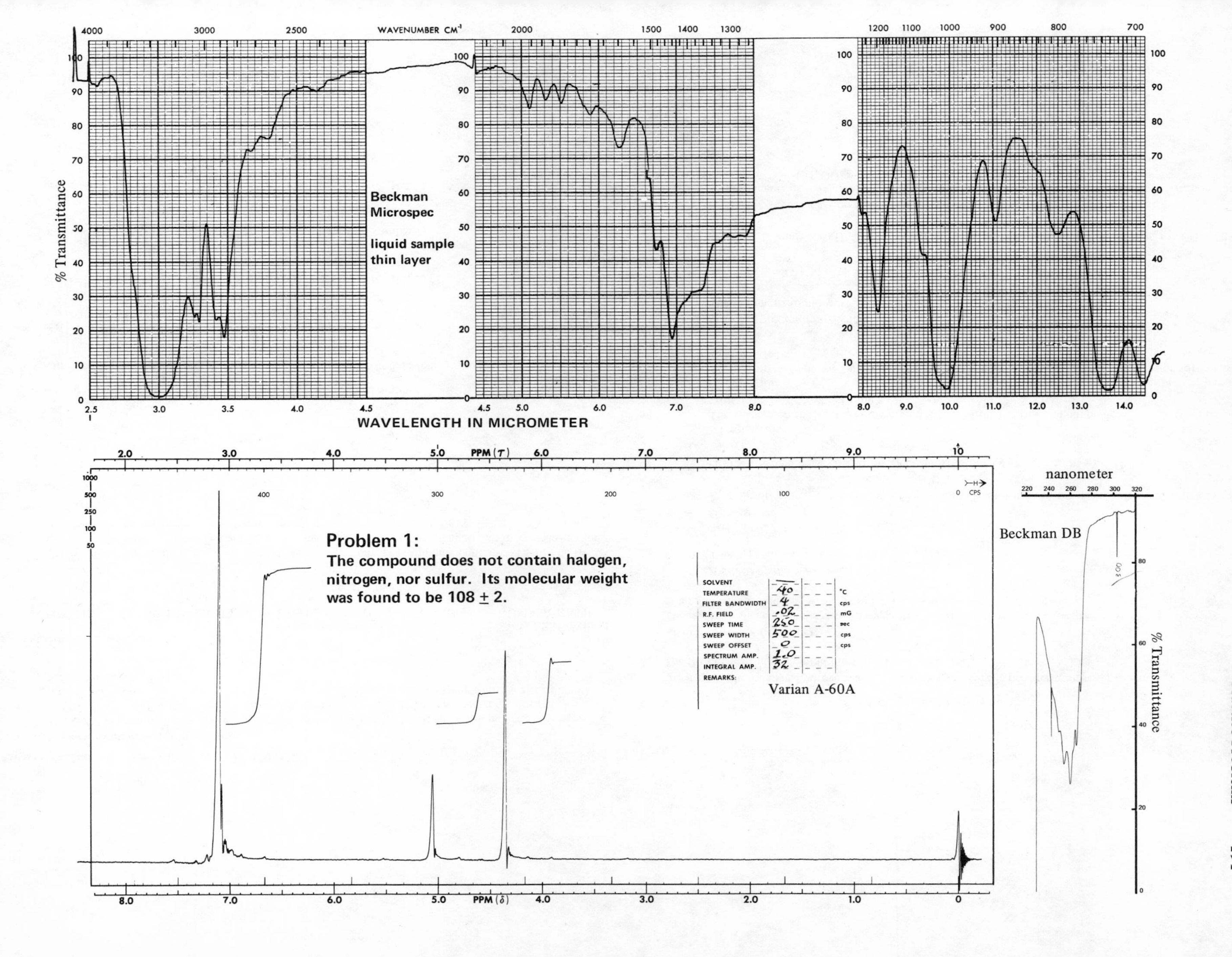
WAVENUMBER CM⁻¹
% Transmittance
Beckman
Microspec
liquid sample
thin layer
WAVELENGTH IN MICROMETER
PPM (τ)
Problem 1:
The compound does not contain halogen, nitrogen, nor sulfur. Its molecular weight was found to be 108 ± 2.
SOLVENT
TEMPERATURE 40 °C
FILTER BANDWIDTH 4 cps
R.F. FIELD .02 mG
SWEEP TIME 250 sec
SWEEP WIDTH 500 cps
SWEEP OFFSET 0 cps
SPECTRUM AMP. 1.0
INTEGRAL AMP. 32
REMARKS:
Varian A-60A
PPM (δ)
nanometer
Beckman DB
% Transmittance

**Answer Problem 1: Spectra are of $C_6H_5CH_2OH$.**

Step 1: *Cursory Examination of the Data:*

The UV spectrum indicates that a conjugated system is present. The IR spectrum indicates the presence of OH, absence of C=O, and the presence of an aromatic system. The NMR spectrum indicates the presence of aromatic protons (7.1 δ) and protons attached to carbons carrying a strong electron withdrawing group(s) (4.3 δ).

Step 2: *Detailed Analysis of Data:*

*NMR Spectrum*

| POSITION | MULTIPLICITY | INTEGRATION CURVE HEIGHT | RATIO | SMALLEST WHOLE NUMBER RATIO |
|---|---|---|---|---|
| 7.1 δ | Singlet | 9.2 cm | 5.1 | 5 |
| 5.1 δ | Singlet | 1.8 cm | 1 | 1 |
| 4.3 δ | Singlet | 3.7 cm | 2.1 | 2 |

The compound contains eight or some multiple of eight hydrogens and may contain oxygen. The IR indicates OH(3400 $cm^{-1}$ band). Calculations of molecular formulas assuming one or more oxygens and multiples of eight hydrogens yield the following:

$$\#C = \frac{108 \pm 2 - 16 - 8}{12} = 7.2 - 6.8 \quad C_7H_8O$$

$$\#C = \frac{108 \pm 2 - 16 - 16}{12} = 6.5 - 6.2 \quad \text{Does not yield a whole number.}$$

$$\#C = \frac{108 \pm 2 - 16 - 24}{12} = 5.8 - 5.5 \quad \text{Does not yield a whole number.}$$

Other multiples of eight hydrogens and one oxygen produce nonexistent formulas.

$$\#C = \frac{108 \pm 2 - 32 - 8}{12} = 5.8 - 5.5 \quad \text{Does not yield a whole number.}$$

$$\#C = \frac{108 \pm 2 - 32 - 16}{12} = 5.2 - 4.8 \quad C_5H_{16}O_2 \text{ nonexistent } (C_nH_{2n+6})$$

$$\#C = \frac{108 \pm 2 - 48 - 8}{12} = 3.7 - 3.3 \quad \text{Does not yield a whole number.}$$

$$\#C = \frac{108 \pm 2 - 48 - 16}{12} = 3.0 - 2.7 \quad C_3H_{16}O_3 \text{ nonexistent } (C_nH_{2n+10})$$

$$\#C = \frac{108 \pm 2 - 64 - 8}{12} = 3.2 - 2.8 \quad C_3H_8O_4$$

$$\#C = \frac{108 \pm 2 - 80 - 8}{12} = 1.8 - 1.5 \quad \text{Does not yield a whole number.}$$

Other multiples of eight hydrogens and more than five oxygens produce nonexistent formulas. The formula of the compound must be $C_7H_8O$ or $C_3H_8O_4$.

*Analysis of the 7.1 δ singlet:*

The position of the singlet is strongly suggestive of aromatic protons. A five proton singlet is suggestive of a monosubstituted ring with an aliphatic carbon attached to the ring; i.e. $C_6H_5-\overset{|}{\underset{|}{C}}-$

*Analysis of the 5.1 δ singlet:*

The OH proton would account for this peak.

*Analysis of the 4.3 δ singlet:*

A two proton singlet is indicative of ns $-\overset{H}{\underset{H}{C}}-$ ns. The position of the peak indicates that one of the ns groups attached must be a strong electron withdrawing group; i.e., ns $-\overset{H}{\underset{H}{C}}-O-$.

*UV Spectrum*

The absorption substantiates the presence of an aromatic ring and eliminates $C_3H_8O_4$ (no sites of unsaturation) as a possible formula.

*IR Spectrum*

Absorption at 3400 $cm^{-1}$ indicates OH and strong absorptions in the 1600 $cm^{-1}$ region further confirms an aromatic ring.

Step 3: *List of Structural Fragments:*

The fragments found were:

ns$-\overset{H}{\underset{H}{C}}-O-$ , $\underline{-OH}$ , $\underline{C_6H_5-\overset{|}{\underset{|}{C}}-}$

Since the formula is $C_7H_8O$, the oxygen atoms of the fragments must be common, ns$-CH_2-OH$.

Step 4: *Residual Formula and Combination of Fragments*

The fragments ns$-CH_2OH$ and $C_6H_5-$ account for all atoms in the formula $C_7H_8O$. The compound must be $C_6H_5-CH_2OH$.

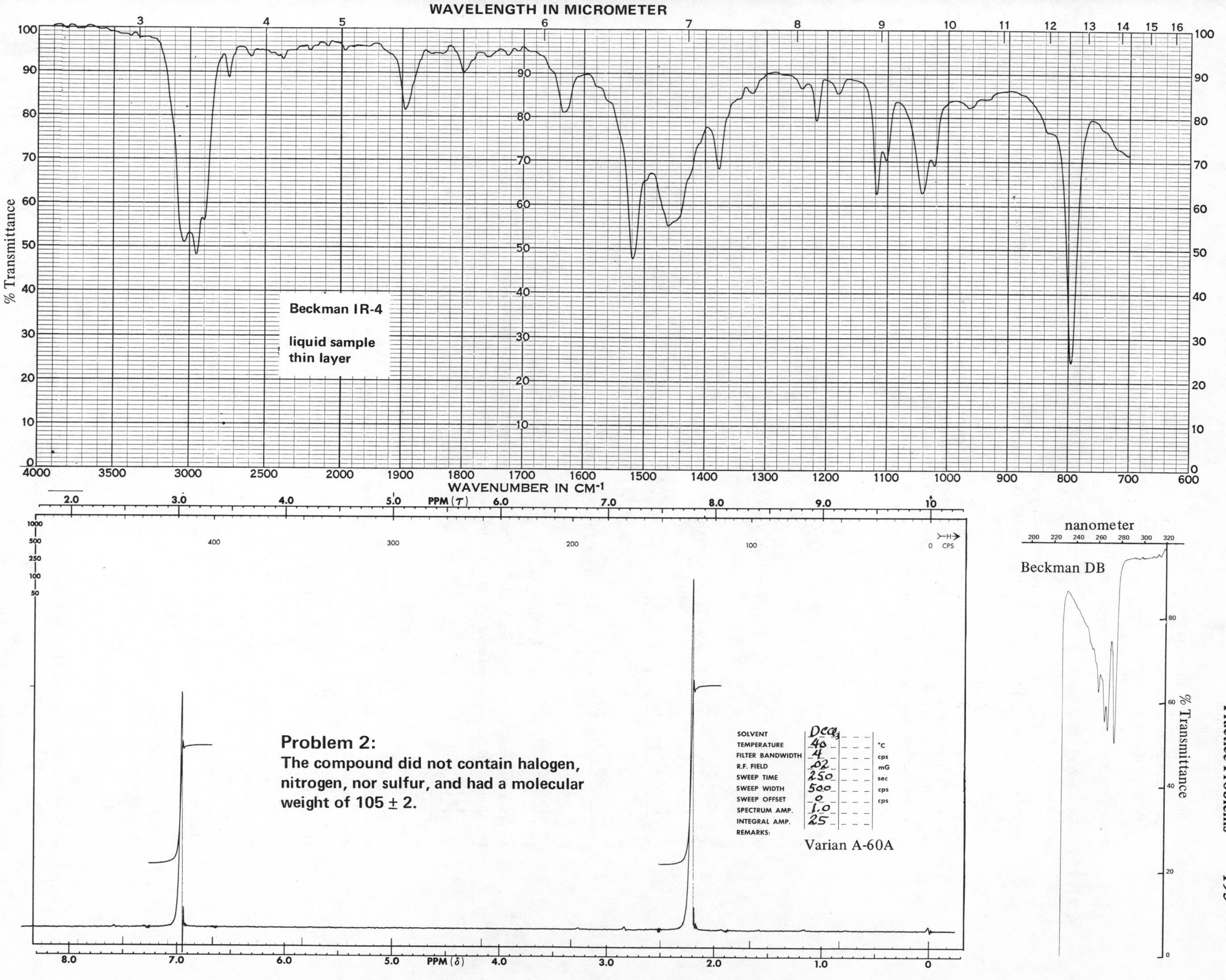
WAVELENGTH IN MICROMETER
% Transmittance
Beckman IR-4
liquid sample
thin layer
WAVENUMBER IN CM-1
PPM (τ)
Problem 2:
The compound did not contain halogen,
nitrogen, nor sulfur, and had a molecular
weight of 105 ± 2.
SOLVENT DCCl3
TEMPERATURE 40 °C
FILTER BANDWIDTH 4 cps
R.F. FIELD .02 mG
SWEEP TIME 250 sec
SWEEP WIDTH 500 cps
SWEEP OFFSET 0 cps
SPECTRUM AMP. 1.0
INTEGRAL AMP. 25
REMARKS:
Varian A-60A
PPM (δ)
nanometer
Beckman DB
% Transmittance

**Answer Problem 2: Spectra are of *p*-xylene.**

Step 1: *Cursory Examination of the Data:*

The UV spectrum indicates a conjugated system. The IR spectrum indicates the lack of OH and C=O and the presence of an aromatic system. The NMR spectrum indicates the presence of aromatic protons and aliphatic protons with non-splitting groups attached.

Step 2: *Detailed Analysis of the Data:*

*NMR Spectrum*

| POSITION | MULTIPLICITY | INTEGRATION CURVE HEIGHT | RATIO | SMALLEST WHOLE NUMBER RATIO |
|---|---|---|---|---|
| 7.0 δ | Singlet | 6.8 cm | 1 | 2 |
| 2.2 δ | Singlet | 10.4 cm | 1.5 | 3 |

Because the compound does not contain halogen or nitrogen, the number of hydrogens must be even. Thus, the hydrogens in the molecule must be a multiple of 10. Calculations of molecular formulas assuming the absence or presence of oxygen (although the IR gives no evidence that oxygen is present) and multiples of ten hydrogens yield the following:

$$\#C = \frac{105 \pm 2 - 10}{12} = 8.1 - 7.8 \quad C_8H_{10}$$

$$\#C = \frac{105 \pm 2 - 20}{12} = 7.3 - 6.9 \quad C_7H_{20} \text{ nonexistent } (C_nH_{2n+6})$$

$$\#C = \frac{105 \pm 2 - 16 - 10}{12} = 6.8 - 6.4 \quad \text{Does not yield a whole number.}$$

$$\#C = \frac{105 \pm 2 - 16 - 20}{12} = 5.9 - 5.6 \quad C_6H_{20}O \text{ nonexistent } (C_nH_{2n+8})$$

$$\#C = \frac{105 \pm 2 - 32 - 10}{12} = 5.4 - 5.1 \quad C_5H_{10}O_2$$

$$\#C = \frac{105 \pm 2 - 32 - 20}{12} = 4.6 - 4.3 \quad \text{Does not yield a whole number.}$$

$$\#C = \frac{105 \pm 2 - 48 - 10}{12} = 4.1 - 3.8 \quad C_4H_{10}O_4$$

All other combinations of oxygen and multiples of ten hydrogens give nonexistent formulas.

The formula of the compound must be one of the following:

$C_8H_{10}$
$C_5H_{10}O_2$
$C_4H_{10}O_4$

*Analysis of the 7.0 δ Singlet:*

The position of the peak is indicative of aromatic protons. The number of protons (4) is indicative of a disubstituted ring. The singlet nature of the peak is indicative of a symmetrical *para* substitution (substituents identical).

*Analysis of the 2.2 δ Singlet:*

The six proton singlet is indicative of $(ns - CH_3)_2$ or $(ns - CH_2 - ns)_3$

*UV Spectrum*

Absorption indicates a conjugated system thus eliminating the formula $C_4H_{10}O_4$ ($\Omega$=O). The UV spectrum supports the NMR evidence that an aromatic ring is present.

*IR Spectrum*

The absorption bands at 1520 and 800 $cm^{-1}$ are further indications that an aromatic ring is present. The substantial evidence that an aromatic ring is present in the molecule (UV, IR, NMR) excludes $C_5H_{10}O_2$ as a possible formula.

Step 3: *List of Structural Fragments:*

The fragments identified are:

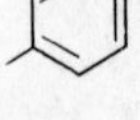

[para-disubstituted benzene ring] and $(ns - CH_3)_2$ or $(ns - CH_2 - ns)_3$

Step 4: *Residual Formula and Combination of Fragments*

Because the formula of the compound is $C_8H_{10}$, the fragments must be [para-disubstituted benzene ring] and $(ns - CH_3)_2$.

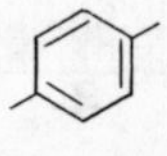

The compound must be $CH_3-C_6H_4-CH_3$.

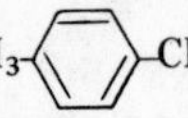

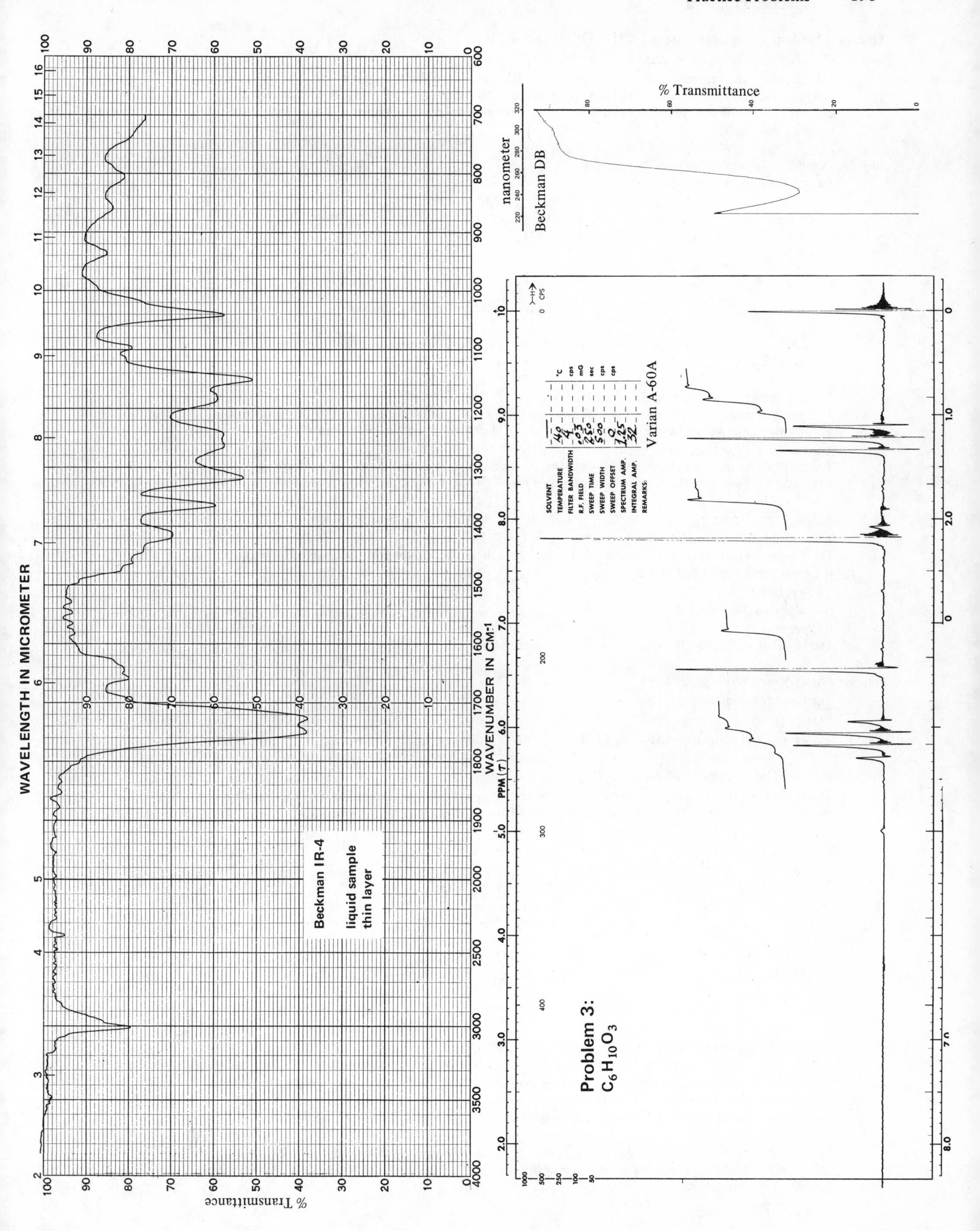
WAVELENGTH IN MICROMETER
WAVENUMBER IN CM-1
% Transmittance
Beckman IR-4
liquid sample
thin layer
nanometer
Beckman DB
% Transmittance
PPM (τ)
SOLVENT
TEMPERATURE
FILTER BANDWIDTH
R.F. FIELD
SWEEP TIME
SWEEP WIDTH
SWEEP OFFSET
SPECTRUM AMP.
INTEGRAL AMP.
REMARKS:
Varian A-60A
Problem 3:
$C_6H_{10}O_3$

**Answer Problem 3: Spectra are of $CH_3COCH_2COOC_2H_5$.**

Step 1: *Cursory Examination of the Data:*

The formula indicates two sites of unsaturation. The UV spectrum indicates conjugation. The IR spectrum indicates the absence of OH but the presence of two different carbonyl groups. The NMR indicates the possible presence of an ethoxyl group (quartet at 4.1 δ and triplet at 1.2 δ with identical coupling constants).

Step 2: *Detailed Analysis of the Data:*

*NMR Spectrum:*

| POSITION | MULTIPLICITY | INTEGRATION CURVE HEIGHT | NUMBER OF PROTONS |
|---|---|---|---|
| 4.1 δ | Quartet | 4.0 cm | 2 |
| 3.5 δ | Singlet | 3.5 cm | 2 |
| 2.2 δ | Singlet | 5.4 cm | 3 |
| 1.2 δ | Triplet | 5.9 cm | 3 |
| | | 18.8 cm/10 protons = 1.9 cm/proton | |

*Analysis of the 4.1 δ Quartet:*
The multiplicity and area under the peaks indicate $-\underline{CH_2}-CH_3$. The position of the quartet indicates $-O-CH_2CH_3$.

*Analysis of the 3.5 δ Singlet:*
The multiplicity and area indicate ns – $CH_2$ – ns.
The position indicates that electron withdrawing groups are attached to the carbon, but the electron withdrawing groups must not be oxygen (oxygen would shift absorption further downfield).

*Analysis of the 2.2 δ Singlet:*
The multiplicity and area indicate $CH_3$ – ns. The position indicates a weak electron withdrawing group attached to the methyl. The position is appropriate for $CH_3\overset{O}{\overset{\|}{C}}-$

*Analysis of the 1.2 δ Triplet:*
The multiplicity, area and position are indicative of $\underline{CH_3}CH_2$ – .
The *J* value is identical to that of the 4.1 δ quartet, thus, these two groups must be coupled. The fragment present must be $CH_3CH_2O-$.

*UV Spectrum:*
A conjugated system is indicated.

*IR Spectrum:*
The bands at 1720 and 1750 cm$^{-1}$ indicate $-\overset{O}{\overset{\|}{C}}-$ and $-\overset{O}{\overset{\|}{C}}-$.

Step 3: *List of Structural Fragments:*

The fragments indicated are:

$-\overset{O}{\overset{\|}{C}}-$, $-\overset{O}{\overset{\|}{C}}-$, $CH_3CH_2O-$, $-CH_2-$, and $CH_3-$

Step 4: *Residual Formula and Combination of Fragments:*

The formula $C_6H_{10}O_3$ indicates two sites of unsaturation, and both are accounted for by the fragments shown above. The above fragments also account for all atoms. Combining the fragments:

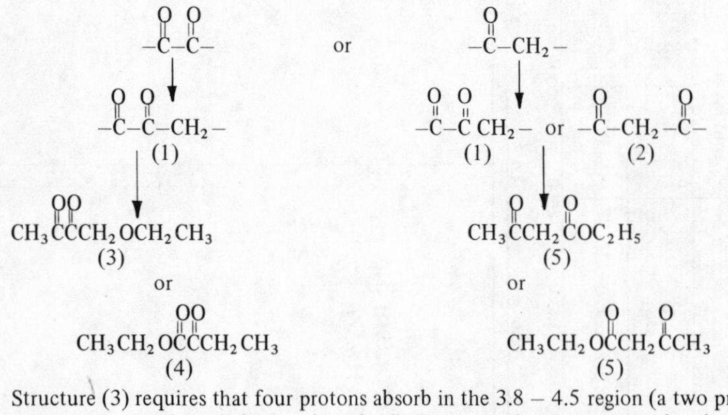

$-\overset{O}{\overset{\|}{C}}-\overset{O}{\overset{\|}{C}}-$ or $-\overset{O}{\overset{\|}{C}}-CH_2-$

↓

$-\overset{O}{\overset{\|}{C}}-\overset{O}{\overset{\|}{C}}-CH_2-$ (1) ; $-\overset{O}{\overset{\|}{C}}-\overset{O}{\overset{\|}{C}}CH_2-$ (1) or $-\overset{O}{\overset{\|}{C}}-CH_2-\overset{O}{\overset{\|}{C}}-$ (2)

↓

$CH_3\overset{O}{\overset{\|}{C}}\overset{O}{\overset{\|}{C}}CH_2OCH_2CH_3$ (3) ; $CH_3\overset{O}{\overset{\|}{C}}CH_2\overset{O}{\overset{\|}{C}}OC_2H_5$ (5)

or ; or

$CH_3CH_2O\overset{O}{\overset{\|}{C}}\overset{O}{\overset{\|}{C}}CH_2CH_3$ (4) ; $CH_3CH_2O\overset{O}{\overset{\|}{C}}CH_2\overset{O}{\overset{\|}{C}}CH_3$ (5)

Structure (3) requires that four protons absorb in the 3.8 – 4.5 region (a two proton quartet at 4.2 δ and a two proton singlet still further downfield). Because only two protons absorb in this region, structure (3) may be rejected. Structure (4) requires a two proton quartet in the 4.2 region ($CH_3\underline{CH_2}O-$) and a two proton quartet in the 2.0 δ region ($\overset{O}{\overset{\|}{C}}\underline{CH_2}CH_3$). Because only a two proton singlet is observed in the 2.0 δ region, structure (4) may be rejected. Only structure (5) is consistent with all the data. (Enolization accounts for conjugation.)

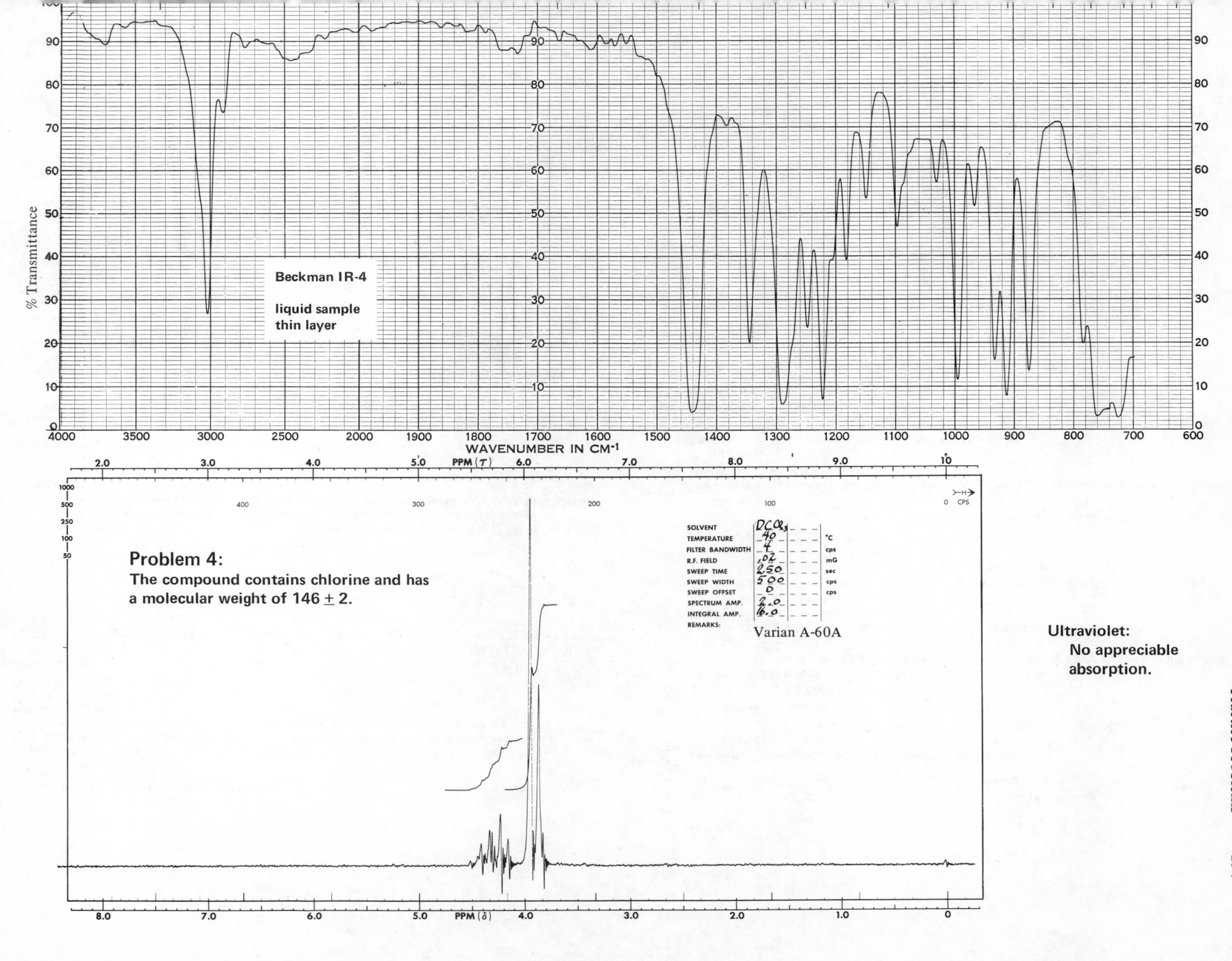
% Transmittance
Beckman IR-4
liquid sample
thin layer
WAVENUMBER IN CM-1
PPM (τ)
0 CPS
SOLVENT DCCl3
TEMPERATURE 40 °C
FILTER BANDWIDTH 4 cps
R.F. FIELD .02 mG
SWEEP TIME 250 sec
SWEEP WIDTH 500 cps
SWEEP OFFSET 0 cps
SPECTRUM AMP. 2.0
INTEGRAL AMP. 16.0
REMARKS:
Varian A-60A
PPM (δ)
Problem 4:
The compound contains chlorine and has
a molecular weight of 146 ± 2.
Ultraviolet:
No appreciable
absorption.

**Answer Problem 4: Spectra are of $ClCH_2CHClCH_2Cl$.**

Step 1: *Cursory Examination of the Data:*

The UV indicates that no conjugated system is present. The IR indicates no OH, C=O, aliphatic, or aromatic double bond stretch. The NMR indicates that no aromatic protons are present and that all protons are either ethylenic (inconsistent with IR) or attached to carbons carrying a strong electron withdrawing group.

Step 2: *Detailed Analysis of the Data:*

*NMR Spectrum:*

| POSITION | MULTIPLICITY | INTEGRATION CURVE HEIGHT | RATIO | SMALLEST WHOLE NUMBER RATIO |
|---|---|---|---|---|
| 4.3 δ | Multiplet | 2.8 cm | 1 | 2 |
| 3.9 δ | Singlet | 6.8 cm | 2.4 | 5 |
| 3.85 δ | Singlet | 4.1 cm | 1.5 | 3 |
| | | OR | | |
| 4.3 δ | Multiplet | 2.8 cm | 1 | 1 |
| 3.9 δ | Doublet | 10.9 cm | 3.9 | 4 |

If there are three sets of NMR peaks of ratio 2:5:3, then there must be an even number of halogens, i.e., $Cl_2$, $Cl_4$, etc. Because of the molecular weight restriction only $Cl_2$ is possible. The calculation of molecular formulas assuming two chlorines and a multiple of ten hydrogens yields only nonstoichiometric formulas, i.e.,

$$\#C = \frac{146 \pm 2 - 71 - 10}{12} = 5.6 - 5.3$$ Does not yield a whole number.

$$\#C = \frac{146 \pm 2 - 71 - 20}{12} = 4.8 - 4.4$$ Does not yield a whole number.

If there are two sets of NMR peaks of ratio 1:4 and the molecule has odd multiples of five hydrogens, then there must be an odd number of halogens, i.e., Cl, $Cl_3$, $Cl_5$, etc. Because of the molecular weight restriction only Cl and $Cl_3$ are possible. The calculation of molecular formulas assuming one or three chlorines and odd multiples of five hydrogens yields the following:

$$\#C = \frac{146 \pm 2 - 35.5 - 5}{12} = 9.0 - 8.6 \quad C_9H_5Cl$$

$$\#C = \frac{146 \pm 2 - 35.5 - 15}{12} = 8.1 - 7.8 \quad C_8H_{15}Cl$$

$$\#C = \frac{146 \pm 2 - 35.5 - 25}{12} = 7.3 - 7.0 \quad C_7H_{25}Cl \text{ nonexistent } (C_nH_{2n+11})$$

$$\#C = \frac{146 \pm 2 - 106.5 - 5}{12} = 3.0 - 2.7 \quad C_3H_5Cl_3$$

$$\#C = \frac{146 \pm 2 - 106.5 - 15}{12} = 2.2 - 1.9 \quad C_2H_{15}Cl_3 \text{ nonexistent } (C_nH_{2n+11})$$

One of the following must be the formula of the compound:

$C_9H_5Cl$ $\quad \Omega = 7$
$C_8H_{15}Cl$ $\quad \Omega = 1$
$C_3H_5Cl_3$ $\quad \Omega = 0$

*Analysis of the 3.9 δ Doublet:*
The four (or 12) proton doublet (3.9 δ) is indicative of $(ns - \underline{CH_2} - \overset{|}{\underset{|}{C}}H-)_{2 \text{ or } 6}$.
The position of the doublet (3.9 δ) suggests an electron withdrawing group attached to the carbon $(Cl - \underline{CH_2} - CH-)_{2 \text{ or } 6}$.
Because six chlorines in the molecule are not possible (mole wt), only two fragments, $(ClCH_2CH-)_2$, are possible.
*Analysis of the 4.3 δ Multiplet:*
The one (or 3) proton multiplet is indicative of $(sp - \overset{H}{\underset{|}{\overset{|}{C}}} - sp)_{1 \text{ or } 3}$ (sp = splitting group).
The position of the multiplet suggests an electron withdrawing group attached to the carbon
$(sp - \overset{H}{\underset{Cl}{C}} - sp)_{1 \text{ or } 3}$.

*UV Spectrum:*
The lack of any significant absorption eliminates the possibility of conjugated systems. This strongly suggests that $C_9H_5Cl(\Omega{=}7)$ is not the molecular formula.
*IR Spectrum:*
The lack of aliphatic and aromatic unsaturation is significant.

Step 3: *List of Structural Fragments:*
The fragments found were:

$$\underset{A}{\underline{ClCH_2}}\underset{H}{C}- + \underset{A}{\underline{ClCH_2}}\underset{H}{C}- + \text{one or three } \underset{B}{\underline{sp - \overset{Cl}{\underset{H}{C}} - sp}}$$

These fragments dictate that at least three chlorines must be present in the molecule. Therefore, the only formula possible is $C_3H_5Cl_3$. Thus, the $-\underset{H}{C}-$ group of each "A" fragment must be common to fragment B.

Step 4: *Residual Formula and Combination of Fragments:*

$$\underset{A}{\underline{ClCH_2}} - \underset{B}{\underline{\overset{H}{\underset{Cl}{C}}}} - \underset{A}{\underline{CH_2Cl}}$$

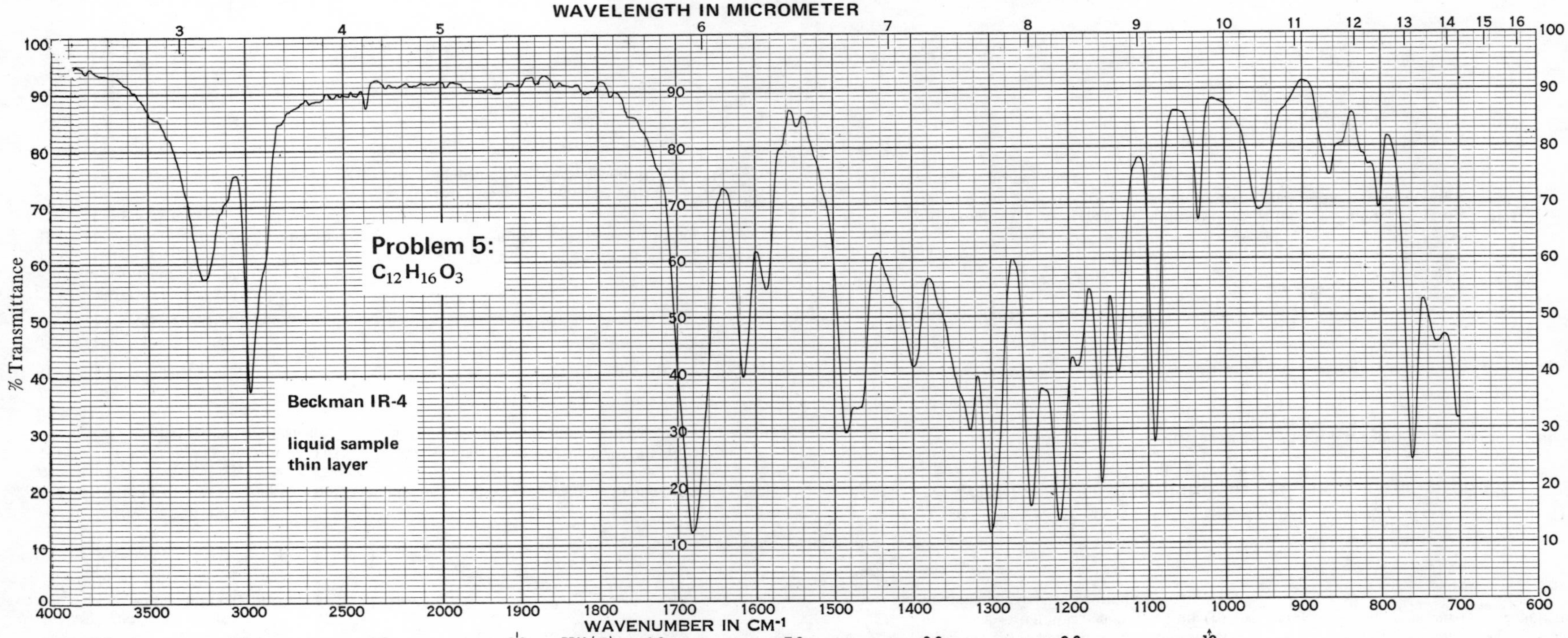
WAVELENGTH IN MICROMETER
Problem 5:
$C_{12}H_{16}O_3$
Beckman IR-4
liquid sample
thin layer
% Transmittance
WAVENUMBER IN CM$^{-1}$

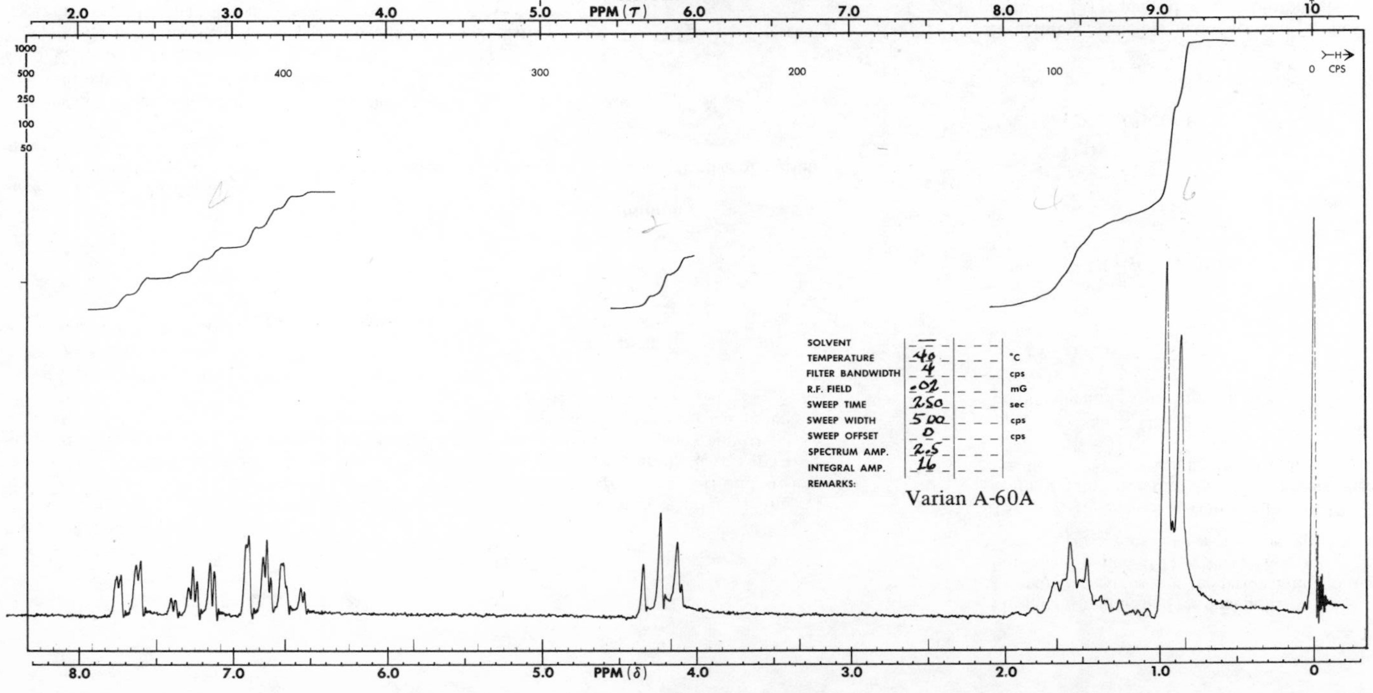
PPM (τ)
PPM (δ)
0 CPS
SOLVENT
TEMPERATURE 40 °C
FILTER BANDWIDTH 4 cps
R.F. FIELD .02 mG
SWEEP TIME 250 sec
SWEEP WIDTH 500 cps
SWEEP OFFSET 0 cps
SPECTRUM AMP. 2.5
INTEGRAL AMP. 16
REMARKS:
Varian A-60A

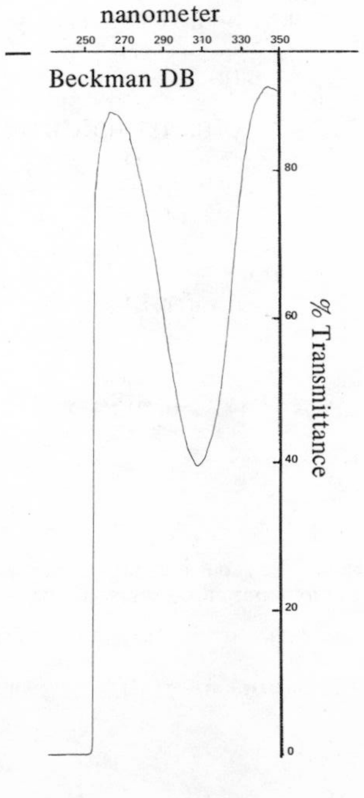
nanometer
Beckman DB
% Transmittance

**Answer Problem 5: Spectra are of** (o-hydroxyphenyl) $C(=O)-OCH_2CH_2CH(CH_3)_2$

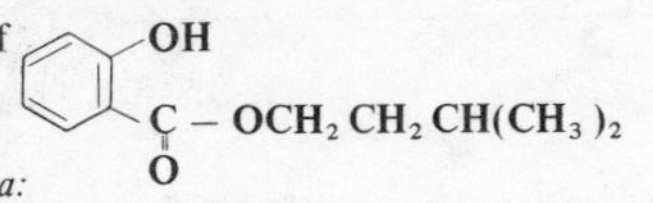

Step 1: *Cursory Examination of the Data:*

The formula $C_{12}H_{16}O_3$ indicates five sites of unsaturation (an aromatic ring may be present). The absorption in the UV indicates a conjugated system. The IR indicates the presence of OH, and the low frequency (3200 $cm^{-1}$) is suggestive of a phenol. The IR is also indicative of an aromatic ring and a conjugated carbonyl. The NMR indicates the presence of aromatic protons (7 δ region), aliphatic protons attached to carbon carrying a strong electron withdrawing group (4.1 δ region), and other aliphatic protons.

Step 2: *Detailed Analysis of the Data:*

*NMR Spectrum:*

| POSITION | MULTIPLICITY | INTEGRATION CURVE HEIGHT | NUMBER OF PROTONS |
|---|---|---|---|
| 7.1 δ | Multiplet | 4.7 cm | 4 |
| 4.2 δ | Triplet | 2.1 cm | 2 |
| 1.5 δ | Multiplet | 3.9 cm | 4 |
| 0.9 δ | Doublet | 6.7 cm | 6 |
| | | 17.4 cm/16 protons = 1.1 cm/proton | |

*Analysis of the 7.1 δ Multiplet:*
The position of the peak is indicative of an aromatic ring. The four proton area is indicative of a disubstituted ring.

*Analysis of the 4.2 δ Triplet:*
The position of the two proton peak is indicative of $-O-\underline{CH_2}-$.
The multiplicity of the peak is indicative of $-O-\underline{CH_2}-CH_2-$.

*Analysis of the 1.5 δ Multiplet:*
The position and multiplicity of the peak are indicative of aliphatic protons. The OH proton (presence confirmed from IR) may also absorb in this region.

*Analysis of the 0.9 δ Doublet:*
The position and multiplicity of the peak is indicative of

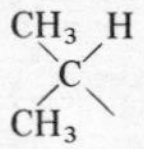

$(CH_3)_2C-H$ or ns $-CH_3$ and ns′ $-CH_3$

*UV Spectrum:*
The absorption indicates a conjugated system and is consistent for the spectrum of an aromatic system.

*IR Spectrum:*
The peak at 3200 $cm^{-1}$ is indicative of OH and its low frequency suggests a phenol. The presence of a conjugated carbonyl (1680 $cm^{-1}$), an aromatic system (1620 and 1580 $cm^{-1}$), and an *ortho* disubstituted benzene ring (760 $cm^{-1}$) are all indicated.

Step 3: *List of Structural Fragments:*

The fragments indicated are:

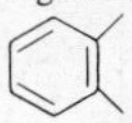

$C_6H_4$ (ortho-disubstituted ring), $-C(=O)-$ (conj), OH (possible phenol)

$O-\underline{CH_2CH_2}-$ and $\underline{(CH_3)_2CH}-$ or $CH_3$–ns plus $CH_3$–ns′.

Step 4: *Residual Formula and Combination of Fragments:*

The formula $C_{12}H_{16}O_3$ indicates five sites of unsaturation. The fragments above account for all five sites. The residual units are found by subtracting the sum of the known fragments from the formula.

SUM OF FRAGMENTS:

$C_6H_4$
C O
H O*
$C_2H_4$ O*
$C_3H_7$ or $C_2H_6$

$C_{12}H_{16}O_3$(A) or $C_{11}H_{15}O_3$(C)
or or
$C_{12}H_{16}O_2$(B) $C_{11}H_{15}O_2$(D)

* These oxygens may be common, i.e., OH and $-OCH_2CH_2-$ may be combined to give the fragment $HOCH_2CH_2-$

Formula A requires no residual unit.
Formula B requires O as the residual unit, and assumes $HOCH_2CH_2-$ is a fragment.
Formula C requires CH as the residual unit.
Formula D requires CHO as the residual unit and assumes $HOCH_2CH_2-$ is a fragment.

The low frequency of the OH band in the IR (3200 $cm^{-1}$) is a strong indication of phenolic OH. If the assumption is made that the OH is, indeed, phenolic, then formula B and D can be rejected and a new fragment $C_6H_4OH$ can be postulated.

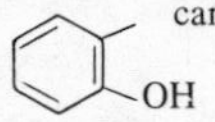

Combining units using formula A:

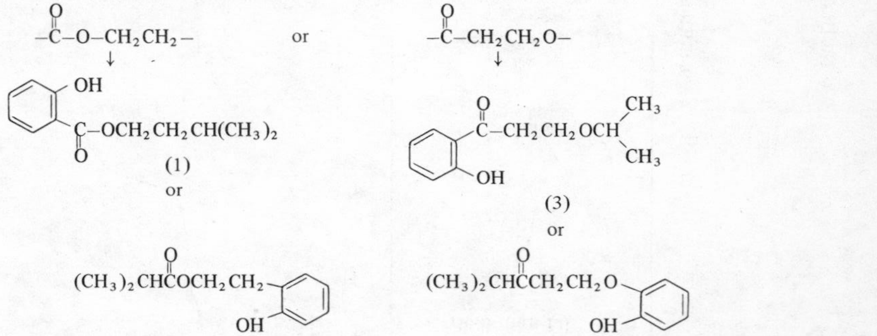

$-C(=O)-O-CH_2CH_2-$ or $-C(=O)-CH_2CH_2O-$

(1) o-HOC$_6$H$_4$–$C(=O)-OCH_2CH_2CH(CH_3)_2$
or
(2) $(CH_3)_2CHC(=O)OCH_2CH_2$–C$_6$H$_4$OH

(3) o-HOC$_6$H$_4$–$C(=O)-CH_2CH_2OCH(CH_3)_2$
or
(4) $(CH_3)_2CHC(=O)CH_2CH_2O$–C$_6$H$_4$OH

Structures (2) and (4) do not account for the conjugated carbonyl (IR spectrum) and can be rejected. Structure (3) requires that three protons absorb in the 3.8 – 4.5 region of the NMR. Because only two protons absorb in this region, structure (3) may be rejected. Only structure (1) accounts for all the data.

Combining units using formula C:

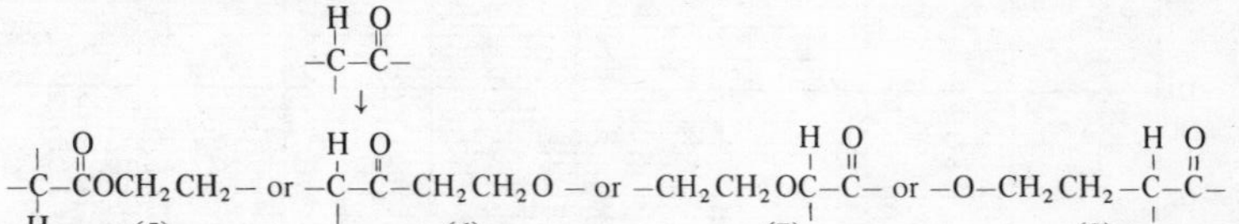

$-CH-C(=O)-$

$-CH(-C(=O)-)OCH_2CH_2-$ (5) or $-CH-C(=O)-CH_2CH_2O-$ (6) or $-CH_2CH_2OC(=O)-CH-$ (7) or $-O-CH_2CH_2-CH-C(=O)-$ (8)

Structures (5) and (6) may be rejected because neither will account for a conjugated carbonyl. Structure (7) may be rejected because the remaining fragments (ns–$CH_3$, ns′ $-CH_3$, and $C_6H_4OH$) cannot be accommodated.

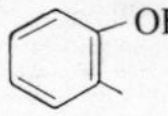

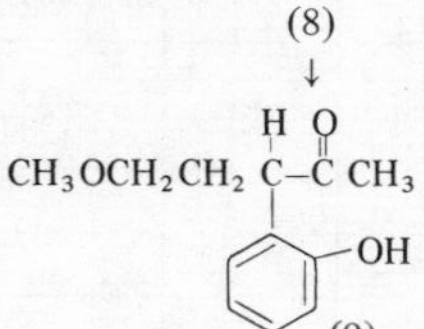

(8) →

$CH_3OCH_2CH_2CH(C_6H_4OH)-C(=O)CH_3$
(9)

Structure (9) requires that five protons absorb in the 3.8 – 4.5 region of the NMR ($\underline{CH_3}-O-\underline{CH_2}-$). Because the NMR spectrum shows only two protons in this region, structure (9) may be rejected. The only structure accounting for all the data is

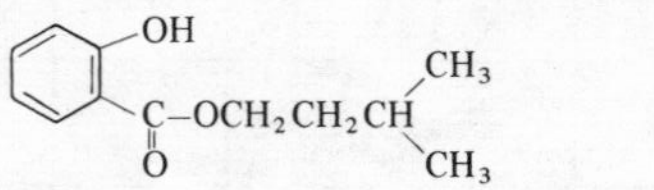

o-HOC$_6$H$_4$–$C(=O)-OCH_2CH_2CH(CH_3)_2$

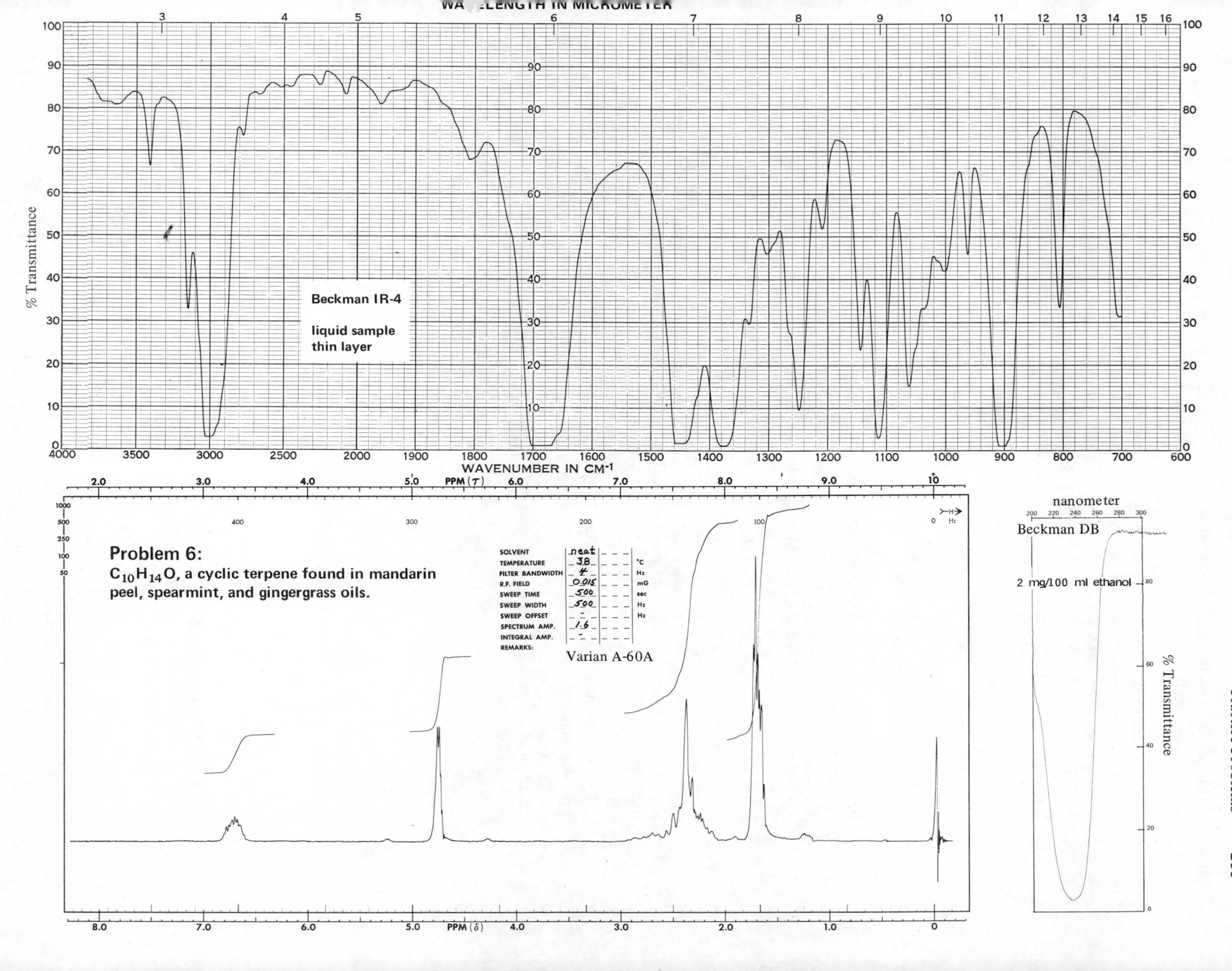


**Problem 6:**
$C_{10}H_{14}O$, a cyclic terpene found in mandarin peel, spearmint, and gingergrass oils.

## Answer Problem 6: Spectra are of *l*-carvone.

Step 1: *Cursory Examination of the Data:*

The IR spectrum indicates a conjugated carbonyl and a geminal disubstituted alkene. The NMR indicates ethylenic hydrogens. The UV indicates a conjugated system. Either a diene system or an $\alpha\beta$-unsaturated ketone must be present. Six membered rings are common cyclic terpene structures.

Step 2: *Detailed Analysis of the Data:*

*NMR Spectrum:*

| POSITION | MULTIPLICITY | INTEGRATION CURVE HEIGHT | PROTONS |
|---|---|---|---|
| 6.8 δ | Multiplet | 2.6 cm | 1 |
| 4.8 δ | Multiplet | 5.2 cm | 2 |
| 2.5 δ | Multiplet | 14.1 cm | 5 |
| 1.8 δ | Multiplet | 16.0 cm | 6 |
| | | 37.9 cm/14 protons = 2.7 cm/proton | |

The formula $C_{10}H_{14}O$ indicates four sites of unsaturation.

*Analysis of 6.8 δ* Multiplet:

The position, area and multiplicity of the peak indicate $\mathrm{>C{=}C<^{H}_{ns}}$ with long-range splitting. The low field position of the proton must be explained by an anisotropic or an inductive effect.

*Analysis of 4.8 δ Multiplet:*

The position, area and multiplicity of this peak indicate

$\left(\mathrm{>C{=}C<^{H}_{ns}}\right)_2$ or, more likely, $\mathrm{>C{=}CH_2}$.

*Analysis of 2.5 δ and 1.8 δ Multiplets:*

The fact that no peaks appear below 1.5 δ is indicative that all methyl groups, if any are present, must be attached to an electron withdrawing group such as C=C or C=O. The six proton multiplet at 1.8 δ is strongly suggestive of

$\mathrm{CH_3}$ >C=CH (*cis* or *trans*) , $\mathrm{CH_3}$ >C=CH (*cis* or *trans*)

with long-range splitting.

*UV Spectrum:*

The absorption maxima of 238 nm is indicative of

H(R)C=C(R)–C(=O)–

| | |
|---|---|
| Parent system | 215 *nm* |
| α substituent | 10 *nm* |
| β substituent | 12 *nm* |
| | 237 *nm* |

Step 3: *List of Structural Fragments:*

The fragments which are strongly indicated from the spectral data are:

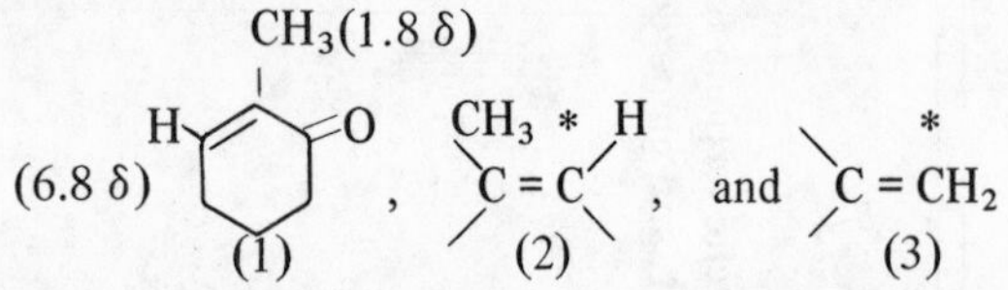

$\mathrm{CH_3(1.8\ \delta)}$

(6.8 δ) H–[cyclohexenone ring] (1) , $\mathrm{CH_3}$ * H >C=C< (2) , and * >C = $\mathrm{CH_2}$ (3)

*Carbons in these groups may be common.

Step 4: *Residual Formula and Combination of Fragments:*

Combination of fragments yields:

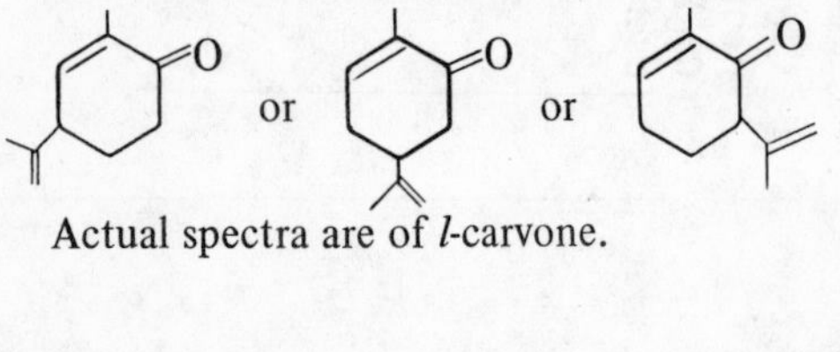

[structure] or [structure] or [structure]

Actual spectra are of *l*-carvone.

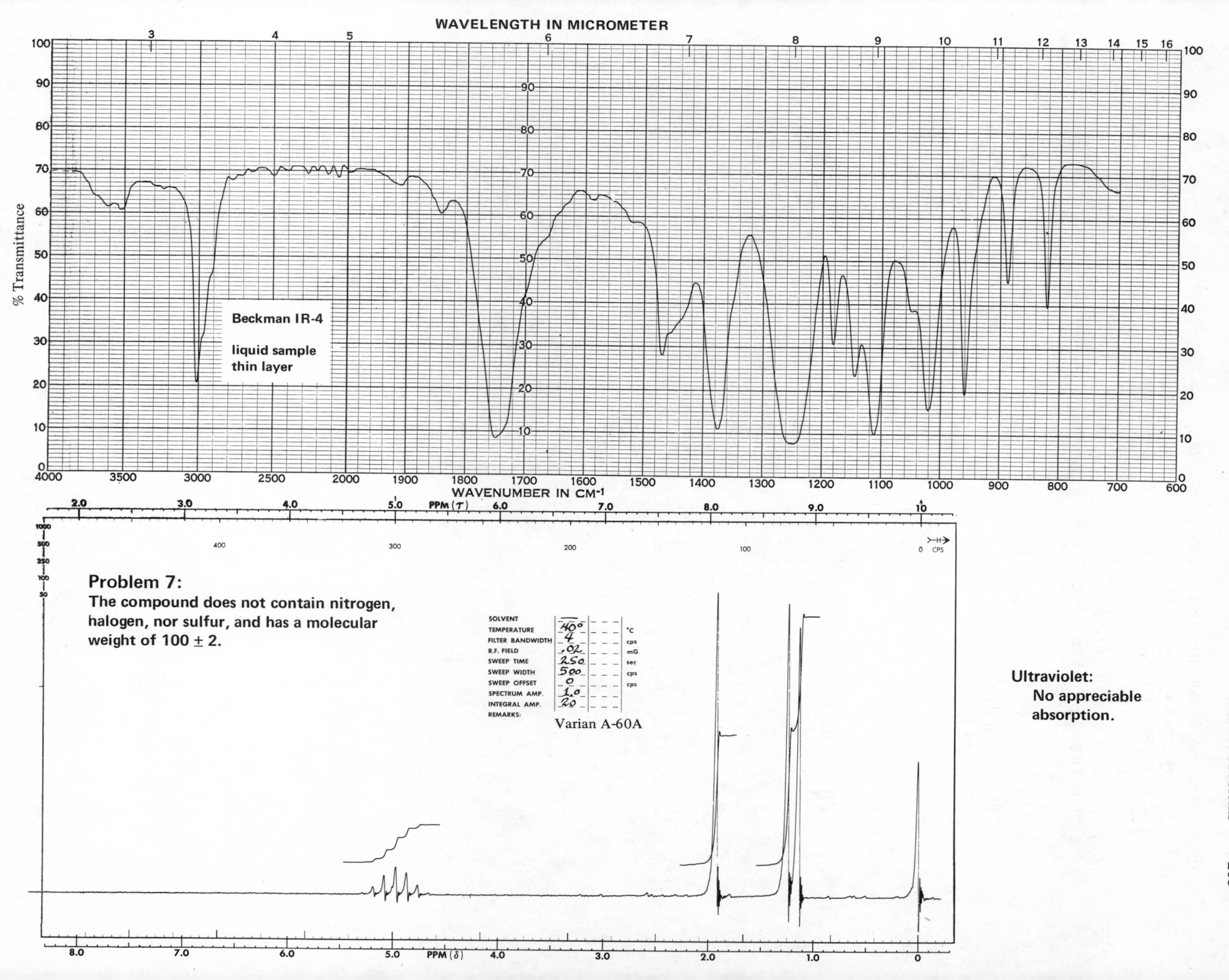


**Problem 7:**
The compound does not contain nitrogen, halogen, nor sulfur, and has a molecular weight of 100 ± 2.

**Answer Problem 7: Spectra are of *i*-propyl acetate.**

Step 1: *Cursory Examination of the Data:*

$CH_3\underset{\mid}{C}H\,CH_3$ (5.0 δ septet, 1.2 δ doublet), $\rangle C=O$ (1750 $cm^{-1}$)
$\quad O-$

Step 2: *Detailed Analysis of the Data:*

| POSITION | PROTON RATIO |
|---|---|
| 5.0 δ | 1 |
| 1.9 δ | 3 |
| 1.2 δ | 6 |

Possible formula: $C_5H_{10}O_2$.

Step 3: *List of Structural Fragments:*

$-\overset{O}{\overset{\|}{C}}-$ (1750 $cm^{-1}$, IR), $CH_3\underset{O-}{\underset{\mid}{C}}HCH_3$ (NMR), $CH_3$ -ns (NMR)

Step 4: *Residual Formula and Combination of Fragments:*

$CH_3\overset{O}{\overset{\|}{C}}-O-CH\,(CH_3)_2$

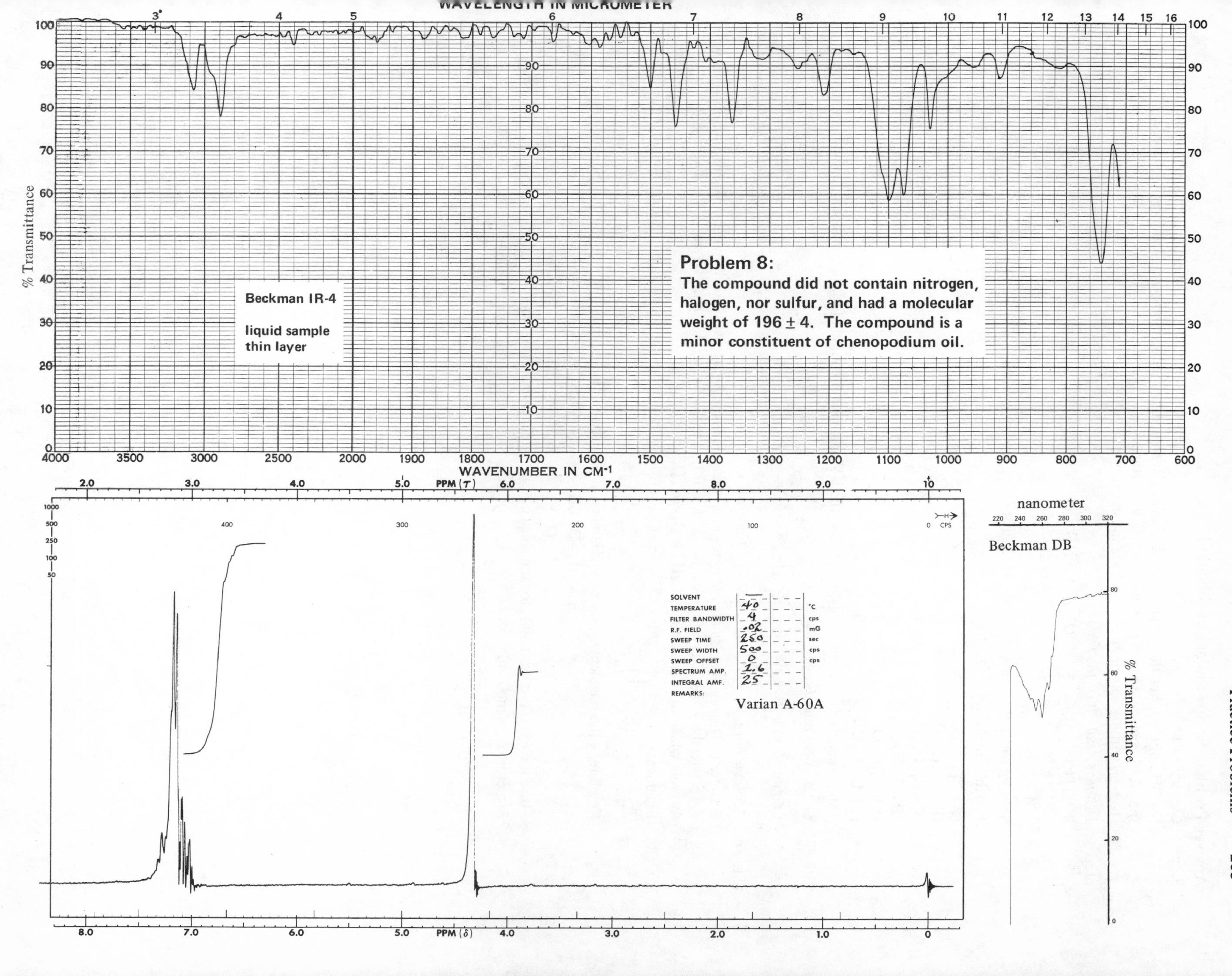
WAVELENGTH IN MICROMETER
% Transmittance
Beckman IR-4
liquid sample
thin layer
Problem 8:
The compound did not contain nitrogen, halogen, nor sulfur, and had a molecular weight of 196 ± 4. The compound is a minor constituent of chenopodium oil.
WAVENUMBER IN CM-1
PPM (τ)
SOLVENT
TEMPERATURE
FILTER BANDWIDTH
R.F. FIELD
SWEEP TIME
SWEEP WIDTH
SWEEP OFFSET
SPECTRUM AMP.
INTEGRAL AMP.
REMARKS:
Varian A-60A
PPM (δ)
nanometer
Beckman DB
% Transmittance

## Answer Problem 8: Spectra are of dibenzyl ether.

Step 1: *Cursory Examination of the Data:*

No OH nor C=O, but the presence of an aromatic system and C–O– (1100 $cm^{-1}$ band) is indicated by the IR. A conjugated system is indicated (UV). Aromatic protons and aliphatic protons attached to strong electron withdrawing groups (4.3 δ) are indicated by the NMR.

Step 2: *Detailed Analysis of the Data:*

| POSITION | PROTON RATIO |
|---|---|
| 7.2 δ | 5 |
| 4.3 δ | 2 |

There must be 14 or a multiple of 14 hydrogens in the compound.

Possible formula:

(A) $C_{14}H_{14}O$ $\Omega = 8$
(B) $C_{13}H_{28}O$ $\Omega = 0$
(C) $C_{13}H_{14}O_2$ $\Omega = 7$
(D) $C_{10}H_{14}O_4$ $\Omega = 4$
(E) $C_9H_{14}O_5$ $\Omega = 3$
(F) $C_6H_{14}O_7$ $\Omega = 0$

The presence of an aromatic system (NMR, IR, UV) eliminates formula (B), (E), and (F).

Step 3: *List of Structural Fragments:*

Fragments indicated are: [phenyl ring], ns –$CH_2$–ns, –O–

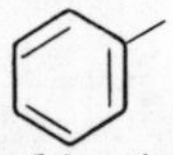

Step 4: *Residual Formula and Combination of Fragments:*

Because the fragments indicate only seven carbons and seven hydrogens and all hydrogens are accounted for in the NMR, the actual C:H ratio must be some multiple of $C_7H_7$, i.e., $C_{14}H_{14}$, $C_{21}H_{21}$, etc. The only formula which fits a multiple of $C_7H_7$ is $C_{14}H_{14}O$.

Fragments: ([phenyl ring])$_2$, (ns – $CH_2$ – ns)$_2$, – O –

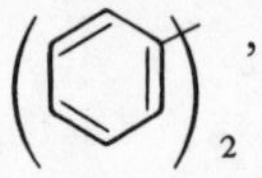

Combining the fragments gives ns – $CH_2$–O– or ns–$CH_2$–$CH_2$–ns →

ns–$CH_2$–O–$CH_2$–ns (1) or ns–$CH_2$–$CH_2$–O– (2)

Structure (2) will not account for the singlet at 4.3 δ and may be rejected.

Structure of compound must be

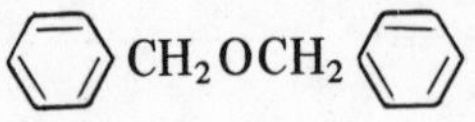

[phenyl ring]–$CH_2OCH_2$–[phenyl ring]

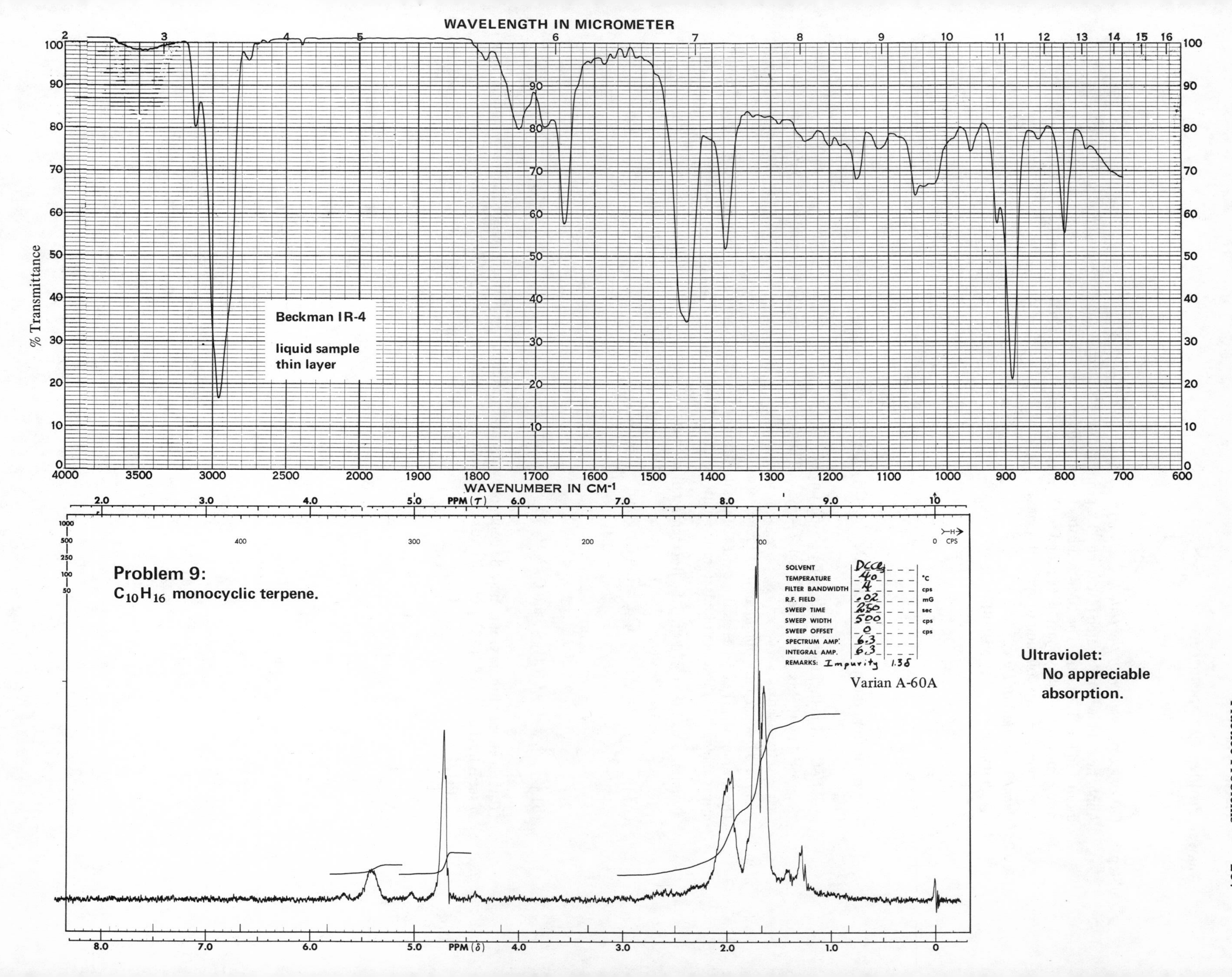


Problem 9:
$C_{10}H_{16}$ monocyclic terpene.

Ultraviolet:
No appreciable absorption.

## Answer Problem 9: Spectra are of

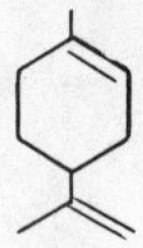

Step 1: *Cursory Examination of the Data:*

Aliphatic double bond stretch (1650 $cm^{-1}$), a geminal disubstituted alkene ($>C=CH_2$, 890 $cm^{-1}$) are indicated in the IR spectra. The NMR indicates two different types of ethylenic hydrogens. If methyl groups are present, they must be attached to C=C (no absorption in the NMR below 1.5 δ). Six membered rings are very common to cyclic terpenes.

Step 2: *Detailed Analysis of Data:*

| POSITION | NUMBER OF PROTONS |
|---|---|
| 5.4 δ | 1 |
| 4.7 δ | 2 |
| 2.0 δ | 7 |
| 1.7-1.5 δ | 6 |

Fragments from the spectrum: $-CH=C<$, $>C=CH_2$, $CH_3-C=C<$ (methyl indicates long-range splitting), $CH_3-C=C<$* (methyl indicates long-range splitting).

*Suggested by the six proton peaks at 1.7 – 1.5 δ. The double bonds may be common to those in the first two fragments.

Step 4: *Residual Formula and Combination of Fragments:*

The formula $C_{10}H_{16}$ indicates three sites of unsaturation. At least two sites are indicated by the fragments above. The third site of unsaturation must be a ring. The compound is a monocyclic terpene.

Assuming a six membered ring, the fragments indicated are:

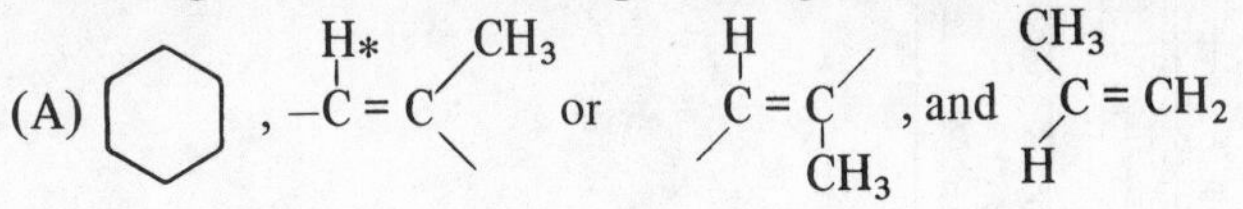

or

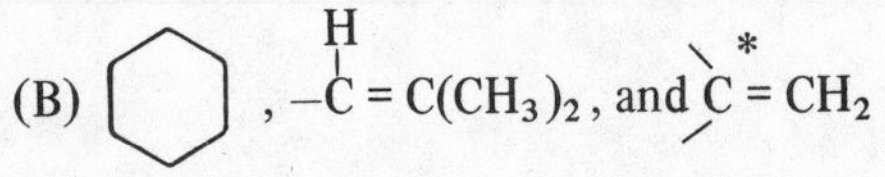

*Carbon atoms of these groups may be common with ring carbons. Note that any possible combination of set B fragments yields a $C_{11}$ compound. Combinations of set A fragments yield:

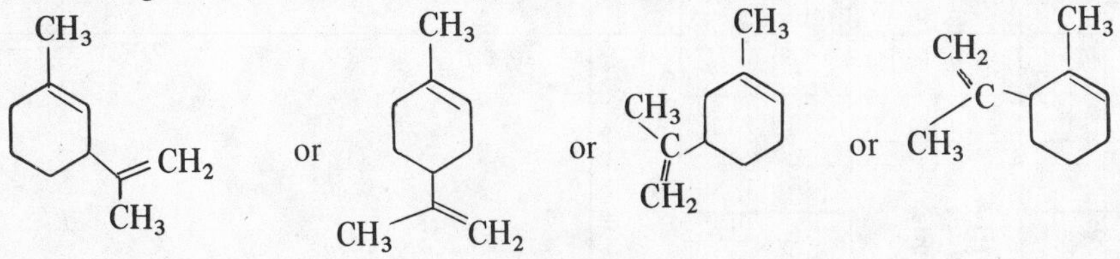

Spectra are of *d*-limonine,

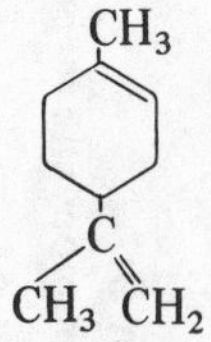

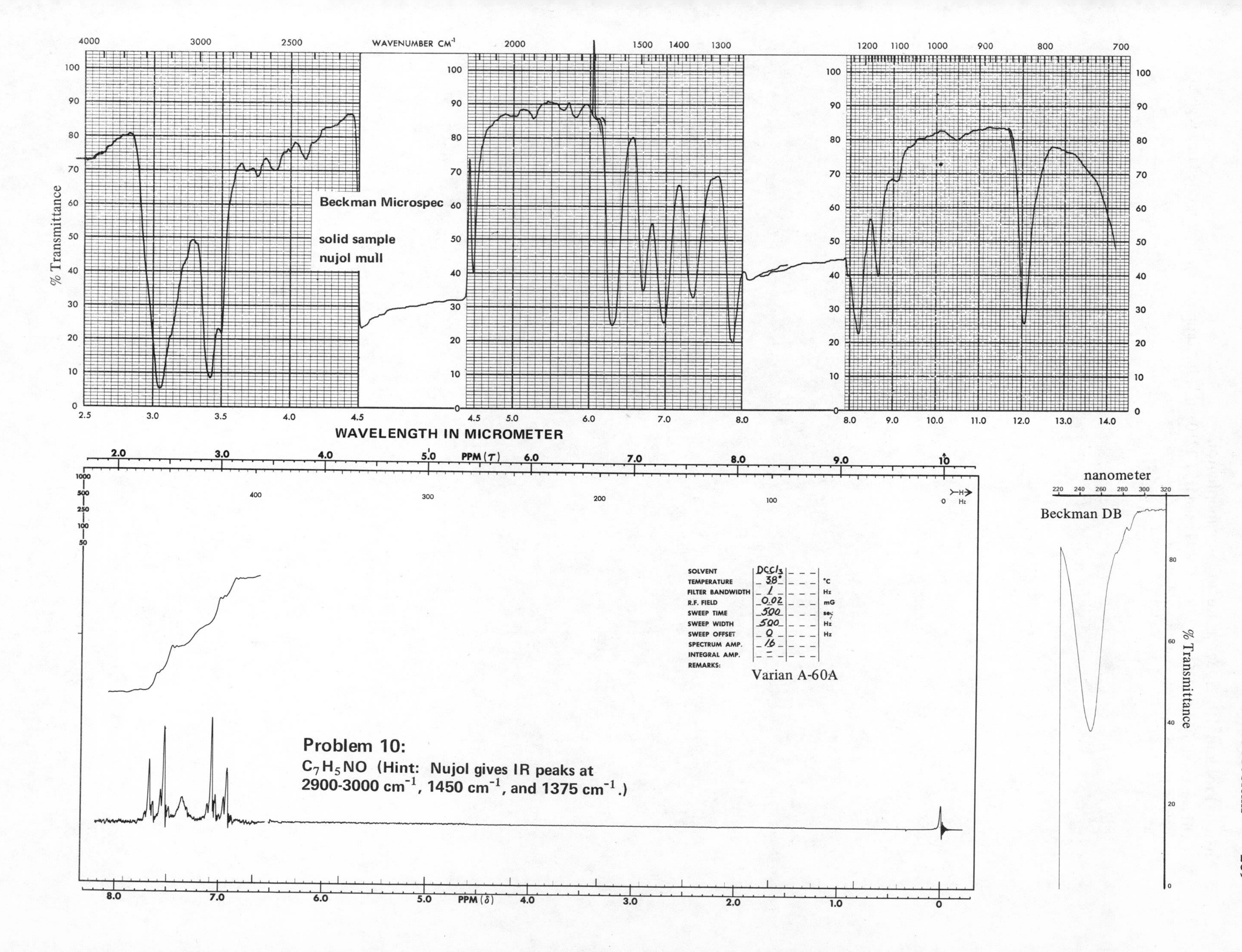

WAVENUMBER CM⁻¹
% Transmittance
Beckman Microspec
solid sample
nujol mull
WAVELENGTH IN MICROMETER
PPM (τ)
SOLVENT DCCl3
TEMPERATURE 38° °C
FILTER BANDWIDTH 1 Hz
R.F. FIELD 0.02 mG
SWEEP TIME 500 sec
SWEEP WIDTH 500 Hz
SWEEP OFFSET 0 Hz
SPECTRUM AMP. 16
INTEGRAL AMP. –
REMARKS:
Varian A-60A
Problem 10:
C7H5NO (Hint: Nujol gives IR peaks at
2900-3000 cm-1, 1450 cm-1, and 1375 cm-1.)
PPM (δ)
nanometer
Beckman DB
% Transmittance

## Answer Problem 10: Spectra are of *p*-cyanophenol.

The OH band is shifted to 2950 $cm^{-1}$. The $C \equiv N$ band is at 2210 $cm^{-1}$. Indication of a *p*-disubstituted ring is given by aromatic proton absorption in the NMR.

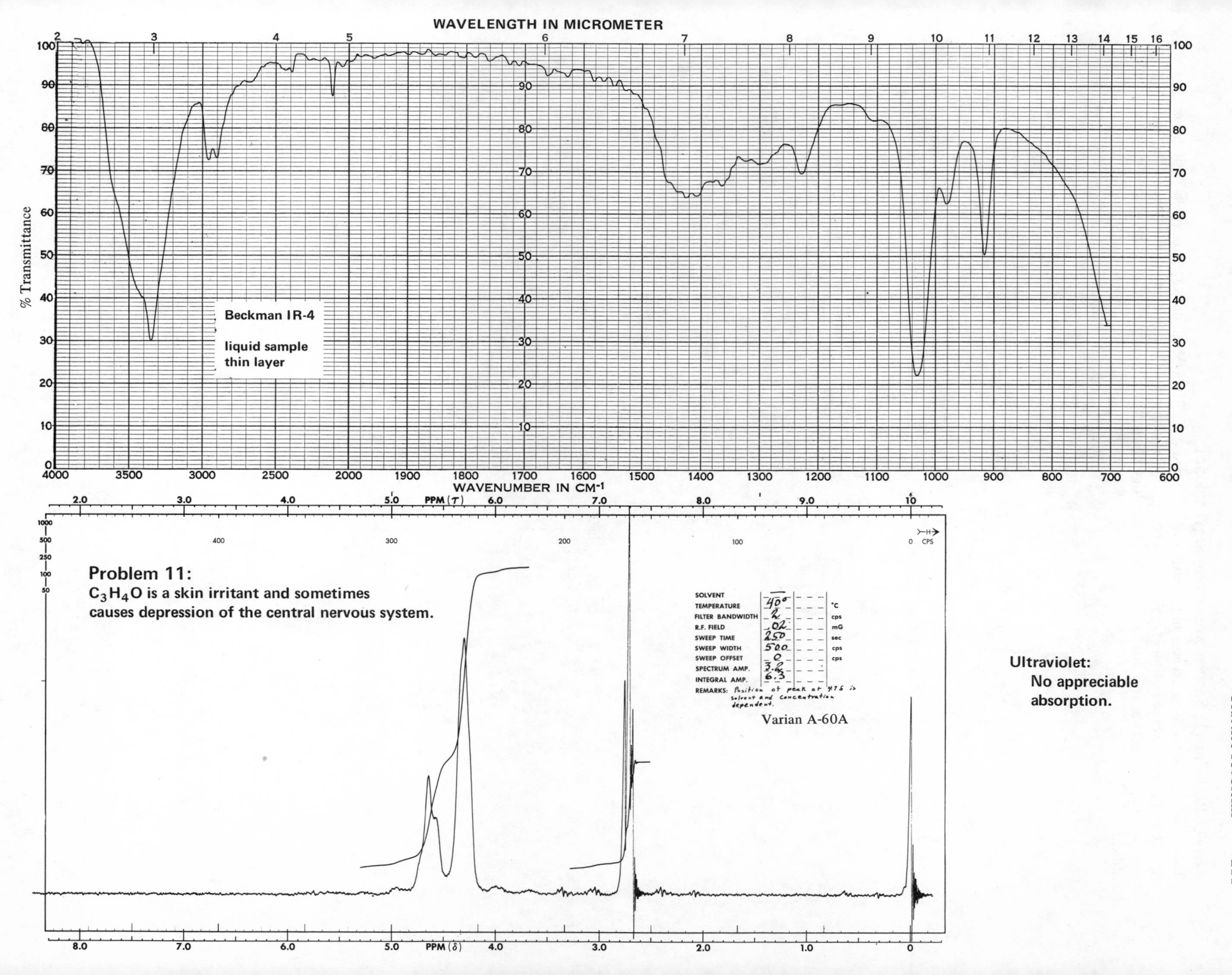
WAVELENGTH IN MICROMETER
% Transmittance
Beckman IR-4
liquid sample
thin layer
WAVENUMBER IN CM-1
PPM (τ)
PPM (δ)
SOLVENT
TEMPERATURE 40° °C
FILTER BANDWIDTH 2 cps
R.F. FIELD .02 mG
SWEEP TIME 250 sec
SWEEP WIDTH 500 cps
SWEEP OFFSET 0 cps
SPECTRUM AMP. 3.2
INTEGRAL AMP. 6.3
REMARKS: Position of peak at 475 is solvent and concentration dependent.
Varian A-60A
Problem 11:
$C_3H_4O$ is a skin irritant and sometimes
causes depression of the central nervous system.
Ultraviolet:
No appreciable
absorption.

## Answer Problem 11: Spectra are of propargyl alcohol.

The IR spectrum should have given an indication of –OH and C≡C. The NMR indicates three peaks with ratios of 1:2:1. The peak whose position is solvent and concentration dependent must be due to the OH proton. The two proton peak must be coupled to the 2.7 δ proton, long range. The system $\underbrace{H}_{1.8\ \delta}-C\equiv C-\underbrace{CH_2}_{4.3\ \delta}-OH$ is strongly indicated.

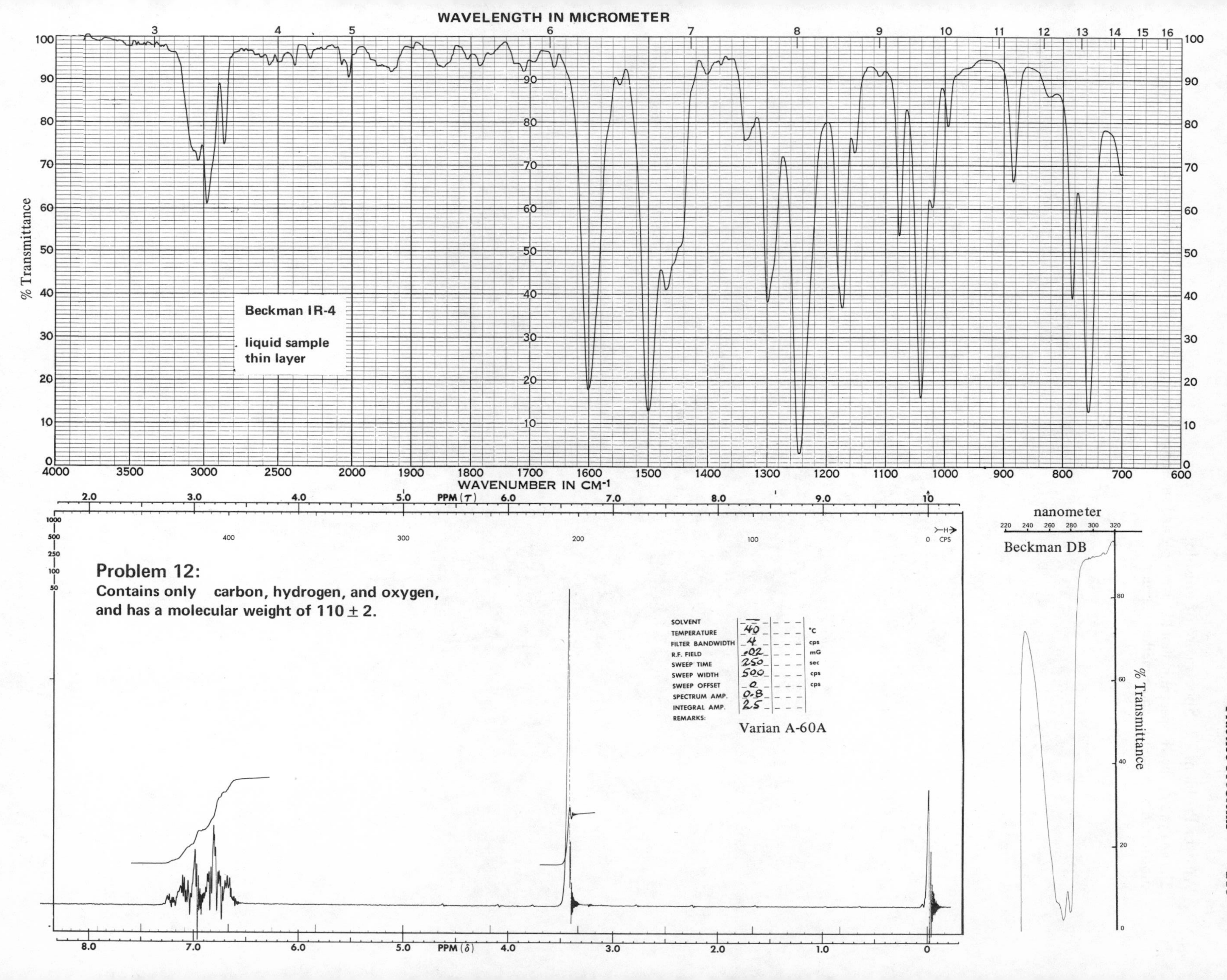
WAVELENGTH IN MICROMETER
% Transmittance
Beckman IR-4
liquid sample
thin layer
WAVENUMBER IN CM-1
PPM (τ)
Problem 12:
Contains only carbon, hydrogen, and oxygen,
and has a molecular weight of 110 ± 2.
SOLVENT
TEMPERATURE 40 °C
FILTER BANDWIDTH 4 cps
R.F. FIELD .02 mG
SWEEP TIME 250 sec
SWEEP WIDTH 500 cps
SWEEP OFFSET 0 cps
SPECTRUM AMP. 0.8
INTEGRAL AMP. 2.5
REMARKS:
Varian A-60A
PPM (δ)
nanometer
Beckman DB
% Transmittance

## Answer Problem 12: Spectra are of anisole.

The analysis of the IR spectrum should have suggested an aromatic system. The NMR gives evidence of a mono-substituted benzene ring and a methoxyl group.

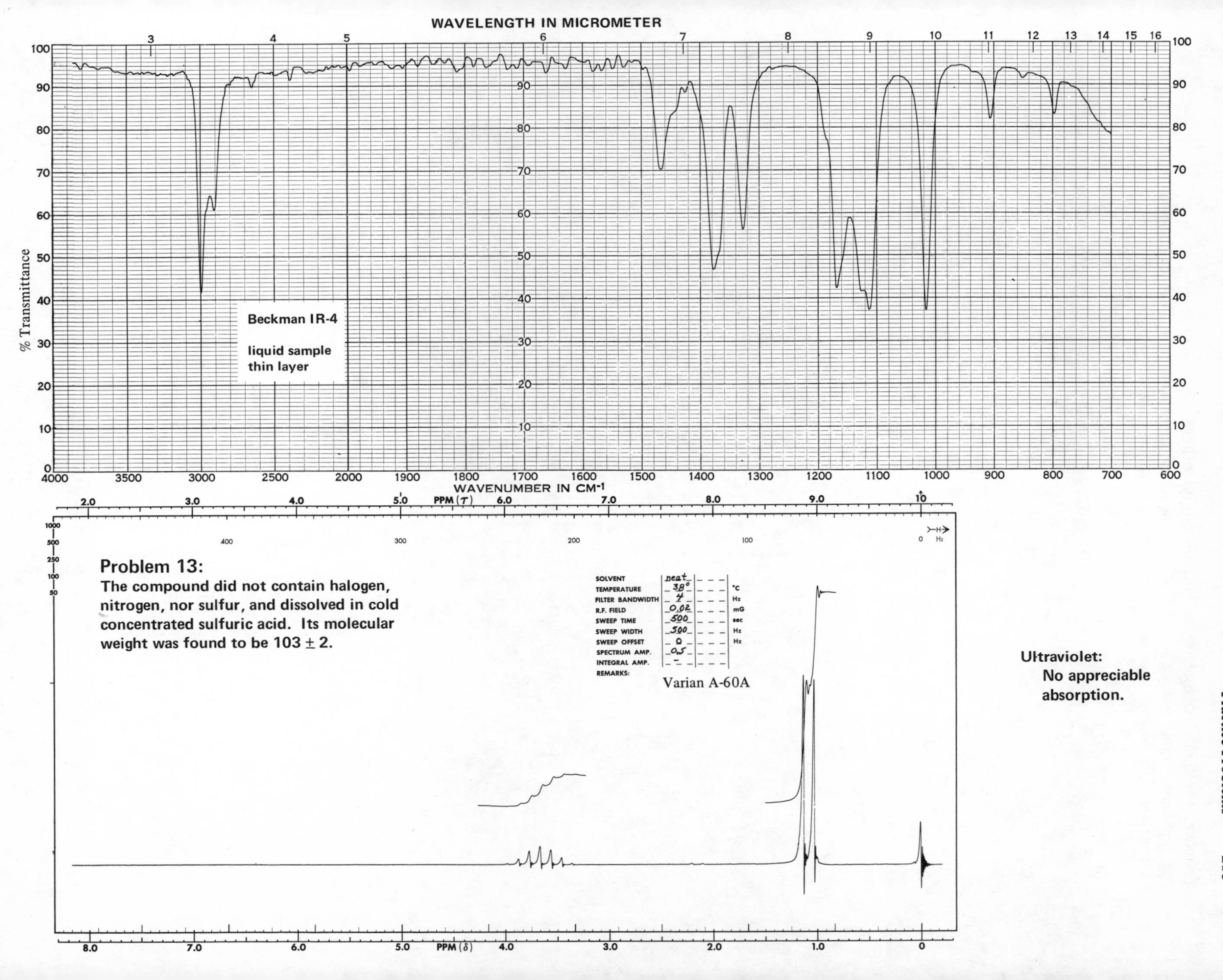
WAVELENGTH IN MICROMETER
WAVENUMBER IN CM-1
% Transmittance
Beckman IR-4
liquid sample
thin layer
Problem 13:
The compound did not contain halogen, nitrogen, nor sulfur, and dissolved in cold concentrated sulfuric acid. Its molecular weight was found to be 103 ± 2.
SOLVENT neat
TEMPERATURE 38° °C
FILTER BANDWIDTH 4 Hz
R.F. FIELD 0.02 mG
SWEEP TIME 500 sec
SWEEP WIDTH 500 Hz
SWEEP OFFSET 0 Hz
SPECTRUM AMP. 0.5
INTEGRAL AMP. –
REMARKS:
Varian A-60A
PPM (τ)
PPM (δ)
Ultraviolet:
No appreciable
absorption.

**Answer Problem 13: Spectra are of di-*i*-propyl ether.**

Strong bands at 1100 $cm^{-1}$ are indicative of C–O bonds. Identification of $(CH_3)_2CH{-}O{-}$ from the NMR and the calculation of the possible formula should have made this an easy one.

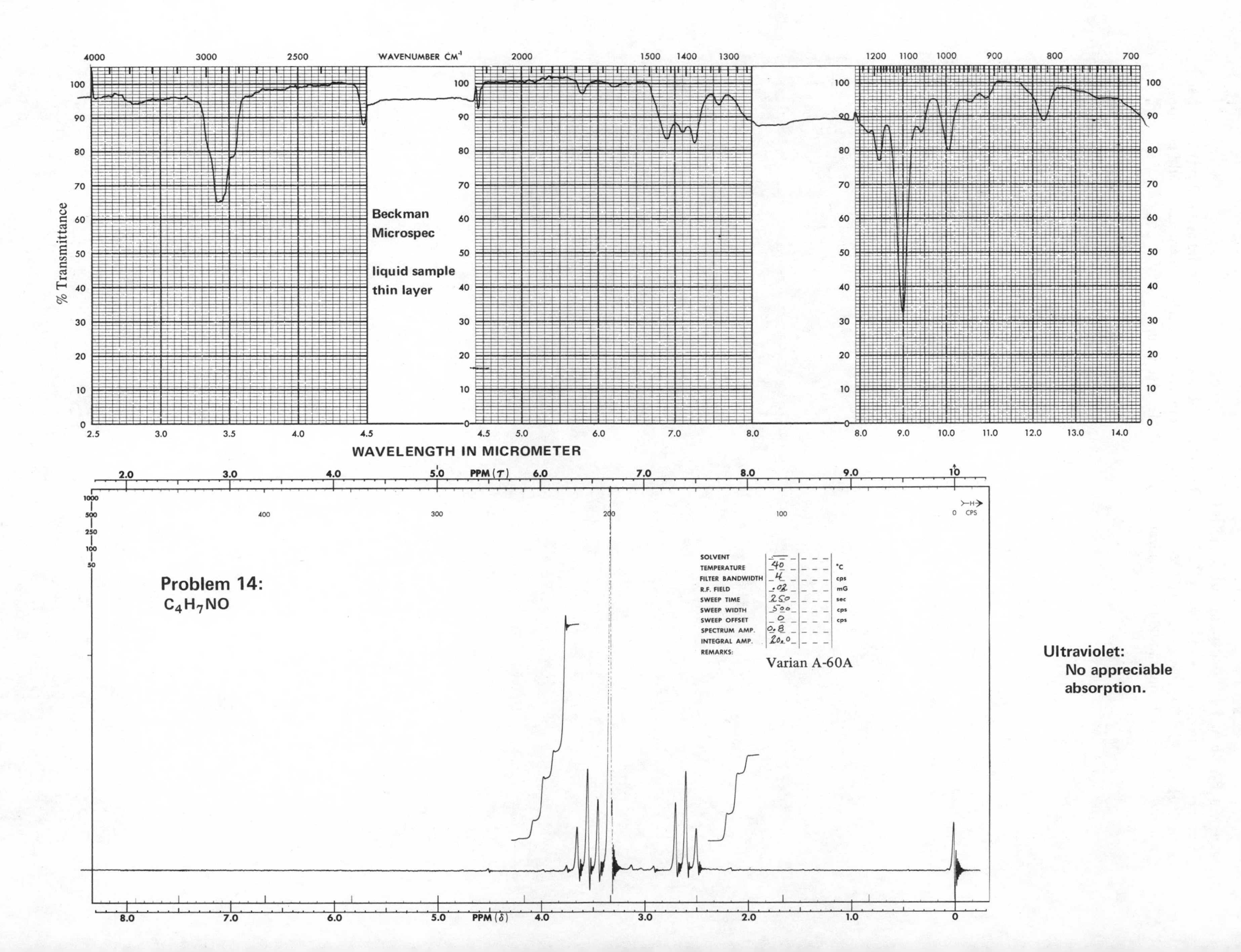
WAVENUMBER CM$^{-1}$
% Transmittance
Beckman
Microspec
liquid sample
thin layer
WAVELENGTH IN MICROMETER
PPM ($\tau$)
Problem 14:
$C_4H_7NO$
SOLVENT
TEMPERATURE 40 °C
FILTER BANDWIDTH 4 cps
R.F. FIELD .02 mG
SWEEP TIME 250 sec
SWEEP WIDTH 500 cps
SWEEP OFFSET 0 cps
SPECTRUM AMP. 0.8
INTEGRAL AMP. 20.0
REMARKS:
Varian A-60A
PPM ($\delta$)
Ultraviolet:
No appreciable
absorption.

## Answer Problem 14: Spectra are of 3-methoxypropionitrile.

A nitrile is indicated by peak at 2250 $cm^{-1}$. The C–O bond is indicated by a strong band at 1100 $cm^{-1}$. The fragments $\underline{CH_3}O$ – (3.3 δ singlet), –O–$\underline{CH_2}CH_2$– (3.5 δ triplet), and ns–$\underline{CH_2}$–$CH_2$–(where ns could be nitrogen or a multiple carbon bond) should be recognizable from the NMR spectrum.

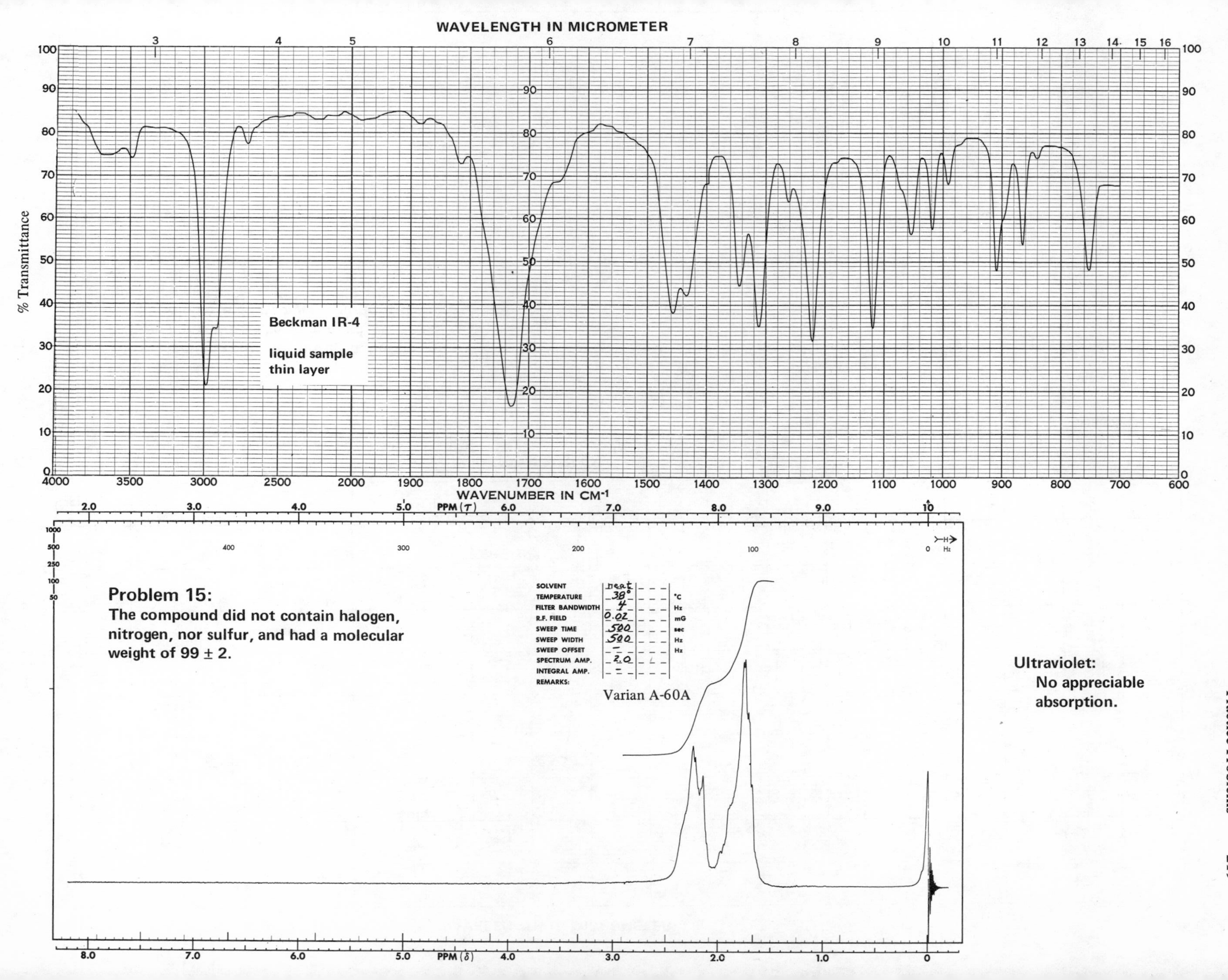


**Problem 15:**
The compound did not contain halogen, nitrogen, nor sulfur, and had a molecular weight of $99 \pm 2$.

**Ultraviolet:**
No appreciable absorption.

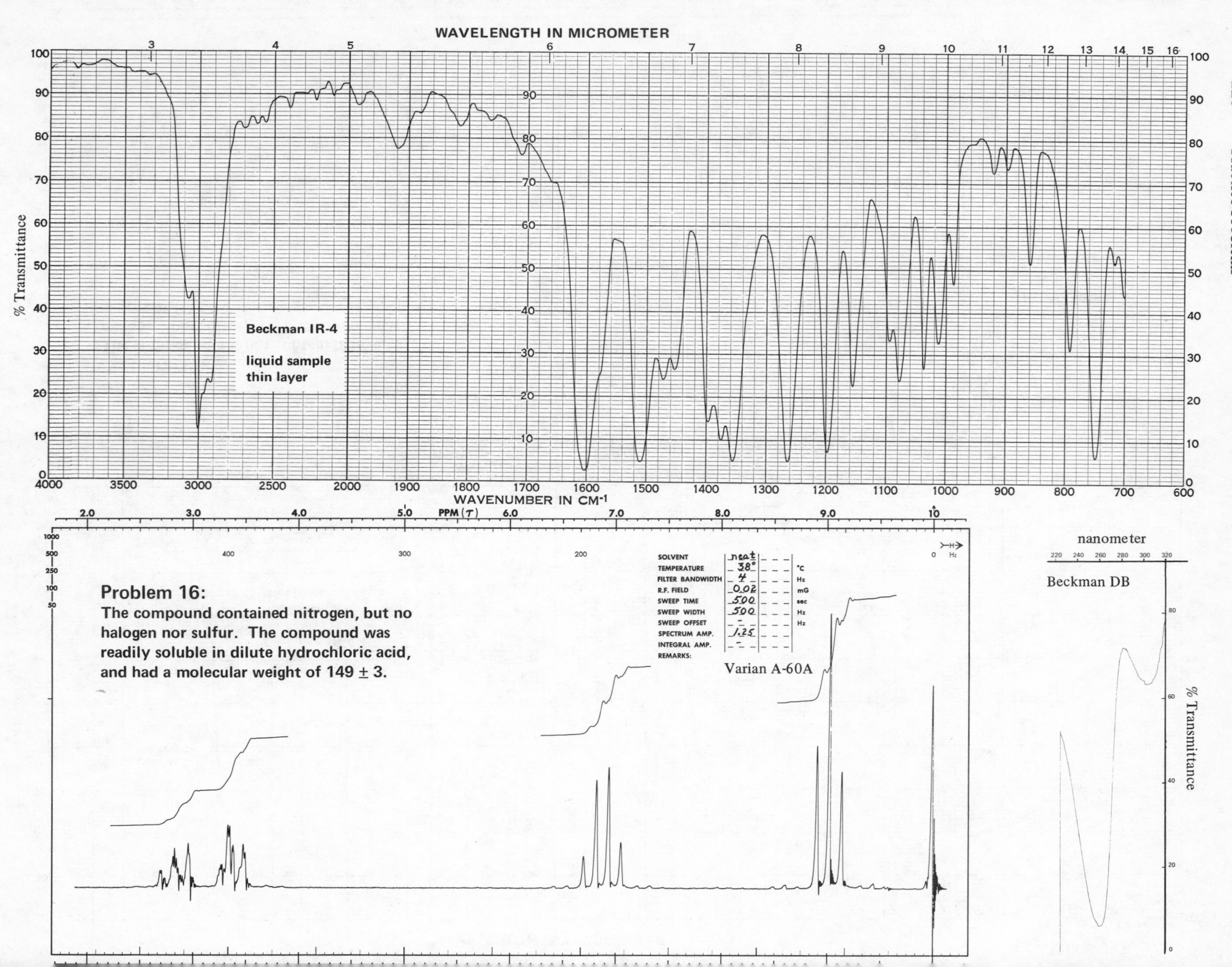


## Problem 16:

The compound contained nitrogen, but no halogen nor sulfur. The compound was readily soluble in dilute hydrochloric acid, and had a molecular weight of 149 ± 3.

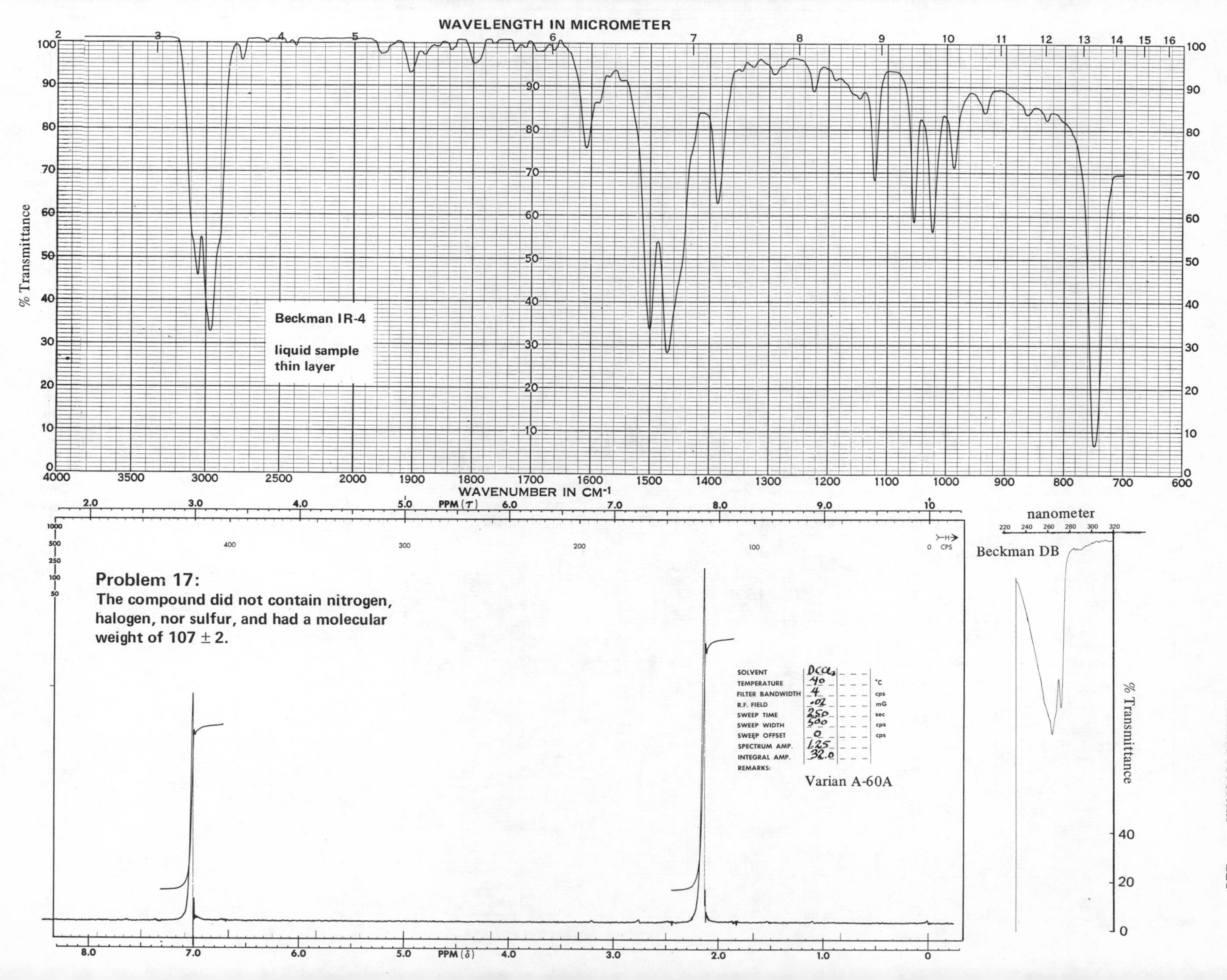


Problem 17:
The compound did not contain nitrogen, halogen, nor sulfur, and had a molecular weight of 107 ± 2.

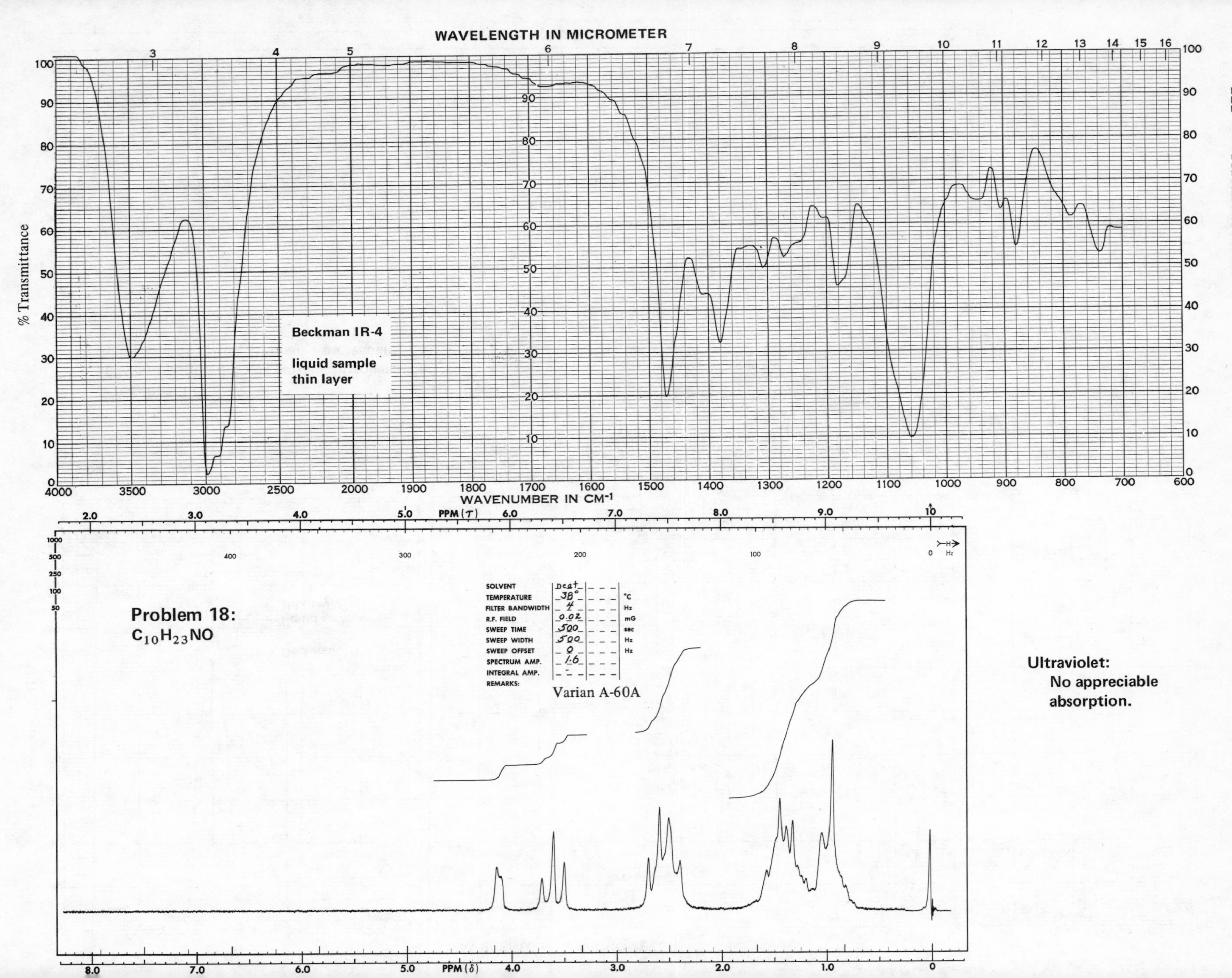
WAVELENGTH IN MICROMETER
WAVENUMBER IN CM-1
% Transmittance
Beckman IR-4
liquid sample
thin layer
Problem 18:
$C_{10}H_{23}NO$
SOLVENT neat
TEMPERATURE 38° °C
FILTER BANDWIDTH 4 Hz
R.F. FIELD 0.02 mG
SWEEP TIME 500 sec
SWEEP WIDTH 500 Hz
SWEEP OFFSET 0 Hz
SPECTRUM AMP. 1.6
INTEGRAL AMP.
REMARKS:
Varian A-60A
PPM (τ)
PPM (δ)
Ultraviolet:
No appreciable
absorption.

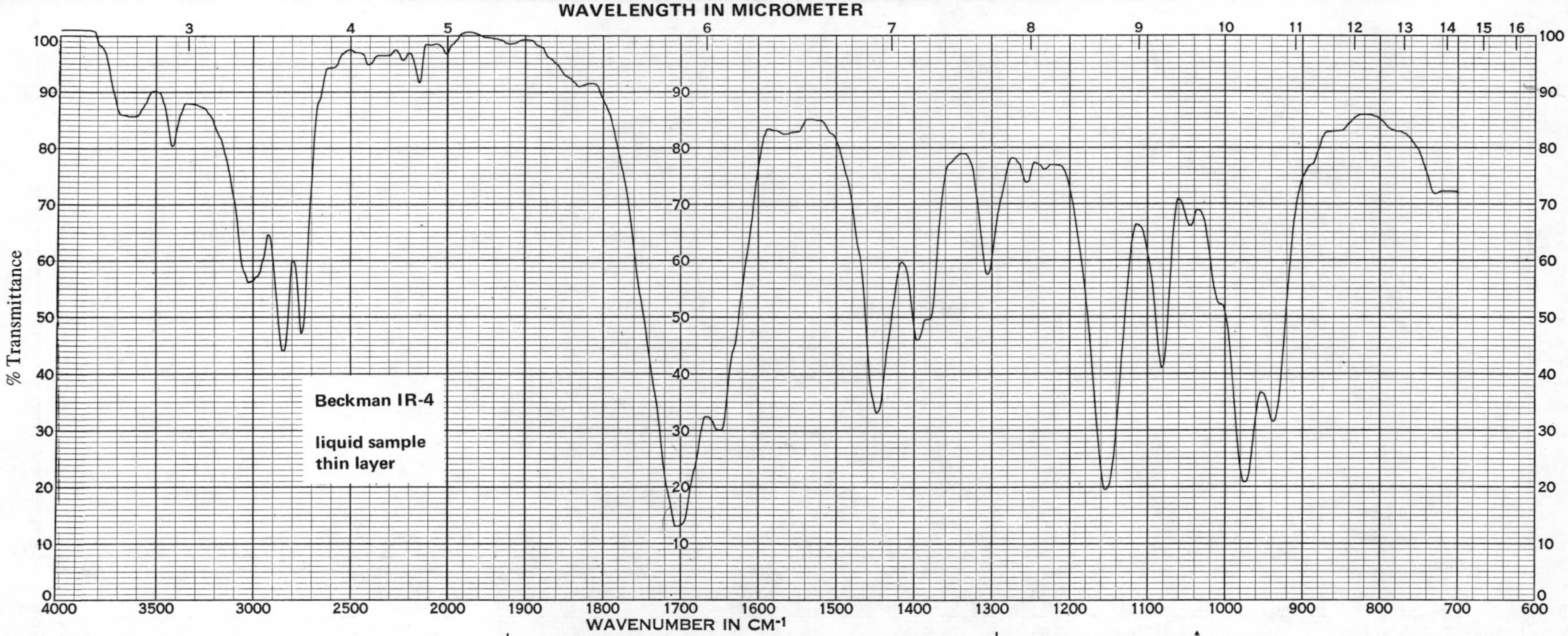

Beckman IR-4

liquid sample
thin layer

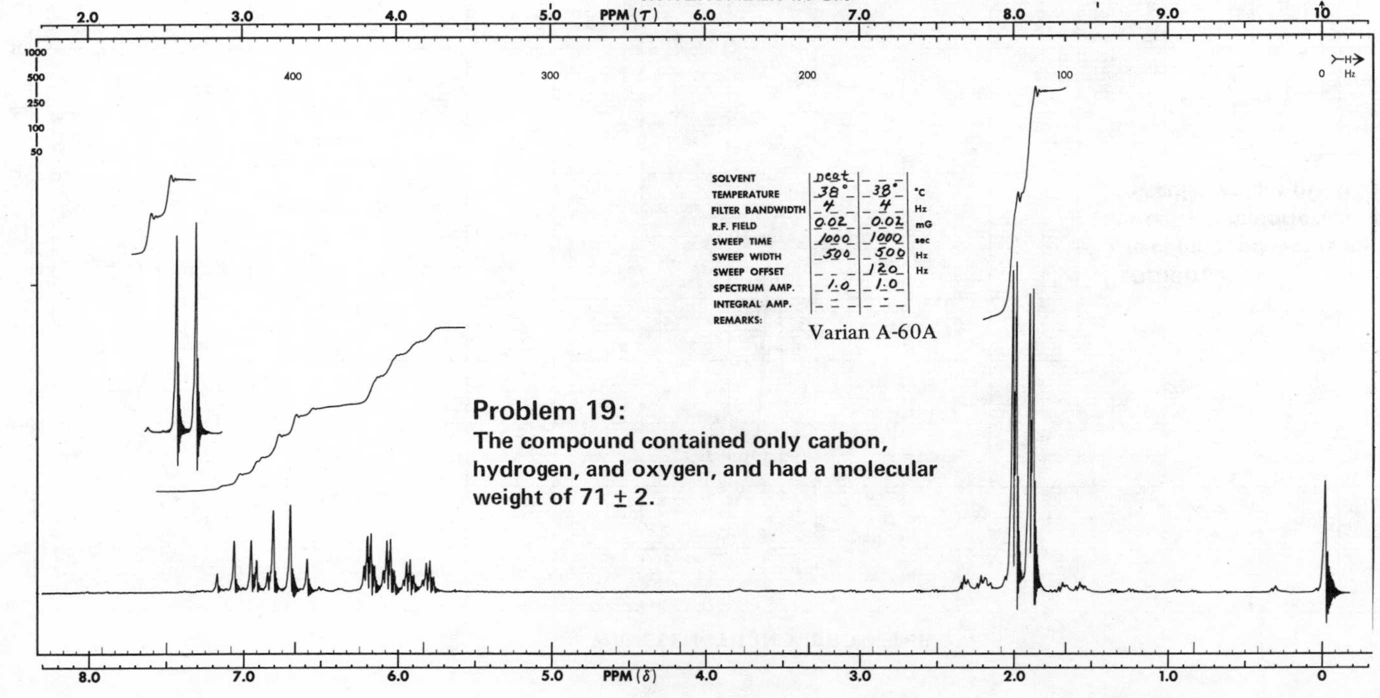

| SOLVENT | neat | | |
| --- | --- | --- | --- |
| TEMPERATURE | 38° | 38° | °C |
| FILTER BANDWIDTH | 4 | 4 | Hz |
| R.F. FIELD | 0.02 | 0.02 | mG |
| SWEEP TIME | 1000 | 1000 | sec |
| SWEEP WIDTH | 500 | 500 | Hz |
| SWEEP OFFSET | – | 120 | Hz |
| SPECTRUM AMP. | 1.0 | 1.0 | |
| INTEGRAL AMP. | – | – | |
| REMARKS: | | | |

Varian A-60A

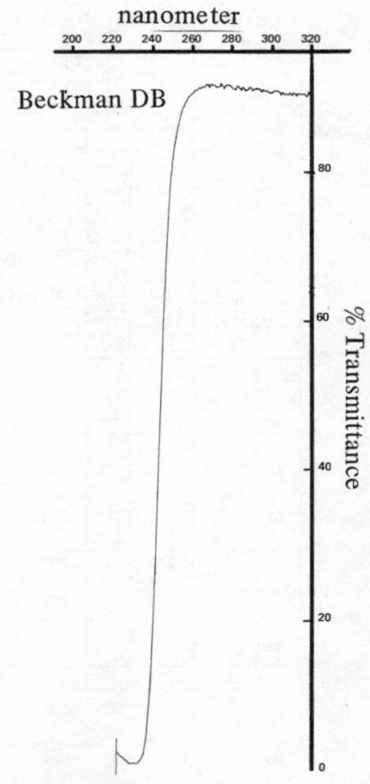

Beckman DB

## Problem 19:

**The compound contained only carbon, hydrogen, and oxygen, and had a molecular weight of 71 ± 2.**

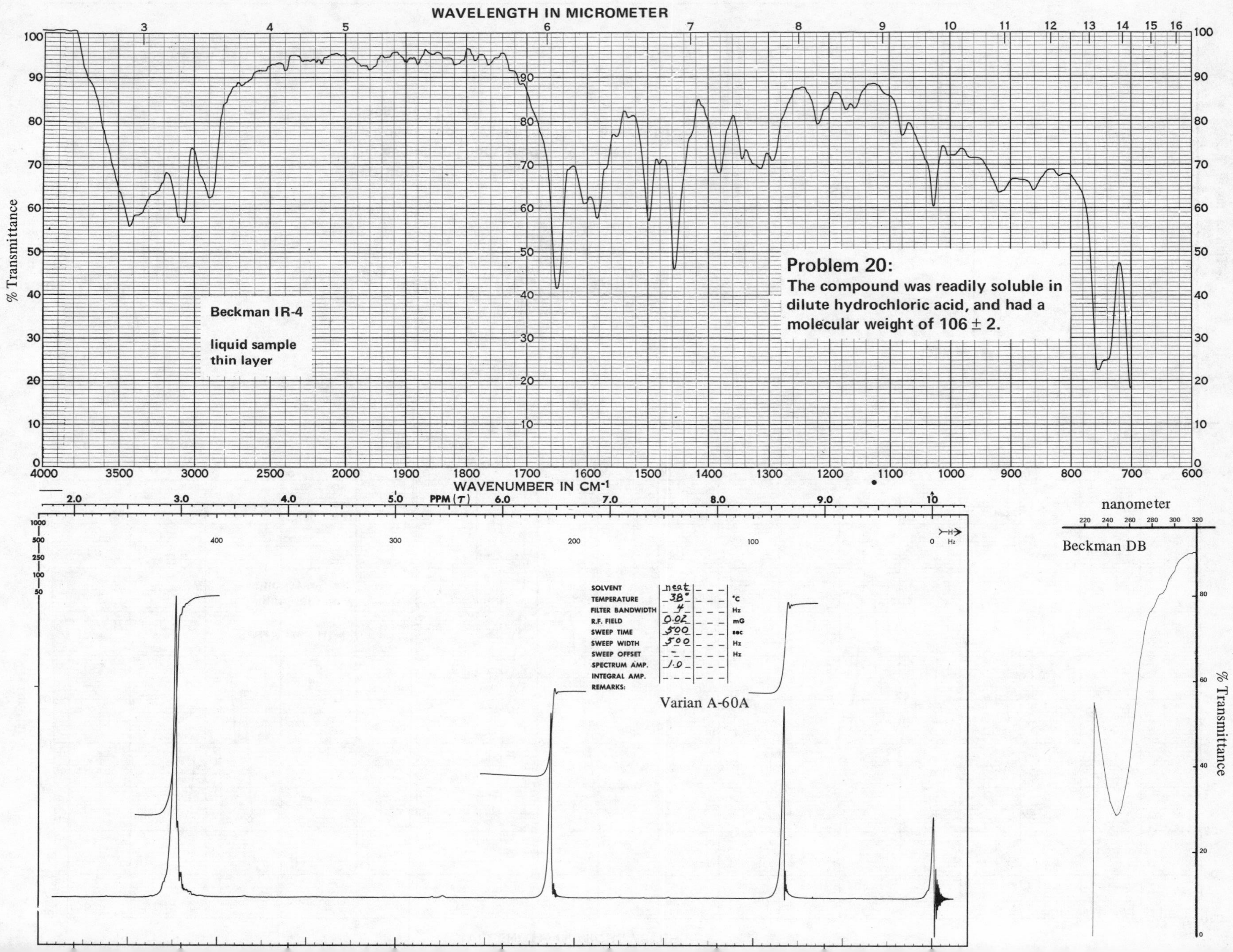

WAVELENGTH IN MICROMETER
% Transmittance
Beckman IR-4
liquid sample
thin layer
Problem 20:
The compound was readily soluble in dilute hydrochloric acid, and had a molecular weight of 106 ± 2.
WAVENUMBER IN CM-1
PPM (τ)
SOLVENT neat
TEMPERATURE 38° °C
FILTER BANDWIDTH 4 Hz
R.F. FIELD 0.02 mG
SWEEP TIME 500 sec
SWEEP WIDTH 500 Hz
SWEEP OFFSET – Hz
SPECTRUM AMP. 1.0
INTEGRAL AMP.
REMARKS:
Varian A-60A
nanometer
Beckman DB
% Transmittance

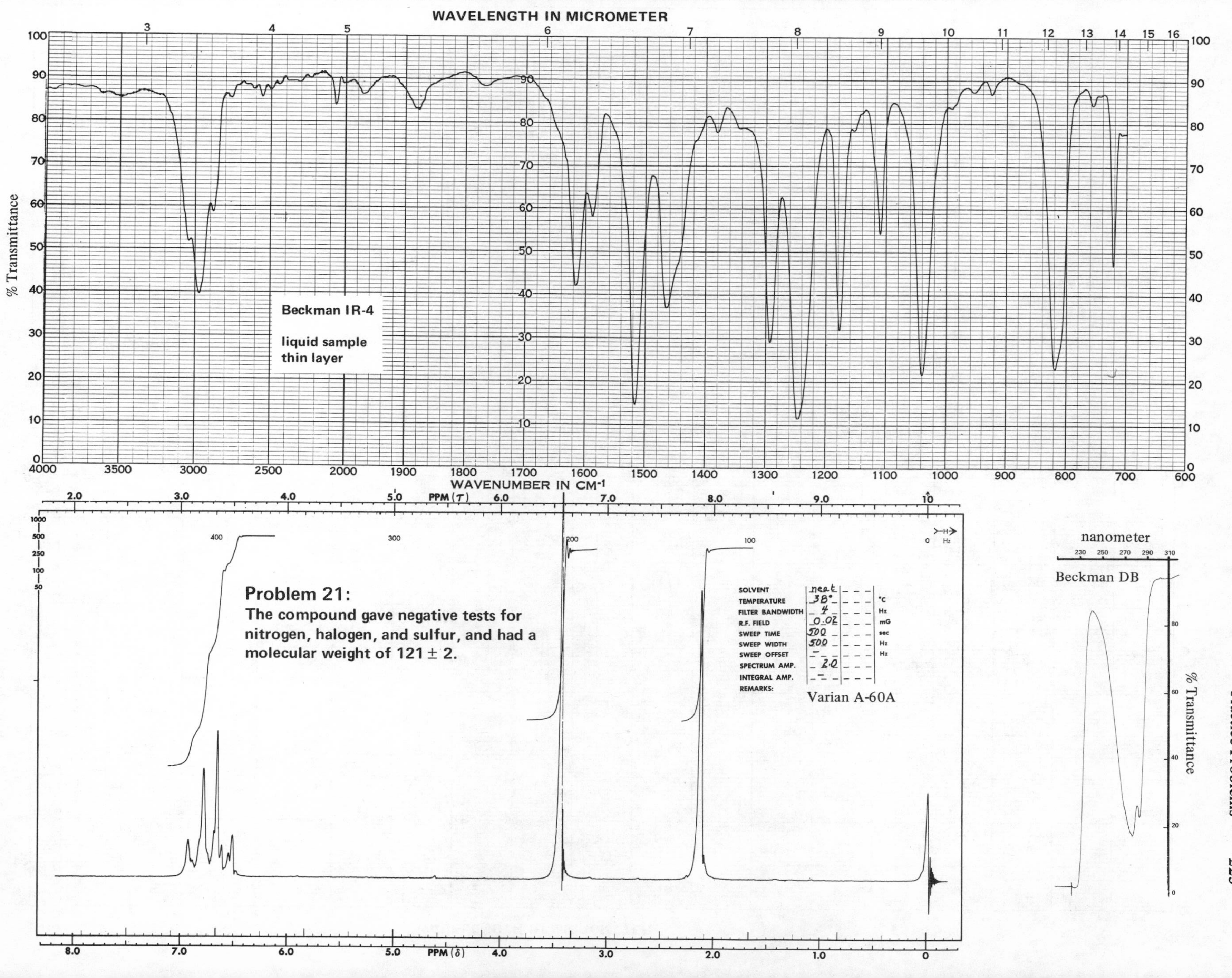
WAVELENGTH IN MICROMETER
WAVENUMBER IN CM-1
% Transmittance
Beckman IR-4
liquid sample
thin layer
Problem 21:
The compound gave negative tests for nitrogen, halogen, and sulfur, and had a molecular weight of 121 ± 2.
SOLVENT neat
TEMPERATURE 38° °C
FILTER BANDWIDTH 4 Hz
R.F. FIELD 0.02 mG
SWEEP TIME 500 sec
SWEEP WIDTH 500 Hz
SWEEP OFFSET – Hz
SPECTRUM AMP. 20
INTEGRAL AMP. –
REMARKS:
Varian A-60A
PPM (τ)
PPM (δ)
nanometer
Beckman DB
% Transmittance

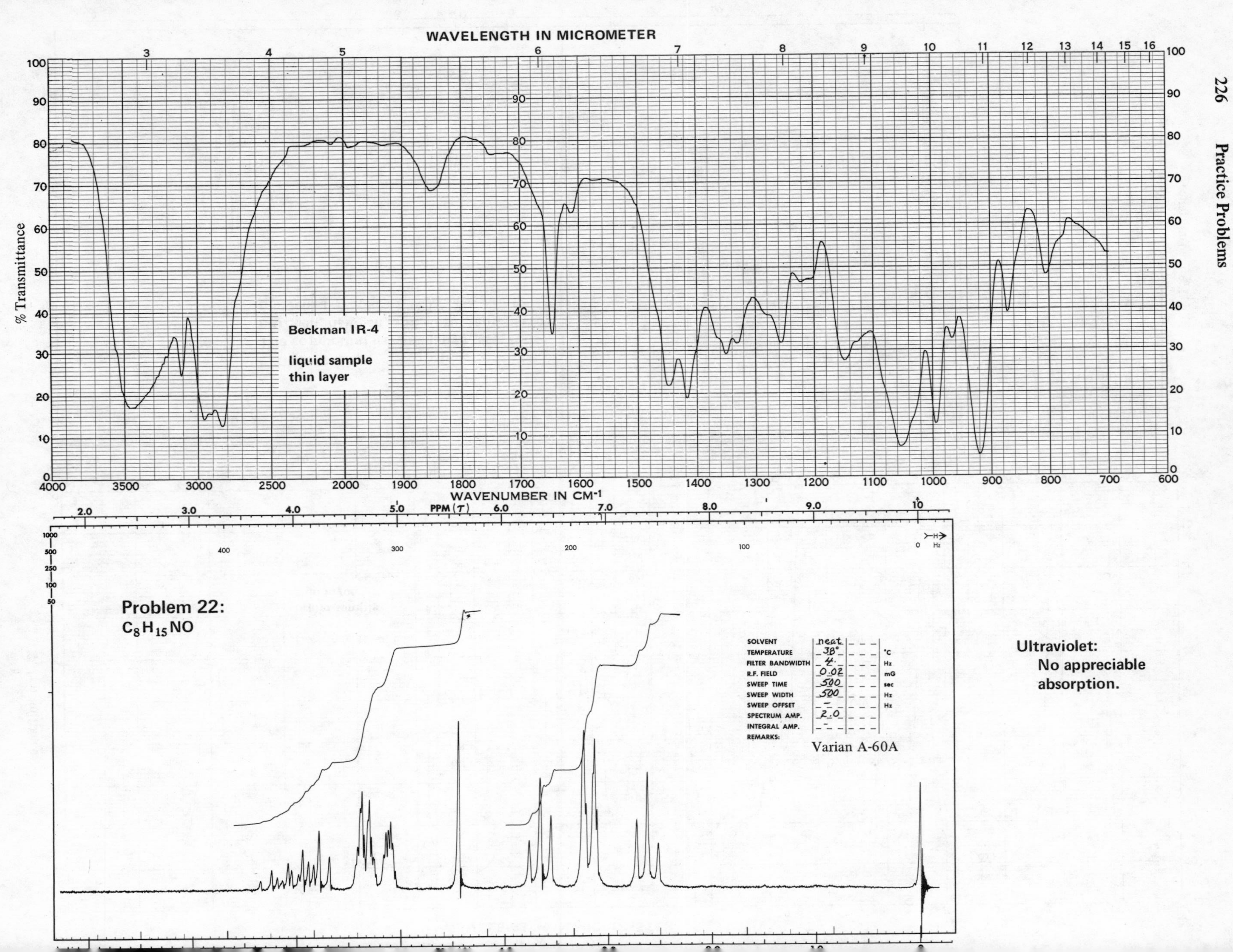
WAVELENGTH IN MICROMETER
WAVENUMBER IN CM-1
% Transmittance
Beckman IR-4
liquid sample
thin layer
PPM (τ)
Problem 22:
C8H15NO
SOLVENT neat
TEMPERATURE 38° °C
FILTER BANDWIDTH 4 Hz
R.F. FIELD 0.02 mG
SWEEP TIME 500 sec
SWEEP WIDTH 500 Hz
SWEEP OFFSET – Hz
SPECTRUM AMP. 2.0
INTEGRAL AMP. –
REMARKS:
Varian A-60A
Ultraviolet:
No appreciable
absorption.

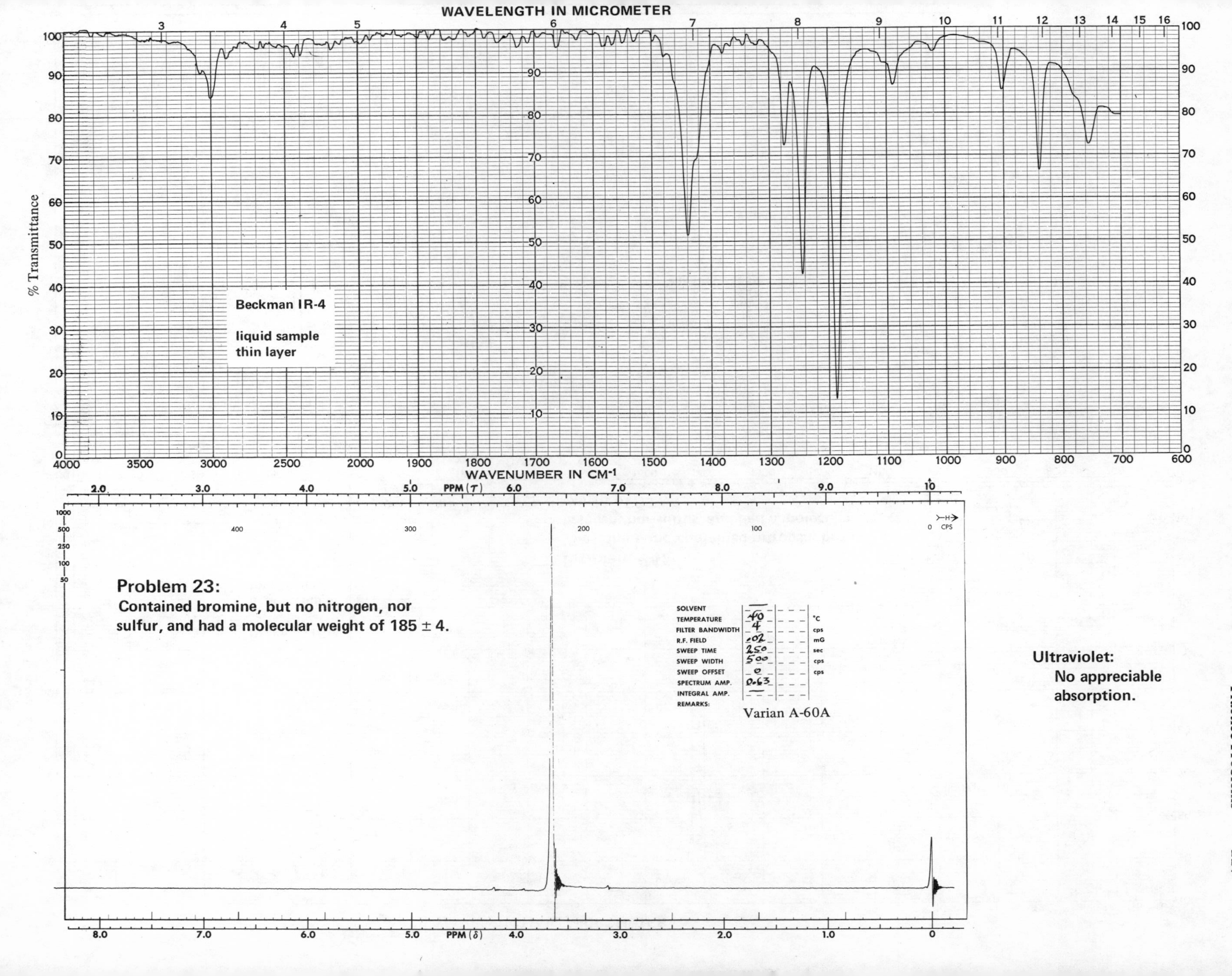


## Problem 23:

Contained bromine, but no nitrogen, nor sulfur, and had a molecular weight of 185 ± 4.

**Ultraviolet:** No appreciable absorption.

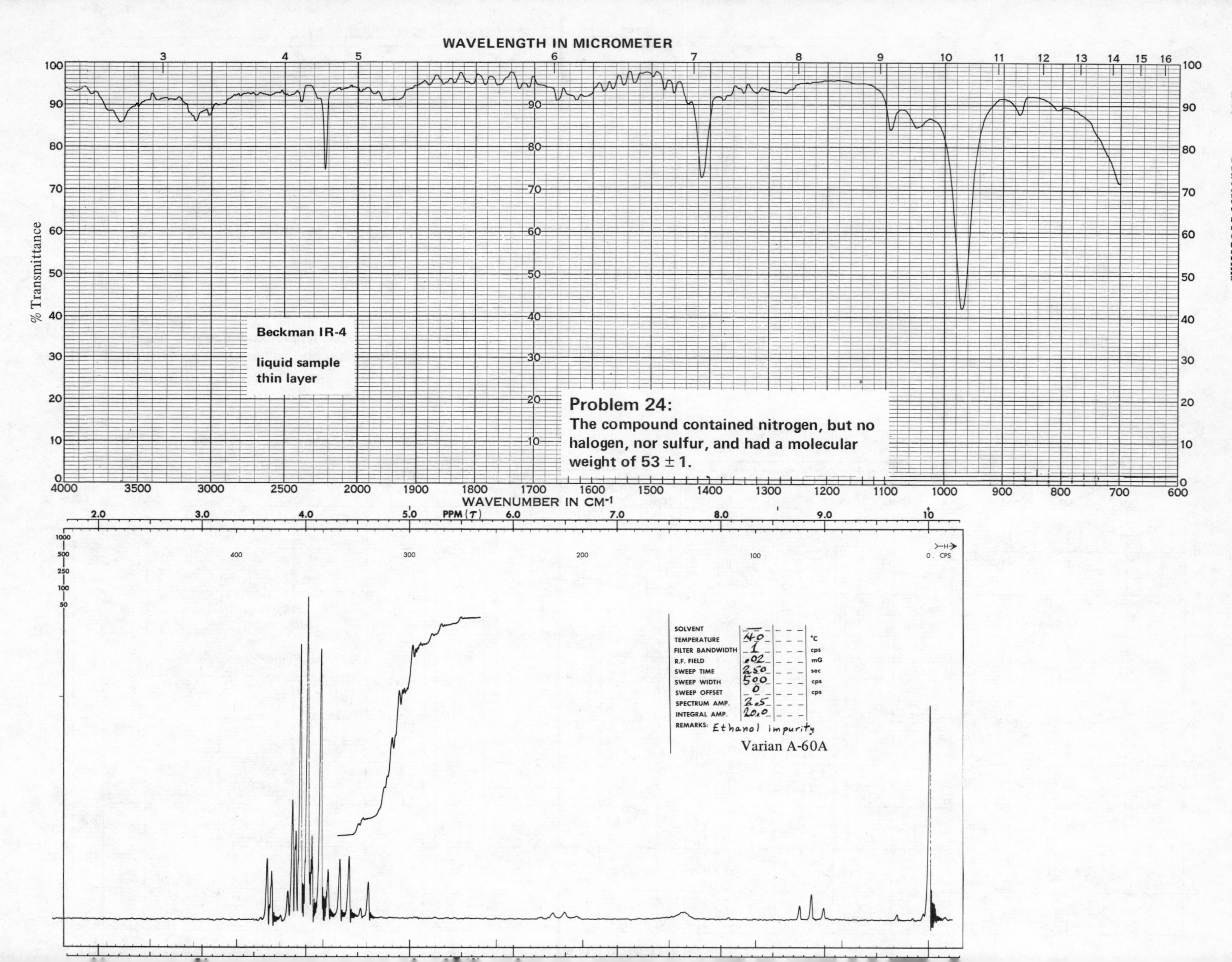
WAVELENGTH IN MICROMETER
% Transmittance
Beckman IR-4
liquid sample
thin layer
Problem 24:
The compound contained nitrogen, but no halogen, nor sulfur, and had a molecular weight of 53 ± 1.
WAVENUMBER IN CM-1
PPM (τ)
SOLVENT
TEMPERATURE 40 °C
FILTER BANDWIDTH 1 cps
R.F. FIELD .02 mG
SWEEP TIME 250 sec
SWEEP WIDTH 500 cps
SWEEP OFFSET 0 cps
SPECTRUM AMP. 2.5
INTEGRAL AMP. 20.0
REMARKS: Ethanol impurity
Varian A-60A

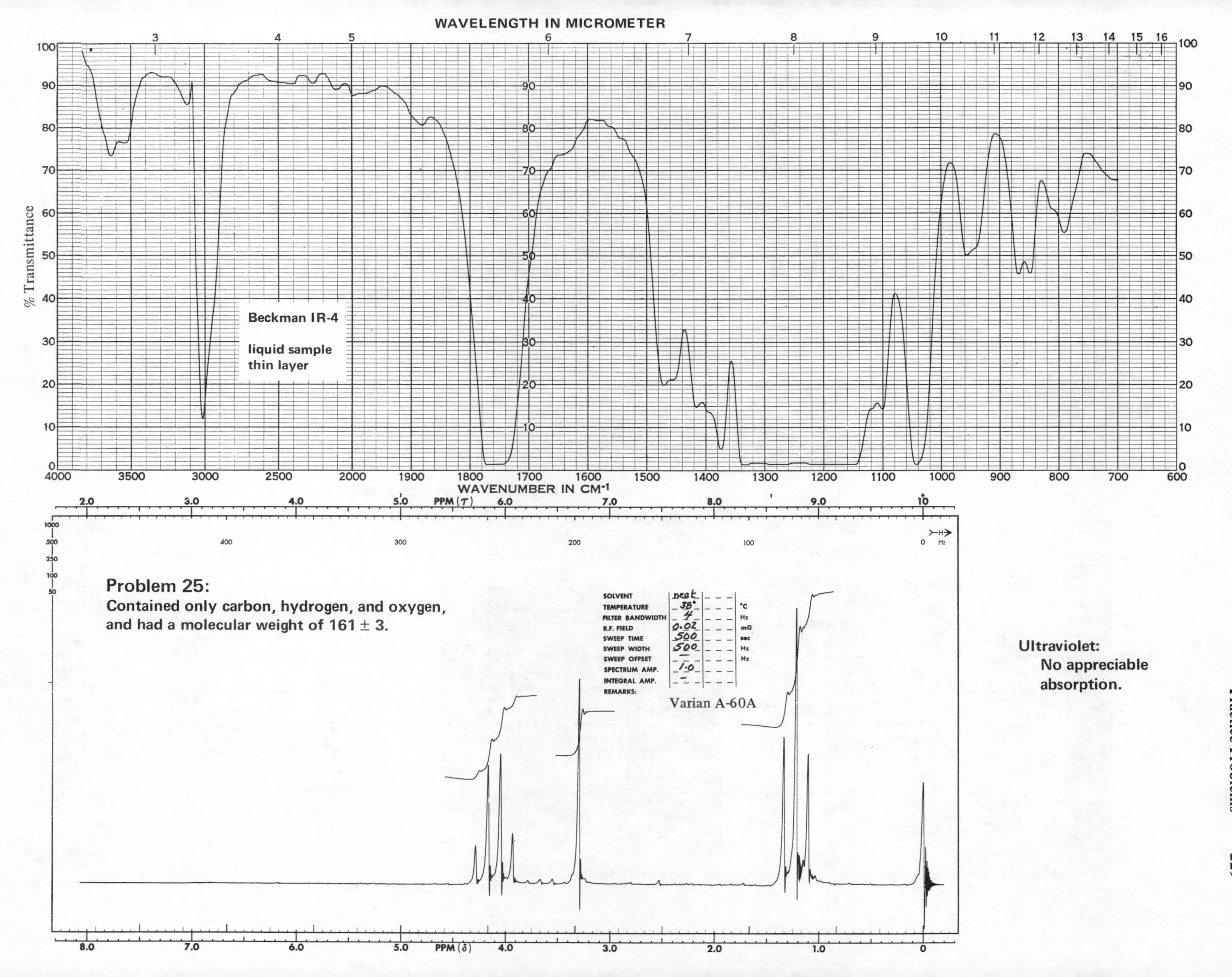
WAVELENGTH IN MICROMETER
WAVENUMBER IN CM-1
% Transmittance
Beckman IR-4
liquid sample
thin layer
Problem 25:
Contained only carbon, hydrogen, and oxygen, and had a molecular weight of 161 ± 3.
SOLVENT neat
TEMPERATURE 38° °C
FILTER BANDWIDTH 4 Hz
R.F. FIELD 0.02 mG
SWEEP TIME 500 sec
SWEEP WIDTH 500 Hz
SWEEP OFFSET — Hz
SPECTRUM AMP. 1.0
INTEGRAL AMP. —
REMARKS:
Varian A-60A
Ultraviolet:
No appreciable absorption.
PPM (τ)
PPM (δ)

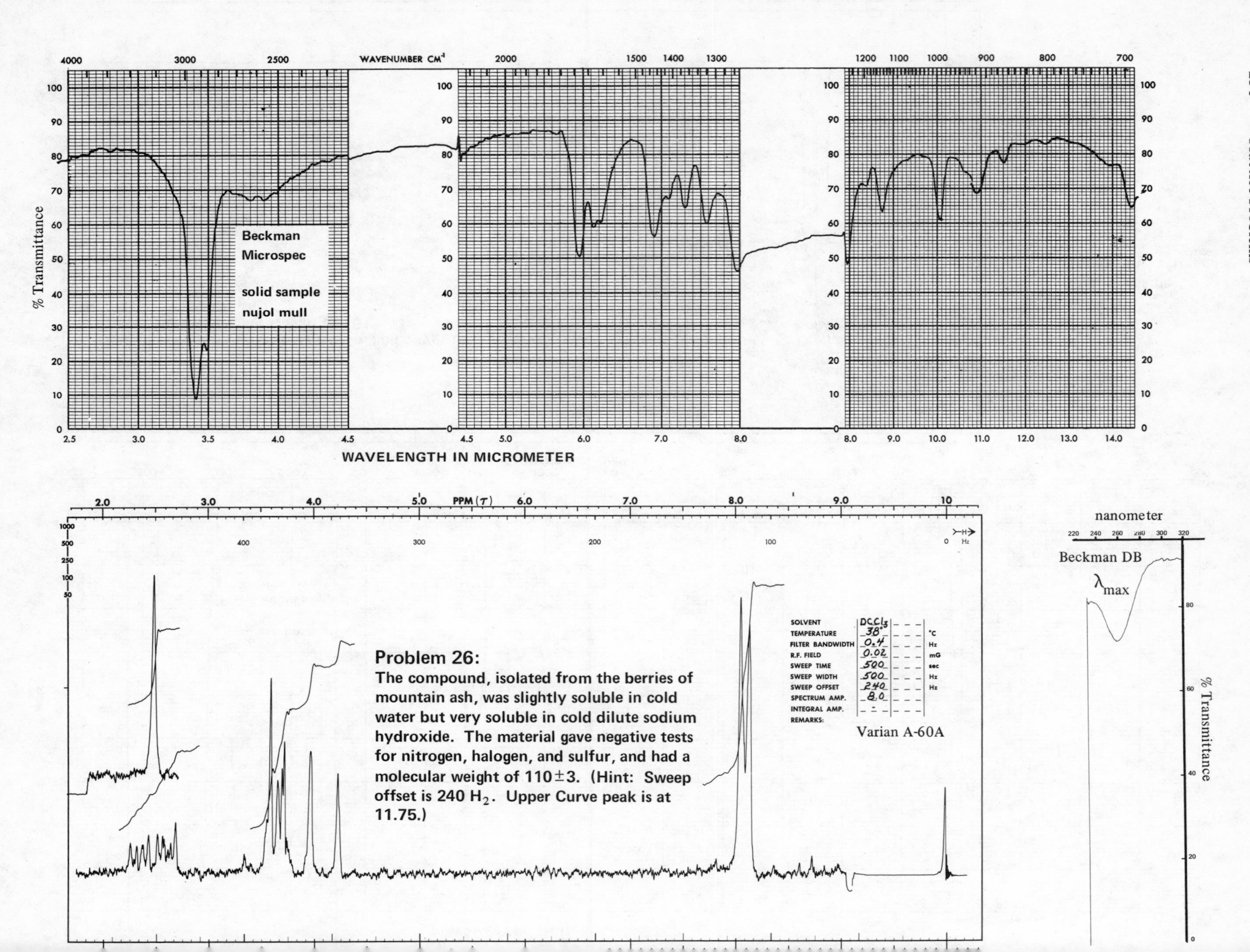


## Problem 26:

The compound, isolated from the berries of mountain ash, was slightly soluble in cold water but very soluble in cold dilute sodium hydroxide. The material gave negative tests for nitrogen, halogen, and sulfur, and had a molecular weight of $110 \pm 3$. (Hint: Sweep offset is 240 $H_2$. Upper Curve peak is at 11.75.)

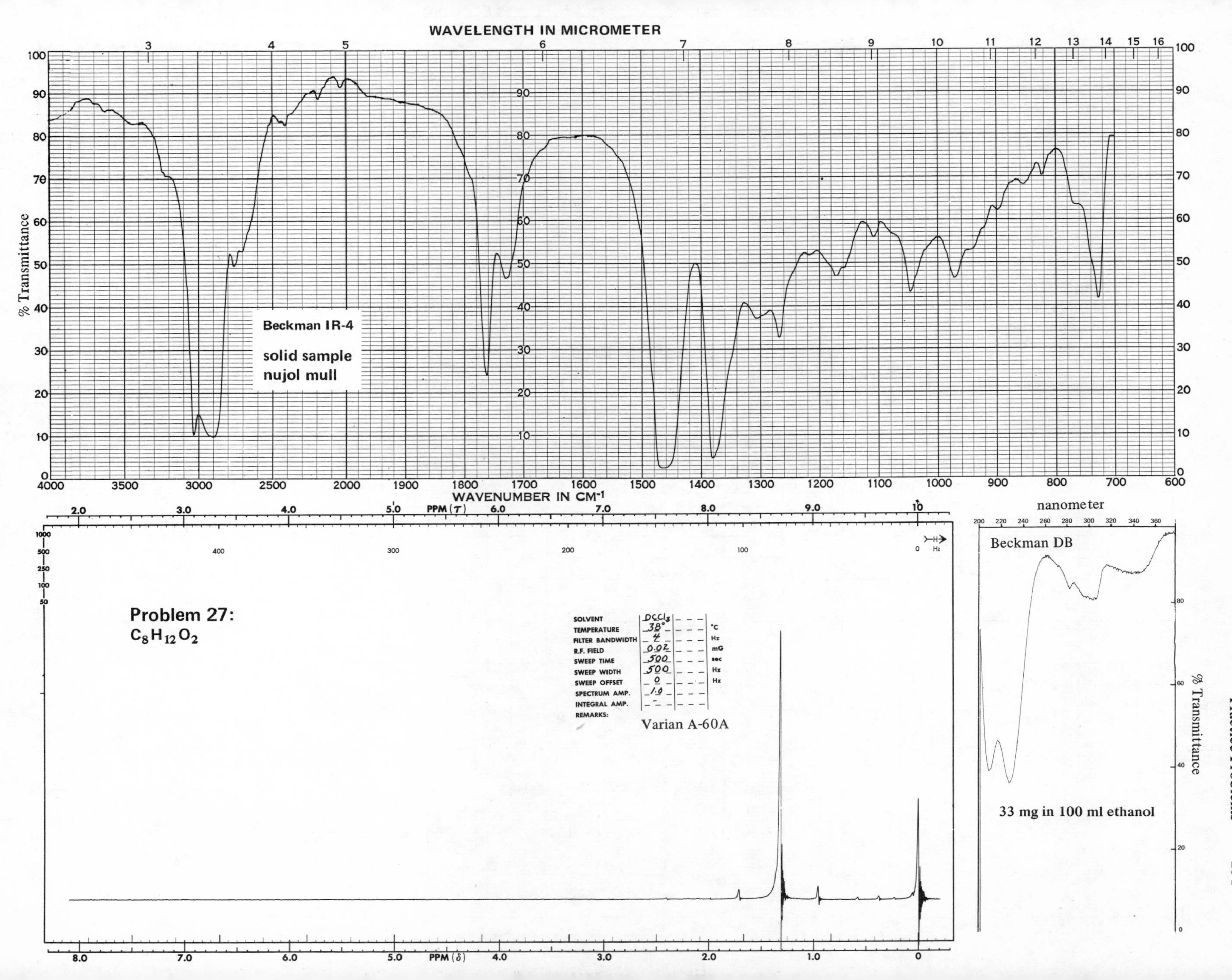
WAVELENGTH IN MICROMETER
WAVENUMBER IN CM-1
% Transmittance
Beckman IR-4
solid sample
nujol mull
nanometer
Beckman DB
33 mg in 100 ml ethanol
% Transmittance
PPM (τ)
PPM (δ)
Problem 27:
C8H12O2
SOLVENT DCCl3
TEMPERATURE 38° °C
FILTER BANDWIDTH 4 Hz
R.F. FIELD 0.02 mG
SWEEP TIME 500 sec
SWEEP WIDTH 500 Hz
SWEEP OFFSET 0 Hz
SPECTRUM AMP. 1.0
INTEGRAL AMP. -
REMARKS:
Varian A-60A

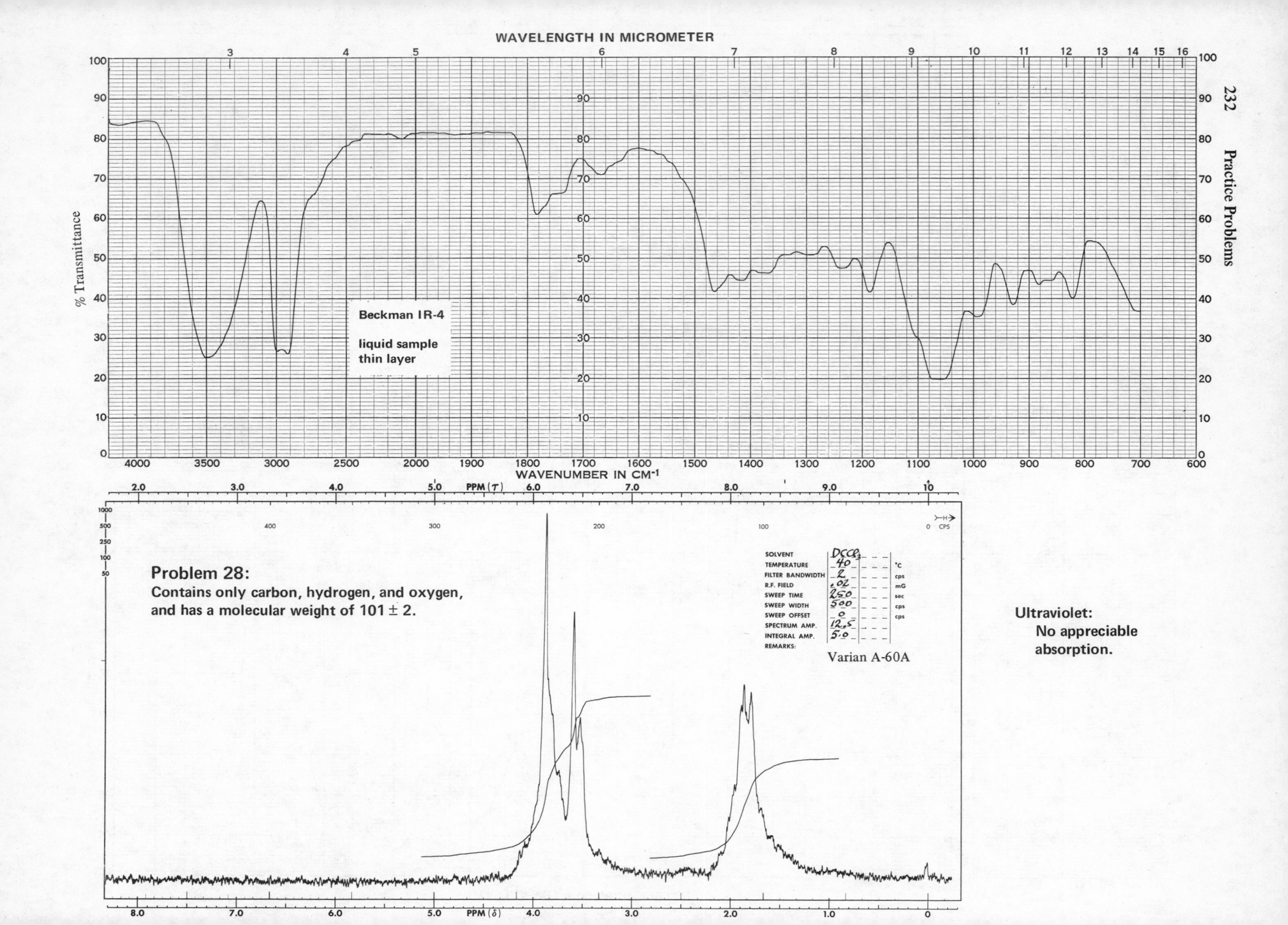


Problem 28:
Contains only carbon, hydrogen, and oxygen, and has a molecular weight of 101 ± 2.

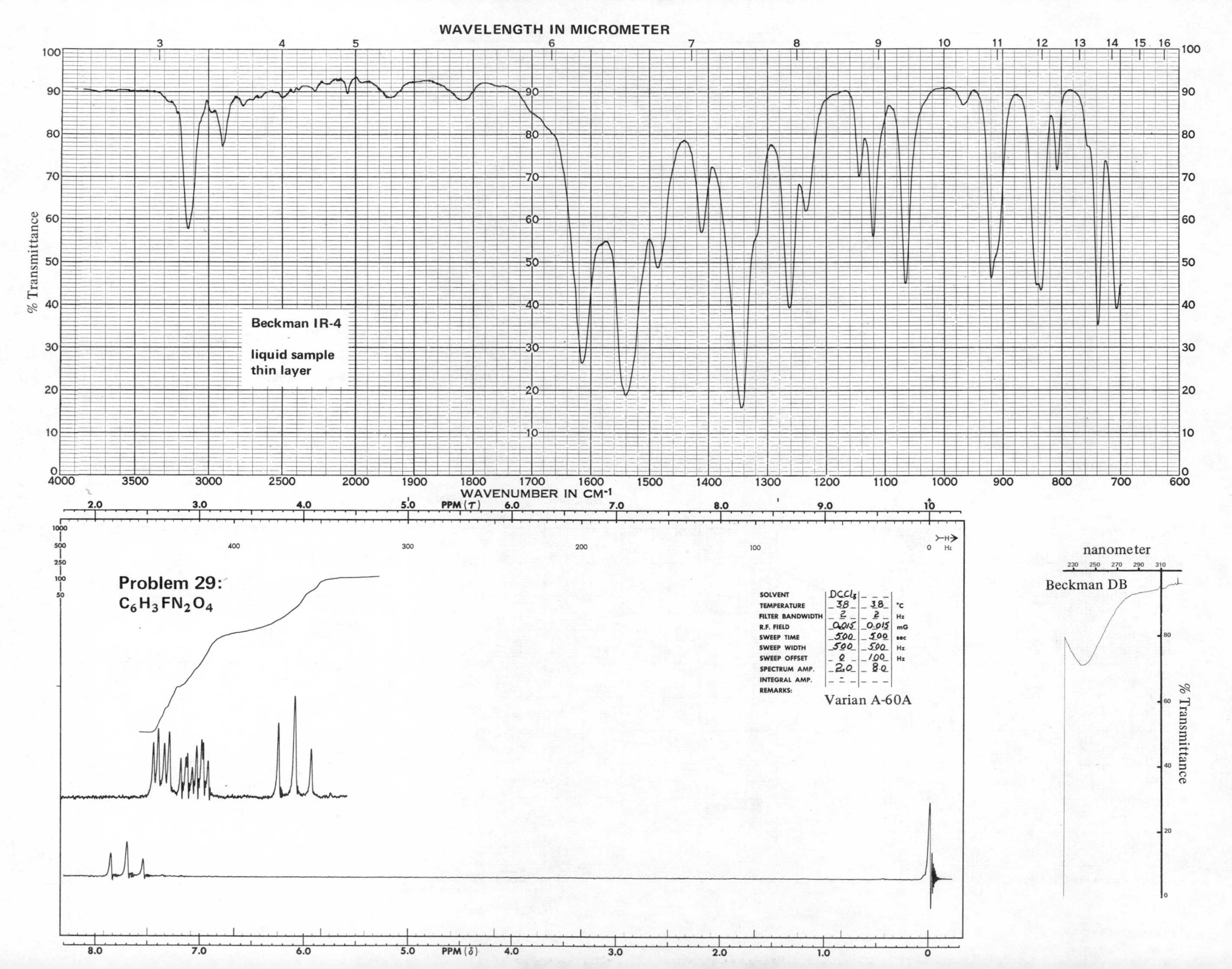
WAVELENGTH IN MICROMETER
WAVENUMBER IN CM-1
% Transmittance
Beckman IR-4
liquid sample
thin layer
PPM (τ)
PPM (δ)
Problem 29:
C6H3FN2O4
SOLVENT DCCl3
TEMPERATURE 38 38 °C
FILTER BANDWIDTH 2 2 Hz
R.F. FIELD 0.015 0.015 mG
SWEEP TIME 500 500 sec
SWEEP WIDTH 500 500 Hz
SWEEP OFFSET 0 100 Hz
SPECTRUM AMP. 2.0 8.0
INTEGRAL AMP.
REMARKS:
Varian A-60A
nanometer
Beckman DB
% Transmittance

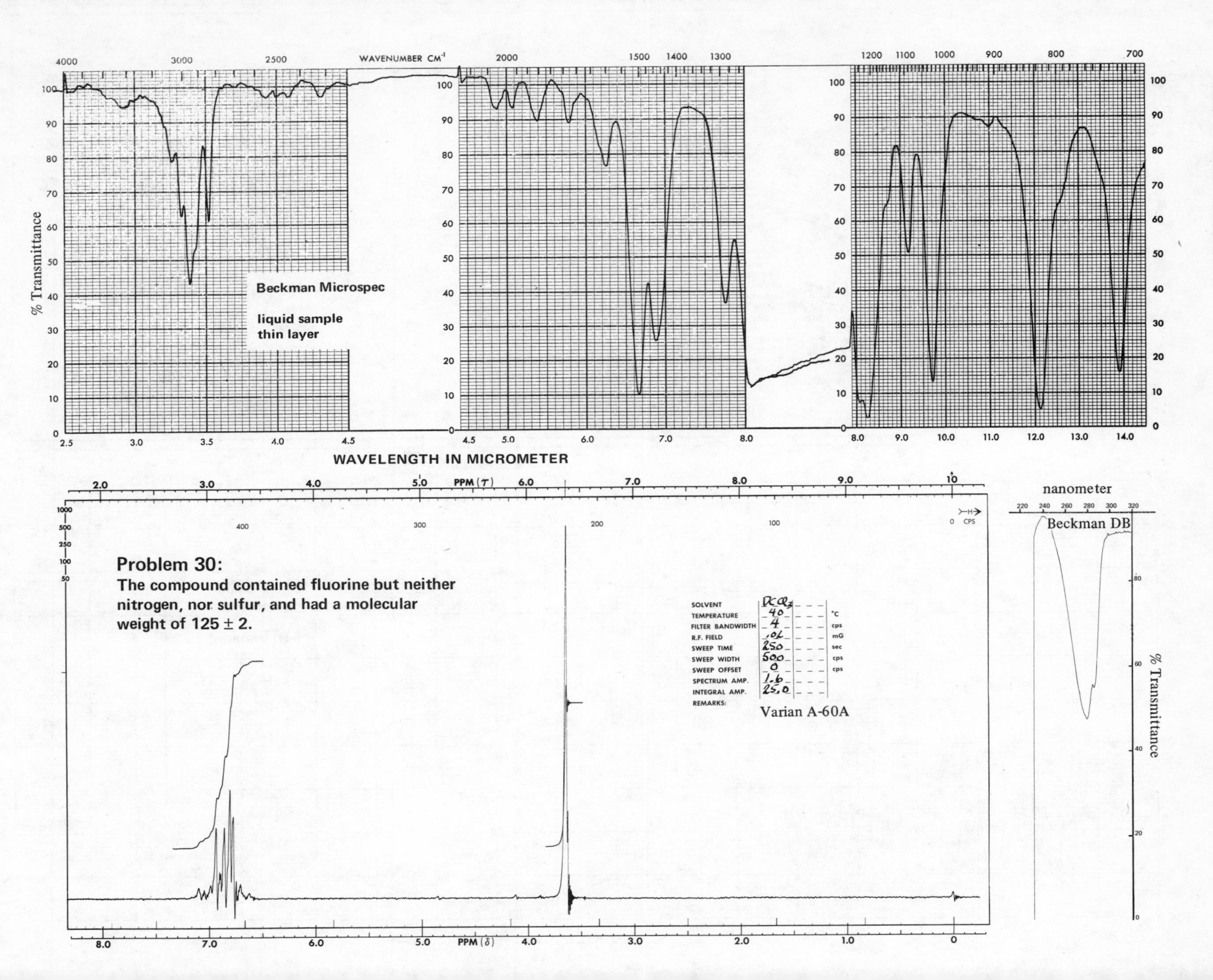
WAVENUMBER CM⁻¹
% Transmittance
Beckman Microspec
liquid sample
thin layer
WAVELENGTH IN MICROMETER
PPM (τ)
PPM (δ)
Problem 30:
The compound contained fluorine but neither nitrogen, nor sulfur, and had a molecular weight of 125 ± 2.
SOLVENT DCCl3
TEMPERATURE 40 °C
FILTER BANDWIDTH 4 cps
R.F. FIELD .06 mG
SWEEP TIME 250 sec
SWEEP WIDTH 500 cps
SWEEP OFFSET 0 cps
SPECTRUM AMP. 1.6
INTEGRAL AMP. 25.0
REMARKS:
Varian A-60A
nanometer
Beckman DB
% Transmittance

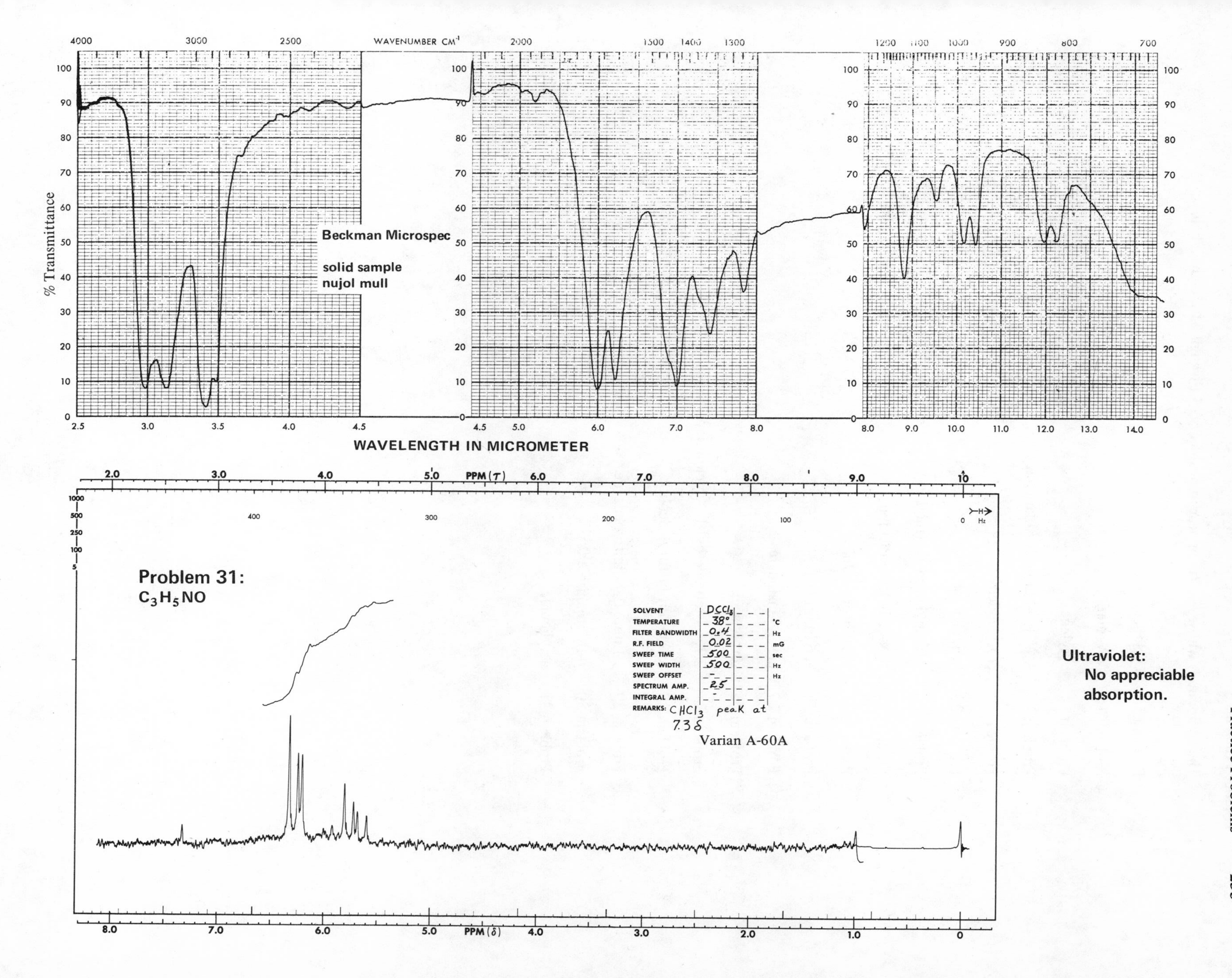
WAVENUMBER CM⁻¹
% Transmittance
Beckman Microspec
solid sample
nujol mull
WAVELENGTH IN MICROMETER
PPM (τ)
Problem 31:
$C_3H_5NO$
SOLVENT DCCl3
TEMPERATURE 38° °C
FILTER BANDWIDTH 0.4 Hz
R.F. FIELD 0.02 mG
SWEEP TIME 500 sec
SWEEP WIDTH 500 Hz
SWEEP OFFSET - Hz
SPECTRUM AMP. 25
INTEGRAL AMP. -
REMARKS: CHCl3 peak at 7.3 δ
Varian A-60A
PPM (δ)
Ultraviolet:
No appreciable
absorption.

**The compounds whose spectra appear in problems 15 through 31 are given below:**

| | |
|---|---|
| Problem 15 | cyclohexanone |
| Problem 16 | *N,N*- diethylaniline. |
| Problem 17 | *o*-xylene |
| Problem 18 | *N,N*-di-*n*-butylethanolamine |
| Problem 19 | crotonaldehyde |
| Problem 20 | benzylamine |
| Problem 21 | *p*-methylanisole (methyl-*p*-tolyl ether) |
| Problem 22 | N,N-diallylethanolamine |
| Problem 23 | 1,2-dibromoethane |
| Problem 24 | acrylonitrile |
| Problem 25 | diethyl malonate |
| Problem 26 | sorbic acid |
| Problem 27 | 2,2,4,4-tetramethyl-1,3-cyclobutanedione |
| Problem 28 | tetrahydrofurfuryl alcohol |
| Problem 29 | 1-fluoro-2,4-dinitrobenzene |
| Problem 30 | *p*-fluoroanisole. |
| Problem 31 | acrylamide |

# Chapter 7 MASS SPECTROSCOPY

The student may experience considerable conceptual difficulty in the process of learning mass spectrometry. A major problem arises because of the diversity of notations which are employed to describe the electronic structure of radical cations and to illustrate fragmentation. The best mechanistic system used in the literature is that originated by Shannon,[1] and more recently employed by Djerassi.[2]

This text will utilize the Shannon convention, but with certain modifications[3] which will be carefully defined. It is anticipated that these modifications, especially when employed to rationalize the mass spectra of hydrocarbons, will facilitate the learning of mass spectral fragmentation processes.

This chapter will develop the theory of mass spectroscopy and the way it is employed in structure elucidation.

**PART I: Separations Based Upon Mass/Charge Ratio**

After completing this section, you should be able to:

a) understand why positive ions of different masses are separated in the mass spectrometer
b) generate formulas from ion masses
c) recognize the significance of odd and even m/e values of molecular ions
d) recognize the significance of the isotopic patterns in the molecular ion region
e) use metastable ion peaks to corroborate fragmentation processes.

**PART II: Fission Processes in Which One Bond is Broken**

After completing this section you should be able to:

a) suggest a logical mechanism to explain fragmentations in which ions of odd m/e result from a parent ion of even m/e value
b) predict the structure of the ions which will result from simple cleavages in hydrocarbons, alcohols, ethers, amines, alkylhalides, aldehydes, ketones, and esters
c) distinguish between various structures on the basis of predictable fragmentations.

**PART III: Fission and Rearrangement Processes in Which Two Bonds are Broken**

After completing this section, you should be able to:

a) suggest a logical mechanism to explain fragmentations in which ions of even m/e result from a parent ion of even m/e value
b) predict the structure of the ions resulting from the McLafferty, retro-Diels Alder, dehydration and other fission processes in which two bonds are broken
c) distinguish between various structures on the basis of ions resulting from fission processes in which two bonds are broken.

**PART IV: Complex Fissions in Which Three Bonds are Broken**

After completing this section you should be able to:

a) predict the structure of ions resulting from complex fissions in cyclic alcohols, amines, and ketones and in acyclic amines and ethers.

**PART V: Analysis of Mass Spectral Data**

After completing this section you should be able to:

a) generate possible molecular formulas and determine the number of sites of unsaturation in each
b) determine structural and residual units as indicated by the mass spectral data
c) arrange structural units into molecular structures.

[1] J. S. Shannon, *Proc. Royal Austral. Chem. Inst.*, 328 (1964).
[2] H. Budzikiewicz, C. Djerassi, and D. H. Williams, *Mass Spectrometry of Organic Compounds,* Holden-Day Inc., San Francisco, (1967).
[3] M. M. Campbell and O. Runquist, *J. Chem. Ed.*, 49, 104 (1972).

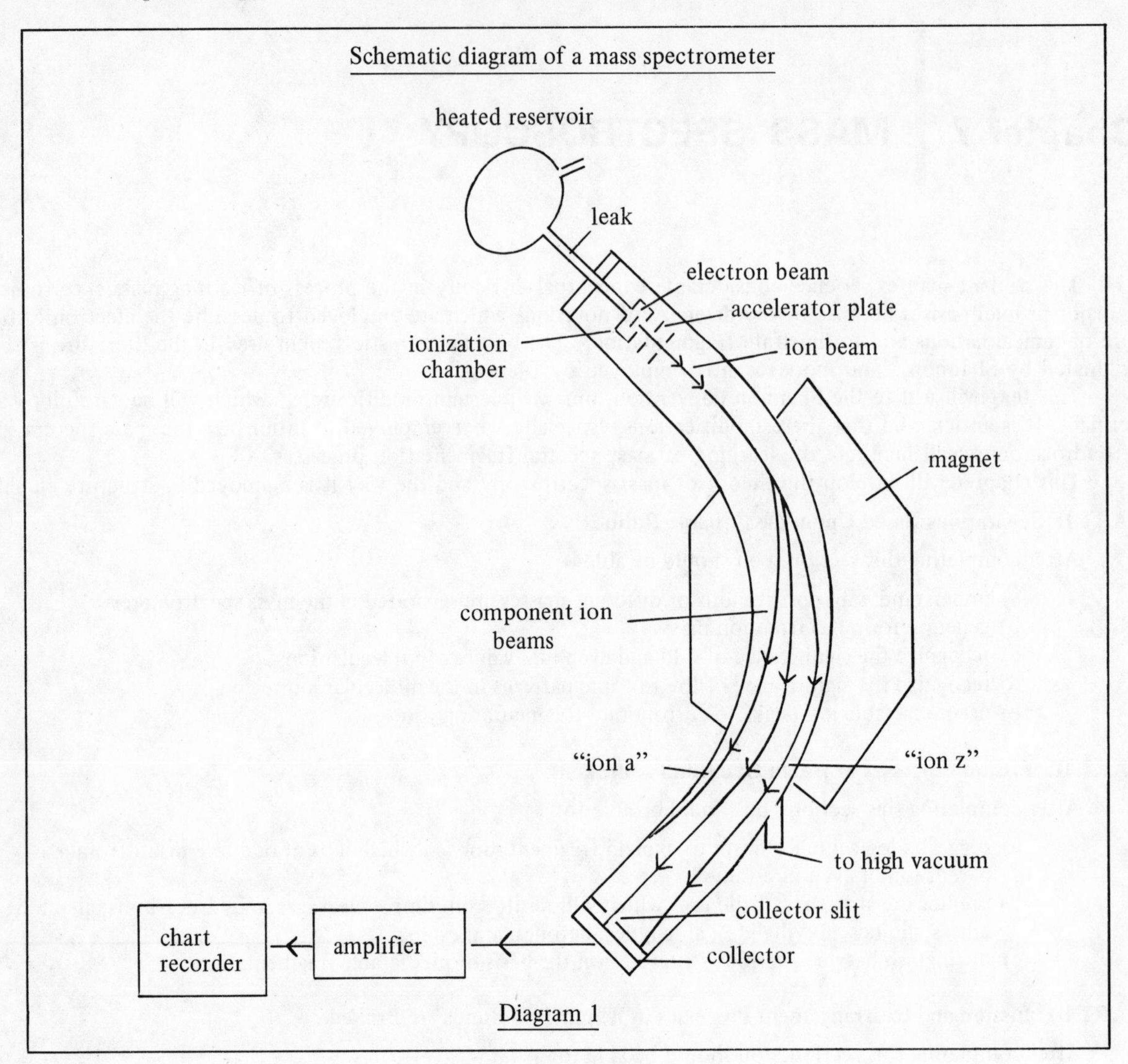
Schematic diagram of a mass spectrometer
heated reservoir
leak
electron beam
accelerator plate
ionization chamber
ion beam
magnet
component ion beams
"ion a"
"ion z"
to high vacuum
collector slit
chart recorder
amplifier
collector

Diagram 1

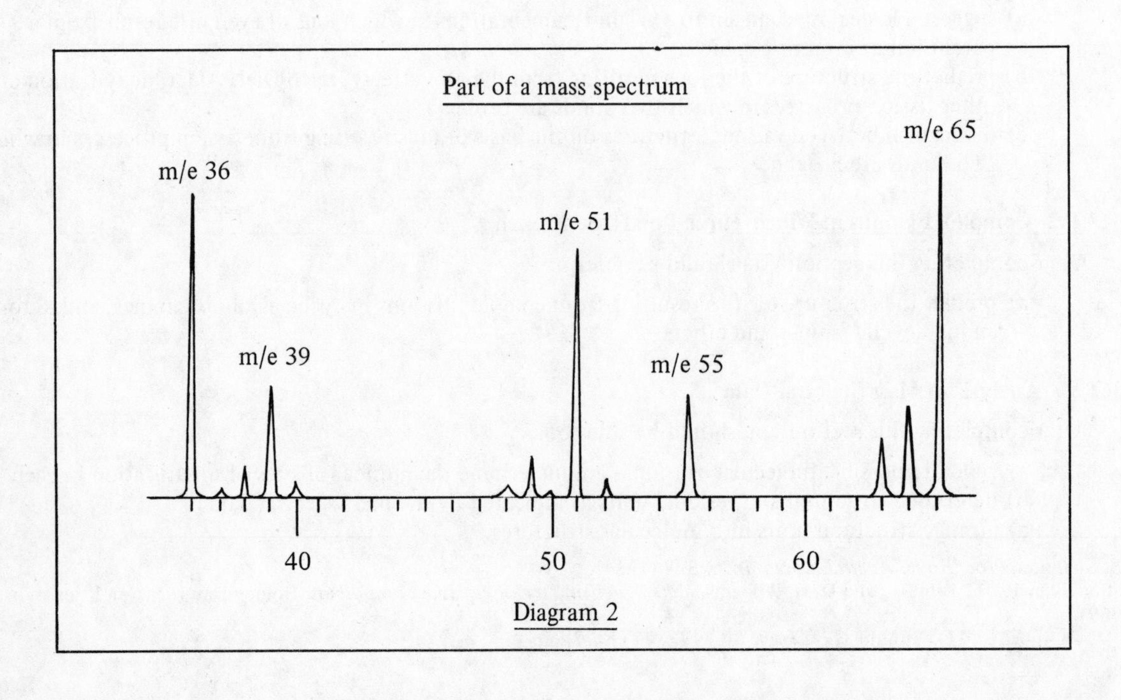
Part of a mass spectrum
m/e 36
m/e 39
m/e 51
m/e 55
m/e 65
40
50
60

Diagram 2

## PART I: Separations Based Upon Mass/Charge Ratio

S-1 The Operation of a Mass Spectrometer:

1. Liquid samples are volatilized under vacuum in the heated reservoir (as little as one microgram will suffice), and the vapor is leaked into the ionization chamber. Heating of the reservoir is often necessary to facilitate volatilization, especially when samples are high boiling. Solid samples are introduced into the ionization chamber on the tip of an insertion probe.
2. In the ion source the sample is bombarded with a stream of electrons of 70 eV energy. The energy absorbed by the molecules promotes ionization by loss of electrons from bonding or non-bonding orbitals. Ions formed by the removal of one electron from the original molecule are called *molecular,* or *parent,* ions. Some of the molecular ions fragment into smaller *daughter* ions and neutral fragments. Both positive and negative ions are formed but only positively charged species will be of concern. A small positive potential is used to repel the positive ions out of the ionization chamber.
3. An accelerator plate with a positive potential of 2000 V is used to accelerate the positive ions down the tube into the magnetic field.
4. The ions are deflected by the magnetic field to different extents depending on their mass/charge ratio, e.g., ion $C_5H_{11}^+$ is deflected to a lesser extent than ion $C_3H_7^+$. The ion beam is thus split into component ion beams of different mass/charge ratio.
5. Each ion beam in turn is made to pass through a collector slit and impinges on a collector plate. Each ion acquires an electron from the plate which neutralizes the positive charge. A flow of current is produced in the collector circuit, amplified, and recorded as a function of mass/charge ratio. The size of each peak is a measure of the relative number of ions in each beam. A typical portion of a mass spectrum is shown in Diagram 2.

---

Q-1 Why are only positive ions of consequence in organic mass spectrometry?

A-1 Only positive ions are repelled into the tube of the mass spectrometer by the positive potential. The negative ions are attracted by the positive potential and neutral fragments are unaffected.

Q-2 Different ions formed by the fragmentation process have different mass/charge (m/e) ratios. Most of the ions have unit positive charge, i.e., e = 1. How will the momentum of an ion travelling into the magnetic field vary with m/e?

A-2 The higher the m/e value the greater the momentum.

Q-3 How will the amount of deflection caused by the magnetic field vary with the momentum of the ion?

A-3 The greater the momentum, the less the deflection will be.

Q-4 What is the kinetic energy of an ion of mass m travelling with velocity v?

A-4 $\frac{1}{2}mv^2$

Q-5 The potential energy of an ion of charge e accelerated by a potential V is eV. If kinetic energy of the ion is equal to potential energy, how will the velocity of an ion change when the potential, V, is doubled?

A-5 $\frac{1}{2} mv^2 = eV$ (equation 1)

therefore, $v = \sqrt{\frac{2eV}{m}}$.

If V is doubled, v will increase by a factor of $\sqrt{2}$.

Q-6 A magnetic field B will deflect an ion of charge e and velocity v into a radial path of radius R. The centripetal force on the ion will be Bev. The centrifugal force on the ion will be $mv^2/R$. What are the conditions necessary for the ion to remain in the radial path of radius R?

A-6 Centrifugal force = centripetal force (otherwise the ion would fly off at a tangent).

$Bev = \frac{mv^2}{R}$ (equation 2)

Q-7 Eliminate v between equation 1 and equation 2 and show that

$m/e = \frac{B^2R^2}{2V}$.

A-7

$eV = \frac{mv^2}{2}$ --- (1)

$Bev = \frac{mv^2}{R}$ --- (2)

from (2) $v = \frac{BeR}{m}$

substituting in (1)

$eV = \frac{mB^2e^2R^2}{2m^2}$

$= \frac{B^2e^2R^2}{2m}$

$V = \frac{B^2eR^2}{2m}$

$m/e = \frac{B^2R^2}{2V}$ (equation 3)

Q-8 A beam of ions containing species with different m/e values is produced in a mass spectrometer which has a fixed slit position and a constant potential V. The magnetic field B is slowly increased. Will the species with the lowest or highest m/e value pass through the slit first? Why?

A-8 The low m/e species will pass through the slit first.

$m/e = \frac{B^2R^2}{2V}$

Because R and V are constant, the m/e value of the species which pass through the slit is directly proportional to $B^2$.

Q-9 In some mass spectrometers B is held constant and V, the accelerating potential, is gradually increased. Consider an ion of mass z and a lighter ion of mass a. Which species will pass through a fixed collector slit first as V is increased?

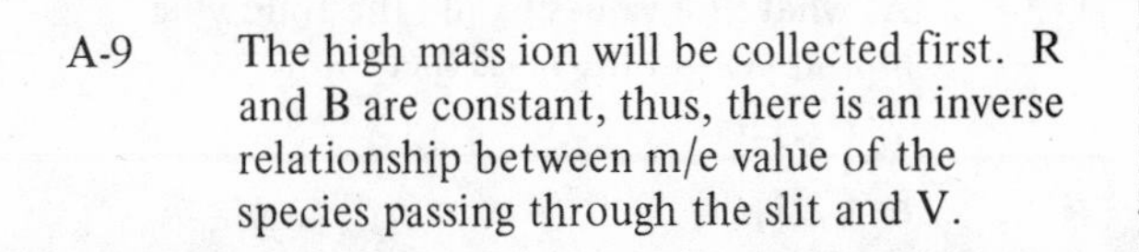

A-9 The high mass ion will be collected first. R and B are constant, thus, there is an inverse relationship between m/e value of the species passing through the slit and V.

Q-10 The current produced by ions entering the collection slit is recorded as a peak on a chart. The intensity (height) of the recorded peak is proportional to the current. Compare the peak intensities of two ions a and z if species z is more abundant.

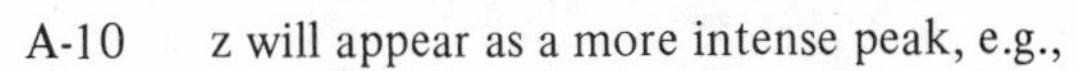

A-10 z will appear as a more intense peak, e.g.,

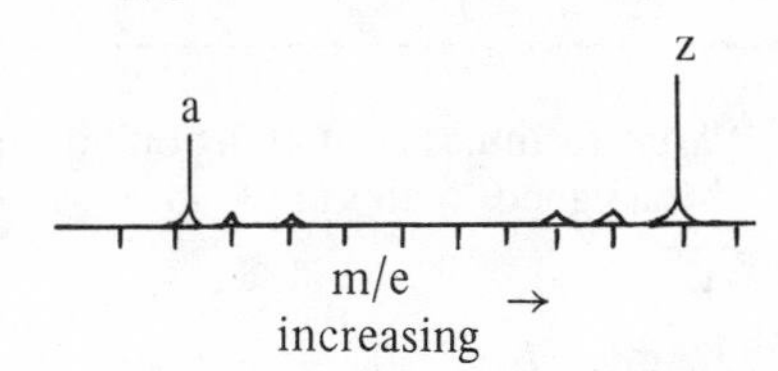


Q-11 What is the relationship between the observed heights of peaks in the mass spectrum and the relative abundance of the ions represented by the peak?

A-11 The more ions of a particular mass that there are, the more intense the corresponding peak in the mass spectrum will be.

Q-12 Comment on the relative abundance of ions, x, y, and z in the partial mass spectrum:

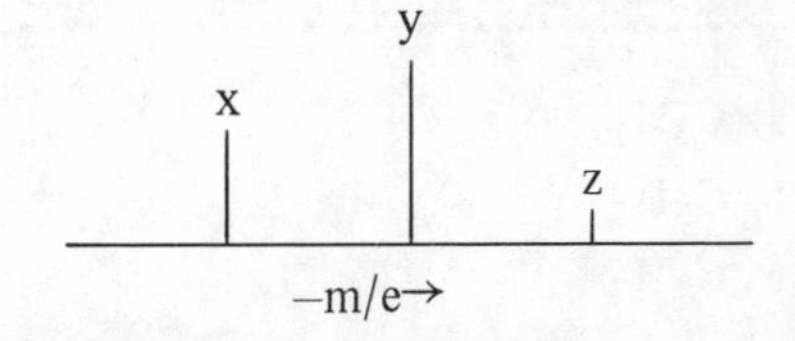


A-12 Ion y is the most abundant ion, followed by ion x and ion z.

**R** Ion beams are resolved according to their m/e values. They are recorded on a chart as a mass spectrum of their m/e values. The intensity of the peak in the mass spectrum is proportional to the number of ions formed.

## S-2 Low Resolution Mass Spectrometry

In low resolution mass spectrometry of organic compounds, the following approximate molecular weights are used to calculate the mass of ions and to determine possible elemental compositions of ions:

| | |
|---|---|
| H – 1 | N – 14 |
| D – 2 | O – 16 |
| C – 12 | S – 32 |

Q-13 What are the masses of the following common ions?

a. $CH_3^+$

b. $C_2H_5^+$

c. $C_6H_5^+$

A-13 a. 15 mass units
b. 29
c. 77

Q-14 At what m/e values would the following ions appear in the mass spectrum?
a. $C_2H_3O^+$
b. $C_6H_{11}O^+$
c. $C_2H_4I^+$ (I = 127)
d. $C_3H_4N^+$
e. $C_8H_{11}NOS^+$
f. $C_{10}H_8{}^{++}$

A-14 a. m/e 43
b. m/e 99
c. m/e 155
d. m/e 54
e. m/e 169
f. m/e 64 (e = 2)

Q-15 What formulas containing only C and H can be assigned to an ion of
a. m/e 43
b. m/e 65
c. m/e 91

A-15 a. $C_3H_7$
b. $C_5H_5$
c. $C_7H_7$

Q-16 What formulas containing C and H *or* C, H, and O can be assigned to an ion at m/e 71.

A-16 $C_5H_{11}$ *or* $C_4H_7O$ *or* $C_3H_3O_2$
$C_2H_{15}O_2$ and $CH_{11}O_3$ are unrealistic since no rational structures can be envisaged.

Q-17 What combinations of C, H, and N could account for the following ions?
a. m/e 43
b. m/e 57

A-17 a. $CH_3N_2$ *or* $C_2H_5N$
b. $CH_3N_3$ *or* $C_2H_5N_2$ *or* $C_3H_7N$

Q-18 An amide exhibits a fragment ion peak in its mass spectrum at m/e 58. What formula containing C and H could account for this peak?

A-18 $C_4H_{10}$ only

Q-19 Do the molecular ions of hydrocarbons occur at *odd* or *even* m/e values?

A-19 Invariably at *even* m/e

Q-20 Do the molecular ions of compounds containing C, H, and O occur at *odd* or *even* m/e values? Devise a series of test compounds.

A-20 *Even* m/e

Q-21 State a general rule governing the m/e value of the molecular ion of any compound containing C, H, and O.

---

A-21 The molecular ion will invariably occur at *even* m/e.

Q-22 Devise a series of compounds containing C, H, and an even number of N atoms. Are the molecular ion m/e values *odd* or *even?*

---

A-22 *Even* m/e

Q-23 Devise a series of compounds containing an *odd* number of N atoms, together with C and H. Are the molecular ions at *odd* or *even* m/e?

---

A-23 *Odd* m/e

Q-24 State the rule which governs the m/e values for molecular ions of compounds containing C, H and N.

---

A-24 An *even* number of N atoms results in an *even* m/e value. An *odd* number of N atoms results in an *odd* m/e value.

Q-25 Does the presence of O in an N compound invalidate the rule just expressed?

---

A-25 No. Devise a series of compounds, if necessary, to convince yourself.

Q-26 An N containing compound has a molecular ion at m/e 201. What immediate conclusion can be drawn?

---

A-26 The compound contains an odd number of N atoms.

Q-27 Is a molecular ion at m/e 151 consistent with structure 1?

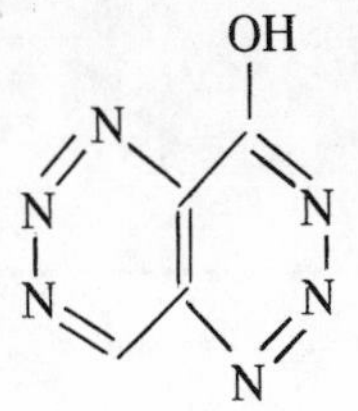

1

---

A-27 No, an even number of N atoms necessitates an even molecular weight.

Q-28 On the basis of the rules postulated above, what immediate conclusion can be drawn about the elemental constitution of an alkaloid of molecular weight 407 (derived by mass spectrometry)?

A-28 The alkaloid contains an odd number of N atoms. The molecular ion (molecular weight) is odd.

**R**

a. Empirical formulas can be assigned to ions in a mass spectrum. Often, several alternative formulaes are possible for a given m/e value. A method for determining which is the correct formula will be described in the following section on high resolution mass spectrometry.

b. Simple rules can be postulated which can be utilized in the initial examination of a mass spectrum to predict whether an *odd* or *even* number of N atoms are present.

S-3 High Resolution Mass Spectrometry

The low resolution mass spectrum affords unit m/e values for molecular ions and fragment ions. However, more than one elemental composition can often be attributed to a given m/e value as was shown in the previous section.

Certain high resolution mass spectrometers are capable of measuring the mass of an ion accurately to several decimal places, and can thus differentiate between alternative possible formulas for the ion. This technique is known as *high resolution mass spectrometry.*

Taking C = 12.000 the masses of the other elements frequently present in organic compounds are:

H = 1.0078
C = 12.0000
N = 14.0031
O = 15.9949
S = 31.9721

Q-29 Using values listed in S-3, calculate the exact mass of the nitrogen molecule ion.

A-29 28.0062

Q-30 In low resolution mass spectrometry, an ion is observed at m/e 28. What are four possible formulas for this ion?

A-30
a. CO
b. $N_2$
c. $CH_2N$
d. $C_2H_4$

Q-31 Using values listed in S-3, calculate the exact mass corresponding to each formula given in A-30.

A-31
a. 27.9949
b. 28.0062
c. 28.0187
d. 28.0312

Q-32 In the low resolution mass spectrum of an organic compound containing C, H, O, and N, an ion is observed at m/e 28. High resolution mass measurement indicates a value of 28.0188. Which of the four ions given in A-30 best corresponds to the high resolution mass measurement?

A-32 $CH_2N$.

Q-33 What elemental compositions are possible for an ion at m/e 66 (low resolution spectrum) from an organic compound which may contain O and/or N?

---

A-33 $C_5H_6$ *or*
$C_4H_2O$ *or*
$C_4H_4N$ *or* $C_3H_2N_2$
($C_3NO$ and $C_2N_3$ are possible but improbable formulas).

Q-34 The measured mass of the ion at m/e 66 is 66.0459. Which formula best corresponds to this mass?

---

A-34
$C_5H_6$ = 66.0468
$C_4H_2O$ = 66.0105
$C_4H_4N$ = 66.0343
$C_3H_2N_2$ = 66.0218
$C_3NO$ = 65.9980
$C_2N_3$ = 66.0093
Ion $C_5H_6$ fits best.

Q-35 Calculate the molecular ion m/e value for compound A (below). A significant ion appears at m/e 114 in the low resolution mass spectrum. How many mass units are lost from the molecular ion to produce m/e 114?

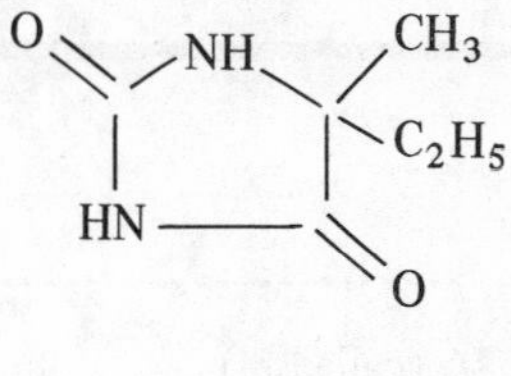

O, NH, $CH_3$, $C_2H_5$, HN, O

A

---

A-35 The molecular ion (or parent ion) has m/e 142.
Loss of 28 mass units produces the ion at m/e 114.

Q-36 The loss of 28 mass units occurs by extrusion of a neutral fragment. What formulas could the fragment have?

---

A-36
CO
$N_2$
$CH_2N$
$C_2H_4$

Q-37 What formulae can be assigned to the ion at m/e 114 described in Q-35.

---

A-37
$C_5H_{10}N_2O$
$C_6H_{10}O_2$
$C_5H_8NO_2$
$C_4H_6N_2O_2$

Q-38 High resolution mass spectral examination of the ion at m/e 114, Q-35, shows that it is a doublet of two ions of mass 114.0793 and 114.0429. Which of the postulated formulas best fit these data.

---

A-38 $C_5H_{10}N_2O$ (114.0791) and $C_4H_6N_2O_2$ (114.0428)

Q-39 Formulate the two extrusion processes which are occurring in the fragmentation of A to produce the ions of m/e 114.

A-39 $C_6H_{10}N_2O_2 \xrightarrow{-CO} C_5H_{10}N_2O$

$C_6H_{10}N_2O_2 \xrightarrow{-C_2H_4} C_4H_6N_2O_2$

**R** High resolution mass spectrometry can be used to assign elemental compositions to ions.

S-4 Isotopic Contributions

Many of the elements common to organic compounds occur naturally as isotopic mixtures. The natural abundance ratios of the elements and their isotopes are:

| | | | |
|---|---|---|---|
| $C^{12}$ | – 98.89% | $Cl^{35}$ | – 75.53% |
| $C^{13}$ | – 1.11% | $Cl^{37}$ | – 24.47% |
| $H^1$ | – 99.99% | $I^{127}$ | – 100% |
| $H^2$ | – 0.01% * | $F^{19}$ | – 100% |
| $N^{14}$ | – 99.64% | $Br^{79}$ | – 50.52% |
| $N^{15}$ | – 0.36% * | $Br^{81}$ | – 49.48% |
| $O^{16}$ | – 99.76% | $S^{32}$ | – 95.02% |
| $O^{17}$ | – 0.04%* | $S^{33}$ | – 0.75%* |
| $O^{18}$ | – 0.20%* | $S^{34}$ | – 4.22% |

* These isotopes are of particularly low abundance and will be ignored.

Q-40 At what m/e values will the molecular ions of $C^{12}H_4$ and $C^{13}H_4$ appear?

A-40 m/e 16 (called M)

m/e 17 (called M + 1)

The symbol M is used to denote the molecular ion containing the most abundant isotopes of the elements present.

Q-41 What are the odds against a molecule of methane containing $C^{13}$? Consult the list of natural isotopic abundances.

A-41 98.89 to 1.11 against the occurrence of $C^{13}H_4$.

Q-42 What relative intensities will the molecular ions M and M + 1 have in the low resolution mass spectrum of methane?

A-42 M : M + 1

= 98.89 : 1.11

this is normally rounded off to 98.9 : 1.1

Q-43 The isotopes of carbon make significant contributions to the M and M+1 peaks. To which peaks will the chlorine isotopes make a significant contribution in compounds containing one chlorine atom? Why?

A-43 M, contribution of $Cl^{35}$ isotope.

M + 2, contribution of $Cl^{37}$ isotope.

Q-44 For compounds containing *one* chlorine atom, what relative intensities will M and M + 2 have?

A-44 3 : 1

Q-45 Consider a compound containing *one* C atom and *one* Cl atom. What isotopic combinations are possible, and to which molecular ion peaks will they contribute?

A-45 $C^{12}Cl^{35}$ (M)
$C^{13}Cl^{35}$ (M + 1)
$C^{12}Cl^{37}$ (M + 2)
$C^{13}Cl^{37}$ (M +3)

Q-46 For the compound described in Q-45, what will be the relative heights of M and M + 1? What will be the relative heights of M and M + 2?

A-46 M and M + 1 differ only in C isotopes, therefore they will be in the ratio

98.9 : 1.1

M and M + 2 differ only in Cl isotopes, therefore they will be in the ratio 3 : 1, i.e.,

98.9 : 33.3

Q-47 For the compound described in Q-45, what are the relative heights of M+2 and M+3? What are the intensities of M+2 and M+3 relative to an M intensity of 98.9?

A-47 M+2 and M+3 differ only in C isotope content, therefore they will be in the ratio of

98.9 : 1.1

M+2 has been shown to have an intensity of 33.3 relative to an M intensity of 98.9. Therefore, M: M+2: M + 3 will be

$$98.9:33.3:\ 33.3 \times \frac{1.1}{98.9}$$

= 98.9:33.3:0.33.

Q-48 Write the relative intensities which have been derived for the M, M + 1, M + 2 and M + 3 peaks in a compound containing one C and one Cl atom.

Draw in the other molecular ion peaks.

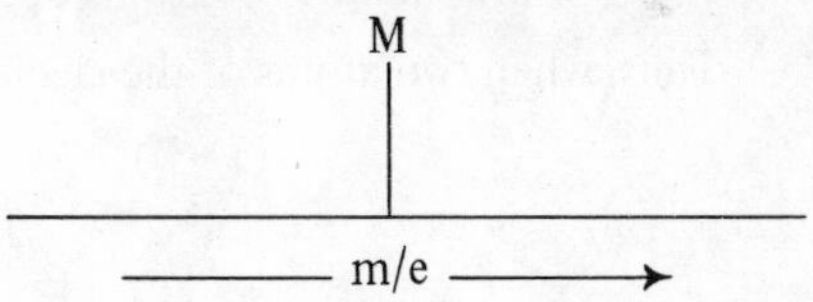

M

m/e →

A-48 M : M + 1 : M + 2 : M + 3
= 98.9 : 1.1 : 33.3 : 0.3

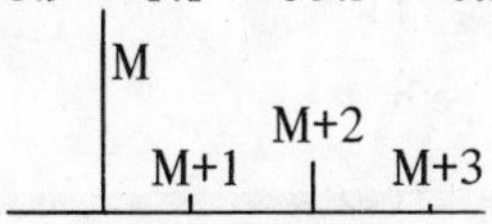

M
M+2
M+1 M+3

Q-49 Draw the molecular ion region expected for methyl bromide. Insert m/e values for the appropriate peaks. Label the isotopic peaks.

A-49 M M+2

— m/e →

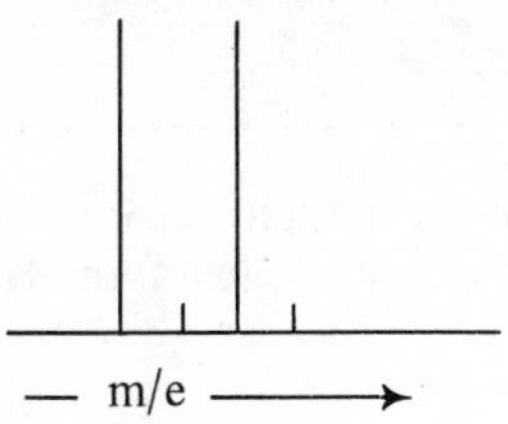

M = $C^{12}H_3Br^{79}$

M + 1 = $C^{13}H_3Br^{79}$

M + 2 = $C^{12}H_3Br^{81}$

M + 3 = $C^{13}H_3Br^{81}$ .

Note that $Br^{79}$ and $Br^{81}$ are equally abundant and contribute equally to M and M + 2 respectively.

Q-50 Draw the molecular ion region for methyl fluoride.

A-50

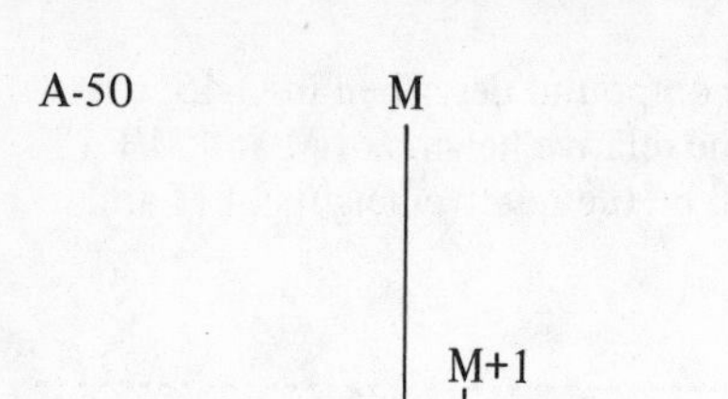


The peaks are caused by $C^{12}H_3F$ and $C^{13}H_3F$.

There are no F isotopes.

The presence of isotopes, especially for C, Cl, Br, and S may give rise to more than one molecular ion.

M is the symbol used to denote the molecular ion. The molecular ion contains the most abundant isotopes of the elements present.

S-5 The molecular ion region is complex for those molecules containing more than one atom which has a significant isotope, e.g., C, Cl, S, Br. The relative intensities of the peaks in the molecular ion region for these molecules may be calculated from the expression

$$(a + b)^m$$

where a = relative abundance of the lighter isotope

b = relative abundance of the heavier isotope

m = number of atoms of the element present in the molecule.

Thus, when two atoms of the element are present, the expression becomes

$$(a + b)^2$$

$$= \underset{\text{Term 1}}{a^2} + \underset{\text{Term 2}}{2ab} + \underset{\text{Term 3}}{b^2}$$

The appropriate values of a and b are substituted.

Term 1 gives the relative intensity of the peak containing only isotope a.

Term 2 gives the relative intensity of the peak containing isotopes a and b.

Term 3 gives the relative intensity of the peak containing only isotope b.

Q-51 What formulas are possible for ethane, taking into consideration the occurrence of $C^{12}$ and $C^{13}$?

A-51
a. $C^{12}H_3\ C^{12}H_3$
b. $C^{12}H_3\ C^{13}H_3$
c. $C^{13}H_3\ C^{13}H_3$

Q-52 What molecular ion values will correspond to each of the isotopic formulas listed in A-51.

A-52
a. m/e 30 (M)
b. m/e 31 (M + 1)
c. m/e 32 (M + 2)

Q-53 Apply the expression $(a + b)^m$, and calculate the relative intensities of M, M + 1, and M + 2 for ethane.

A-53 $(a + b)^m$

$= (a + b)^2$

$= a^2 + 2ab + b^2$

$= 98.9^2 + 2 \times 98.9 \times 1.1 + 1.1^2$

$= 9781 + 218 + 1.2$

Term 1 refers to
$C^{12}H_3C^{12}H_3$ (M)

Term 2 refers to
$C^{12}H_3C^{13}H_3$ (M + 1)

Term 3 refers to
$C^{13}H_3C^{13}H_3$ (M + 2)

M : M+1 : M+2

= 9781 : 218 : 1.2

= 97.8 : 2.2 : 0.01

Q-54 Draw the expected molecular ion region for ethane.

---

A-54

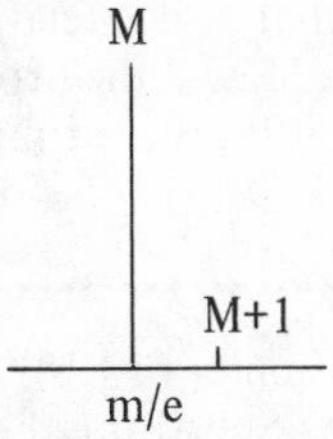

M

M+1

m/e

Q-55 As a rule of thumb, for every C atom in a molecule there will be a contribution of 1.1% to M + 1. This empirical rule is derived from the $(a + b)^m$ expression. What will the relative heights of M and M + 1 be in the mass spectrum of propane?

---

A-55 If M is regarded as 100 then M + 1 will be 3 x 1.1% = 3.3%

Q-56 What is the ratio of M to M + 1 in the mass spectrum of butane?

---

A-56 100 : 4.4

Q-57 The mass spectrum of a hydrocarbon exhibits a molecular ion peak at m/e 142 and a further peak at m/e 143. Why are there two peaks?

---

A-57 The peak at m/e 142 (M) is the molecular ion due to molecules containing $C^{12}$. The peak at m/e 143 (M + 1) is the isotope peak caused by those molecules which contain an atom of $C^{13}$

Q-58 What formulae can be ascribed to the hydrocarbon of m/e 142? What approximate ratios should M and M+1 have?

---

A-58 a. $C_{10}H_{22}$

M : M + 1 = 100 : 11

b. $C_{11}H_{10}$

M : M+1 = 100 : 12.

Q-59 An M : M + 1 ratio in a spectrum was 100 : 24. How many C atoms are present?

---

A-59 Approximately 22.

Q-60 What is the relative intensity of M + 1 to M in $C_{100}H_{202}$?

A-60 110% relative to M, i.e., the probability is that M + 1 is bigger than M. This is, in fact, observed.

**R** $C^{13}$ will contribute approximately 1.1% to M + 1 (relative to M) for every C atom the molecule contains.

S-6 The $(a + b)^m$ expression is especially suitable for calculating the isotopic abundances in poly-halogenated molecules. Polysulfides can be treated similarly.

Q-61 To what peaks, relative to the molecular ion peak, M, will the bromine isotopes make a significant contribution in compounds containing one bromine atom? What are the relative intensities of the contributions made by the bromine isotopes?

$Br^{79}$ = 50.52% $Br^{81}$ = 49.48%

A-61 Contributions will be made to the M (due to $Br^{79}$) and the M+2 (due to $Br^{81}$) peaks. The relative intensity of the contributions will be 50.52: 49.48 or approximately 1 : 1.

Q-62 What isotopic combinations are possible in a dibromide? To what peaks, relative to the molecular ion peak, M, will the bromine isotopes make a significant contribution.

A-62 a. $Br^{79}$ $Br^{79}$ M

b. $Br^{79}$ $Br^{81}$ M+2

c. $Br^{81}$ $Br^{81}$ M+4

Q-63 Consider the expansion of $(a + b)^m$ in the case of a dibromide. Substitute the value for m and explain the significance of the terms containing a and b.

A-63 $(a + b)^m = (a + b)^2 = a^2 + 2\,ab + b^2$

$a^2$ refers to those molecules containing $Br^{79}$ $Br^{79}$.

2ab refers to those molecules containing $Br^{79}$ $Br^{81}$.

$b^2$ refers to those molecules containing $Br^{81}$ $Br^{81}$.

Q-64 For the expression derived in A-63, substitute the appropriate values for a and b. Find the relative intensities of M, M + 2 and M + 4.

A-64 $a^2 + 2ab + b^2$

$= 1^2 + 2 \times 1 \times 1 + 1^2$

$= 1 + 2 + 1$

M : M + 2 : M + 4 =

1 : 2 : 1.

Q-65 To what peaks, relative to the molecular ion peak, M, will the chlorine isotopes make a significant contribution in a compound containing two chlorine atoms? What are the relative intensities of the contributions due to the chlorine isotopes?

$Cl^{35}$–75.53%, $Cl^{37}$–24.47%

(approximately, $Cl^{35} : Cl^{37}$ = 3:1)

A-65 M : M+2 : M+4 = 9:6:1

$(a + b)^m$

$= (a + b)^2$

$= a^2 + 2\,ab + b^2$

$= 3^2 + 2 \times 3 \times 1 + 1^2$

$= 9 + 6 + 1$

M : M + 2 : M + 4

= 9 : 6 : 1

Q-66 To what peaks relative to the molecular ion peak, M, will the chlorine isotopes make a significant contribution in a compound containing three chlorine atoms? What are the relative intensities of the contributions by the chlorine isotopes?

A-66 M : M+2 : M+4 : M+6 = 27 : 27 : 9 : 1

$(a + b)^m$

$= (a + b)^3$

$= a^3 + 3a^2b + 3ab^2 + b^3$

$= 3^3 + 3 \times 3^2 \times 1 + 3 \times 3 \times 1^2 + 1^3$

$= 27 + 27 + 9 + 1$

M : M + 2 : M + 4 : M + 6

= 27 : 27 : 9 : 1

Q-67 Write down all the isotopic combinations of $C^{12}$, $C^{13}$, $Cl^{35}$, and $Cl^{37}$ possible for chloroform. To which isotope peaks do they contribute in the molecular ion region of the mass spectrum?

A-67

| | |
|---|---|
| $C^{12}Cl^{35}Cl^{35}Cl^{35}$ | M |
| $C^{13}Cl^{35}Cl^{35}Cl^{35}$ | M + 1 |
| $C^{12}Cl^{37}Cl^{35}Cl^{35}$ | M + 2 |
| $C^{13}Cl^{37}Cl^{35}Cl^{35}$ | M + 3 |
| $C^{12}Cl^{37}Cl^{37}Cl^{35}$ | M + 4 |
| $C^{13}Cl^{37}Cl^{37}Cl^{35}$ | M + 5 |
| $C^{12}Cl^{37}Cl^{37}Cl^{37}$ | M + 6 |
| $C^{13}Cl^{37}Cl^{37}Cl^{37}$ | M + 7 |

Q-68 For chloroform the intensities of M and M+1 peaks differ only because of the contributions of the carbon isotopes. What are the relative intensities of the M and M+1 peaks in chloroform? What are the relative intensities of M+2 : M+3; M+4 : M+5; M+6 : M+7?

A-68 M : M + 1

= 98.9 : 1.1

The other ratios are also 98.9 : 1.1.

Q-69 In trichlorides the relative contributions made by the chlorine isotopes to the M:M+2 : M+4 : M+6 peaks are 27 : 27 : 9 : 1 respectively. For chloroform, what are the relative intensities of peaks M : M+1 : M+2 : M+3 : M+4 : M+5 : M+6 : M+7?

A-69 27 : 0.27 : 27 : 0.27 :
9 : 0.09 : 1 : 0.01

Q-70 When mixed halogen atoms are present, e.g., Cl and Br, the binomial expansion

$$(a + b)^m \ (c + d)^n$$

is used. The first bracket refers to the Cl isotopes (a = 3, b = 1) and the second bracket refers to the bromine isotopes ( c = 1, d = 1); m and n refer to the number of each of the halogen atoms present.

Expand the binomial expression, when m = 1 and n = 1.

To which isotope peak will each of the four terms obtained contribute?

A-70 $(a + b)^1 (c + d)^1$
$= ac + ad + bc + bd$

ac refers to $Cl^{35}Br^{79}$ M

ad refers to $Cl^{35}Br^{81}$ M + 2

bc refers to $Cl^{37}Br^{79}$ M + 2

bd refers to $Cl^{37}Br^{81}$ M + 4

Note that ad and bc both contribute to M + 2 and should therefore be summed when substituting for a, b, c, and d.

Q-71 Substitute the appropriate values of a, b, c, and d and calculate the relative intensities of contributions to M, M+2, M+4 by the halogen isotopes in a mono-bromo, mono-chloro-compound.

A-71 M = ac = 3 x 1 = 3

M + 2 = ad + bc = 3 x 1 + 1 x 1 = 4

M + 4 = bd = 1 x 1 = 1

Therefore M : M + 2 : M + 4 = 3 : 4 : 1.

Q-72 What isotopic contribution would sulfur make in $CH_3SH$? What are the relative intensities of contributions made by the S isotopes?

A-72 The $S^{34}$ atom will contribute to the M + 2 peak. M : M + 2 = 95 : 4.

Q-73 Calculate the sulfur isotope pattern in the mass spectrum of a disulfide.

A-73 $(a + b)^m$

$= (a + b)^2$

$= a^2 + 2ab + b^2$

$= 95^2 + 2 \times 95 \times 4 + 4^2$

$= 9025 + 760 + 16$

approximate ratio

= 100 : 8 : 0.2

= M : M + 2 : M + 4.

**R** From the binomial expansion of $(a + b)^m$, the isotope patterns can be predicted. Information can be derived about the elemental composition of molecules from isotopic patterns.

S-7 Ionization of an organic molecule normally requires approximately 10-15 eV. In mass spectrometry, however, molecules are subjected to electronic bombardment of energy 70 eV. One electron is removed from the molecule, forming a high energy radical cation which has a large probability of fragmenting in order to dissipate its excess energy.

As a rule, the electron will be lost from the most easily ionizable site in the molecule, e.g., from a lone pair of electrons on an atom such as O, N, S or halogen, or from an unsaturated bond. If a molecule has no lone pairs or sites of unsaturation, then the electron will be lost from a sigma bond. C-C sigma bonds are more easily ionized than C-H sigma bonds.

---

Q-74 How many electrons surround the nucleus in
a. hydrogen cation ($H^+$)
b. hydrogen radical (H·)
c. hydrogen anion ($H^-$)?

---

A-74
a. none
b. one
c. two

Q-75 Draw the electronic structure of
a. methyl cation, $CH_3^+$.
b. methyl radical, $CH_3^{\bullet}$.

How many valence electrons surround the C-atom in each case? Put the appropriate charge outside brackets in each case.

---

A-75
a. 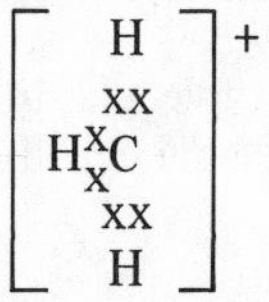
```
[    H    ]+
[   xx    ]
[  HxC    ]     6 electrons
[   x     ]     (+ charge)
[   xx    ]
[    H    ]
```
b. 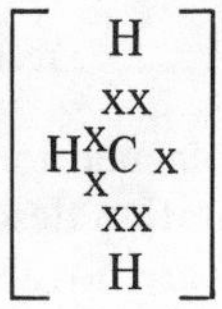
```
[    H    ]
[   xx    ]
[  HxC x  ]     7 electrons
[   x     ]     (0 charge)
[   xx    ]
[    H    ]
```

Q-76 Draw the electronic structure of ethane, showing the valence electrons.

Draw the structure obtained by ionization of the C-C sigma bond. Do not yet write in any charge.

---

A-76
```
   H  H
  xx xx
 HxC x CxH        ethane
  x  x  x
  xx xx
   H  H

   H  H
  xx xx
 HxC x CxH        ionized ethane
  x     x
  xx xx
   H  H
```

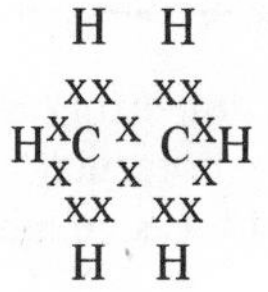

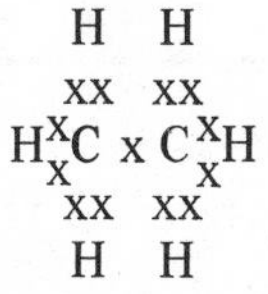

Q-77 Consider the single sigma electron in ionized ethane to be localized on one of the methyl groups.

a. How many valence electrons are associated with that methyl group? Can the methyl group be considered as a cation, a radical or an anion?

b. How many valence electrons are associated with the other methyl group? Is it a cation, radical or anion?

---

A-77
a. 7 electrons, i.e., *radical.*
b. 6 electrons, i.e., *cation.*

Q-78 What is the net charge associated with the ethane molecular ion?

A-78 +1

It is both a radical (no charge associated with the unpaired electron) and a cation, i.e., a *radical cation.*

Q-79 In the chemical literature, ethane molecular ion is commonly written as

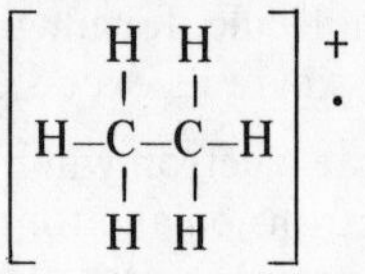

```
┌  H H  ┐+
   | |   •
   H–C–C–H
   | |
└  H H  ┘
```

Criticize this formal representation.

---

A-79 The sigma bonds are drawn as lines, implying two electrons in each bond. However, the radical cation charge implies that one of the sigma bonds contains only *one* electron.

The site of ionization is thus not clearly defined.

Q-80 Assuming the C–C bond to be ionized in ethane, draw the molecule with the symbol +• joining the C atoms instead of a line. Enclose the structure in brackets.

How many electrons are in the carbon-carbon sigma bond?

---

A-80

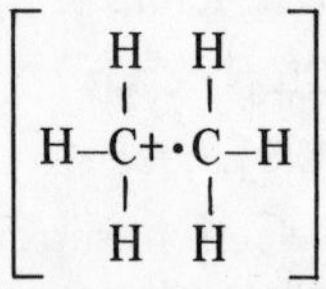

```
┌    H  H    ┐
     |  |
   H–C+•C–H
     |  |
└    H  H    ┘
```

There is a single electron in the carbon-carbon sigma bond.

Q-81 Draw the electronic structure of methane radical cation indicating all valence electrons.

---

A-81

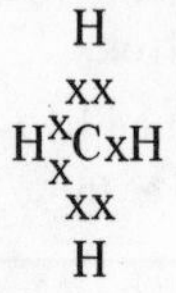

```
   H
   xx
 HˣCxH
  x
   xx
   H
```

Q-82 Draw the methane molecular ion as it might appear in the chemical literature.

---

A-82

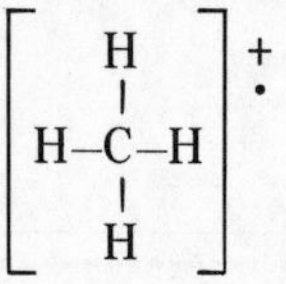

```
┌   H   ┐+
    |    •
  H–C–H
    |
└   H   ┘
```

Q-83 Draw the methane molecular ion using the more precise bond notation described in A-80. Two structures are possible.

---

A-83

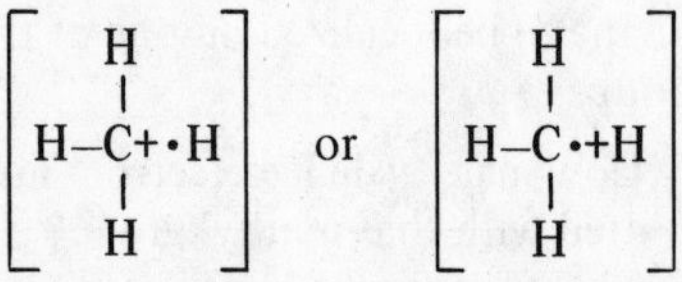

```
┌   H    ┐          ┌   H    ┐
    |                   |
  H–C+•H      or      H–C•+H
    |                   |
└   H    ┘          └   H    ┘
```

Q-84 Draw the two radical cation structures possible for the propane molecular ion. Assume ionization of a C–C sigma bond.

A-84

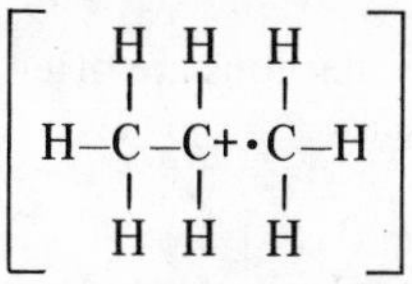

```
 H  H  H
 |  |  |
H–C–C+•C–H
 |  |  |
 H  H  H
```

and

```
 H  H  H
 |  |  |
H–C–C•+C–H
 |  |  |
 H  H  H
```

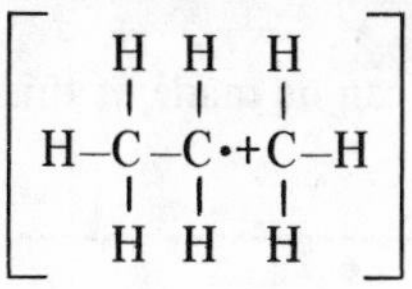

Q-85 What radical cation structures could result from ionization of butane? Assume that C–C sigma bonds are ionized.

A-85

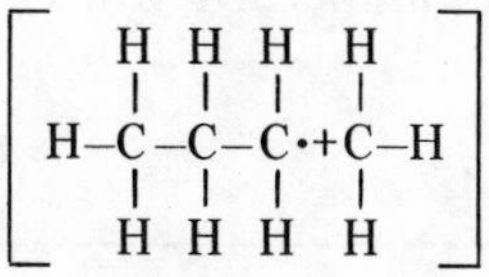

a.

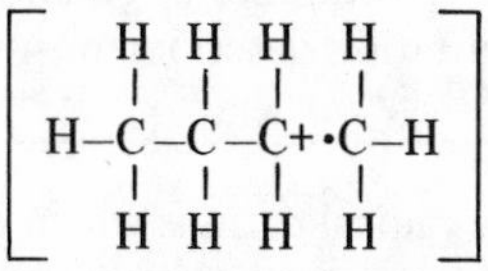

```
 H  H  H  H
 |  |  |  |
H–C–C–C+•C–H
 |  |  |  |
 H  H  H  H
```

b.

```
 H  H  H  H
 |  |  |  |
H–C–C–C•+C–H
 |  |  |  |
 H  H  H  H
```

c.

```
 H  H  H  H
 |  |  |  |
H–C–C+•C–C–H
 |  |  |  |
 H  H  H  H
```

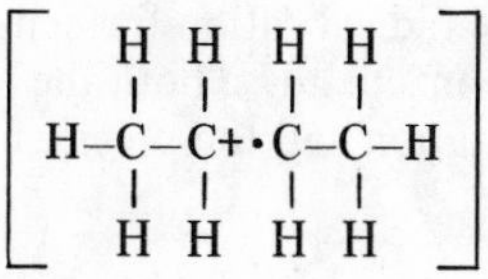

Q-86 Draw the molecular orbitals (bonding and anti-bonding) for ethylene, showing the relative energies of the sigma and pi orbitals.

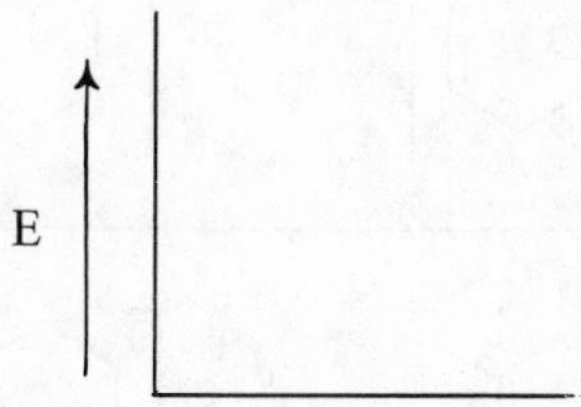

E

A-86

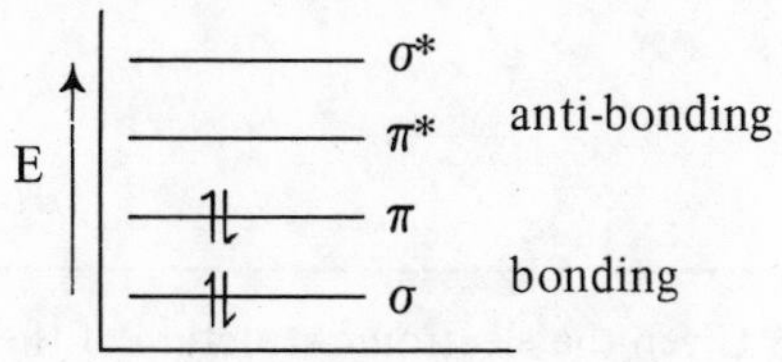

E

$\sigma^*$

$\pi^*$ anti-bonding

$\pi$

$\sigma$ bonding

Q-87 Which electron is most easily lost during the process of ionization of ethylene?

A-87 One of the two pi electrons.

Q-88 Draw the ethylene molecular ion. Indicate a two electron sigma bond with a line and the ionized bond with the convention given in Q-80.

A-88

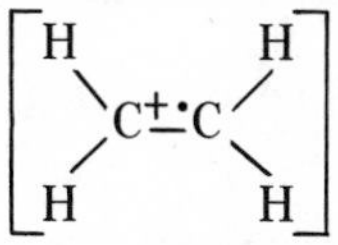

```
H       H
 \     /
  C+•C
 /     \
H       H
```

Q-89 How many electrons are in the bonding pi orbital in the ionized ethylene structure?

A-89 One.

A-90 The sigma and pi bonds are drawn intact. The site of ionization, although implied to be in the pi-bond, is not clearly shown.

A-91

$[\text{cyclohexene}]^{+\cdot}$ or $[\text{cyclohexene}^{+\cdot}]$

A-92

1-methylcyclohexene $\xrightarrow{-e}$ 1 or 2

A-93

literature or literature

precise

A-94 An n electron, i.e., one of the lone pair electrons.

Q-90 Ethylenic molecular ions are often depicted in the literature as

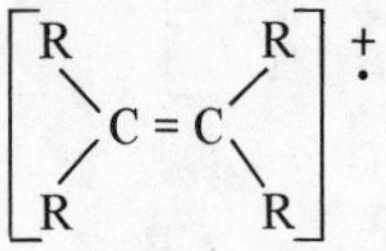

$$\left[\begin{matrix} R & & R \\ & C=C & \\ R & & R \end{matrix}\right]^{+\cdot}$$

What criticism can be made of this representation?

Q-91 Draw the cyclohexene molecular ion, using both the literature notation and the more precise notation described in A-80.

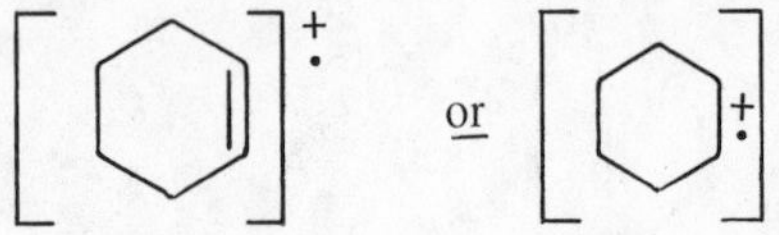

Q-92 Draw the radical cations which can be formed from 1-methylcyclohexene.

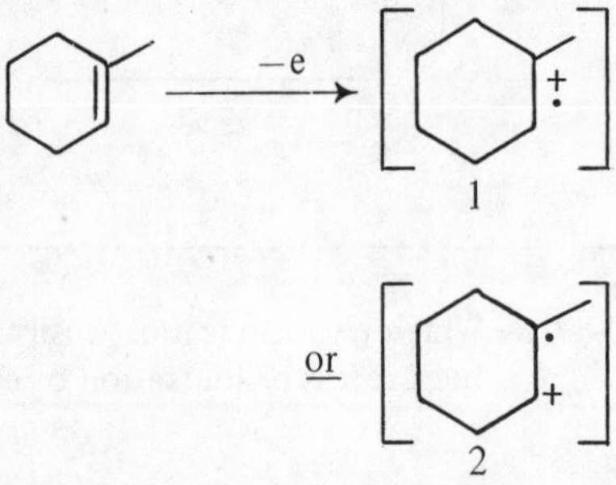

Q-93 Draw the radical cation formed by ionization of benzene. Use both the literature representation and the precise notation.

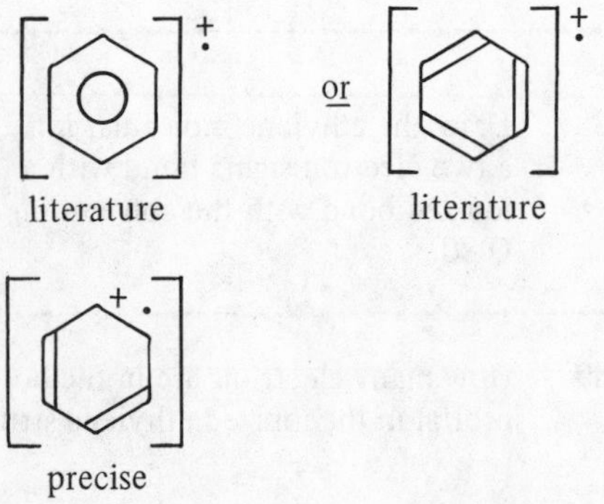

Q-94 Given the electronic structure of the carbonyl group

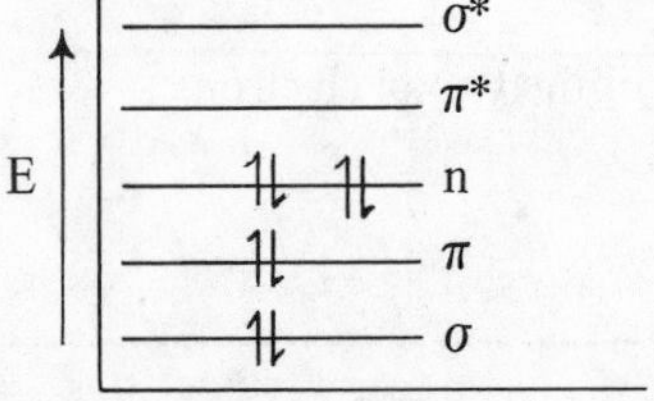

E

$\sigma^*$

$\pi^*$

n

$\pi$

$\sigma$

which electron will be lost during formation of the molecular ion?

Q-95 Draw the acetone molecular ion with the radical cation charge at the point of ionization. Include all other lone pair electrons. Draw a resonance form with the + charge on C.

A-95 [Resonance structures of the acetone radical cation: $(CH_3)_2C{=}\dot{O}^+$ ⟷ $(CH_3)_2C^+{-}\ddot{O}\cdot$]

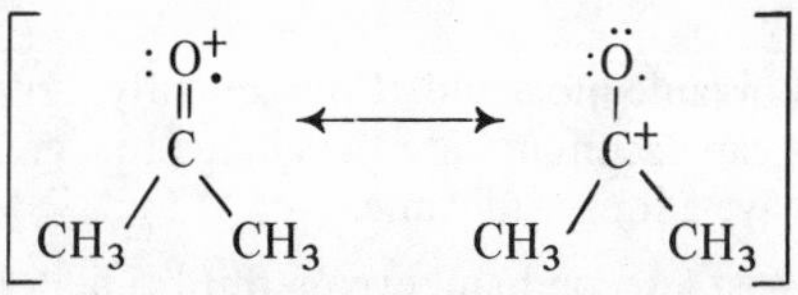

Q-96 Draw the cyclohexanone radical cation.

A-96 [Resonance structures of the cyclohexanone radical cation: $\dot{O}^+{=}$cyclohexane ⟷ $\cdot\ddot{O}{-}C^+$ cyclohexyl]

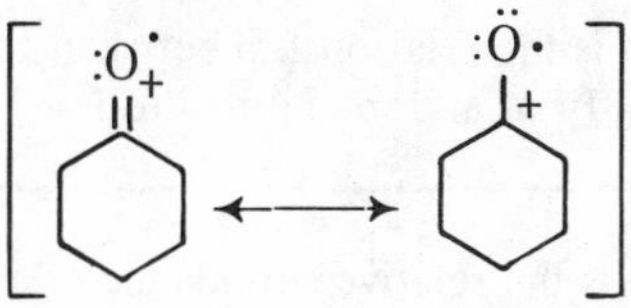

Q-97 Which electrons are most easily ionized in an alcohol?

A-97 The lone pair electrons on oxygen.

Q-98 Draw the propanol radical cation. Include all lone pair electrons.

A-98 $[CH_3{-}CH_2{-}CH_2{-}\dot{O}^+{-}H]$

Q-99 Which electrons are most easily ionized in dialkyl ethers?

A-99 The lone pair electrons on oxygen.

Q-100 Draw the diethyl ether molecular ion.

A-100 $[CH_3{-}CH_2{-}\dot{O}^+{-}CH_2{-}CH_3]$

Q-101 Draw the molecular ion formed from cyclohexyl chloride.

A-101 [Cyclohexyl–$\ddot{Cl}^{+\cdot}$]

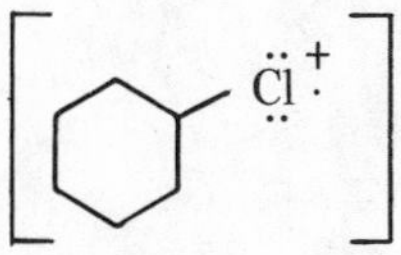

Q-102 Draw the molecular ions which may be formed from methyl vinyl ether.

A-102
a. $CH_3{-}\dot{O}^+{-}CH{=}CH_2$
b. $CH_3{-}\ddot{O}{-}\overset{+\cdot}{CH}{-}CH_2$
c. $CH_3{-}\ddot{O}{-}\overset{\cdot+}{CH}{-}CH_2$

Q-103 What molecular ions can be formed from 4-methoxycyclohexanone?

A-103 a. [4-methoxycyclohexanone with $:\dot{O}^+{=}$ ⟷ $:\ddot{O}\cdot$ with $C^+$; substituent $:\ddot{O}CH_3$]

[4-methoxycyclohexanone with $:O:$ carbonyl and $\overset{\cdot}{O}^+CH_3$ substituent]

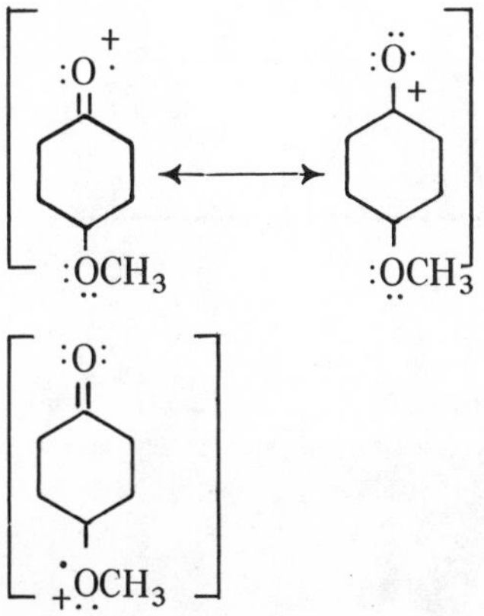

**R** Ionization of an alkane will generally occur at a C-C sigma bond. Alkenes will be ionized at the olefinic bond.

When an atom containing a lone pair is present, one of the lone pair electrons tends to be ionized.

Ease of ionization is
lone pair >C=C >C-C >C-H

The molecular ion is drawn inside square brackets at all times throughout this text.

S-8 A potential of 70 eV is more than is required to ionize an organic molecule. Consequently, ionized molecules tend to fragment. Some classes of compounds can accommodate the charge to a greater extent than others by virtue of delocalization, and thus have a longer lifetime.

Certain molecular ions are so long lived that they are ionized a second and even a third time.

Q-104 What is the relationship between the stability of an ion and its lifetime?

A-104 The more stable the ion, the longer it will live before fragmentation.

Q-105 How is the relative abundance of an ion in the mass spectrum related to its stability?

A-105 The more stable the ion, the longer it will live and thus the more abundant it will be in the mass spectrum.

Q-106 If an ion has low stability, what will be the relative intensity of its peak in the mass spectrum?

A-106 The relative peak height will be small as compared to peak heights of more stable ions.

Q-107 What factor would increase the stability of an ion?

A-107 The ability to delocalize the charge.

Q-108 Draw the radical cation from butadiene. Why does it exhibit a strong molecular ion?

A-108

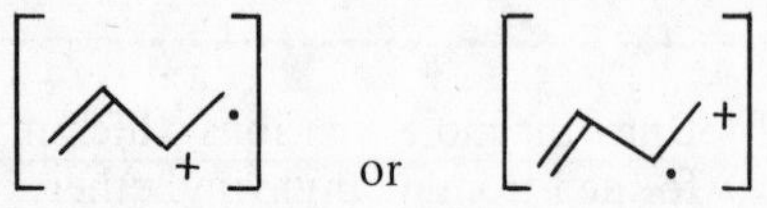

or

Resonance is possible, with delocalization of the charge, e.g.,

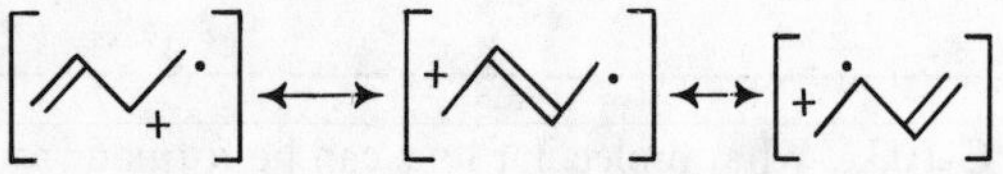

Q-109 Why would you expect the benzene radical cation to be stable?

A-109 Several resonance forms can be invoked to delocalize the charge around the ring, e.g.,

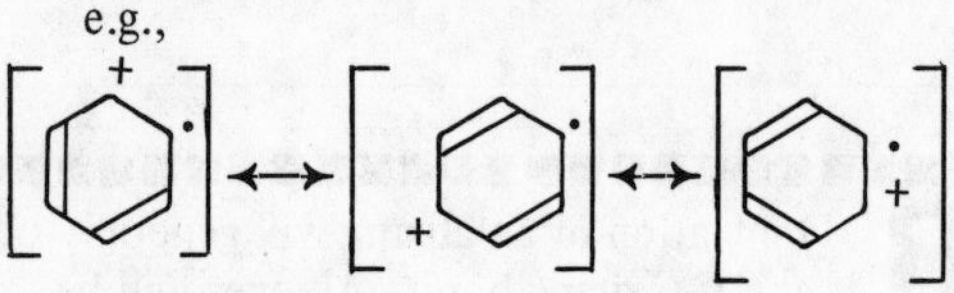

Note that the radical charge can also be delocalized.

Q-110 Explain the very high intensity of the molecular ion of naphthalene.

A-110 Several canonical forms are possible, thus stabilizing the charged species.

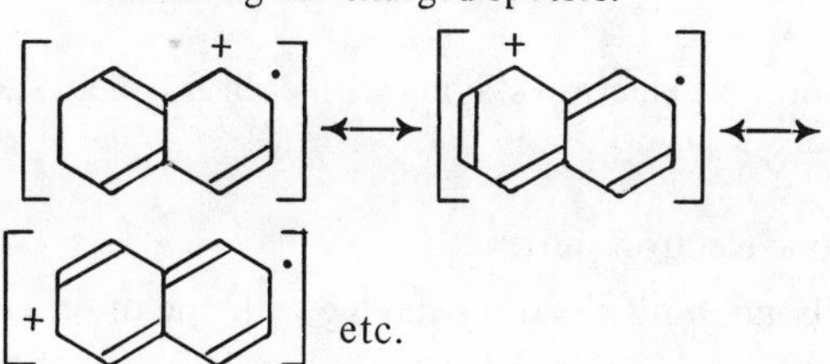

etc.

Q-111 Rank the following in the order of molecular ion stability:

a. n-pentane

b. benzene

Explain the difference.

A-111 In order of stability

b > a

The molecular ion of benzene is stabilized by resonance.

Q-112 The naphthalene molecular ion is very stable. It lives long enough in the ion source that the probability of a *second* ionization occurring is quite high. Draw the structure obtained by ionizing bond C(4)–C(5) in the ion.

and show that a di-cation is obtained. Draw some of the resonance structures of the di-cation.

A-112

Q-113 At what m/e value does the singly charged molecular ion of naphthalene appear? Where does the doubly charged molecular ion appear?

A-113 m/e 128

m/e 64 (e = 2)

a. Delocalization of charge tends to stabilize molecular ions, especially in aromatic systems.

b. Highly stable molecular ions can be ionized a second and even a third time. The resultant multiple charged molecular ions can appear at unit m/e values or at fractional m/e values.

S-9 The fission process is one in which a bond is ruptured in the molecule. Three types of fission can be defined:

a. HOMOLYSIS–cleavage of a two electron sigma bond, one electron staying with each fragment, i.e.,

$$X - Y \longrightarrow X\cdot + Y\cdot$$

Note the use of a single headed arrow, implying a one electron shift.

b. HETEROLYSIS–cleavage of a two electron sigma bond, both electrons staying with one or other fragment, i.e.,

$$X - Y \longrightarrow X^{+} + Y^{:}$$

The double headed arrow implies a two electron shift. Note that X remains electron deficient, as a cation.

c. HEMI-HETEROLYSIS–cleavage of an ionized sigma bond, i.e.,

$$X +\cdot Y \longrightarrow X^{+} + Y\cdot$$

There is only one electron in the bonding sigma orbital, hence the use of a single headed arrow. Note that all molecular ions (cation radicals) are depicted inside square brackets.

---

Q-114 Draw the ethane radical cation in which the C–C bond has been ionized.

Show by "arrow-pushing" how the ethane molecular ion may fragment to yield the methyl cation ($CH_3^+$, m/e 15). Is the methyl radical observed in the mass spectrum? Why?

---

A-114

$$\left[ H_3C + \cdot CH_3 \right] \longrightarrow H_3C^{+} + \cdot CH_3$$

The methyl radical is not seen. Only cations are seen in the spectrum.

Q-115 Classify the fragmentation mechanism of the ethane cation radical as

a. homolysis,

b. heterolysis, or

c. hemi-heterolysis.

---

A-115 c. hemi-heterolysis, i.e., a one-electron bond has been fragmented.

Q-116 Draw the two radical cations which can be formed by ionization of a C–C bond in propane.

Show by "arrow-pushing"

a. how an ion at m/e 29 can be formed.

b. how an ion at m/e 15 can be formed.

Classify the fragmentation mechanisms.

A-116 a.

$$\left[ H{-}\overset{H}{\underset{H}{C}}{-}\overset{H}{\underset{H}{C}}{+}\cdot\overset{H}{\underset{H}{C}}{-}H \right] \longrightarrow$$

$$H{-}\overset{H}{\underset{H}{C}}{-}\overset{H}{\underset{H}{C}}{+} \quad + \quad \cdot\overset{H}{\underset{H}{C}}{-}H$$

m/e 29

hemi-heterolysis

b.

$$\left[ H{-}\overset{H}{\underset{H}{C}}{-}\overset{H}{\underset{H}{C}}\cdot{+}\overset{H}{\underset{H}{C}}{-}H \right] \longrightarrow$$

$$H{-}\overset{H}{\underset{H}{C}}{-}\overset{H}{\underset{H}{C}}\cdot \quad + \quad {}^{+}\overset{H}{\underset{H}{C}}{-}H$$

m/e 15

hemi-heterolysis

Q-117 The origin of cations at m/e 15 and m/e 29 in the mass spectrum of propane has been rationalized by hemi-heterolysis.

There is a possible second route to the ion of m/e 15, by *heterolytic* scission of the C-C bond in the ion of m/e 29. Explain by "arrow-pushing."

(Hint: A neutral species of mass 14 is extruded.)

A-117

$$H{-}\overset{H}{\underset{H}{C}}{-}\overset{H}{\underset{H}{C}}{+} \longrightarrow H{-}\overset{H}{\underset{H}{C}}{+} \quad + \quad CH_2$$

m/e 29 m/e 15

The double headed arrow implies heterolysis.

Q-118 Explain the formation of ions at m/e 15 and m/e 43 in the mass spectrum of 2-methylpropane.

A-118

a.

$$\left[ CH_3{-}\overset{CH_3}{\underset{H}{C}}{+}\cdot CH_3 \right] \xrightarrow{-CH_3^{\bullet}} CH_3{-}\overset{CH_3}{\underset{H}{C}}{+}$$

m/e 43

b.

$$\left[ CH_3{-}\overset{CH_3}{\underset{H}{C}}\cdot{+}CH_3 \right] \longrightarrow CH_3{-}\overset{CH_3}{\underset{H}{C}}\cdot + CH_3^{+}$$

m/e 15

Q-119 Draw the ethyl bromide molecular ion. Include all lone pair electrons.

Illustrate by "arrow-pushing" the formation of the ethyl cation (m/e 29) in the mass spectrum of ethyl bromide. Classify the fragmentation.

A-119 $\left[CH_3-CH_2-\overset{+\cdot}{Br}:\right] \xrightarrow{\text{heterolysis}}$

$CH_3-CH_2^+ + \cdot\ddot{Br}:$

m/e 29

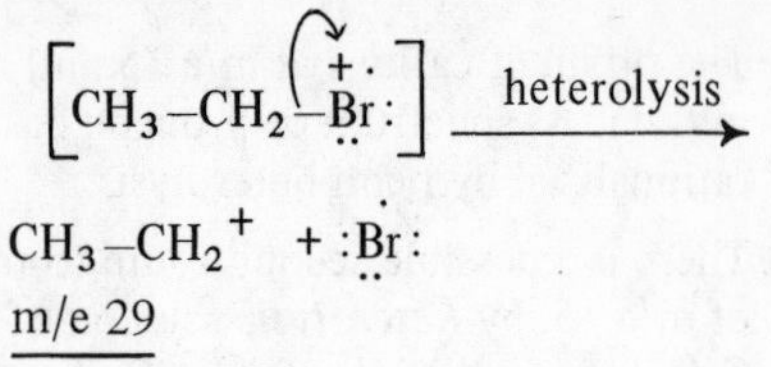

Q-120 What are the charges on the products formed in the heterolytic cleavage of ethyl bromide? How does the net charge compare with the molecular ion charge?

---

A-120 A *cation* and a *radical* are formed. The net charge is (+•). This is the same as the charge on the molecular ion. Note the conservation of charge.

Q-121 In the mass spectrum of ethyl bromide, in addition to the ion of m/e 29, two ions of equal intensity are observed at m/e 93 and m/e 95. Assign formulas to those ions. (Hint: remember the isotope patterns for bromo-compounds.)

---

A-121 $^+CH_2Br^{79}$ (m/e 93)

$^+CH_2Br^{81}$ (m/e 95)

Q-122 Consider the homolytic fission of the C-C sigma bond in the ethyl bromide molecular ion. Draw the fragments obtained. (Hint: note that the unpaired electron on bromine can be paired with an electron from the homolytic fission, forming a multiple bond to bromine.)

---

A-122 $\left[CH_3-CH_2-\overset{\cdot+}{Br}:\right] \xrightarrow{-CH_3^{\bullet}} CH_2=\overset{+}{Br}:$

m/e 93 ($Br^{79}$) and m/e 95 ($Br^{81}$)

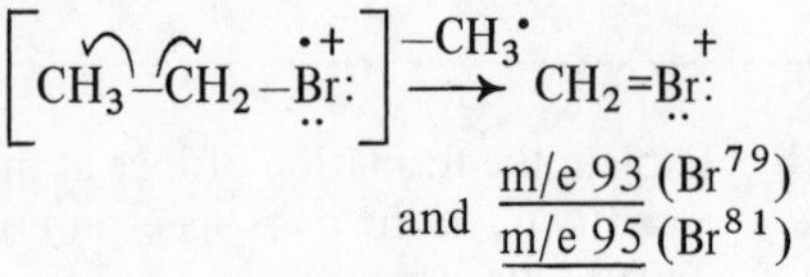

Q-123 Invoke a homolytic fission to account for the formation of the fragment

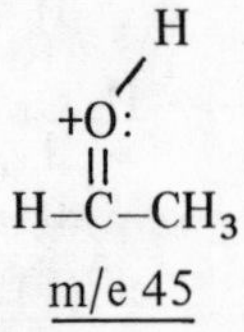

$H-C(=\overset{+}{O}:-H)-CH_3$

m/e 45

detected in the mass spectrum of 2-propanol.

---

A-123 $\left[CH_3-C(\overset{+}{\ddot{O}}H)(H)-CH_3\right] \xrightarrow{-CH_3^{\bullet}} H-C(=\overset{+}{\ddot{O}}H)-CH_3$

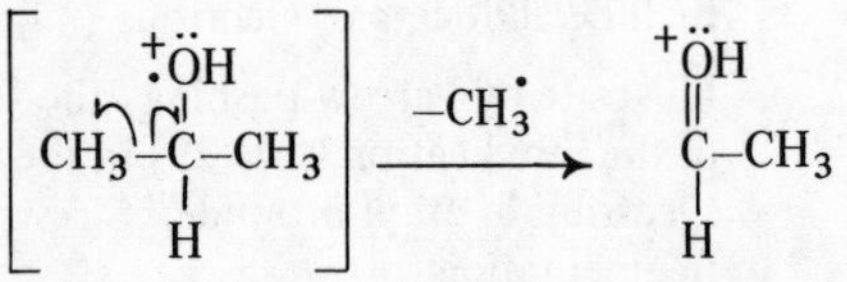

Q-124 Predict the structure of the fragment ion produced by homolysis of bond C(2) – C(3) in the spectrum of

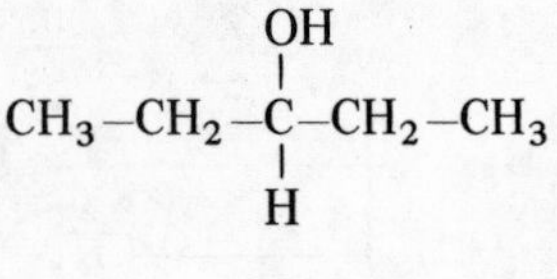

$CH_3-CH_2-C(OH)(H)-CH_2-CH_3$

A-124

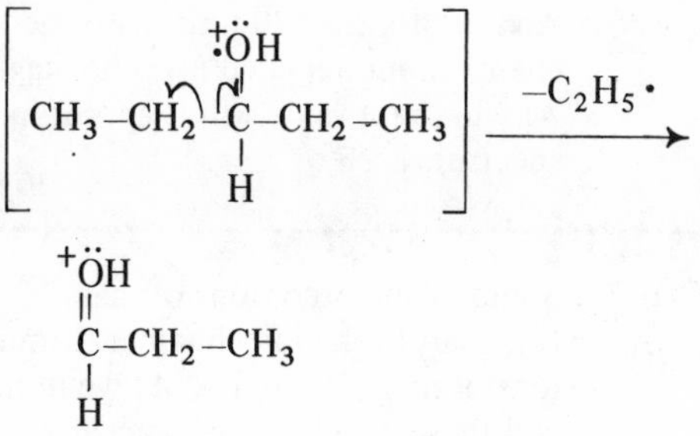

$$\left[ CH_3-CH_2-\overset{\overset{+\ddot{O}H}{|}}{\underset{\underset{H}{|}}{C}}-CH_2-CH_3 \right] \xrightarrow{-C_2H_5\cdot} \overset{\overset{+\ddot{O}H}{\|}}{\underset{\underset{H}{|}}{C}}-CH_2-CH_3$$

m/e 59

Q-125 In the mass spectrum of 3-hexene, peaks corresponding to the following ions appear:

a. $CH_3^+$ (m/e 15)

b. $CH_3-CH_2-\overset{+}{C}H-CH=CH_2$ (m/e 69)

Explain the formation of the ion of m/e 15. (Hint: consider fission of C(1)–C(2).)

Explain the formation of the ion of m/e 69. (Hint: consider fission of C(5)–C(6).)

A-125 a. $\left[CH_3-CH_2-\overset{+}{C}H-\overset{\bullet}{C}H-CH_2-CH_3\right] \longrightarrow$

$CH_3^+ + CH_2=CH-\overset{\bullet}{C}H-CH_2-CH_3$

m/e 15

b. $\left[CH_3-CH_2-CH\overset{+}{-}\overset{\bullet}{C}H-CH_2-CH_3\right] \longrightarrow$

$CH_3-CH_2-\overset{+}{C}H-CH=CH_2 + \overset{\bullet}{C}H_3$

M-15

where M is the molecular ion mass

**R** Fission may occur by the following processes:

a. homolysis,

b. heterolysis or

c. hemi-heterolysis.

S-10 Consider a number of ions of mass $m_1$ in the ion chamber of a mass spectrometer. Some of these ions will appear at the appropriate m/e value ($m_1$) in the mass spectrum. Some others will fragment in the ion chamber, affording *daughter ions* of mass $m_2$ which appear at m/e $m_2$, i.e.,

$$m_1^+ \longrightarrow m_2^+ \quad + \text{ neutral fragment.}$$

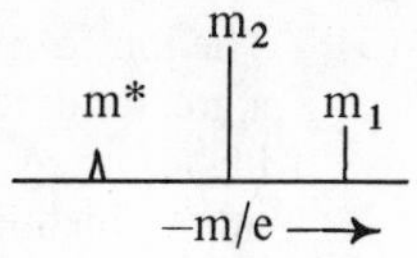

m2 m* m1 —m/e →

However, it is often possible for *some* of the $m_1$ ions to decompose *during flight* rather than in the ion chamber. The daughter ion formed in flight (called a metastable ion) does not appear at m/e $m_2$. Instead a diffuse peak at lower m/e value (m*) is observed (see diagram).

Whereas $m_1$ and $m_2$ are observed at unit m/e values, m* may appear at a fractional m/e value. A mathematical relationship exists, defining the interdependence of $m_1$, $m_2$ and m*:

$$\frac{(m_2)^2}{m_1} = m^*$$

Thus, if ions are observed at mass $m_1$ and $m_2$ and it is suspected that fragmentation of $m_1$ affords $m_2$, it is possible to confirm this by seeking the metastable ion at the value of m* defined by the equation. The presence of the metastable ion therefore corroborates the suspected relationship between the two ions, $m_1$ and $m_2$. Not all fragmentations produce a metastable ion, therefore, the absence of a metastable ion peak is *not* evidence against a fragmentation relationship.

The theoretically calculated m/e value for a metastable ion is sometimes 0.1–0.4 mass units lower than its observed m/e value.

| | |
|---|---|
| | Q-126 An ion of mass 120 units is formed in the ion chamber and repelled into the magnetic field. At what m/e value will it appear in the mass spectrum? |
| A-126 m/e 120 | Q-127 Some of the other ions of mass 120 units fragment in the ion chamber, forming fragment ions of mass 105 units. At what m/e value will these daughter ions appear? |
| A-127 m/e 105 | Q-128 Some other ions of mass 120 units are repelled out of the ion chamber and decompose during flight forming ions of mass 105 units. What effect does loss of the neutral fragment have on the kinetic energy of the daughter ion? |
| A-128 The kinetic energy of the daughter ion is lessened, because the neutral fragment will carry away a portion of the energy. | Q-129 How will the kinetic energy of the daughter ion thus formed compare with the kinetic energy of a daughter ion formed in and repelled from the ion chamber? |
| A-129 The kinetic energy of the daughter ion formed during flight will be *less.* | Q-130 Will the daughter ion formed during flight be deflected by the magnetic field to a lesser or greater extent than a daughter ion formed in the ion chamber? |
| A-130 The daughter ion formed during flight has less kinetic energy, therefore, it is deflected to a *greater* extent.<br>Another way of picturing this is to consider the ion formed during flight to have less momentum than the ion formed in the ion chamber. The former ion will then be more easily deflected. | Q-131 The ion formed during flight is deflected to a greater extent. Will it appear at lower or higher m/e value than the ion of similar mass which is formed in the ion chamber? |
| A-131 At lower m/e value. | Q-132 Fragmentation in the ion chamber afforded a daughter ion of mass 105 units.<br>What is the mass of the corresponding daughter ion formed during flight? |
| A-132 105 mass units. | Q-133 The parent ion has mass 120 and the daughter ion has mass 105. What is the relationship defining the m/e value of the metastable ion? |

A-133 $\frac{(m_2)^2}{m_1} = m^* = \frac{105^2}{120} = 91.9$

where m* is the m/e value of the metastable daughter ion.

Q-134 For $m_1 = 150$ and $m_2 = 122$, calculate the m/e value of m*.

Will the metastable ion peak appear exactly at this calculated value of m*?

A-134 $\frac{(m_2)^2}{m_1} = \frac{122^2}{150} = m^*$

$m^* = 99.2$

The observed metastable ion peak may appear slightly higher than this value.

Q-135 What shape will the metastable ion peak have?

A-135 It will be a broad or diffuse peak.

Q-136 What two important differences are there between a metastable ion peak and a normal fragment ion peak?

A-136 A metastable ion produces a *broad peak* which is often at a non-integral, or fractional, m/e value..

Q-137 Are there any other ion species which appear at fractional m/e value in low resolution mass spectra?

A-137 Doubly or triply-charged ions may appear at non-integral m/e value.

Q-138 In the mass spectrum of a hydrocarbon, peaks are observed at m/e 57 and m/e 43. A diffuse peak is observed at m/e 32.5. Determine the molecular formulas corresponding to the two peaks which appear at integral m/e value. Suggest a possible relationship between these two ions.

A-138 m/e 57 – $C_4H_9^+$

m/e 43 – $C_3H_7^+$

possible parent-daughter relationship is

$C_4H_9^+ \longrightarrow C_3H_7^+ + CH_2$

Q-139 Account for the diffuse peak which occurs at m/e 32.5 in the mass spectrum of the hydrocarbon described in Q-138.

A-139 The ion at m/e 32.5 is the metastable ion corresponding to the fragmentation described in A-138. Calculation shows that the metastable ion should appear at

$$\frac{(m_2)^2}{m_1} = \frac{43^2}{57} = 32.4$$

Q-140 What is the mass of the metastable ion at m/e 32.5?

A-140 43 mass units.

Q-141 At what m/e value will the metastable ion formed by the following fragmentation appear?

$$C_6H_5\text{-}C\equiv\overset{+}{O} \longrightarrow C_6H_5^+ + CO$$

A-141 The metastable ion should appear at m/e 56.5.

$m_1 = 105 \qquad m_2 = 77$

$$m^* = \frac{(m_2)^2}{m_1} = \frac{77^2}{105} = 56.5$$

Q-142 Interpret the following partial mass spectrum:

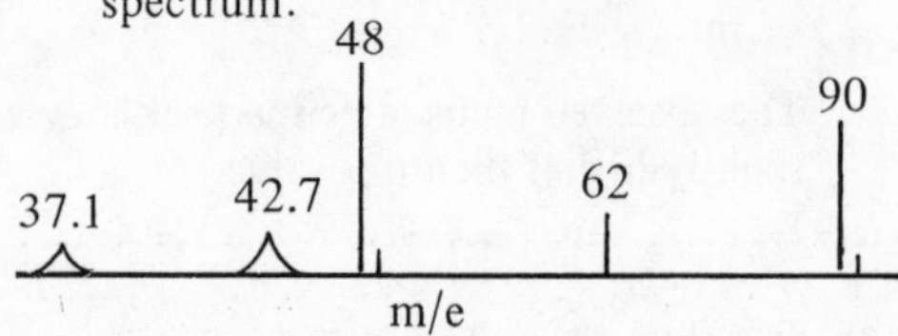


A-142 The ion at m/e 90 is fragmenting to form the ion at m/e 62 which in turn breaks down to give the ion at m/e 48, i.e.,

m/e 90 ⟶ m/e 62 ⟶ m/e 48

These fragmentations are supported by the existence of metastable ions at the appropriate values:

a. m/e 90 ⟶ m/e 62

$$m^* = \frac{62^2}{90} = 42.7$$

b. m/e 62 ⟶ m/e 48

$$m^* = \frac{48^2}{62} = 37.1$$

Q-143 In the last example, no metastable ion is observed which corresponds to the process

m/e 90 ⟶ m/e 48.

Does this mean that the ion at m/e 48 cannot be formed from the ion at m/e 90 by a single fragmentation step?

A-143 No. The absence of a metastable ion does not necessarily mean that the fragmentation is not occurring.

**R** The presence of metastable ion peaks can be used to corroborate suspected fragmentations. Since some fragmentations do not produce metastable ions, the absence of metastable ion peaks cannot be used as evidence against a fragmentation.

## PART II: Fission Processes in Which One Bond Is Broken. The Formation of Odd m/e Fragments When M is Even.

S-1 When saturated hydrocarbons are subjected to mass spectrometry, radical cations form. Since there is no functional group in which to localize the charge, the charge may reside in any of the sigma bonds. Simple fission processes predominate, yielding odd m/e fragments, and can be generalized:

a. In straight chain alkanes an alkyl group is lost from one end of the molecule, and then successive losses of 14 mass units ($CH_2$) are observed. A homologous series of peaks is observed with the typical pattern:

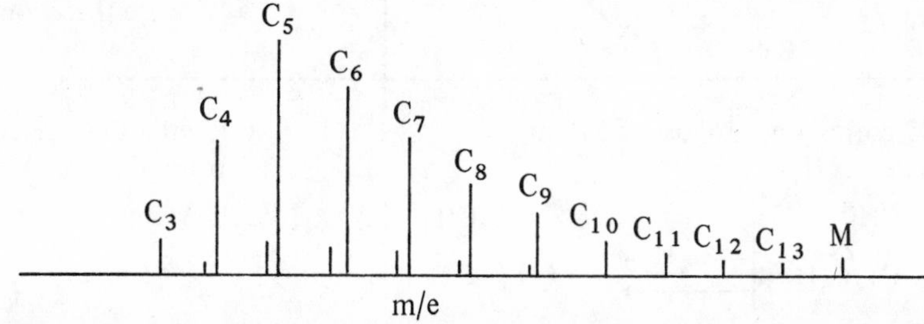


An accompanying elimination of hydrogen forms the smaller peaks which are observed two mass units below these hydrocarbon ions. Note the maximum which occurs at $C_3$-$C_6$.

b. In branched hydrocarbons cleavage tends to occur at the site of branching. The locus of branching is thus readily discerned because preferential cleavage will afford relatively abundant fragment ions. When alternative fissions at a branched site are possible, the group with largest mass tends to be lost preferentially.

---

Q-1 a. What type of hydrocarbon will give a spectrum with the peak pattern shown below?

b. What is the formula of the hydrocarbon?

c. What are the differences in mass between the major peaks?

43 57 71 85 29 99 113 127 M 142

m/e

---

A-1 a. A straight chain saturated hydrocarbon.

b. $C_{10}H_{22}$.

c. The molecular ion M is separated by 15 mass units from the ion of m/e 127. Other peaks in the spectrum differ by 14 mass units.

Q-2 Account for the formation of the ion which appears at m/e 127 (the M-15 ion) in the spectrum shown in Q-1.

---

A-2 m/e 127 corresponds to loss of $CH_3$ from M, i.e.,

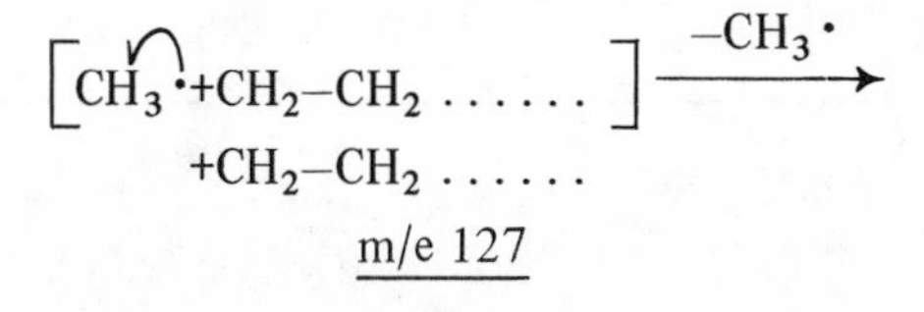


Q-3 Interpret the following mass spectrum.

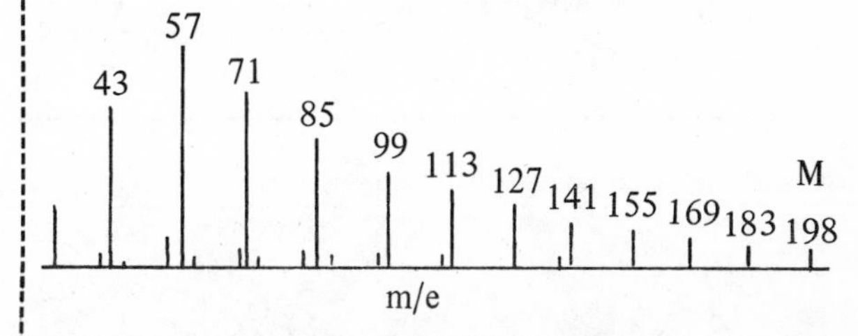

A-3 The molecular weight is consistent with a saturated hydrocarbon of structure $C_{14}H_{30}$. The pattern of peaks suggests an unbranched, saturated structure, i.e., a straight chain molecule.

M-15 occurs by hemi-heterolysis as described in A-2. Subsequent fragmentations may be due to consecutive losses of $CH_2$ by heterolysis.

Q-4 Explain the formation of the abundant ions at m/e 57 and m/e 155 in the mass spectrum of

$C_{12}H_{26}$ (M = 170)

Why are these ions particularly abundant?

A-4 m/e 57 corresponds to loss of 113 mass units, i.e., $C_8H_{17}$, from M.

$$\left[CH_3-\overset{CH_3}{\underset{CH_3}{C}}{+}\cdot C_8H_{17}\right] \xrightarrow{-C_8H_{17}^{\cdot}} CH_3-\overset{CH_3}{\underset{CH^3}{C}}{+}$$

m/e 57

m/e 155 corresponds to loss of 15 mass units, i.e., $CH_3$, from M

$$\left[CH_3-\overset{\overset{\cdot}{C}H_3}{\underset{CH_3}{\overset{+}{C}}}-C_8H_{17}\right] \xrightarrow{-CH_3^{\cdot}} CH_3-\underset{CH_3}{\overset{+}{C}}-C_8H_{17}$$

m/e 155

m/e 57 is a stable tertiary carbonium ion.
m/e 155 is a tertiary carbonium ion, therefore, relatively stable.

Q-5 Predict the fragment ions in the mass spectrum of

$$C_2H_5-\overset{CH_3}{\underset{CH_3}{C}}-C_3H_7$$

A-5

a.

$$\left[C_2H_5-\overset{\overset{\cdot}{C}H_3}{\underset{CH_3}{\overset{+}{C}}}-C_3H_7\right] \xrightarrow{-CH_3^{\cdot}} C_2H_5-\underset{CH_3}{\overset{+}{C}}-C_3H_7$$

m/e 99

b.

$$\left[C_2H_5\cdot{+}\overset{CH_3}{\underset{CH_3}{C}}-C_3H_7\right] \xrightarrow{-C_2H_5^{\cdot}} {+}\overset{CH_3}{\underset{CH_3}{C}}-C_3H_7$$

m/e 85

c.

$$\left[C_2H_5-\overset{CH_3}{\underset{CH_3}{C}}{+}\cdot C_3H_7\right] \xrightarrow{-C_3H_7^{\cdot}} C_2H_5-\overset{CH_3}{\underset{CH_3}{C}}{+}$$

m/e 71

Q-6 The mass spectrum of a compound believed to be either methylcyclopentane or ethylcyclobutane showed an intense peak at M-15. Which structure best fits this data? Suggest a mechanism for the fragmentation process.

A-6 methylcyclopentane

$CH_3$ $\xrightarrow{-e}$ $[CH_3^{+\bullet}]$ $\xrightarrow{-CH_3^{\bullet}}$ +

M-15

Q-7 Explain the abundant ion at M-113 in the mass spectrum of cholestane.

A-7 M-113 corresponds to loss of $C_8H_{17}$, i.e., the side-chain.

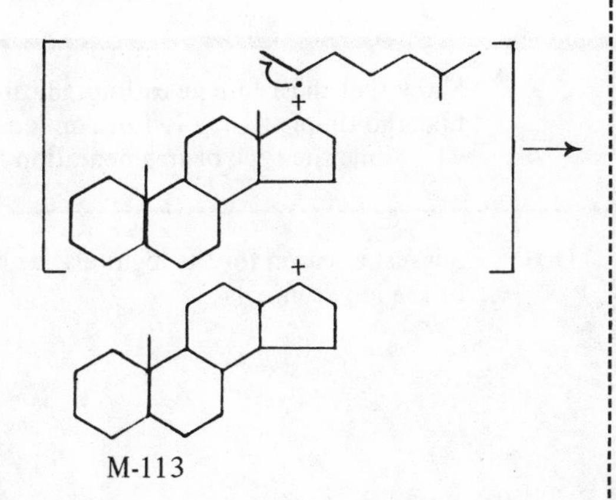


Q-8 The mass spectrum of a compound believed to be either A or B (below) had peaks corresponding to m/e 97 and m/e 111. Which structure best fits the data?

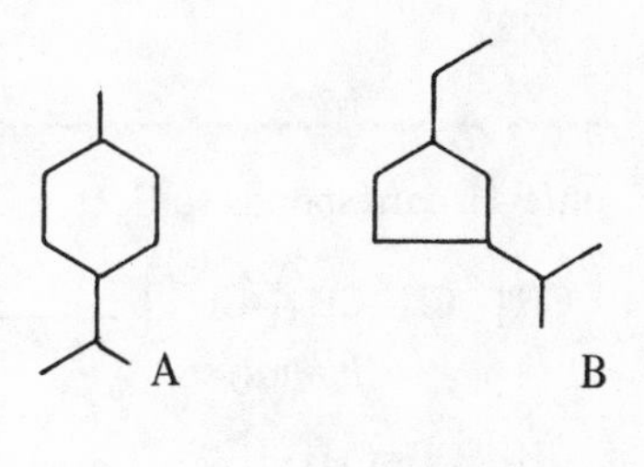


A-8 Structure **B.**
These peaks correspond to side-chain cleavage of the ethyl and propyl groups respectively.

**R** Saturated acyclic alkanes are ionized by the removal of a bonding sigma electron. The resultant ionized bond can be cleaved by a *hemi-heterolysis* process, affording a cation and a neutral radical.

The cation may further fragment by *heterolytic* fission, affording another cation at lower m/e value.

Saturated cyclic hydrocarbons are ionized by abstraction of an electron from a sigma bond. The resultant radical cation can fragment by hemi-heterolytic cleavage of the side-chain.

S-2 The notation which will be used in the description of unsaturated hydrocarbons will be:

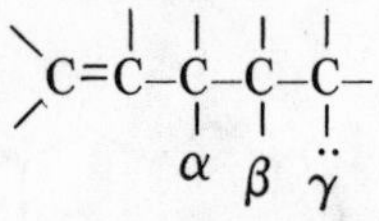

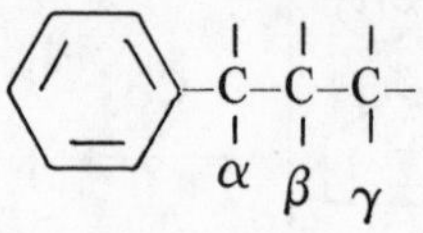

$$>C=\underset{\alpha}{C}-\underset{\beta}{C}-\underset{\gamma}{C}-$$

$$C_6H_5-\underset{\alpha}{C}-\underset{\beta}{C}-\underset{\gamma}{C}-$$

a. C atoms will be numbered $\alpha, \beta, \gamma$ as shown,

b. the bond between the unsaturated group and the $\alpha$-C will be termed vinylic, and

c. the bond between the $\alpha$- and $\beta$- carbon atoms will be termed allylic (in the aliphatic systems) or benzylic (in the aromatic system).

The common simple fission processes which occur in unsaturated hydrocarbons are:

a. cleavage of the vinylic bond—this occurs relatively infrequently, and

b. cleavage of the allylic bond—this process is most commonly encountered.

---

Q-9 Show that the 1-butene cation radical can undergo simple fission to form an ion at m/e 41. Name the type of fragmentation invoked.

---

A-9 m/e 41 corresponds to $C_3H_5+$

$$\left[\dot{C}H_2^{+}-CH-CH_2-CH_3\right] \xrightarrow{-CH_3^{\bullet}} \overset{+}{C}H_2-CH=CH_2$$

*homolysis*

m/e 41

Q-10 Suggest a reason for the high relative intensity of the ion at m/e 41.

---

A-10 Delocalization of the cation charge is possible, i.e.,

$$\overset{+}{C}H_2-CH=CH_2 \longleftrightarrow CH_2=CH-\overset{+}{C}H_2$$

The ion is consequently relatively stable.

Q-11 Show that 1-butene cation radical may also fragment to give an ion at m/e 15. Name the fission type.

---

A-11 $$\left[\overset{\bullet +}{C}H_2-CH-CH_2-CH_3\right] \longrightarrow \overset{\bullet}{C}H_2-CH=CH_2 + CH_3^{+}$$

*heterolysis*

m/e 15

Q-12 The mass spectrum of 5-methyl-3-heptene has a molecular ion peak at m/e 112. Explain the peaks at

a. m/e 97.

b. m/e 83.

A-12 a. m/e 97 corresponds to loss of $CH_3$, i.e.,

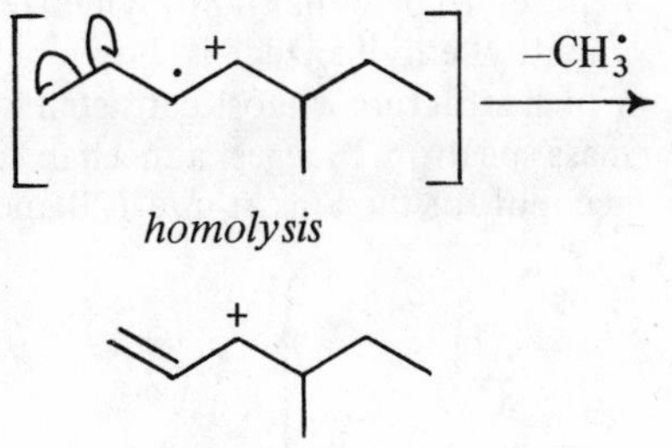

$\xrightarrow{-CH_3^{\bullet}}$

*homolysis*

m/e 97

or:

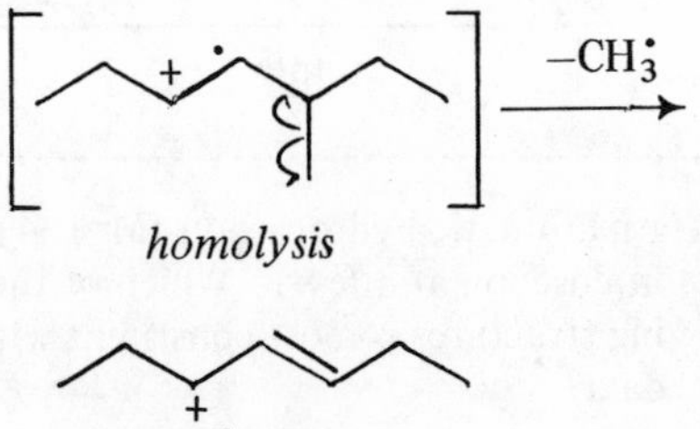

$\xrightarrow{-CH_3^{\bullet}}$

*homolysis*

m/e 97

b. m/e 83 corresponds to loss of 29 mass units, i.e., $C_2H_5$ from M, i.e.,

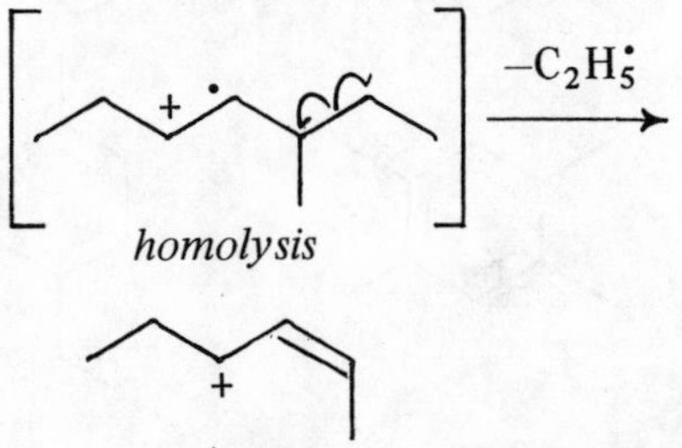

$\xrightarrow{-C_2H_5^{\bullet}}$

*homolysis*

m/e 83

Q-13 A compound could have either structure A or B shown below. Intense ions at m/e 83 and m/e 57 were observed in the mass spectrum of the compound. Which structure is most consistent with the data?

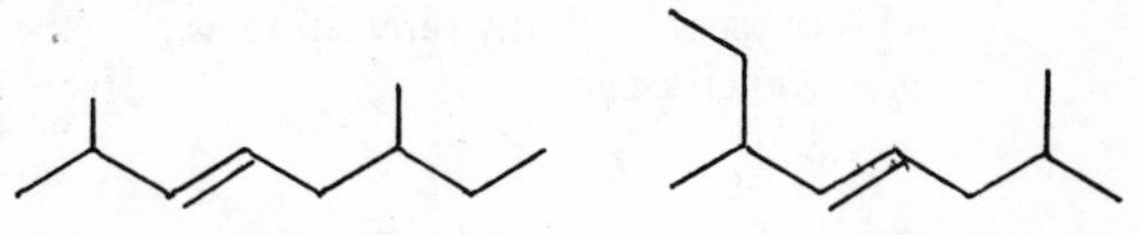

A
M = 140

B
M = 140

A-13 Structure A.
m/e 83 corresponds to loss of $C_4H_9$ from M.

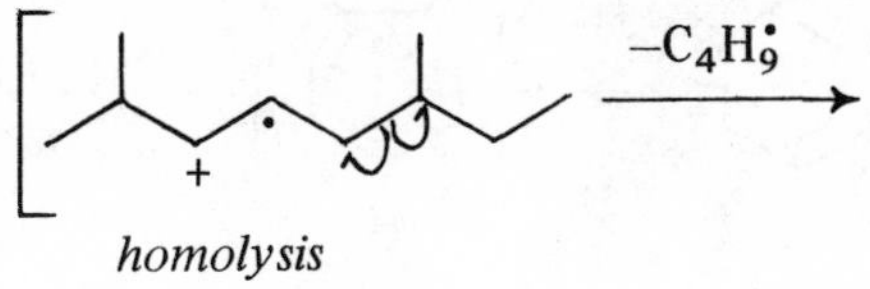

$\xrightarrow{-C_4H_9^{\bullet}}$

*homolysis*

$\longleftrightarrow$

m/e 83

m/e 57

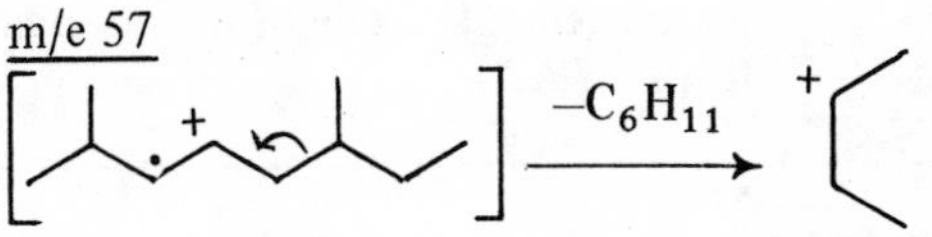

$\xrightarrow{-C_6H_{11}}$

m/e 57

Q-14 The aromatic nucleus is a good site for stabilization of the cation radical charge. Consider the cation radical which may be formed by ionization of ethylbenzene.

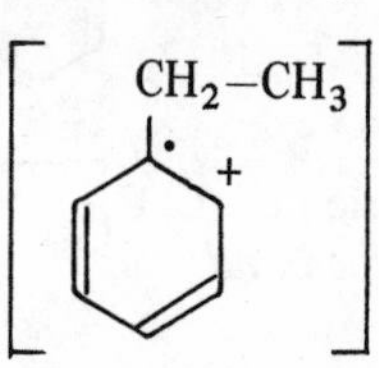

$CH_2{-}CH_3$

M = 106

Depict the homolytic fission process by which a methyl radical is lost forming an ion of m/e 91.

A-14

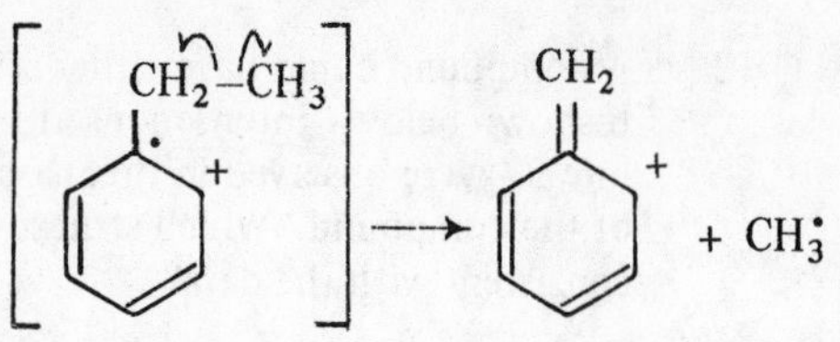

$$\left[ \text{CH}_2\text{—CH}_3 \text{ (radical cation, ring)} \right] \longrightarrow \text{CH}_2{=}\text{(ring)}^+ + \text{CH}_3^{\bullet}$$

m/e 106 m/e 91*

This process is usually referred to as a *benzylic* cleavage.

Q-15 The partial mass spectrum of a compound believed to be either n-propylbenzene or *p*-methyl ethylbenzene is shown below. Which structure is most consistent with the mass spectrum? Suggest a mechanism to account for the ions at m/e 120 and m/e 91.

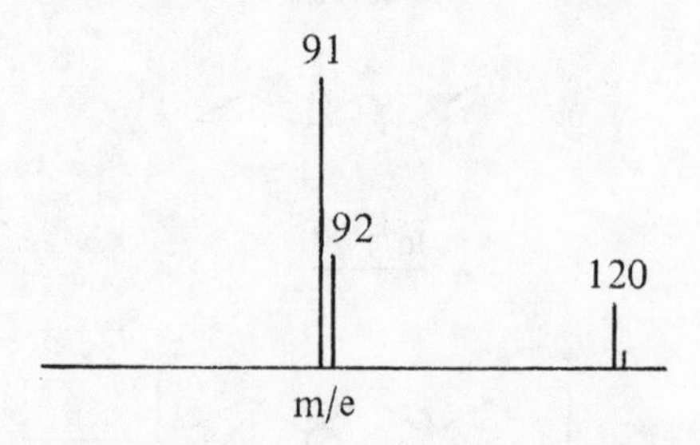


A-15 N - propylbenzene

The ion at m/e 120 corresponds to $C_9H_{12}$, i.e., the molecular ion. The relative abundance of M is due to the ability of the aromatic ring to stabilize charge.

The ion at m/e 91 corresponds to $C_7H_7^+$, i.e., loss of $C_2H_5$.

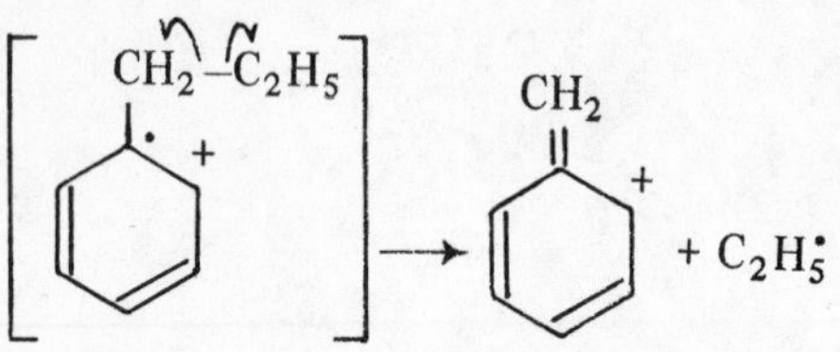

$$\left[ \text{CH}_2\text{—C}_2\text{H}_5 \text{ (radical cation, ring)} \right] \longrightarrow \text{CH}_2{=}\text{(ring)}^+ + \text{C}_2\text{H}_5^{\bullet}$$

m/e 120 m/e 91

Q-16 An aromatic hydrocarbon (M=134) had an intense ion at m/e 91. Which of the following structures is most consistent with the data?

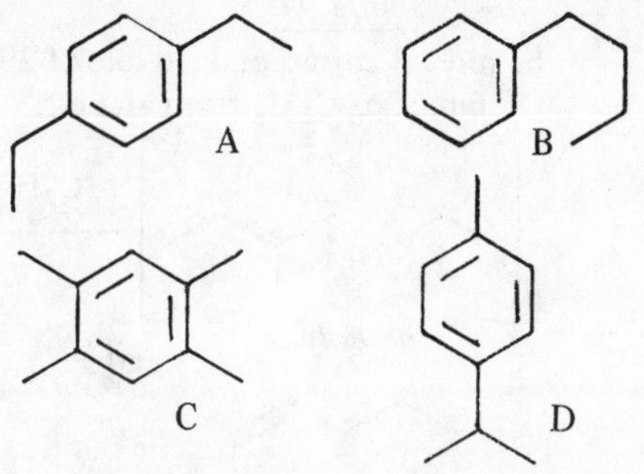

A B C D

A-16 Structure B.

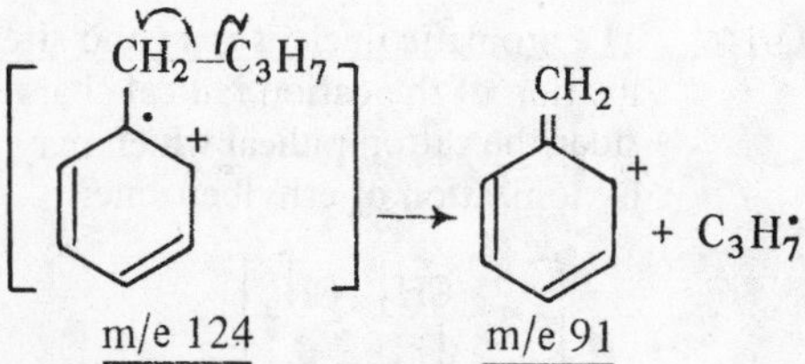

$$\left[ \text{CH}_2\text{—C}_3\text{H}_7 \text{ (radical cation, ring)} \right] \longrightarrow \text{CH}_2{=}\text{(ring)}^+ + \text{C}_3\text{H}_7^{\bullet}$$

m/e 124 m/e 91

*Homolytic* fission of the $C_{(\alpha)}-C_{(\beta)}$ bond, i.e., the *benzylic bond.*

Q-17 The ion formed by benzylic cleavage of alkylbenzenes undergoes ring expansion:

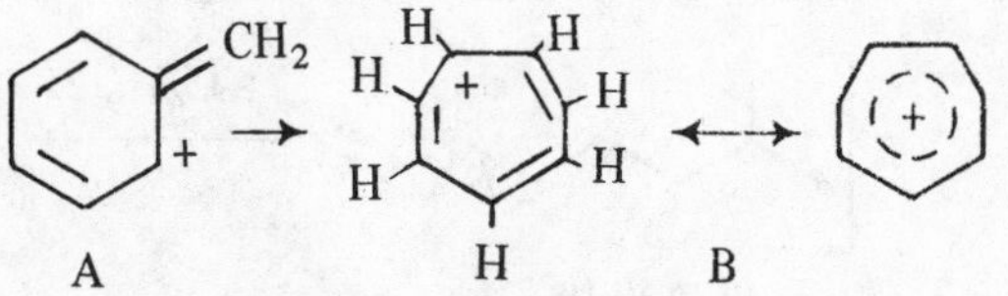

A → B

Suggest a mechanism by which the benzylic fission product A rearranges to form the tropylium ion B.

(Hint: Consider a Wagner-Meerwein type of rearrangement of one of the resonance forms of A.)

*The ion of m/e 91 undergoes further rearrangement as will be described later.

A-17

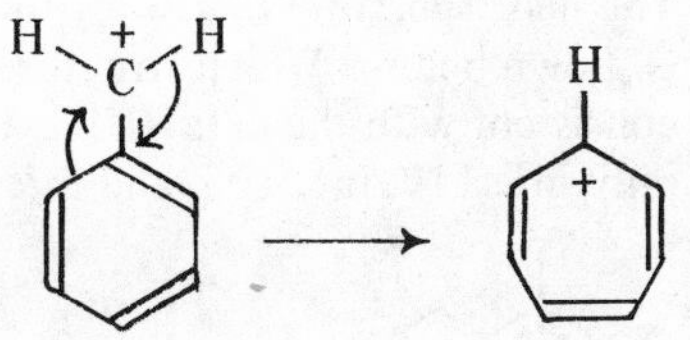

$\xrightarrow{}$ (tropylium ion)

Q-18 Depict the formation of the tropylium ion from the propylbenzene radical cation.

A-18

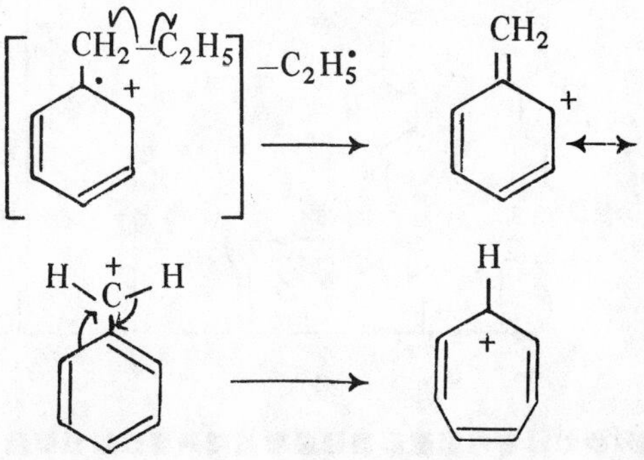

$\left[ \text{C}_6\text{H}_5\text{–CH}_2\text{–C}_2\text{H}_5 \right]^{\cdot +} \xrightarrow{-\text{C}_2\text{H}_5^{\cdot}}$ $\text{C}_6\text{H}_5=\text{CH}_2^{+} \longleftrightarrow$

$\xrightarrow{}$

m/e 91 m/e 91

Q-19 Vinylic cleavage between an aromatic ring and a substituent is commonly encountered but is less favorable than benzylic cleavage.

Show that the cation radical

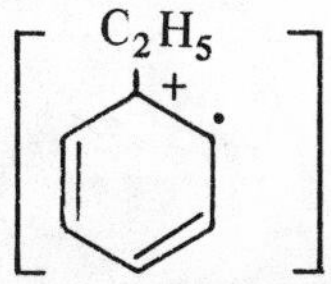

$\left[ \text{C}_6\text{H}_5\text{C}_2\text{H}_5 \right]^{\cdot +}$

can undergo homolytic fission of the vinylic bond, affording benzene cation.

A-19

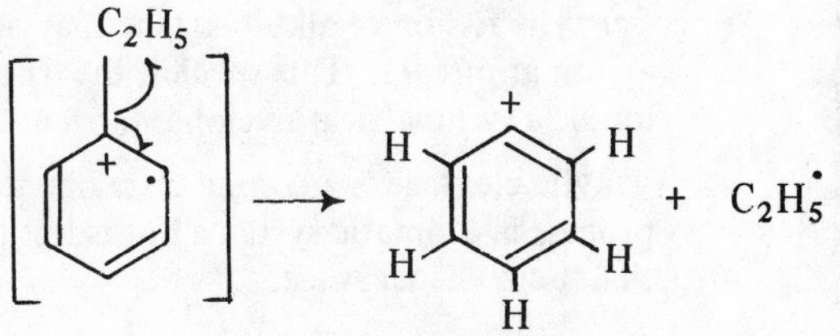

$\left[ \text{C}_6\text{H}_5\text{C}_2\text{H}_5 \right]^{\cdot +} \longrightarrow \text{C}_6\text{H}_5^{+} + \text{C}_2\text{H}_5^{\cdot}$

m/e 77

Q-20 Explain the formation of an ion at m/e 77 in the mass spectrum of chlorobenzene.

A-20

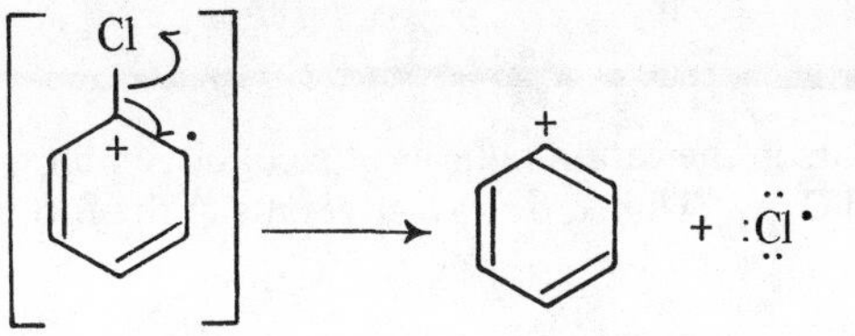

$\left[ \text{C}_6\text{H}_5\text{Cl} \right]^{\cdot +} \longrightarrow \text{C}_6\text{H}_5^{+} + :\ddot{\text{Cl}}^{\cdot}$

m/e 77

Q-21 Show that the ion of mass m/e 77 in A-20 can also be obtained from the chlorobenzene cation radical in which chlorine is ionized. Name the type of cleavage invoked.

A-21

$:\ddot{Cl}:^{+}$ (phenyl) $\longrightarrow$ $C_6H_5^+$ + $:\ddot{Cl}\cdot$

m/e 77

*Heterolytic cleavage*

Q-22 The mass spectrum of a substituted benzene is shown below. Which structure is most consistent with the data? (Hint: Consider only m/e 119, m/e 105, and m/e 77.)

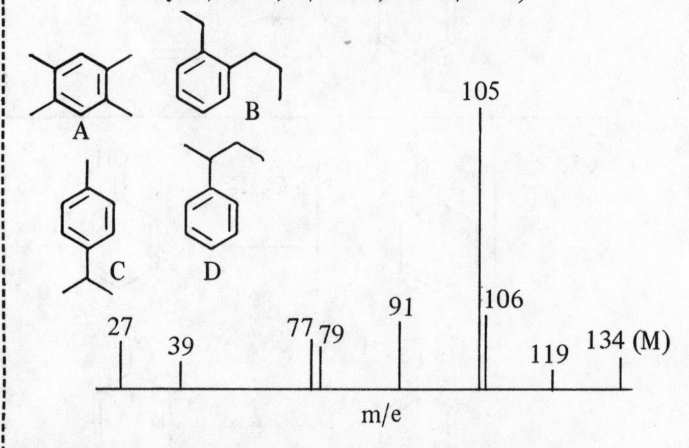


A-22

M = 134

D

m/e 119 corresponds to loss of $CH_3$ by benzylic cleavage.

m/e 105 corresponds to loss of $C_2H_5$ by benzylic cleavage.

m/e 77 represents the vinylic cleavage product, $C_6H_5^+$.

**R** Allylic and benzylic cleavage is a favored process in the mass spectral fragmentation of unsaturated compounds.

Benzylic fission of alkylbenzene may afford an ion at m/e 91. This species, the tropylium ion, is a symmetrical cycloheptatriene cation.

Vinylic cleavage is a common fragmentation process in aromatic systems but is less favored than benzylic cleavage.

S-3 A common simple fragmentation process which occurs in the cation radicals of alcohols, amines, and ethers is the cleavage of the bond between $C_{(\alpha)}$ and $C_{(\beta)}$. This $\alpha, \beta$ cleavage results in the formation of an oxonium or immonium ion.

$\overset{+\dot{O}H}{|}$ —C—C—C— ($\beta$ $\alpha$ $\beta$)  $\overset{+\dot{N}R}{|}$ —C—C—C— ($\beta$ $\alpha$ $\beta$)  $\overset{+\dot{O}R}{|}$ —C—C—C— ($\beta$ $\alpha$ $\beta$)

Q-23 Explain the formation of the ion at m/e 31 which is the base peak (most intense peak) in the mass spectrum of ethanol.

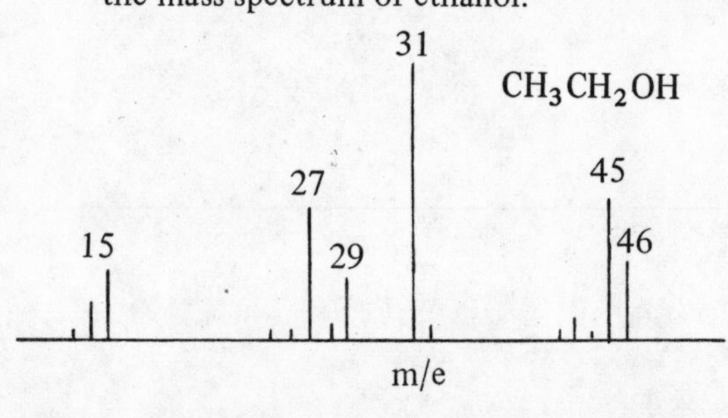

A-23 m/e 31 corresponds to loss of $CH_3$ from M.

$$\left[CH_3-CH_2-\ddot{O}H\right]^{\cdot+} \xrightarrow{-CH_3^{\cdot}} CH_2=\overset{+}{O}H$$

m/e 46 → m/e 31

Q-24 Is the mass spectrum shown below consistent with the structure 2-methyl-2-butanol (M = 88)?

(Hint: Consider the ions of m/e 73 and m/e 59.)

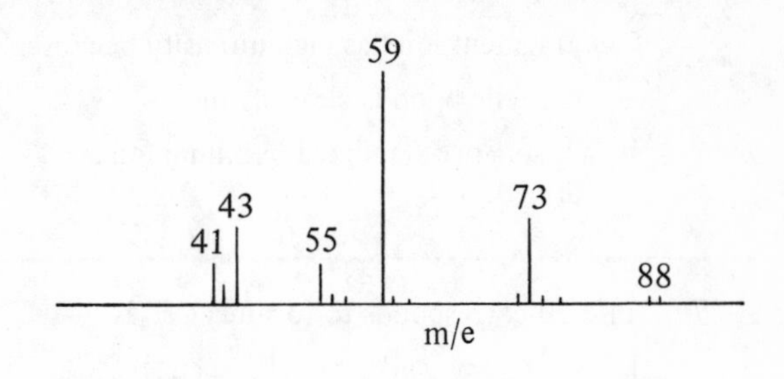


A-24 Yes, the mass spectrum is consistent with the postulated structure. Favorable fragmentation processes in the proposed structure would be loss of $CH_3^{\cdot}$ and $C_2H_5^{\cdot}$.

m/e 88 $\xrightarrow{-CH_3^{\cdot}}$ m/e 73

$\xrightarrow{-C_2H_5^{\cdot}}$ m/e 59

Q-25 A compound suspected of being 3,3-dimethyl-2-butanol (M = 102) or its isomer 2-ethyl-3-pentanol, exhibits intense ions at m/e 87 (30%) and m/e 45 (80%). A weak ion is observed at m/e 102 (2%).

Which structure is consistent with the data? (The per cent figures give the intensity of a peak relative to the most intense peak.)

A-25 3,3-dimethyl-2-butanol.

m/e 102 corresponds to formula $C_6H_{13}OH$, and could, therefore, be the molecular ion.

m/e 87 corresponds to loss of $CH_3$.

$\longrightarrow$ m/e 87 + $CH_3^{\cdot}$

m/e 54 corresponds to loss of $C_4H_9$, by the alternative $C_{(\alpha)}$- $C_{(\beta)}$ homolysis.

Q-26 Comment on the formation of an intense ion of m/e 71 ($C_4H_7O$) in the mass spectrum of the sesquiterpene alcohol below.

OH

A-26

$$\left[\text{(dienol radical cation, } \overset{+}{\ddot{O}}H)\right] \xrightarrow{-C_{11}C_{19}^{\bullet}} \text{(+OH enone)} \quad m/e\ 71$$

The fragment ion has high intensity because:

a. an allylic bond is cleaved, and

b. a resonance-stabilized oxonium ion is formed.

Q-27 In the mass spectrum of an octylamine ($C_8H_{19}N$) the base peak appears at m/e 30. Explain how this ion might be formed. What octylamines would be consistent with this data?

A-27 m/e 30 corresponds to formula $CH_4N$.

$$\left[CH_3(CH_2)_6-CH_2-\overset{\bullet+}{N}H_2\right] \xrightarrow{-C_7H_{15}^{\bullet}} CH_2=\overset{+}{N}H_2 \quad m/e\ 30$$

Any octylamine of the type $C_7H_{15}-CH_2NH_2$ will fit the data given.

Q-28 A compound is thought to be either 3-aminooctane or 4-aminooctane. In its mass spectrum, the most abundant peaks occur at m/e 58 (100%) and m/e 100 (40%). Which structure is more likely? Why?

A-28 3-aminooctane.

m/e 100 corresponds to loss of $C_2H_5$ from M, and m/e 58 corresponds to loss of $C_5H_{11}$.

The two possible $\alpha,\beta$-cleavages of 3-aminooctane account for these ions.

Similar fragmentation of 4-aminooctane would give ions of m/e 86 and m/e 72.

Q-29 A *primary* amine affords the following mass spectrum:

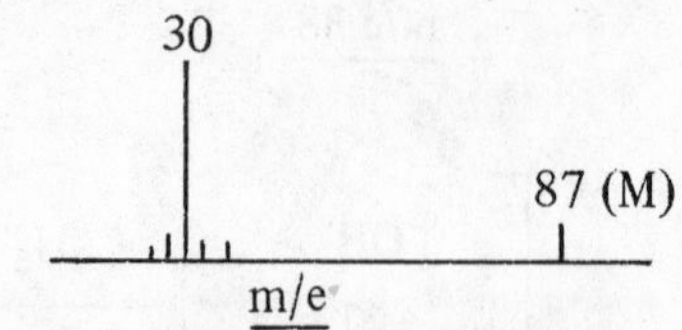


Which of the following structures best fits the mass spectral data?

A: $(CH_3)_2CHCH_2CH_2NH_2$

B: $CH_3CH_2-C(CH_3)_2-NH_2$

A-29 Structure A best fits the mass spectral data. m/e 30 corresponds to a loss of $C_4H_9$ from M; $\alpha,\beta$-cleavage of structure A would account for such a loss.

Q-30 In the mass spectrum of di-*n*-propyl ether (M = 102), ions of m/e 43 and m/e 59 are observed. Explain how these ions might be formed.

A-30 m/e 43 corresponds to $C_3H_7^+$,

m/e 43

m/e 59 corresponds to $C_3H_7O^+$,

m/e 59 2%

m/e 73 corresponds to loss of 29 mass units, i.e., $C_2H_5$.

$+ C_2H_5^{\cdot}$

m/e 102 m/e 73

Q-31 Are the peaks indicated in the following mass spectrum consistent with the structure ethyl butyl ether or methyl n-pentyl ether? Justify your answer.

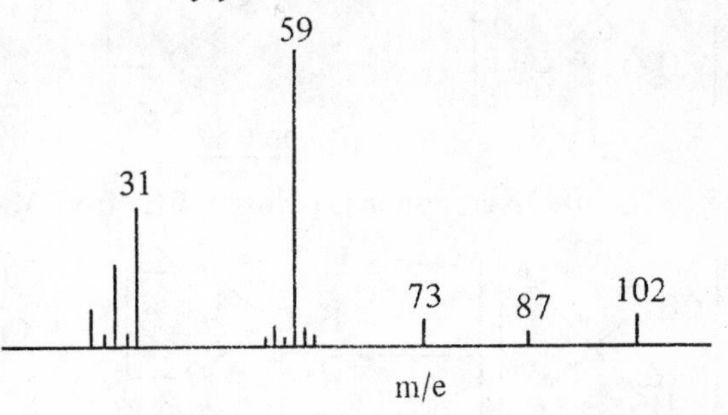


A-31 Ethyl butyl ether.

m/e 102 corresponds to $C_6H_{14}O$, the molecular ion.

m/e 87 corresponds to loss of $CH_3$ from M ($\alpha,\beta$-fission process).

$-CH_3^{\cdot}$

m/e 102 m/e 87

m/e 73 corresponds to loss of $C_2H_5$ from M. This fragment could be formed by alkyl-oxygen fission.

$-C_2H_5^{\cdot}$

m/e 102 m/e 73

m/e 59 corresponds to a loss of $C_3H_7$ from M; this could be formed by the alternative $\alpha,\beta$-fission process:

$-C_3H_7^{\cdot}$

m/e 59

Q-32 In the mass spectrum of the ketal shown below, fragment ions are observed at m/e 87 (15%) and m/e 73 (100%). Explain the formation of each ion.

$CH_3$ O H O

M = 88

A-32 m/e 87 corresponds to loss of H from M.

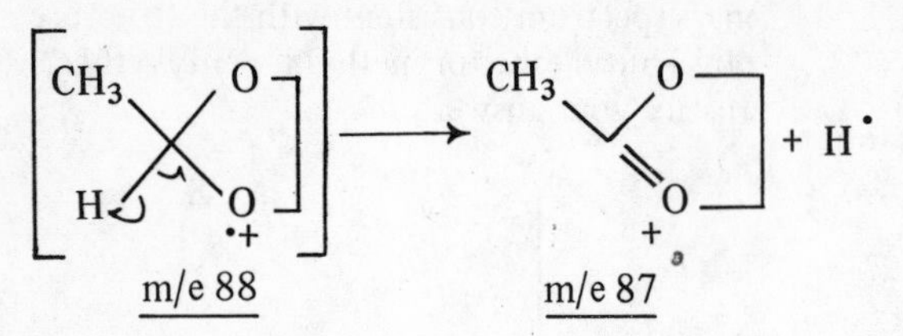


m/e 73 corresponds to loss of $CH_3$ from M.

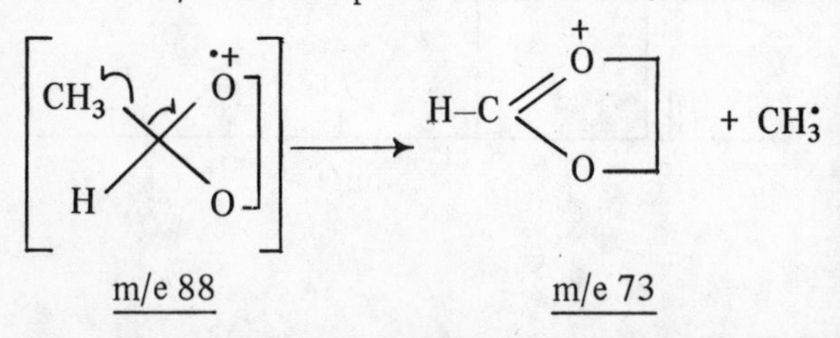


**R** Alcohols, ethers and amines fragment primarily by $\alpha, \beta$-cleavage with formation of oxonium or immonium ions.

The base peak in a mass spectrum is the most intense peak. Per cent figures listed for m/e values indicate the intensity of the peak relative to the base peak.

S-4 Common simple fragmentation processes of halides are:

a. cleavage of the C–X bond,

b. $\alpha, \beta$-fission with formation of halonium ions, i.e., and

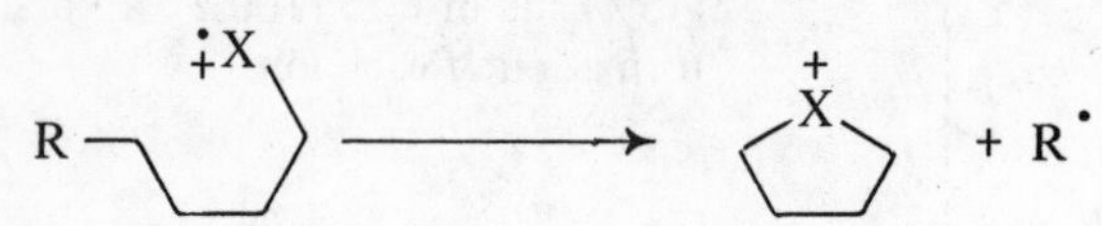


c. fragmentation of a remote alkyl group with formation of a cyclic halonium ion



The atomic masses of the halogens are

F 19
Cl 35 and 37 (3:1)
Br 79 and 81 (1:1)
I 127

Q-33 The mass spectrum of iodoethane (M = 156) is shown below. Suggest mechanisms which would account for the formation of the indicated peaks.

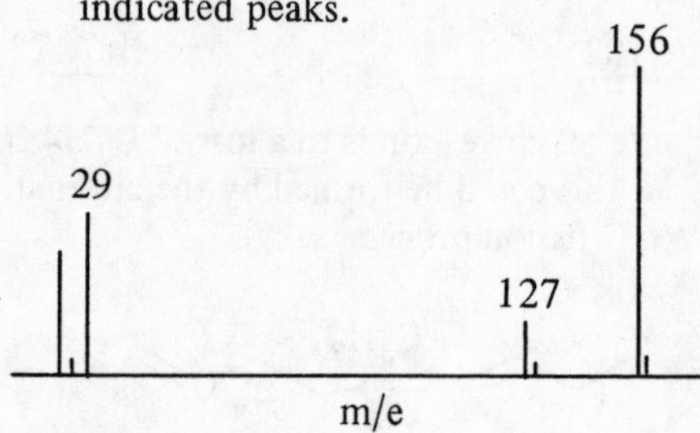

A-33 m/e 156 corresponds to $C_2H_5I^+$

m/e 127 corresponds to $I^+$, formed by C–I cleavage:

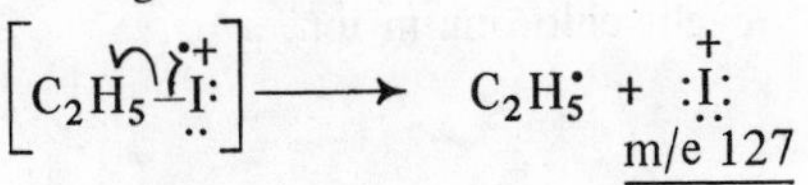

m/e 29 corresponds to ethyl cation $C_2H_5{}^+$ formed by C–I cleavage:

$$\left[C_2H_5-\ddot{I}:\right]^{\cdot+} \longrightarrow \underset{m/e\ 29}{C_2H_5{}^+} + \cdot\ddot{I}:$$

A-34 The isotope pattern of 1:1 of the M and M + 2 peaks suggests a mono-bromo compound. The probable formula is $C_2H_5Br$ (M = 108, M + 2 = 110).

m/e 79 and m/e 81 are due to $(Br^{79})^+$ and $(Br^{81})^+$, formed by homolytic heteroatom-cleavage.

$$\left[C_2H_5-Br\right]^{\cdot+} \longrightarrow C_2H_5^{\cdot} + Br^+$$

m/e 29 ($C_2H_5{}^+$) is formed by heterolytic heteroatom cleavage.

$$\left[C_2H_5-Br\right]^{\cdot+} \longrightarrow C_2H_5{}^+ + Br^{\cdot}$$

A-35 The spectrum is that of fluoroethane.

m/e 48 is the molecular ion.

m/e 47 corresponds to loss of H• by α, β-cleavage:

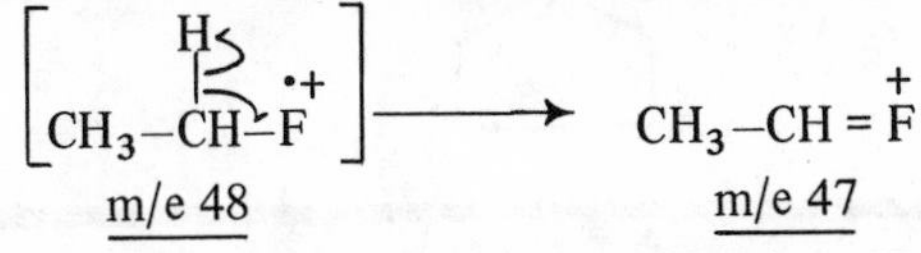

m/e 33 corresponds to loss of $CH_3^{\cdot}$ from M by α, β-cleavage:

$$\left[CH_3-CH_2-F\right]^{\cdot+} \xrightarrow{-CH_3^{\cdot}} \underset{m/e\ 33}{CH_2=\overset{+}{F}}$$

Q-34 Suggest a structure for the molecule which gives the following mass spectrum.

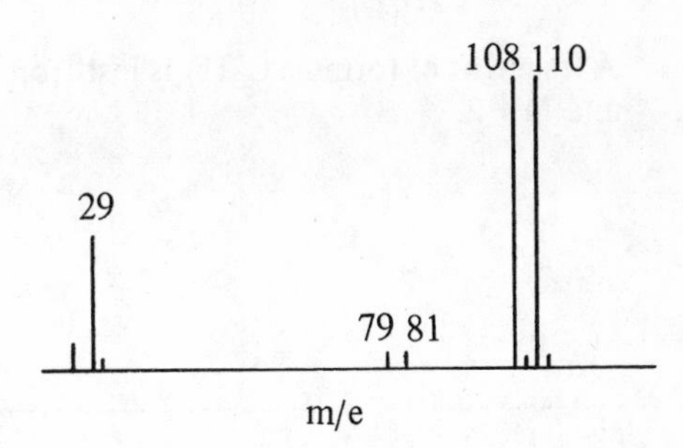


Q-35 Suggest a structure for the compound which affords the following mass spectrum, and explain the formation of the fragment ions.

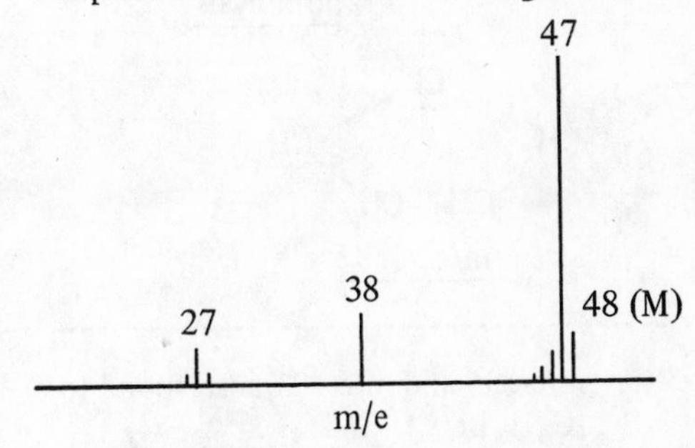


(Do not attempt to explain the origin of the ion of m/e 27.)

Q-36 The mass spectrum of heptyl chloride is shown below. Assign formulas to the ions of m/e 91 (100%) and m/e 93 (35%). What fragment is lost during formation of these ions.

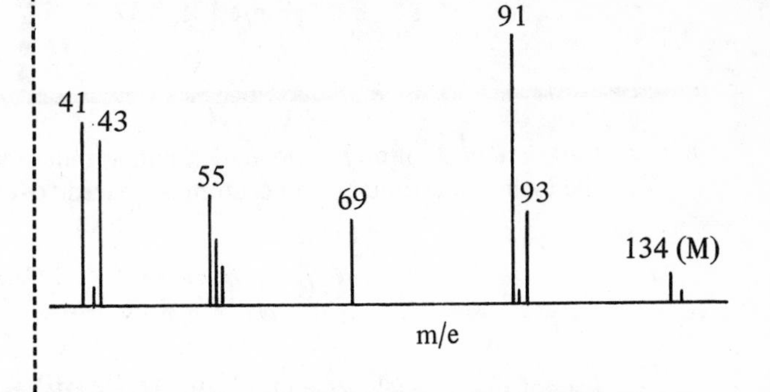

A-36 m/e 91 – $C_4H_8Cl^{35}$
m/e 93 – $C_4H_8Cl^{37}$

A fragment of formula $C_3H_7$ is lost from M and M + 2.

Q-37 Illustrate the loss of the alkyl fragment ($C_3H_7$) from the heptyl chloride cation radical and demonstrate the formation of a cyclic chloronium ion.

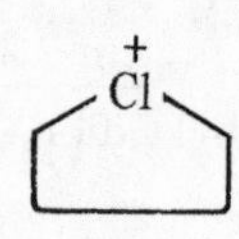


m/e 91, 93

A-37 $\left[ C_3H_7{-}CH_2 \quad \overset{\cdot\cdot}{\underset{}{+Cl:}} \quad CH_2 \atop CH_2{-}CH_2 \right] \xrightarrow{-C_3H_7^{\cdot}}$

homolysis

$\overset{\cdot\cdot}{Cl}^+$ bridging $CH_2$ and $CH_2$, with $CH_2{-}CH_2$

m/e 91, 93

Q-38 Predict the mechanistic pathway which will afford equiabundant ions of m/e 135 and m/e 137 in the mass spectrum of long chain alkyl bromides.

A-38 m/e 135 and m/e 137 correspond to $(C_4H_8Br^{79})^+$ and $(C_4H_8Br^{81})^+$ respectively. By analogy with A-37 cyclic bromonium ions are formed.

$[R{-}\text{ring with } \overset{+}{Br}]$ $\xrightarrow{-R^{\cdot}}$ cyclic $\overset{\cdot\cdot}{Br}^+$

m/e 135, 137

**R** Common simple fragmentation processes of organic halides are:

a. C–X cleavage

b. $\alpha$, $\beta$-fission with the formation of resonance-stabilized halonium ion.

$$CH_2 = \overset{+}{X} \longleftrightarrow \overset{+}{C}H_2{-}X$$

c. remote alkyl group scission with formation of cyclic five-membered halonium ions.

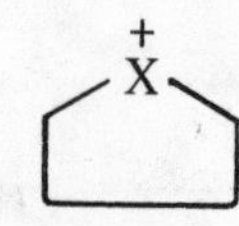


S-5 Cation radicals of esters, aldehydes, and ketones may undergo fragmentations in which the bond between the carbonyl carbon and an $\alpha$ atom is cleaved ($\alpha$-cleavage).

$$\underset{\alpha \qquad \alpha}{R{-}\overset{O}{\overset{\|}{C}}{-}O{-}R'} \qquad \underset{\alpha \qquad \alpha}{R{-}\overset{O}{\overset{\|}{C}}{-}H(R')}$$

Ions of the type $R^+$, $RC{\equiv}O^+$, $^+OR'$, $^+O{\equiv}COR'$ may form. In esters, ions of the type $R'^+$ may be formed.

$$R{-}\overset{:O\dot{+}}{\overset{\|}{C}}{-}O{-}R' \longrightarrow R{-}\overset{:\dot{O}:}{\overset{|}{C}}{=}O + R'^+$$

Q-39 The mass spectrum of methyl butyrate (M = 102) contains peaks at m/e 71 (55%) and m/e 43 (100%). Suggest a mechanism to explain the formation of these ions.

A-39 Both ions arise by $\alpha$-cleavage. m/e 71 represents a loss of $CH_3O$ from M:

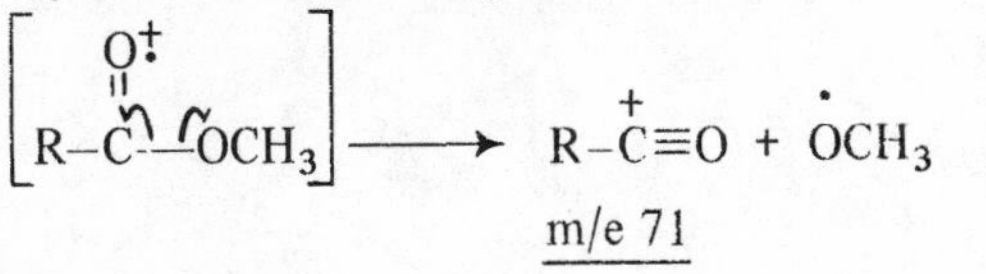

$$\left[R-\overset{O^{\dot{+}}}{\overset{\|}{C}}-OCH_3\right] \longrightarrow R-\overset{+}{C}\equiv O + \dot{O}CH_3$$

m/e 71

alternatively,

$$\left[R-\overset{O}{\overset{\|}{C}}-\overset{\cdot +}{O}CH_3\right] \longrightarrow R-\overset{+}{C}=O + :\dot{O}CH_3$$

m/e 71

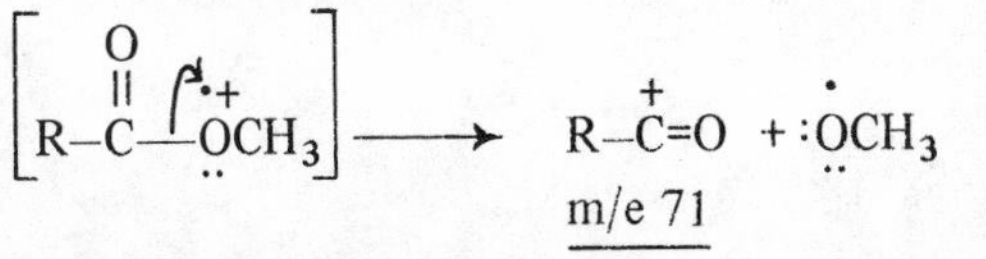

m/e 43 corresponds to $C_3H_7$:

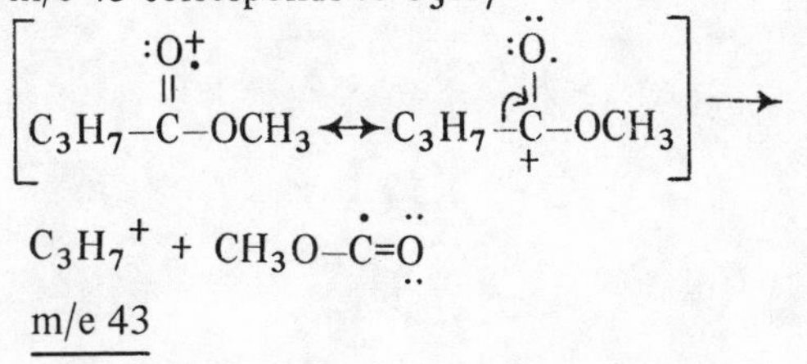

$$\left[C_3H_7-\overset{:O^{\dot{+}}}{\overset{\|}{C}}-OCH_3 \leftrightarrow C_3H_7-\overset{:\ddot{O}\cdot}{\overset{|}{\underset{+}{C}}}-OCH_3\right] \longrightarrow$$

$$C_3H_7^+ + CH_3O-\dot{C}=\ddot{O}$$

m/e 43

alternatively,

$$C_3H_7-\underset{+}{C}\equiv O \xrightarrow{-CO} C_3H_7^+ + CO$$

m/e 71     m/e 43

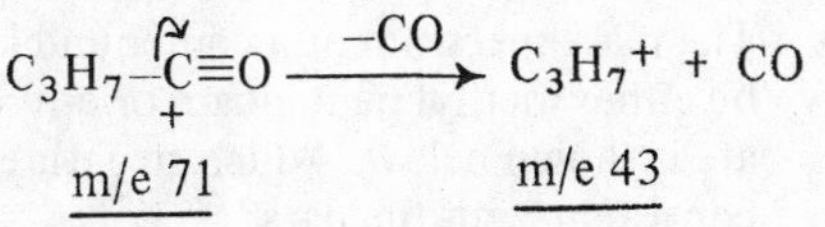

Q-40 The mass spectrum of methyl butyrate also contains ions at m/e 31 (43%) and m/e 59 (25%). Explain the formation of these ions.

A-40 Both ions are formed by $\alpha$-cleavage.
m/e 31 corresponds to $CH_3O^+$

$$\left[R{-}\overset{:\ddot{O}^{\cdot +}}{\overset{\|}{C}}{-}OCH_3 \longleftrightarrow R{-}\overset{:\ddot{O}^{\cdot}}{\overset{|+}{C}}{-}OCH_3\right] \longrightarrow R{-}\dot{C}{=}O + \overset{+}{O}CH_3$$

m/e 31

or, alternatively,

$$\left[R{-}\overset{O}{\overset{\|}{C}}{-}\overset{\cdot +}{O}CH_3\right] \longrightarrow R{-}\dot{C}{=}O + \overset{+}{O}CH_3$$

m/e 31

m/e 59 corresponds to loss of $C_3H_7$ from M

$$\left[C_3H_7{-}\overset{\dot{+}O}{\overset{\|}{C}}{-}OCH_3\right] \longrightarrow C_3H_7^{\cdot} + \overset{O^+}{\overset{|||}{C}}{-}OCH_3$$

m/e 59

or, alternatively,

$$\left[C_3H_7{-}\overset{O}{\overset{\|}{C}}{-}\overset{+\cdot}{O}CH_3\right] \longrightarrow C_3H_7^{\cdot} + \overset{O}{\overset{\|}{C}}{=}\overset{+}{O}CH_3$$

m/e 59

Q-41 The mass spectrum of an ester (M = 116) has prominent ions at m/e 57 (100%), m/e 29 (57%) and m/e 43 (27%). Which of the following esters is consistent with the data?

$(CH_3)_2CHCOOC_2H_5$ A

$CH_3CH_2COOCH_2CH_2CH_3$ B

$CH_3CH_2CH_2COOCH_3$ C

A-41 B $CH_3CH_2COOCH_2CH_2CH_3$

$\alpha$-cleavage products are:

$C_2H_5C{\equiv}O^+$ (m/e 57)

$C_2H_5^+$ (m/e 29)

$C_3H_7^+$ (m/e 43)

A would not yield m/e 57.

C can be eliminated by molecular weight.

Q-42 The mass spectrum of a compound known to be either methyl pentanoate or ethyl butanoate is shown below. Which structure is most consistent with the data?

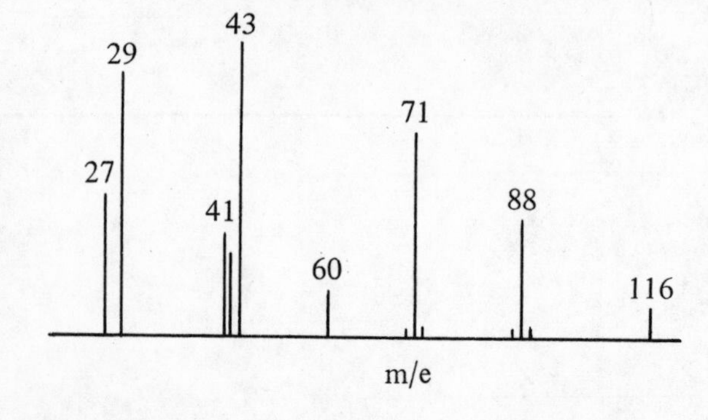

A-42 Ethyl butanoate.

$\alpha$-cleavage products are:

$C_3H_7C{\equiv}O^+$ (m/e 71)

$C_3H_7^+$ (m/e 43)

$C_2H_5^+$ (m/e 29)

Q-43 The mass spectrum of a compound known to be either 3- or 4-heptanone is shown below. Which ketone is most consistent with the data?

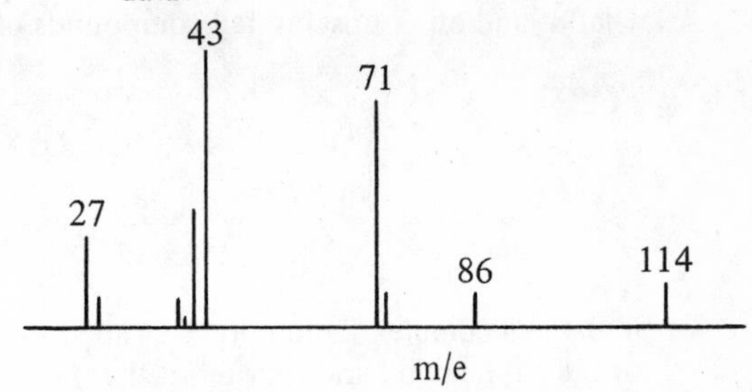


A-43 4-heptanone.

$$\left[ C_3H_7{-}\overset{:\overset{\bullet}{O}{+}}{\overset{\|}{C}}{-}C_3H_7 \right] \xrightarrow{-C_3H_7^{\bullet}} C_3H_7{-}C{\equiv}\overset{+}{O}:$$

*homolysis* m/e 71

Q-44 Illustrate the formation of the ion of m/e 43 from the oxonium ion of m/e 71 (A-43). Name the fragmentation type.

A-44

$$C_3H_7{-}\overset{+}{C}{=}O \longrightarrow C_3H_7^+ + CO$$

m/e 71 m/e 43

heterolysis

Q-45 The mass spectrum of benzaldehyde is shown below.

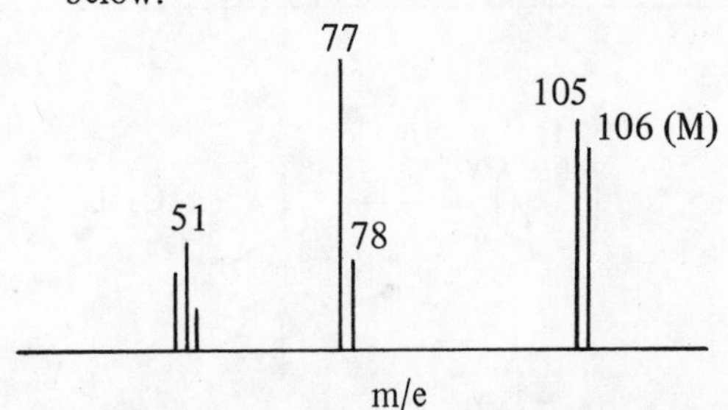


Account for the formation and abundance of m/e 105.

A-45

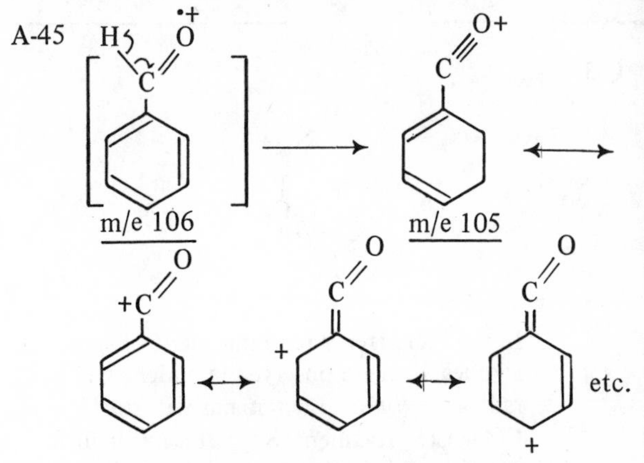


m/e 105 is formed by $\alpha$-cleavage. Resonance stabilization explains the abundance of the fragment ion.

**R** Esters, aldehydes, and ketones undergo cleavage of the bonds between the carbonyl carbon and an $\alpha$ atom ($\alpha$-cleavage).

## PART III: Fission and Rearrangement Processes in Which Two Bonds Are Broken. The Formation of Even m/e Fragments When M is Even.

S-1 Olefins and other unsaturated compounds of the general type

$X^{\alpha}$ — C(=Q) … $Y^{\beta}$ — $Z^{\gamma}$ — H

where Q, X, Y and Z may be almost any combination of C, O, N and S

undergo a complex fission process called the *McLafferty* rearrangement. This process involves cleavage of the allylic bond and transfer of the $\gamma$–H to the ionized double bond. The radical cation charge stays within one or other of the fragments.

---

Q-1 Suggest two mechanistic pathways for the fragmentation of the 1-pentene radical cation to yield ethylene.

H, $\gamma$, $\beta$, $\alpha$ ⟷ H, $\gamma$, $\beta$, $\alpha$

---

A-1

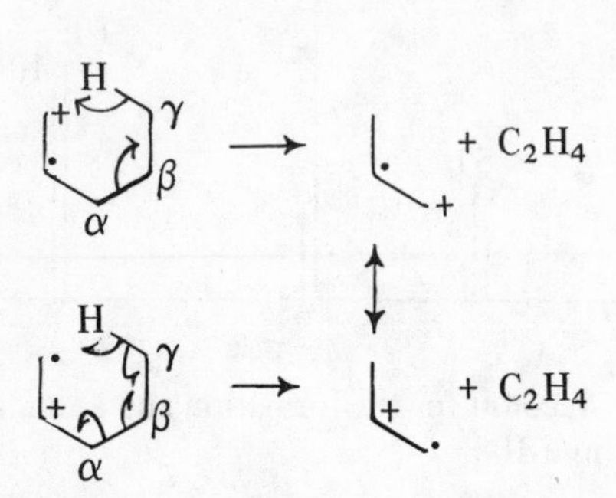


---

Q-2 Illustrate the McLafferty rearrangement by which 2-hexene can afford a fragment at m/e 56.

---

A-2 m/e 56 represents loss of $C_2H_4$ from M

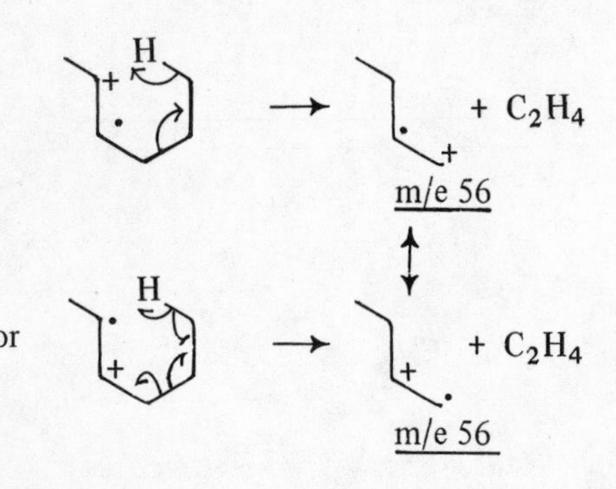


---

Q-3

H, $_1C$, $_2C$, $_3C$, $_4C$, $_5C$

In the McLafferty rearrangement of some alkenes, ions are observed in which the radical cation charge remains with the C(4)–C(5) fragment. Suggest a mechanism to account for this fragmentation.

A-3

a. 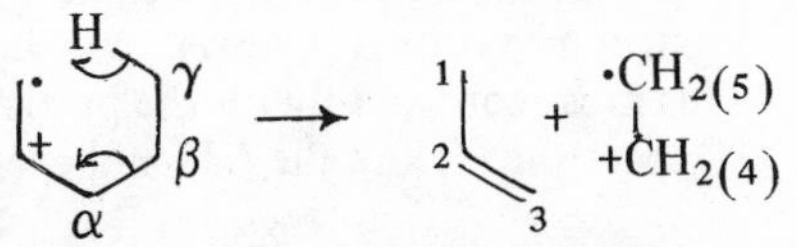 $\longrightarrow$ + $\cdot CH_{2(5)}$ / $+CH_{2(4)}$

b. 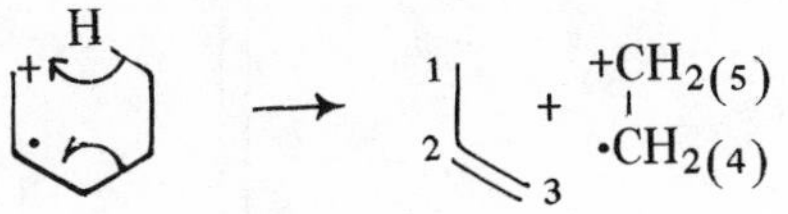 $\longrightarrow$ + $+CH_{2(5)}$ / $\cdot CH_{2(4)}$

Q-4 The mass spectrum of a compound which could be either A or B had an ion peak at m/e 84. Which compound best fits the data. Suggest a mechanism to account for ion m/e 84.

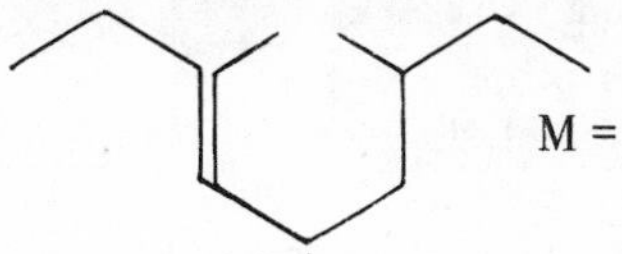

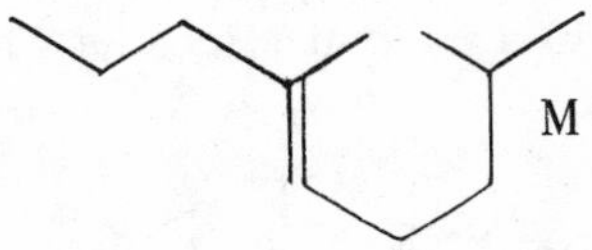

A: M = 154

B: M = 154

A-4 Compound A.
m/e 84 corresponds to loss of 70 mass units, i.e., $C_5H_{10}$, from M.

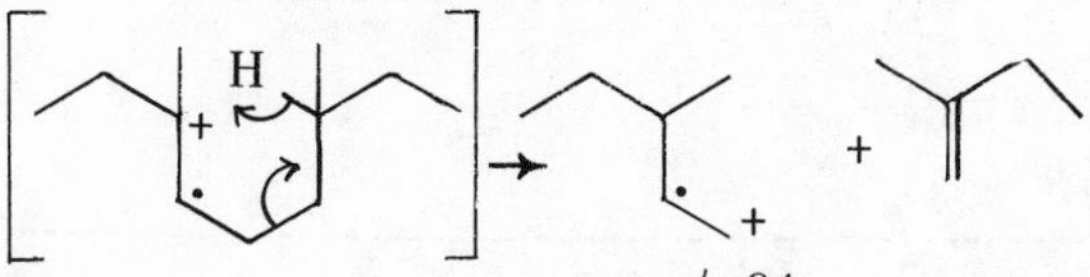

m/e 84

or

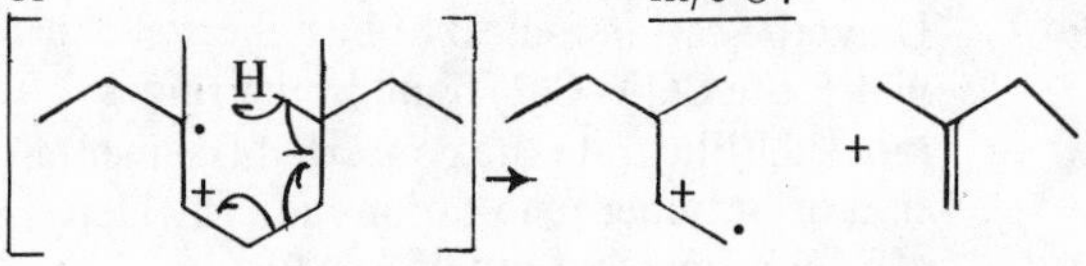

Compound B would extrude

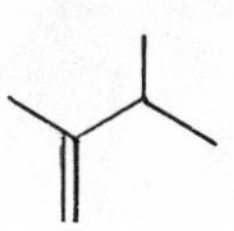

, to give an ion of m/e 70.

Q-5 Which of the following will undergo McLafferty rearrangement?

a.

b.

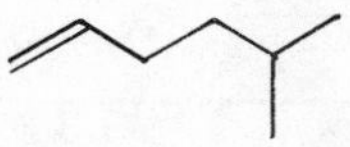

c.

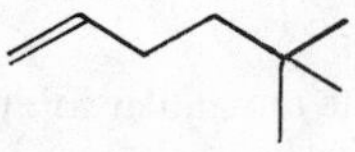

d.

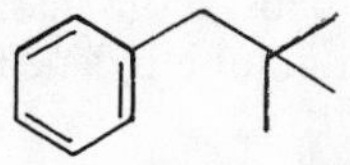

e. $CH_3CH_2CH_2\overset{O}{\overset{\|}{C}}OCH_3$

A-5 a., b., d. and e. will undergo McLafferty rearrangement. c. has no $\gamma$-H; therefore, it will not.

Q-6 Explain the formation of ions at

a. m/e 69

b. m/e 70

in the mass spectrum of 3-heptene (M = 98).

**A-6** a. m/e 69 corresponds to loss of 29 mass units from M, i.e., loss of $C_2H_5$. m/e 69 is an *odd* mass and arises by allylic cleavage, i.e.,

homolysis m/e 69

b. m/e 70 (even) corresponds to loss of 28 mass units from M, i.e., loss of $C_2H_4$. m/e 70 arises from a McLafferty rearrangement.

heterolysis heterolysis

$+ C_2H_4$

m/e 70

**Q-7** A compound was known to have either structure A or B (below). Which structure is most consistent with the partial mass spectrum? Note the odd and even m/e values.

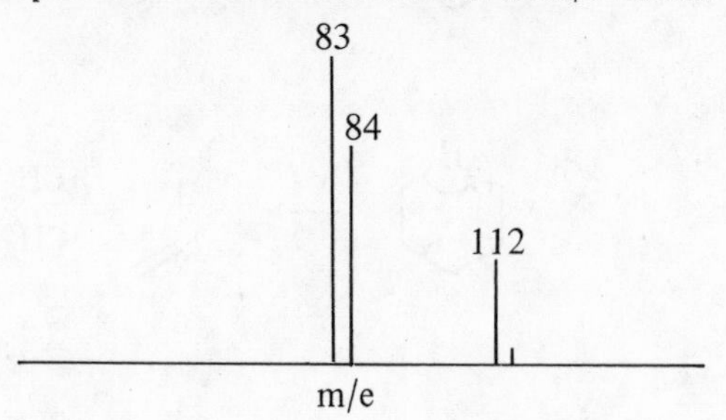


A B

**A-7** Compound B.
m/e 112 is the molecular ion peak. m/e 84 corresponds to loss of 28 mass units from M, i.e., extrusion of ethylene by McLafferty rearrangement.

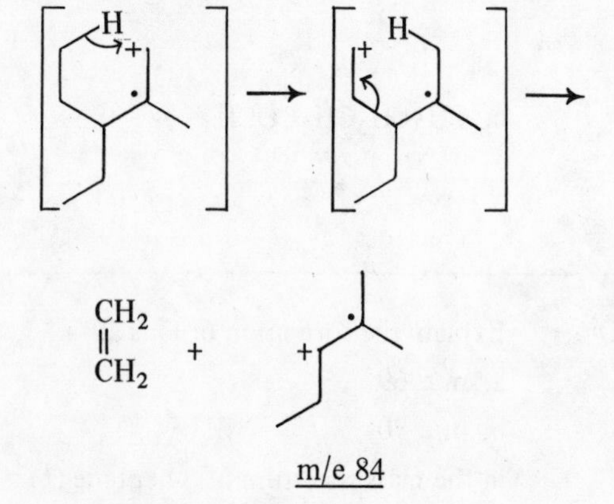


m/e 83 corresponds to loss of 29 mass units, i.e., $C_2H_5$ from M by simple allylic fission.

**Q-8** Draw the cation radical of butylbenzene in which the C(1)–C(2) bond in the ring is ionized. Illustrate the concerted fragmentation of butylbenzene cation radical which affords a fragment ion of m/e 92.

A-8

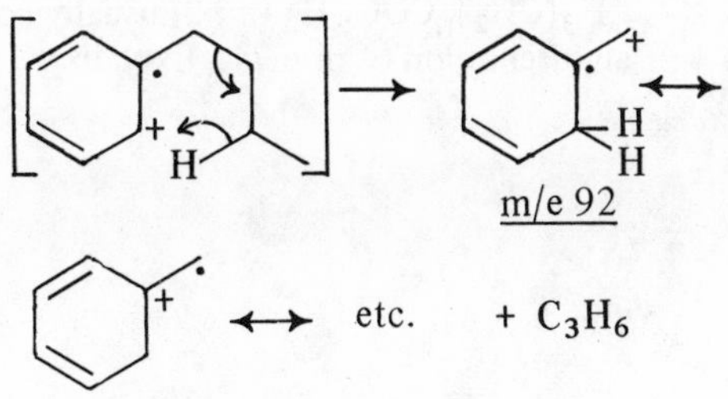

H H H

m/e 92

$\longleftrightarrow$ etc. + $C_3H_6$

Q-9 Account for the major peaks in the mass spectrum of butylbenzene.

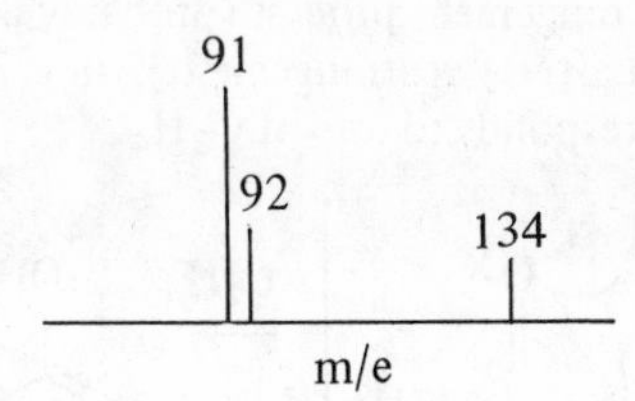


A-9 a. m/e 134: molecular ion corresponding to $C_{10}H_{14}$.

b. m/e 92: corresponds to $C_7H_8^{+\cdot}$, i.e., the McLafferty rearrangement product.

c. m/e 91: corresponds to $C_7H_7^+$, i.e., the tropylium ion formed by benzylic cleavage followed by rearrangement.

Q-10 An ether could have one of the structures given below. The compound gave an intense ion peak at m/e 92. Which structure is most consistent with this data?

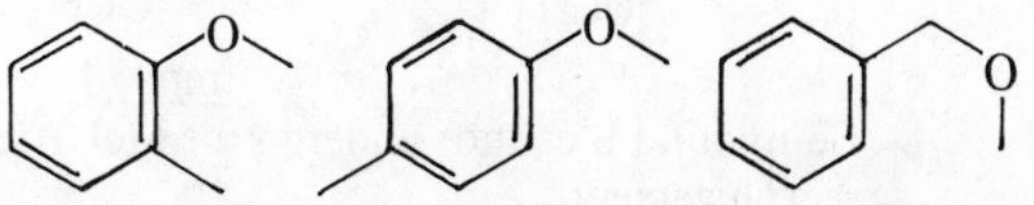

A B C

A-10 Compound C.
The ion of m/e 92 is a McLafferty rearrangement product. Compounds A and B would not undergo a McLafferty.

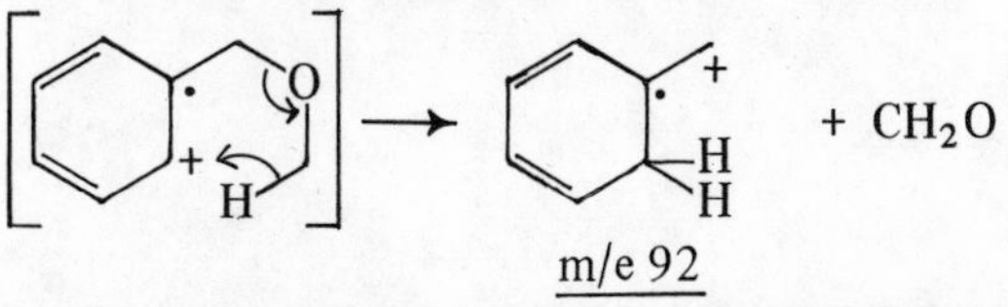

+ $CH_2O$

m/e 92

Q-11 Which of the following ethers would be expected to yield a peak at m/e 94?

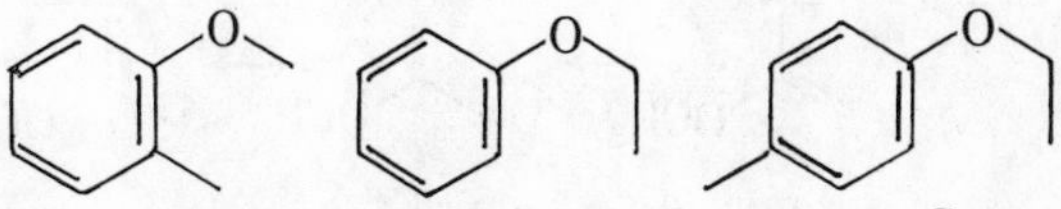

A B C

A-11 Compound B.
The ion of m/e 94 is a McLafferty rearrangement product, i.e.,

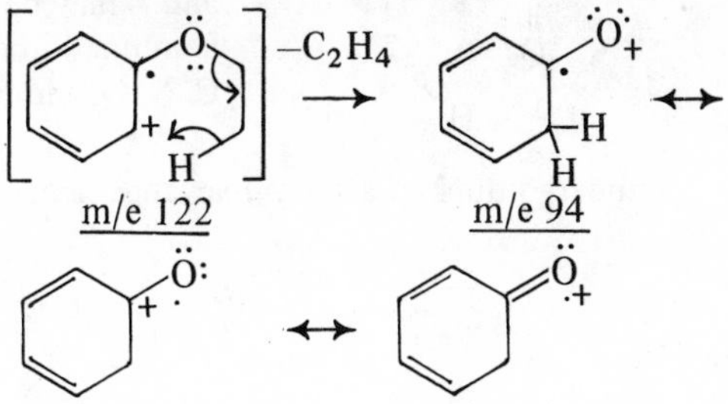

$-C_2H_4$

m/e 122 m/e 94

or, alternatively, if the site of ionization is on the heteroatom

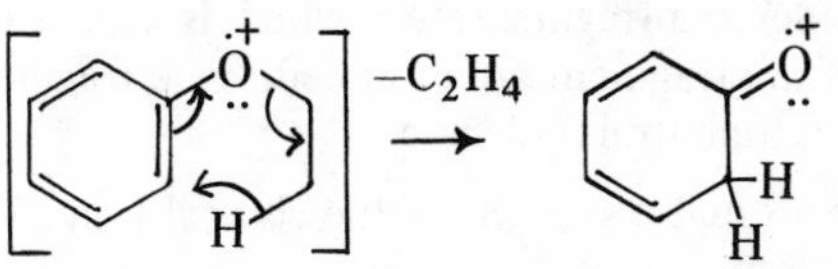

$-C_2H_4$

Compound A would not undergo the McLafferty, and Compound C would yield an ion at m/e 108.

Q-12 The mass spectrum of an ester contains an intense ion at m/e 74 (70%). Which of the two structures given below is most consistent with this observation?

$CH_3CH_2CH_2COOCH_3$

A

$(CH_3)_2CHCOOCH_3$

B

A-12 Structure A.
The even mass of the ion indicates a rearrangement process (probably a McLafferty rearrangement). m/e 74 corresponds to loss of $C_2H_4$.

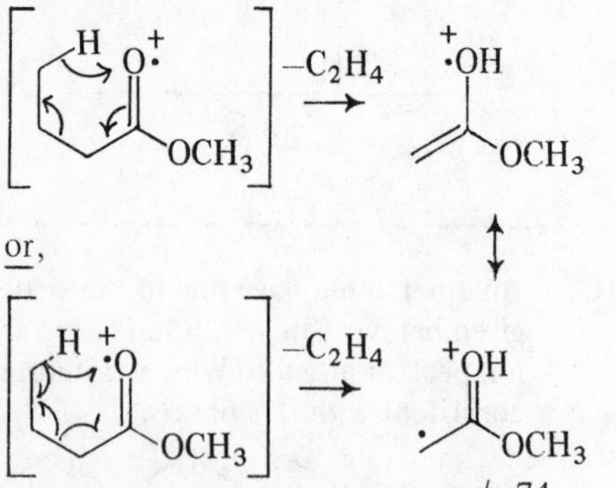

$-C_2H_4$ → $\overset{+}{\cdot}OH$ … $OCH_3$

or,

$-C_2H_4$ → $^{+}OH$ … $OCH_3$

m/e 74

Compound B cannot undergo the McLafferty rearrangement.

Q-13 The mass spectra of esters of the type $CH_3(CH_2)_nCOOCH_3$ ($n \geqslant 2$) usually contain an intense ion of m/e 74. Explain.

A-13 All such esters may undergo McLafferty rearrangement to form the fragment ion of m/e 74:

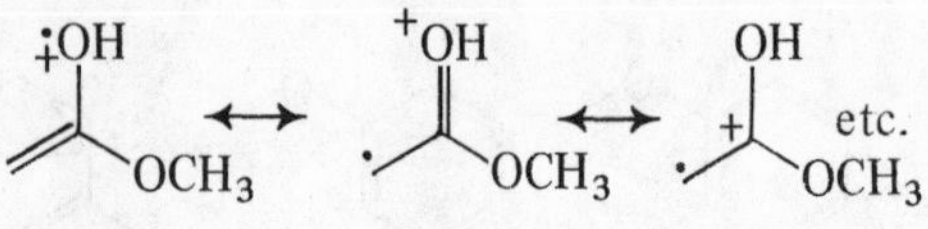

$\overset{+}{:}OH$ … $OCH_3$ ↔ $^{+}OH$ … $OCH_3$ ↔ OH … $OCH_3$ etc.

m/e 74

Q-14 The mass spectrum of 2-methylpentanal (M = 100, 5%) contains ions at m/e 71 (15%), 58 (95%) and 43 (100%). Are these ions consistent with the structure of the aldehyde?

A-14 Yes.

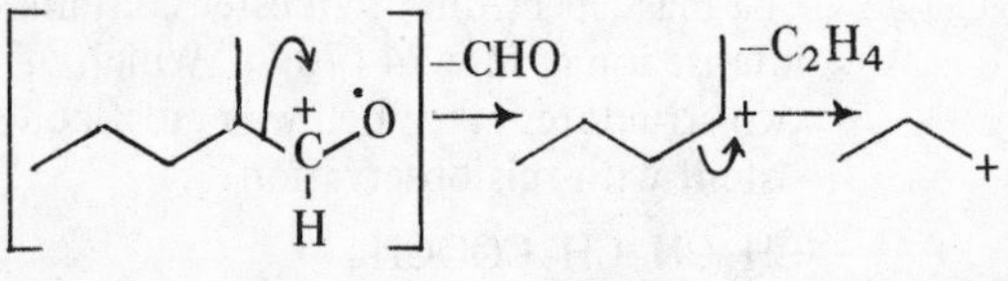

−CHO → $-C_2H_4$ →

m/e 100 m/e 71 m/e 43

→ + $\overset{+}{\cdot}OH$

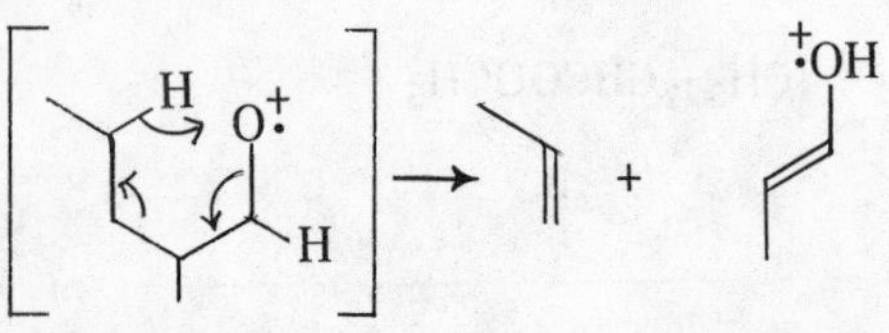

m/e 58

**R** Compounds of the general type

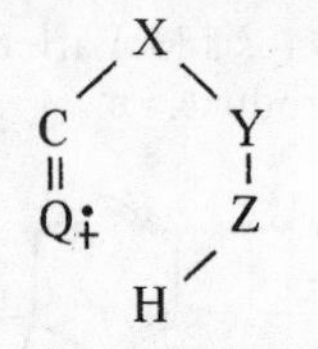


where Q, X, Y, and Z may be any combination of C, N, O, and S

undergo the McLafferty rearrangement.

S-2 Cyclic six-membered ring olefins of the cyclohexene type undergo a fragmentation which is similar to the retro-Diels Alder fragmentation. A butadiene and an olefinic fragment are obtained. One fragment, with even m/e, bears the radical cation charge, and the other is a neutral molecule.

The retro-Diels Alder fragmentation is also a facile process in aromatic systems of the general type.

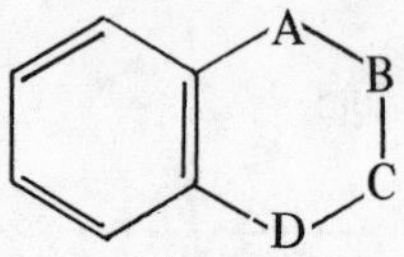

The rearrangement can be easily understood by assuming that an aromatic cation radical is formed. Alternative mechanisms can be postulated, however, when A, B, C or D are ionizable heteroatoms.

---

Q-15 Draw the radical cation formed by ionization of cyclohexene. Suggest two mechanisms which will account for a fragmentation in which C(4) and C(5) are extruded as neutral ethylene.

A-15

$\rightarrow$ + $CH_2CH_2$ (4 5)

m/e 54

$\rightarrow$ + $CH_2CH_2$ (4 5)

m/e 54

Q-16 Explain the formation of an ion at m/e 68 in the mass spectrum of 1-methylcyclohexene (note even m/e).

M = 96

A-16 m/e 68 corresponds to loss of 28 mass units, i.e., $C_2H_4$, from M.

$\rightarrow$ + $C_2H_4$

m/e 68

or

$\rightarrow$ + $C_2H_4$

m/e 68

Similar mechanisms can be derived by considering the initial cleavage of the C(3)–C(4) bond.

Q-17 Which of the following compounds would be expected to undergo a retro-Diels Alder to yield an intense ion at m/e 66?

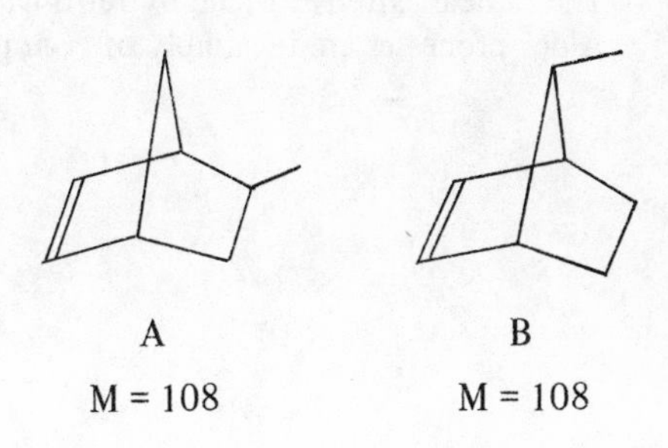


A-17 Compound A.
m/e 66 corresponds to loss of 42 mass units, i.e., $C_3H_6$, from M.

$\rightarrow$ + $CHCH_3{=}CH_2$

m/e 66

$\rightarrow$ + $C_2H_4$

m/e 80

Q-18 Explain the formation of an ion at m/e 222 in the mass spectrum of the steroid,

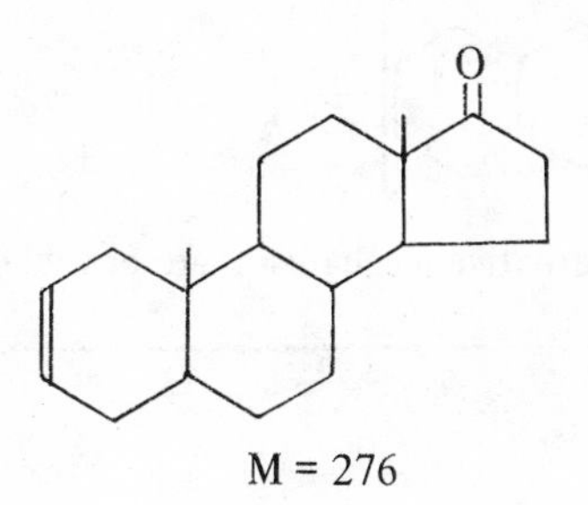


M = 276

A-18 m/e 222 corresponds to loss of 54 mass units, i.e., $C_4H_6$, from M.

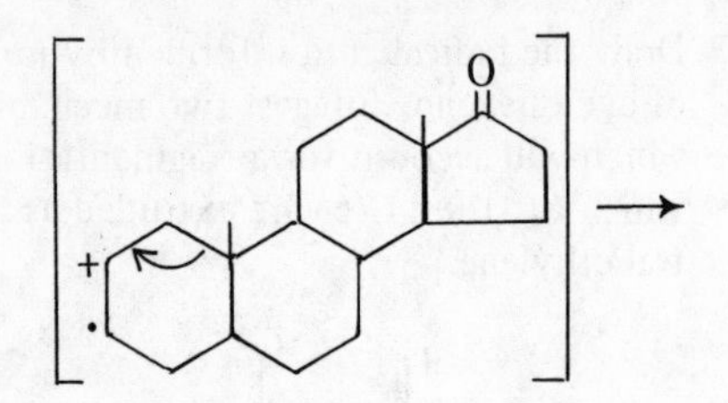

m/e 222

Q-19 Consider the m/e values of the charged fragments formed by the retro-Diels Alder processes described in the previous examples. Are the m/e values of the ions formed *odd* or *even?*

---

A-19 The radical cations formed by retro-Diels Alder processes are invariably of *even* mass.

Q-20 The mass spectrum of tetralin (M = 132) contains an intense ion of m/e 104. What is the structure of the neutral fragment which has been extruded?

Suggest a mechanism for the fragmentation.

---

A-20 m/e 104 corresponds to loss of 28 mass units, i.e., $C_2H_4$, from M.

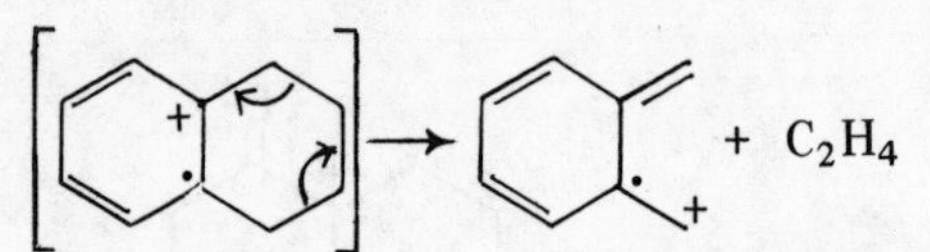


Alternative mechanisms are possible.

Q-21 Which of the following compounds would be expected to undergo a retro-Diels Alder? What m/e values might be expected?

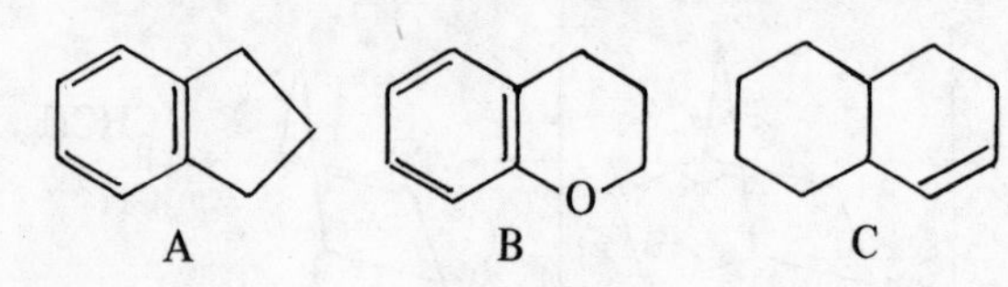

A-21 Only B.

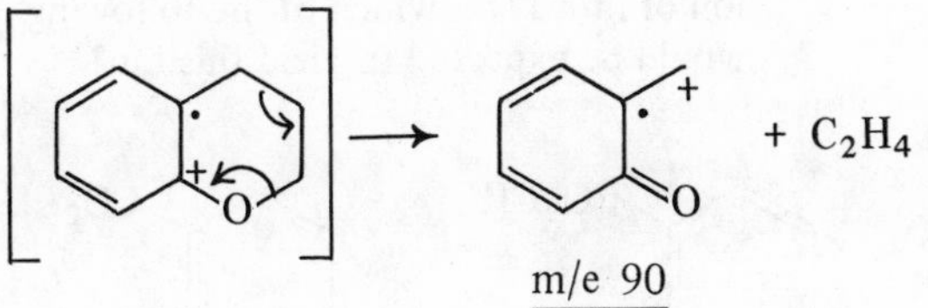

m/e 90

$+ C_2H_4$

or,

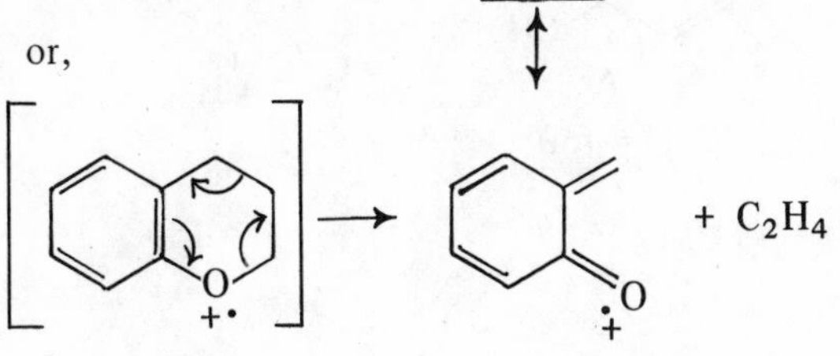

$+ C_2H_4$

also

$+ \cdot CH_2{-}{}^+CH_2$

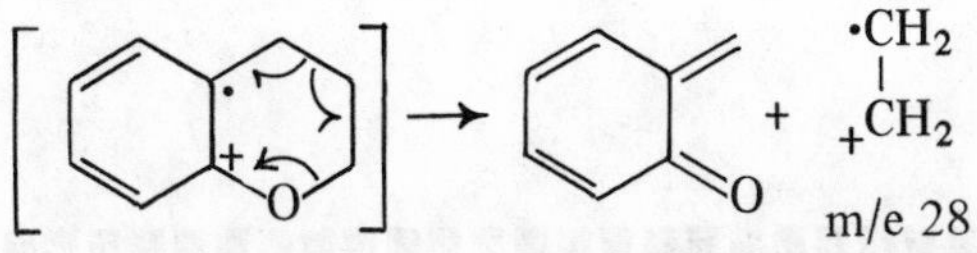

m/e 28

Q-22 Cation radicals of certain *ortho*-disubstituted compounds fragment by a mechanism similar to the retro-Diels Alder. These rearrangements are most easily understood by assuming that an aromatic cation radical is formed, or, that the site of ionization is on heteroatom A.

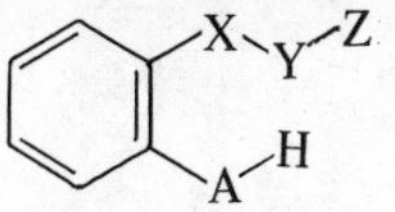

X Y Z A H

The mass spectrum of

$OCH_3$ OH

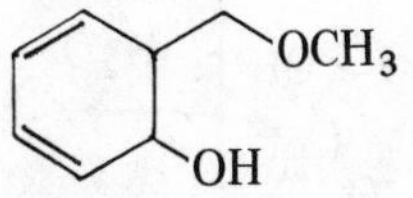

$M = 138$

contains an ion of m/e 106. What neutral fragment has been extruded to form this ion?

A-22 m/e 106 corresponds to loss of 32 mass units, i.e., $CH_3OH$, from M. ($C_2H_8$ is a non-existent molecule and the loss of $O_2$ is very unlikely.)

Q-23 Invoke a concerted mechanism to explain the loss of methanol described in A-22.

A-23

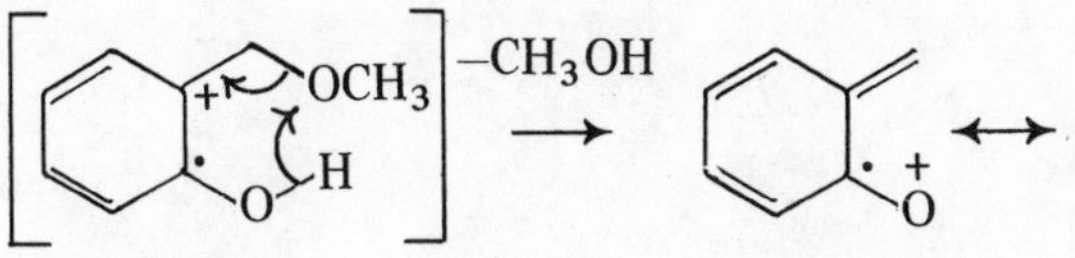

$OCH_3$ H O

$-CH_3OH$

m/e 136 m/e 106

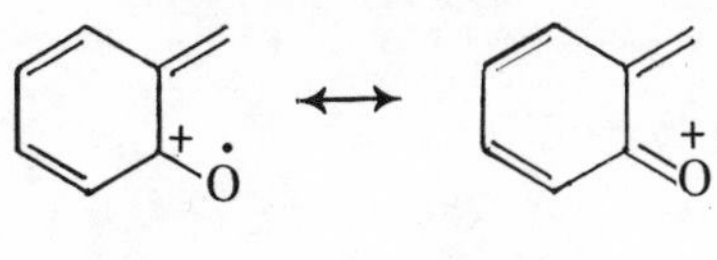

alternatively,

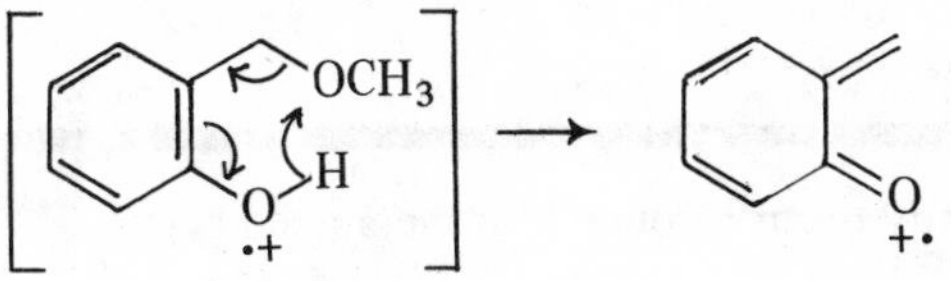

$OCH_3$ H O

Q-24 The mass spectrum of a compound contains a significant ion of m/e 120.

Which structure is most consistent with this data?

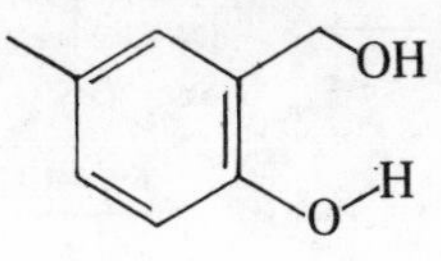

OH O H O H

A $M = 138$

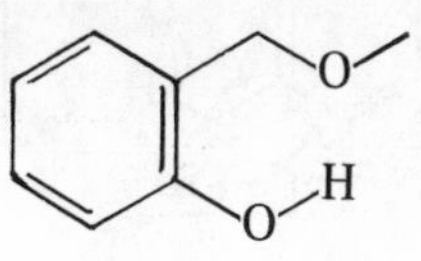

B $M = 138$

**A-24** Compound A.
m/e 120 corresponds to a loss of 18 mass units, i.e., $H_2O$, from M.

[ OH, +, •, O, H ] $-H_2O$ → (•, +, O) m/e 120

↕

or alternatively,

[ OH, O, H, •+ ] $-H_2O$ → (O, +•) m/e 120

Compound B would give a corresponding peak at m/e 106.

**Q-25** The mass spectrum of an ester contains an ion of m/e 118. Which of the following would be expected to yield this ion?

$CO_2CH_3$, $CH_3$

A

M = 150

$CO_2CH_3$, $CH_3$

B

M = 150

**A-25** Compound A.
m/e 118 corresponds to a loss of 32 mass units, i.e., $CH_3OH$, from M.

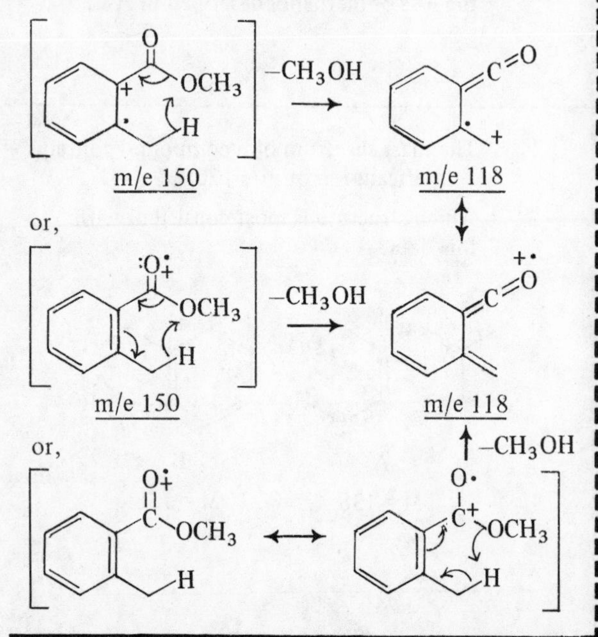


**R** Retro-Diels Alder fragmentations occur in cyclohexene type ring systems and in some aromatic compounds.

*Ortho*-disubstituted aromatic compounds may eliminate a neutral fragment through a cyclic six-membered transition state.

**S-3** Some rearrangements of aromatic cation radicals apparently occur through four-membered cyclic transition states. The general case is:

[ H, H, A, B, +, •, H, H, H, H ] → [ H, H, H, A, B, +, •, H, H, H ] → (H, H, H, +, •, H, H, H) + A–B

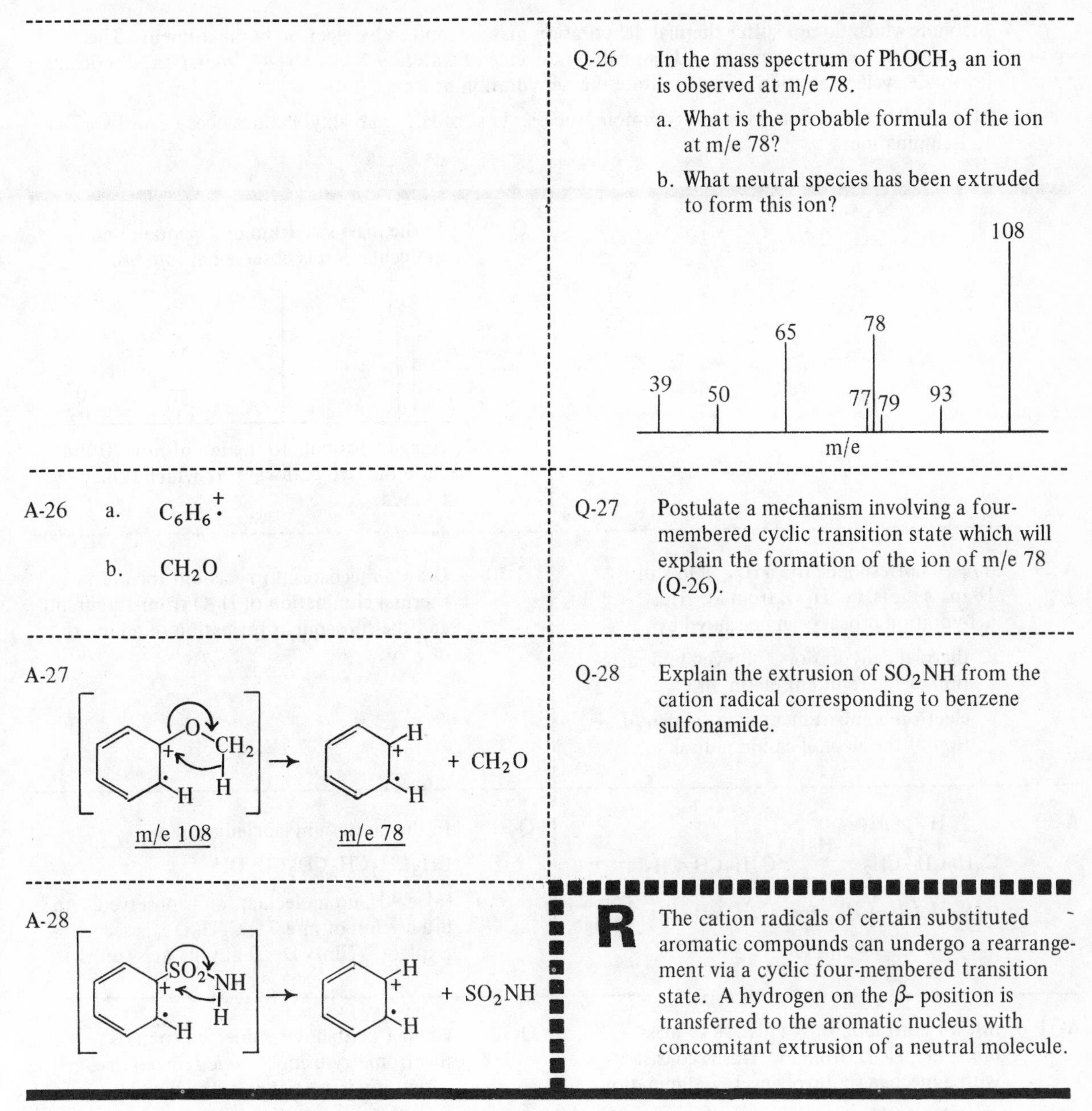

Q-26 In the mass spectrum of $PhOCH_3$ an ion is observed at m/e 78.

a. What is the probable formula of the ion at m/e 78?

b. What neutral species has been extruded to form this ion?



A-26 a. $C_6H_6^{+\cdot}$

b. $CH_2O$

Q-27 Postulate a mechanism involving a four-membered cyclic transition state which will explain the formation of the ion of m/e 78 (Q-26).

A-27

m/e 108 → m/e 78 + $CH_2O$

Q-28 Explain the extrusion of $SO_2NH$ from the cation radical corresponding to benzene sulfonamide.

A-28

→ + $SO_2NH$

**R** The cation radicals of certain substituted aromatic compounds can undergo a rearrangement via a cyclic four-membered transition state. A hydrogen on the $\beta$- position is transferred to the aromatic nucleus with concomitant extrusion of a neutral molecule.

S-4 Dehydration, i.e., loss of 18 mass units from the molecular ion, is a significant process in the mass spectrum of many aliphatic alcohols. Loss of water can occur by

a. thermal dehydration

b. electron-bombardment-induced dehydration.

Thermal dehydration may occur in the ion chamber *prior* to electron bombardment ionization, in which case the resultant olefin will be ionized. This type of dehydration is a common source of difficulty in interpreting the mass spectra of alcohols. As a result of this undesirable side reaction, little or no molecular ion may be observed. The mechanism of thermal dehydration may be depicted as:

$$\underset{4\ \ 3}{CH_3CH_2}-\underset{2}{\overset{H}{\overset{|}{CH}}}-\underset{1}{\overset{OH}{\overset{|}{CH_2}}} \xrightarrow{-H_2O} CH_3CH_2CH{=}CH_2 \xrightarrow{-e} \text{molecular ion of } \textit{olefin.}$$

This dehydration process involves 1, 2-elimination of $H_2O$, i.e., the hydroxyl group and a vicinal hydrogen are eliminated.

Alcohols which do not suffer thermal dehydration may be ionized by electron bombardment. The resultant cation radicals can extrude a neutral molecule of water by 1, 3- or 1, 4-elimination. Sterically favorable cyclic transition states facilitate the dehydration process.

Loss of $NH_3$ from amines is a less common process. Loss of HX from alkyl halides does occur by a 1, 2-elimination.

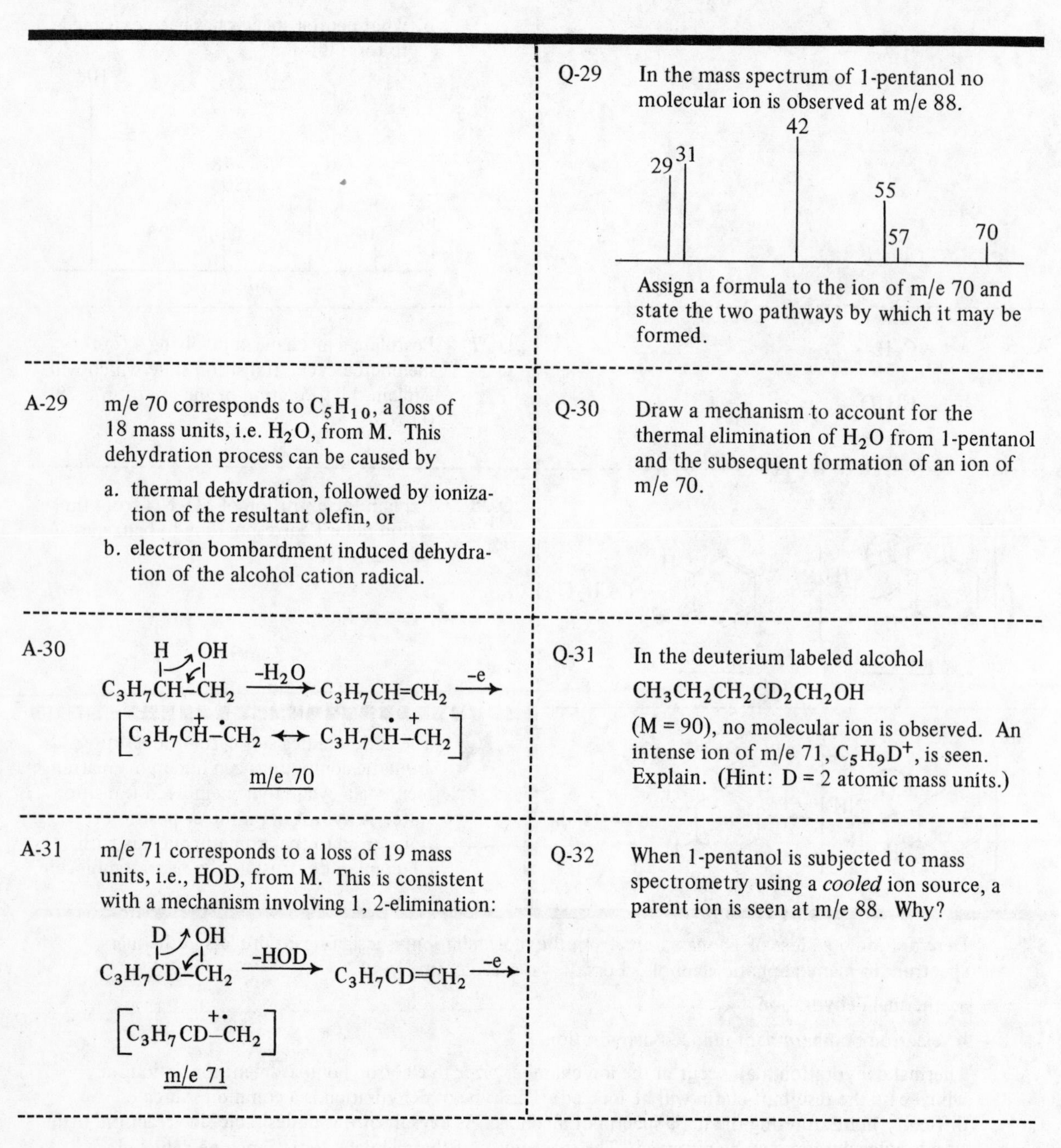

**Q-29** In the mass spectrum of 1-pentanol no molecular ion is observed at m/e 88.



Assign a formula to the ion of m/e 70 and state the two pathways by which it may be formed.

**A-29** m/e 70 corresponds to $C_5H_{10}$, a loss of 18 mass units, i.e. $H_2O$, from M. This dehydration process can be caused by

a. thermal dehydration, followed by ionization of the resultant olefin, or

b. electron bombardment induced dehydration of the alcohol cation radical.

**Q-30** Draw a mechanism to account for the thermal elimination of $H_2O$ from 1-pentanol and the subsequent formation of an ion of m/e 70.

**A-30**

$$C_3H_7\overset{H}{C}H-\overset{OH}{C}H_2 \xrightarrow{-H_2O} C_3H_7CH{=}CH_2 \xrightarrow{-e}$$

$$\left[C_3H_7\overset{+}{C}H-\overset{\bullet}{C}H_2 \leftrightarrow C_3H_7\overset{\bullet}{C}H-\overset{+}{C}H_2\right]$$

m/e 70

**Q-31** In the deuterium labeled alcohol

$CH_3CH_2CH_2CD_2CH_2OH$

(M = 90), no molecular ion is observed. An intense ion of m/e 71, $C_5H_9D^+$, is seen. Explain. (Hint: D = 2 atomic mass units.)

**A-31** m/e 71 corresponds to a loss of 19 mass units, i.e., HOD, from M. This is consistent with a mechanism involving 1, 2-elimination:

$$C_3H_7\overset{D}{C}D-\overset{OH}{C}H_2 \xrightarrow{-HOD} C_3H_7CD{=}CH_2 \xrightarrow{-e}$$

$$\left[C_3H_7CD-\overset{+\bullet}{C}H_2\right]$$

m/e 71

**Q-32** When 1-pentanol is subjected to mass spectrometry using a *cooled* ion source, a parent ion is seen at m/e 88. Why?

A-32 At lower temperatures the amount of thermal dehydration decreases. There is, thus, a greater probability of ionizing the intact hydroxy compound.

Q-33 The electron bombardment dehydration process may occur by a 1, 4- or a 1, 3-elimination.

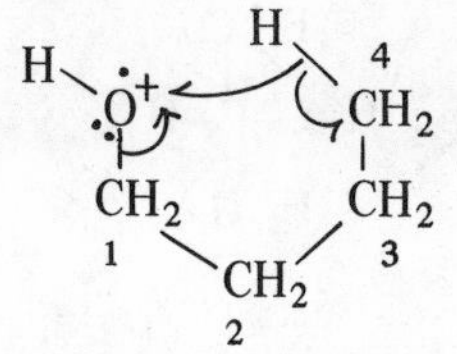

1, 4-elimination

Illustrate the 1, 3-elimination process by which the butanol cation radical may extrude $H_2O$.

A-33 $\left[ HO^{+\cdot}\ CH_2\text{–}CH_2\text{–}CH(H)CH_3 \right] \xrightarrow{-H_2O} \dot{C}H_2^{+}\ CH_2\ \dot{C}HCH_3$

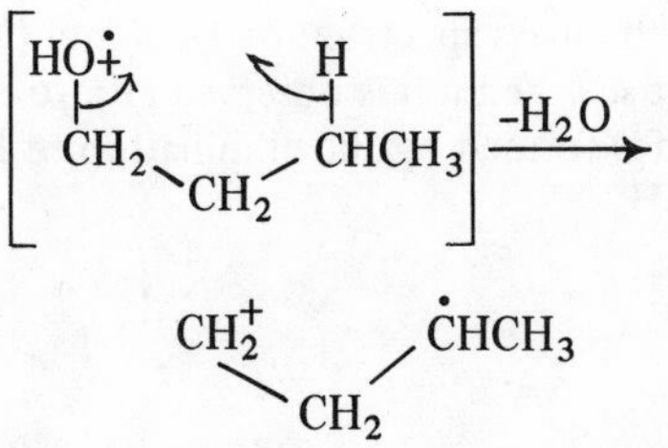

Q-34 A metastable ion is observed at m/e 42.4 when butanol (m/e 74) is subjected to electron bombardment at low ion-source temperature. To what process can this metastable ion be attributed?

A-34 The metastable ion of m/e 42.4 is consistent with the fragmentation

$$[C_4H_9OH]^{+\cdot} \longrightarrow C_4H_8^{+\cdot} + H_2O$$

m/e 74 m/e 56

The calculated metastable ion peak for this process occurs at

$$m^* = \frac{56^2}{74} = 42.4$$

Q-35 Would a metastable ion corresponding to the thermal extrusion shown below be expected? Why?

$$C_5H_{11}OH \xrightarrow{\text{heat}} C_5H_{10} + H_2O$$

A-35 No. The thermal dehydration occurs prior to ionization. No metastable peak will thus be observed for thermal elimination of $H_2O$.

Q-36 In the mass spectrum of an alcohol no molecular ion is observed, but an ion is observed at M-18. Suggest a way of determining whether the elimination of $H_2O$ is thermal or electron bombardment induced.

A-36 a. If a metastable ion corresponding to the transition M→M-18 is observed, then an electron bombardment dehydration is occurring.

b. Alternatively, the ion source can be cooled in order to minimize thermal dehydration. A molecular ion peak corresponding to the alcohol may be observed.

Q-37 In the mass spectrum of ethyl chloride a peak at m/e 28 (95%) occurs. Suggest a mechanism to account for this peak.

A-37 m/e 28 corresponds to elimination of HCl.

H Cl
H–C–C–H
H H

**R** Thermal dehydration of alcohols in the mass spectrometer occurs by a 1, 2-elimination process.

Electron bombardment induced dehydration of alcohols occurs by 1, 3 or 1, 4 elimination. A cation radical and neutral $H_2O$ molecule are formed.

Dehydrohalogenation of alkyl halides occurs by a 1, 2-elimination process.

S-5 In alcohols with more than four C atoms in the alkyl chain, extrusion of $H_2O$ may also be accompanied by extrusion of a neutral olefinic moiety.

Q-38 In the mass spectrum of 1-octanol (M = 130, not seen) an ion is observed at m/e 84. This ion is formed by the elimination of $H_2O$ and $C_2H_4$.

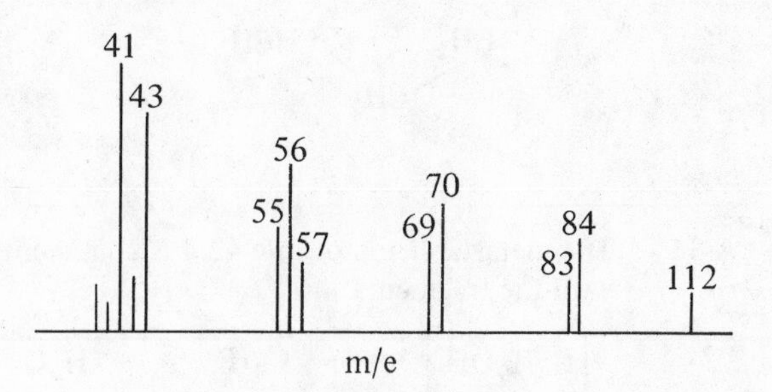


The loss of water is believed to involve the OH group and the hydrogen on C(4). Illustrate the loss of water by invoking a cyclic transition state.

A-38 [ $\overset{+}{\cdot OH}$ H ; $CH_2$ CH $C_4H_9$ ; $CH_2$–$CH_2$ ] $\xrightarrow{-H_2O}$

$\overset{+}{CH_2}$ $\dot{C}H$ $C_4H_9$
$CH_2$–$CH_2$

Q-39 Show that the cation radical dehydration product in A-38 can eliminate a neutral molecule of ethylene by a heterolytic fission process.

A-39 $\overset{+}{CH_2}$ $\dot{C}HC_4H_9$ ; $CH_2$–$CH_2$ $\xrightarrow{-C_2H_4}$ $\cdot CHC_4H_9$ ; $^+CH_2$

m/e 84

Q-40 Explain the formation of an ion of m/e 42 in the mass spectrum of 3-methylbutanol (M = 88).

A-40 m/e 42 corresponds to $C_3H_6$.

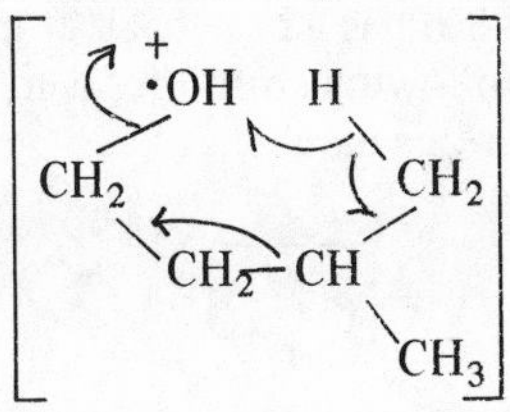

$$\left[\begin{array}{c} \overset{+}{\cdot OH} \quad H \\ CH_2 \qquad CH_2 \\ CH_2-CH \\ CH_3 \end{array}\right] \rightarrow H_2O + C_2H_4 + \overset{+}{\dot{C}H_2}-CHCH_3$$

m/e 42

**R** Acyclic alcohols with an alkyl chain containing at least four C atoms can eliminate neutral $H_2O$ and neutral olefin in a concerted manner. The product is a cation radical of *even* mass.

S-6 Certain cyclic systems fragment by the cleavage of two bonds to form a neutral moiety and a cation radical of *even* m/e (when M is even). These systems are:

Cyclic saturated hydrocarbons: → + $C_2H_4$

Cyclic ethers: → + $CH_2O$

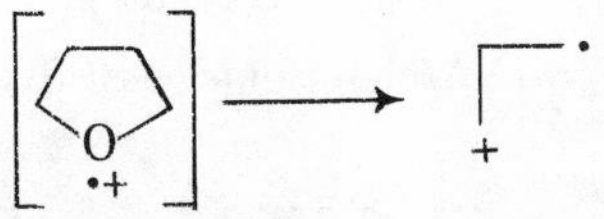

Phenols: → + CO

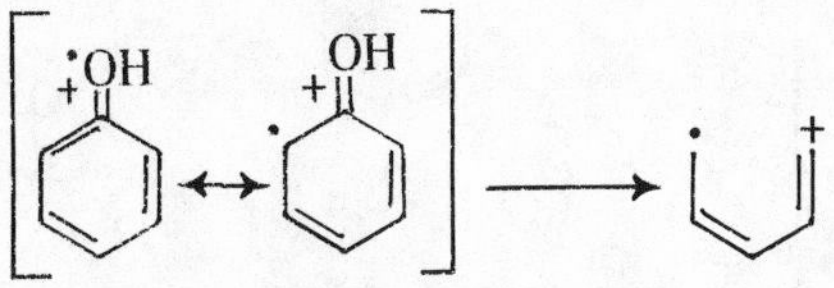

Aromatic systems with bridged carbonyls: → + CO

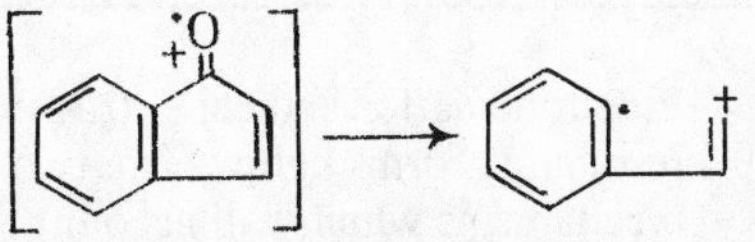

Q-41 Explain the origin of an ion at m/e 42 (25%) in the mass spectrum of methylcyclopentane (M = 84). Note that there may be more than one mechanism.

A-41 m/e 42 corresponds to loss of 42 mass units, i.e., $C_3H_6$, from M. Two possible mechanisms are:

a. → + $C_3H_6$

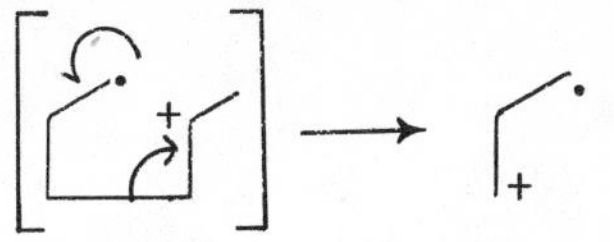

m/e 84 m/e 42

b. → + $C_3H_6$

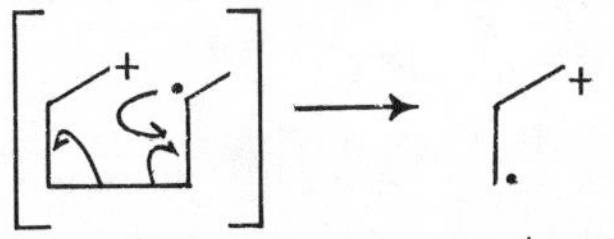

m/e 84 m/e 42

The ion at m/e 42 could also arise from methylcyclopentane molecules ionized in other ring bonds.

Q-42 Suggest a mechanism to account for the fragmentation of tetrahydrofuran cation radical as shown in S-6.

A-42

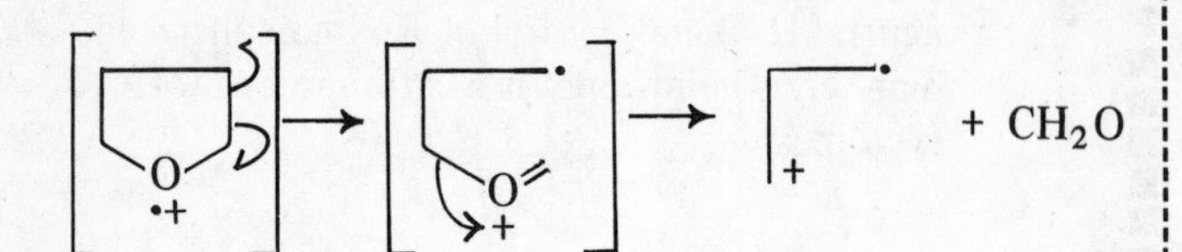


Q-43 The mass spectrum of a methyltetrahydrofuran (M = 86) has a base peak at m/e 56. Which of the following structures is most consistent with this data?

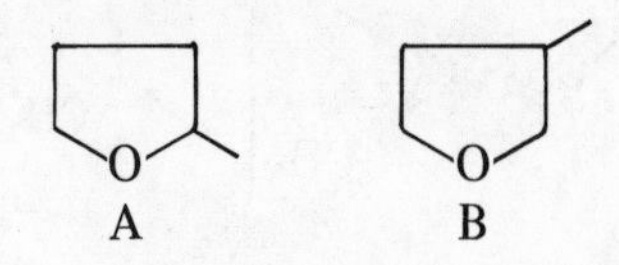


A-43 Structure B.
m/e 56 corresponds to a loss of $CH_2O$ from M.

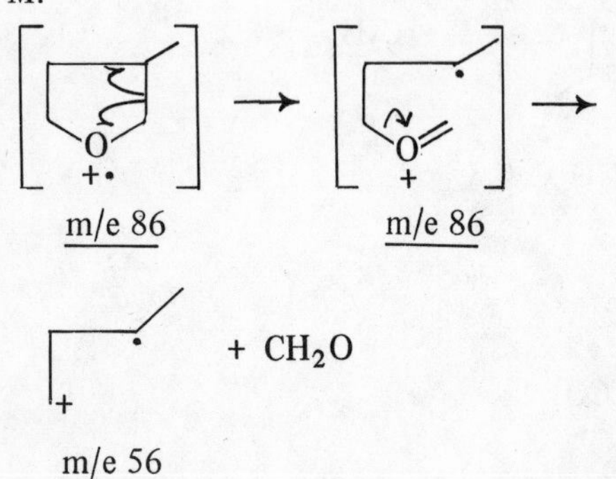


Q-44 In the mass spectrum of a dimethyltetrahydrofuran (M = 100, 9%) ions are observed at

a. m/e 85 (67%)

b. m/e 56 (100%)

Which isomer best fits these data?

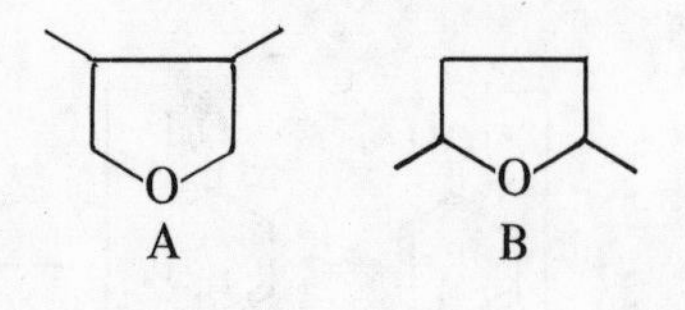


A-44 Structure B.

a. m/e 85 corresponds to loss of $CH_3$ from M by simple $\alpha$, $\beta$-fission.

b. m/e 56 corresponds to loss of 44 mass units, i.e., $C_2H_4O$, from M.

m/e 100 → m/e 100 →

$+ CH_3CHO$

m/e 56

Q-45 Indenone cation radical extrudes CO. Using resonance forms below, give two alternative mechanisms which will account for the elimination of CO.

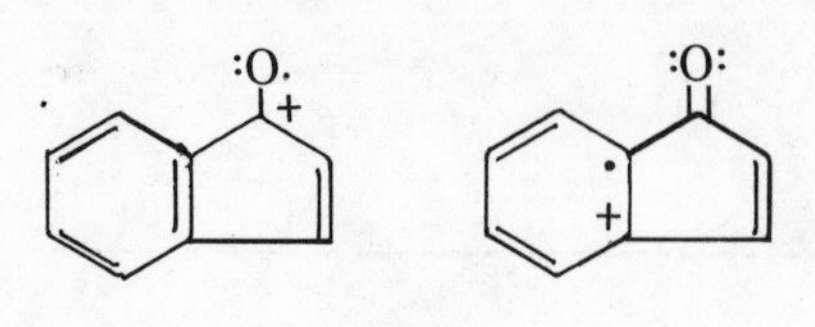

A-45

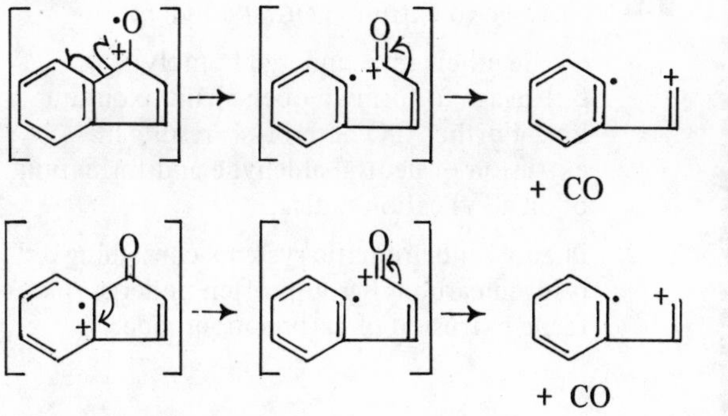

+ CO

+ CO

Q-46 Anthraquinone (M = 208) affords peaks at m/e 180 and m/e 152.

Suggest a mechanism to account for these two ions. (Hint: After the extrusion of the first fragment, consider a ring closure.)

A-46

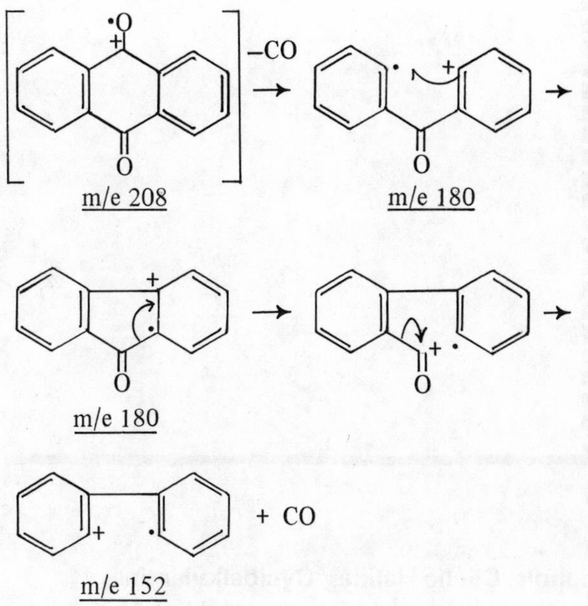

−CO

m/e 208

m/e 180

m/e 180

+ CO

m/e 152

Q-47 The mass spectrum of phenol (MW = 94) has a peak at m/e 66 indicating the extrusion of CO.

Suggest a mechanism to account for this process.

A-47

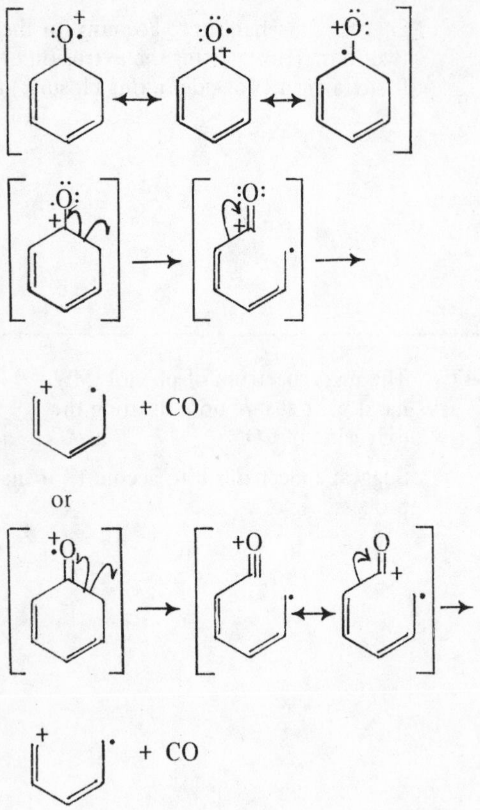

:Ö· (+) ↔ :Ö· (+) ↔ +Ö:

+ CO

or

+ CO

**R** Cyclic saturated hydrocarbons may undergo cleavage to extrude neutral ethylene.

Cyclic ethers may undergo homolytic α, β-cleavage to form an open-chain oxonium ion. Further C-O heterolysis results in extrusion of neutral aldehyde and formation of an alkyl cation radical.

Phenols and aromatic systems containing bridged carbonyl groups often undergo facile extrusion of carbon monoxide.

## PART IV: Complex Fission of Cyclic Alcohols, Cyclic Halides, Cycloalkylamines and Cyclic Ketones. The Formation of Odd m/e Fragments When M is Even, Following Cleavage of Two Bonds and H Transfer.

S-1 *Complex fission* is a common process in the mass spectral fragmentation of cyclic alcohols, cyclic halides, cycloalkylamines and cyclic ketones. This type of fragmentation process involves two ring-bond cleavages and a hydrogen transfer, with formation of a stable oxonium or immonium ion, i.e.,

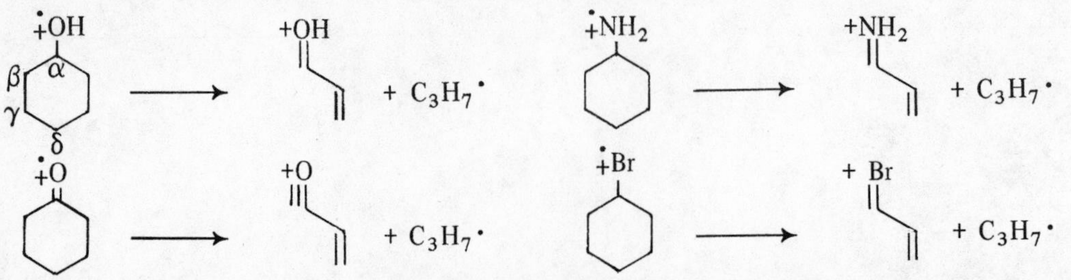

+OH (α, β, γ, δ) ⟶ +OH + $C_3H_7\cdot$

$+NH_2$ ⟶ $+NH_2$ + $C_3H_7\cdot$

+O ⟶ +O + $C_3H_7\cdot$

+Br ⟶ + Br + $C_3H_7\cdot$

Q-1 Using cyclohaxanol cation radical, illustrate the ring-opened oxonium species formed by α, β-fission.

A-1

Q-2 The second step in the complex fission process is usually illustrated as a homolytic transfer of an H atom from the $C_{(\beta)}$ carbon to the radical site.

Depict this step. Why is the H transfer a facile process in the intermediate shown in A-1?

A-2

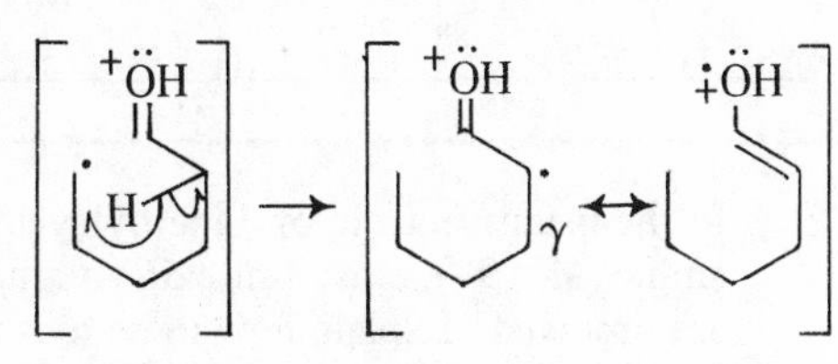


The H transfer is a facile process because:

a. the radical formed is resonance stabilized and

b. a sterically favourable six-membered cyclic transition state is involved.

Q-3 Depict the final homolytic step in which the $C_{(\alpha)}-C_{(\beta)}$ bond is homolyzed. What is the driving force for this step.

A-3

m/e 57

The driving force for this final step is the formation of a resonance-stabilized oxonium ion, i.e.

Q-4 Suggest a mechanism to account for the formation of an ion of m/e 57 (69%) in the mass spectrum of 2-methylcyclohexanol (M = 114). (m/e 57 corresponds to $C_3H_5O$ as shown by high resolution mass measurement.)

A-4

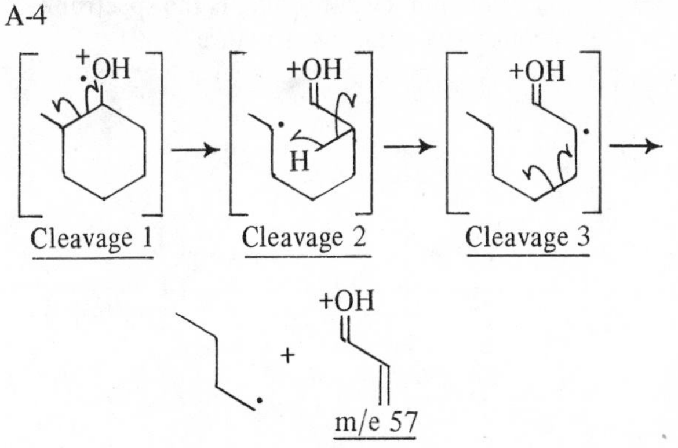


Q-5 Suggest a mechanism to account for the formation of an ion of m/e 71 (31%) in the mass spectrum of 2-methylcyclohexanol. (m/e 71 corresponds to $C_4H_7O$ as shown by high resolution mass measurement.)

A-5

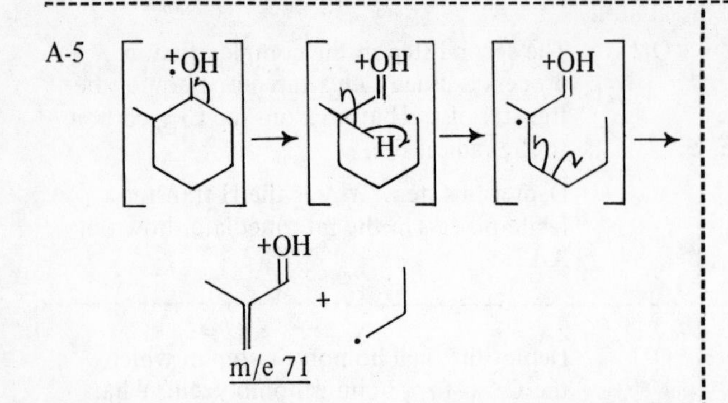


Q-6 The mass spectrum of a compound is shown below. Which structure is most consistent with the spectrum?

57 A OH C OH

B OH

44 68 86 (M)

A-6 Structure B.
m/e 68 corresponds to loss of $H_2O$ from M. This can occur either by thermal elimination or by electron bombardment. m/e 57 arises by a complex fission process, i.e.,

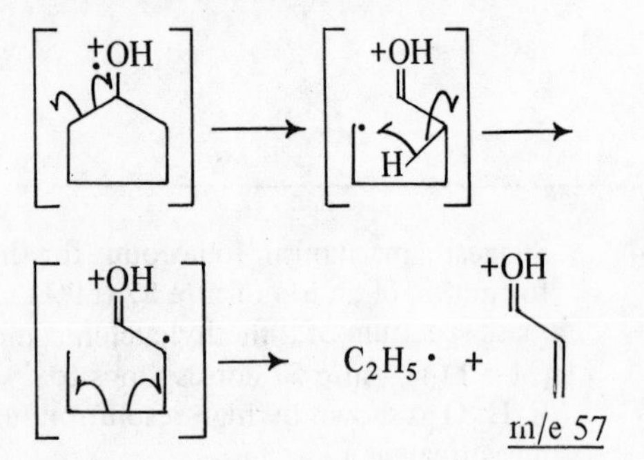


Structure A would not be expected to yield the ion m/e 57.
Structure C has M = 100.

Q-7 In the mass spectrum of 3-methylcyclopentylamine (M = 99), ions of m/e 56 and m/e 70 are observed. Explain how these ions might be formed.

A-7 m/e 70 corresponds to a loss of $C_2H_5$.
m/e 56 corresponds to a loss of $C_3H_7$.

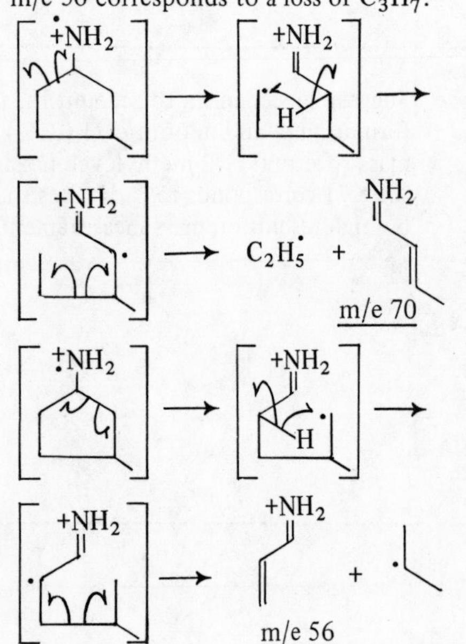


Q-8 In the mass spectrum of a compound, ions at m/e 114 (22%), m/e 97 (51%) and m/e 56 (100%) are observed.

the compound was thought to be 1, 2-diaminocyclohexane. Is the spectrum consistent with this structure?

A-8 Yes.
m/e 114 could be the molecular ion.
m/e 97 corresponds to loss of 17 mass units from M, i.e., $NH_3$. By analogy with the dehydration of alcohols the following mechanism may be invoked:

$H_2N^{+}$ H $-NH_3$ → $NH_2$

m/e 97

m/e 56 corresponds to loss of $C_3H_8N$. This ion is formed by complex fission as in A-7.

$+NH_2$

m/e 56

Q-9 In the mass spectrum of a compound believed to be 2-methylcyclopentanone, intense ions are observed at

a. m/e 98 (54%)

b. m/e 55 (40%)

c. m/e 69 (38%)

Is the data consistent with the suggested structure?

A-9 Yes.

a. m/e 98 could be M.

b. m/e 55 corresponds to loss of 43, i.e., $C_3H_7$, from M. This fragmentation may be illustrated as a complex fission process.

+O → O+ H → +O → + O+

m/e 55

c. m/e 69 arises by the alternative complex fission process.

O+

m/e 69

Q-10 The mass spectrum of a compound believed to be 2-ethyldimedone is shown below. Is the base peak (m/e 83) consistent with the proposed structure?

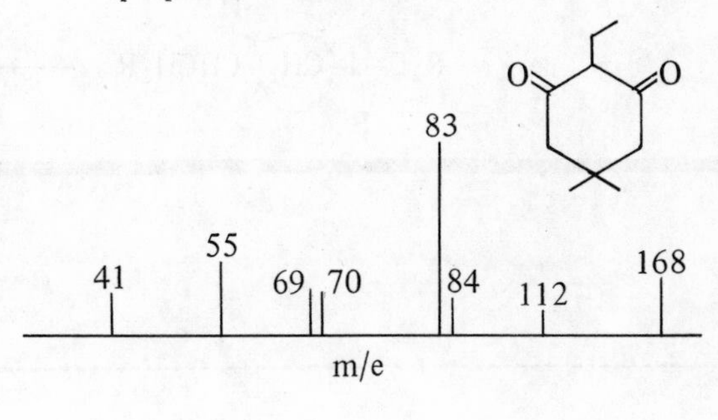

A-10 Yes.
m/e 83 corresponds to $C_5H_7O$, and is formed by the complex fission process:

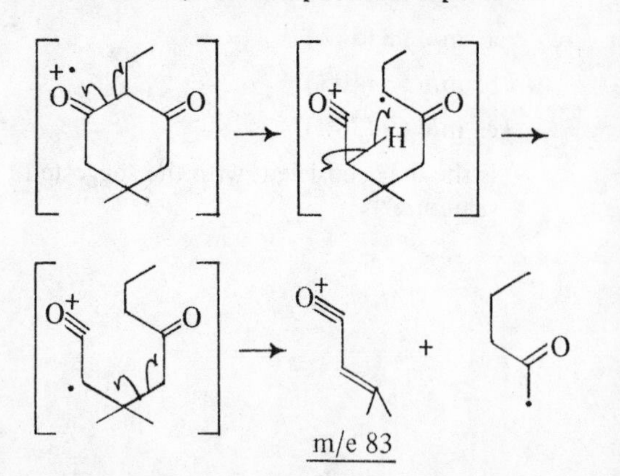

m/e 83

**R** Cyclic alcohols fragment by a complex fission process. Initial $\alpha, \beta$-cleavage is followed by H transfer and C–C homolysis yielding a resonance-stabilized oxonium ion. Cycloalkyl halides, cycloalkylamines and cyclic ketones undergo similar complex fissions.

S-2 The oxonium ions formed by the $\alpha, \beta$-cleavage of ethers and the immonium ions formed by the $\alpha, \beta$-cleavage of amines may undergo further cleavage of the alkyl-oxygen or alkyl-nitrogen bonds with concomitant transfer of H, e.g.,

$$R_2C{=}\overset{+}{O}{-}CH_2{-}CH(H)CH_2R \longrightarrow R_2C{=}\underset{+}{O}H + CH_2{=}CHCH_2R$$

$$R_2C{=}\overset{+}{N}(R){-}CH_2{-}CH(H)CH_2R \longrightarrow R_2C{=}\overset{+}{N}(R){-}H + CH_2{=}CHCH_2R$$

Q-11 The mass spectrum of di-*n*-propyl ether has peaks at m/e 31 and 73. Suggest a mechanism for the formation of these ions.

A-11 $[CH_3CH_2CH_2\overset{\cdot+}{O}CH_2CH_2CH_3]$ (M = 102) → $CH_3CH_2CH_2\overset{+}{O}{=}CH_2$ (m/e 73) + $C_2H_5\cdot$

$CH_2{=}\overset{+}{O}CH_2CH(H)CH_3$ (m/e 73) ⟶ $CH_2{=}\overset{+}{O}H$ (m/e 31) + $CH_2{=}CHCH_3$

Q-12 Draw the ion of m/e 86 which is formed by $\alpha, \beta$-fission of triethylamine (M = 101). Depict the subsequent fission and H rearrangement which affords an ion of m/e 58.

A-12 m/e 86 corresponds to loss of $CH_3$ from M.

$$\left[CH_3{-}CH_2{-}\overset{C_2H_5}{\underset{\bullet+}{N}}{-}C_2H_5\right] \longrightarrow CH_2{=}\overset{C_2H_5}{\underset{+}{N}}{-}C_2H_5 + CH_3^{\bullet}$$

m/e 86

m/e 58 corresponds to loss of $C_2H_4$ from ion m/e 86.

$$CH_2{=}\overset{C_2H_5}{\underset{+}{N}}{-}CH_2{-}\underset{H}{CH_2} \longrightarrow CH_2{=}\overset{C_2H_5}{\underset{+}{N}}{-}H + C_2H_4$$

m/e 86 m/e 58

Q-13 A molecule thought to be di-*i*-propylmethylamine affords the mass spectrum shown below. Are the mass spectral data consistent with the proposed structure?

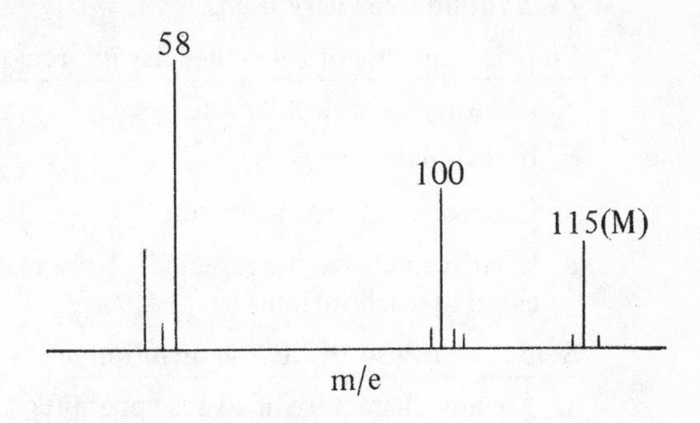


A-13 Yes. The mass spectral data fit the proposed structure. Di-*i*-propylmethylamine (M = 115) should undergo facile $\alpha, \beta$-fission with loss of $CH_3$, i.e.,

$$\left[CH_3{-}\underset{CH_3}{CH}{-}\overset{CH_3}{\underset{\bullet+}{N}}{-}\underset{CH_3}{CH}{-}CH_3\right] \xrightarrow{-CH_3^{\bullet}} CH_3{-}CH{=}\overset{CH_3}{\underset{+}{N}}{-}\underset{CH_3}{CH}{-}CH_3$$

m/e 115 m/e 100

The ion of m/e 100 might further fragment with H rearrangement yielding an ion of m/e 58, i.e.,

$$CH_3{-}CH{=}\overset{CH_3}{\underset{+}{N}}{-}\underset{H{-}CH_2}{CH}{-}CH_3 \xrightarrow{-C_3H_6} CH_3{-}CH{=}\overset{CH_3}{\underset{+}{N}}{-}H$$

m/e 100 m/e 58

**R** Oxonium and immonium ions formed by the $\alpha, \beta$-cleavage of ethers and amines may undergo further cleavage of the alkyl-oxygen or nitrogen bond with concomitant transfer of H.

## PART V: Analysis of Mass Spectral Data.

S-1 While each individual is likely to develop his own unique method of analyzing a mass spectrum in an attempt to determine the structure of an unknown compound, the following systematic approach has been found to be very useful.

Step 1: Analysis of the molecular ion region.

a. Intensity of molecular ion?

b. Is M odd or even?

c. Is isotope pattern significant?

d. What formula can be assigned? How many units of unsaturation (multiple bonds or rings) are indicated by each formula?

Step 2: Analysis of the fragment ions.

a. Are any characteristic losses apparent?

| Ion | Fragment lost | Structural or fragmentation types indicated |
|---|---|---|
| M-1 | H | aldehyde (some ethers and amines) |
| M-15 | $CH_3$ | methyl substituents |
| M-18 | $H_2O$ | alcohols |
| M-28 | $C_2H_4$, CO, $N_2$ | $C_2H_4$, McLafferty Rearrangement, CO, (extrusion from cyclic ketone) |
| M-29 | CHO, $C_2H_5$ | aldehydes, ethyl substituents |
| M-34 | $H_2S$ | thiols |
| M-35, M-36 | Cl, HCl | chlorides |
| M-43 | $CH_3CO$, $C_3H_7$ | methyl ketones, propyl substituents |
| M-45 | COOH | carboxylic acids |
| M-60 | $CH_3COOH$ | acetates |

b. What formulas can be assigned to significant ions?

| Ion | Fragment | Structural types indicated |
|---|---|---|
| 29 | CHO | aldehyde |
| 30 | $CH_2NH_2$ | primary amines |
| 43 | $CH_3CO$, $C_3H_7$ | $CH_3CO$, propyl substituents |
| 29, 43, 57, 71 etc. | $C_2H_5$, $C_3H_7$, etc. | n-alkyl |
| 39, 50, 51, 52, 65, 77 | aromatic fragmentation products | aromatic: Most of these ions will be present if an aromatic system is part of the structure |
| 60 | $CH_3COOH$ | carboxylic acids, acetates, methyl esters |
| 91 | $C_6H_5CH_2$ | benzoyl |
| 105 | $C_6H_5CO$ | benzoyl |

c. Does the odd-even character of significant ions suggest a rearrangement process, i.e., McLafferty, retro-Diels-Alder?

d. Does high resolution analysis allow alternative formulas to be written for the ions?

Step 3: Listing of partial structural units.

a. What are the partial structural units indicated?

b. Are there relations between major ions (metastable peaks)?

c. How many units of unsaturation and atoms are accounted for by the partial structural units? What residual fragments are possible?

Step 4: Postulating structures.

a. Combine partial and residual structures in all possible ways.

b. Can any structures be eliminated on the basis of mass spectra or other data?

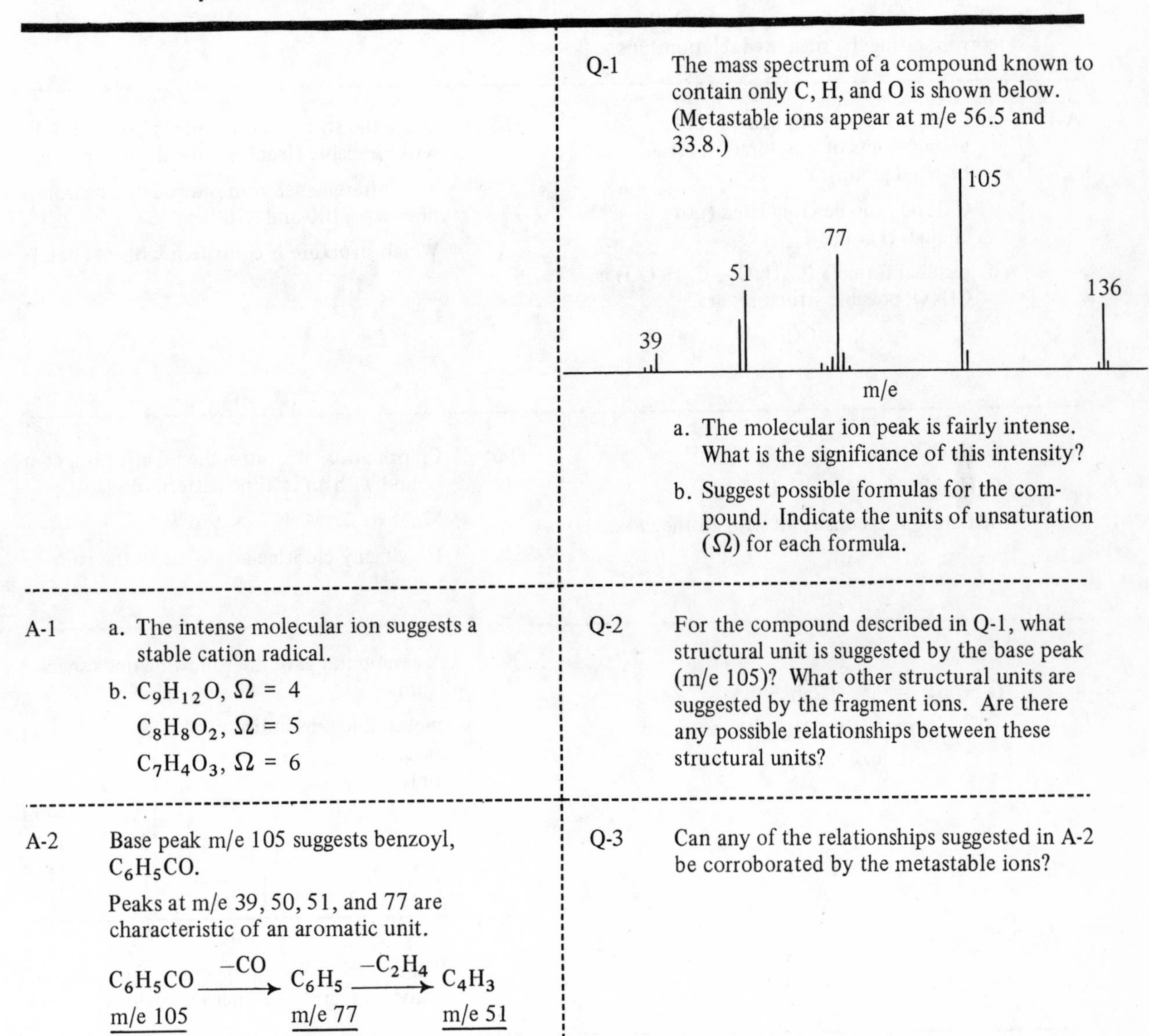

Q-1 The mass spectrum of a compound known to contain only C, H, and O is shown below. (Metastable ions appear at m/e 56.5 and 33.8.)



a. The molecular ion peak is fairly intense. What is the significance of this intensity?

b. Suggest possible formulas for the compound. Indicate the units of unsaturation ($\Omega$) for each formula.

A-1 a. The intense molecular ion suggests a stable cation radical.

b. $C_9H_{12}O$, $\Omega = 4$
$C_8H_8O_2$, $\Omega = 5$
$C_7H_4O_3$, $\Omega = 6$

Q-2 For the compound described in Q-1, what structural unit is suggested by the base peak (m/e 105)? What other structural units are suggested by the fragment ions. Are there any possible relationships between these structural units?

A-2 Base peak m/e 105 suggests benzoyl, $C_6H_5CO$.

Peaks at m/e 39, 50, 51, and 77 are characteristic of an aromatic unit.

$$\underset{m/e\ 105}{C_6H_5CO} \xrightarrow{-CO} \underset{m/e\ 77}{C_6H_5} \xrightarrow{-C_2H_4} \underset{m/e\ 51}{C_4H_3}$$

Q-3 Can any of the relationships suggested in A-2 be corroborated by the metastable ions?

A-3 Yes.

m/e 105 ⟶ m/e 77
would give a metastable ion at

$$\frac{(77)^2}{105} = 56.5$$

m/e 77 ⟶ m/e 51
would give a metastable ion at

$$\frac{(51)^2}{77} = 33.8$$

These metastable ions are observed, thus corroborating the suggested relationships.

Q-4 a. Can any of the formulas in A-1 be eliminated on the basis of the structure postulated for m/e 105?

b. What atoms are not accounted for by the m/e 105 peak? Suggest structures for the "residual" formula.

A-4 a. $C_9H_{12}O$ can be eliminated (not enough units of unsaturation for a benzoyl group).

$C_7H_4O_3$ can be eliminated (not enough H atoms).

b. residual formula ($C_8H_8O_2-C_6H_5CO$) is $CH_3O$; possible structures are:

$-OCH_3$

$-CH_2OH$

Q-5 Using the structural fragments listed in A-4, write possible structures for the compound.

The infrared spectrum showed no absorption between 3100 and 3700 cm$^{-1}$.

Which structure is consistent with the data?

A-5 $C_6H_5COOCH_3$ A

$C_6H_5COCH_2OH$ B

Structure A is consistent with all the data.

Q-6 Chlorination of a nitro-phenol affords a compound with an isotope pattern of

M: M + 2: M + 4 = 9:6:1

How many chlorine atoms are in the compound?

A-6 Two.

$$(a + b)^2 = a^2 + 2ab + b^2$$
$$= 3^2 + (2)(3)(1) + 1^2$$
$$= 9 + 6 + 1$$

Q-7 A compound gave the following mass spectrum:

metastable ions indicate m/e 154 → m/e 139 → m/e 111.

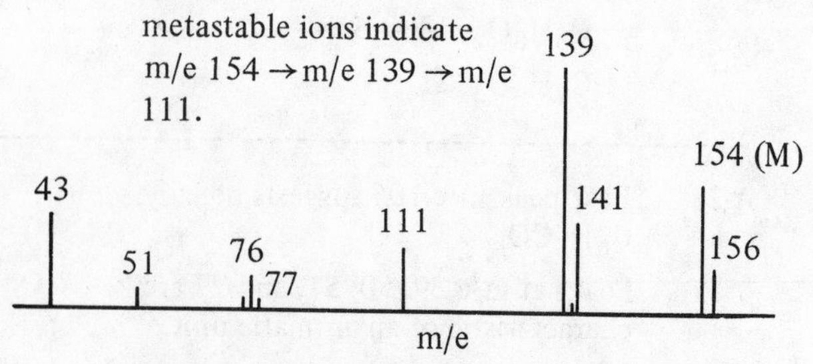


What conclusions can be drawn from an analysis of the molecular ion region?

A-7 The molecular ion is fairly intense and may suggest an aromatic system. The even value of M suggests no N or an even number of N atoms. The isotope pattern suggests a monochloride.

Q-8 For the compound described in Q-7:

a. Are there any characteristic losses from M?

b. Are any of the fragment ions characteristic of particular functional groups?

**A-8**

a. m/e 139 (M-15) suggests a loss of $CH_3$.

b. m/e 43 indicates the presence of $C_3H_7$ or $CH_3CO$.

m/e 51, 76, 77, indicate an aromatic ring.

**Q-9** Using the data derived in A-7 and A-8, suggest possible structures.

**A-9** The compound contains:

Cl, $C_6H_5$, (or $C_6H_4$)

$CH_3CO$ (or $C_3H_7$)

Possible combinations are:

$CH_3CO-C_6H_4-Cl$ A

$n$-$C_3H_7-C_6H_4-Cl$ B

$i$-$C_3H_7-C_6H_4-Cl$ C

**Q-10** Which of the three possible structures given in A-9 is most consistent with the mass spectrum given in Q-7? Why?

**A-10** Structure B should give an intense M-29 peak (benzylic cleavage) and an M-28 peak (McLafferty rearrangement). Neither peak is apparent.

Structure C should afford an intense M-15 peak (benzylic cleavage). Although this is observed in the mass spectrum, it is difficult to account for the subsequent transition m/e 139 → m/e 111 in terms of structure C.

Structure A best fits the data, i.e.,

$[ClC_6H_4COCH_3]^+$ (m/e 154) ⟶ $ClC_6H_4CO^+$ (m/e 139) ⟶ $ClC_6H_4^+$ (m/e 111)

**Q-11** A compound which was known to contain no halogen or nitrogen gave the following mass spectrum.

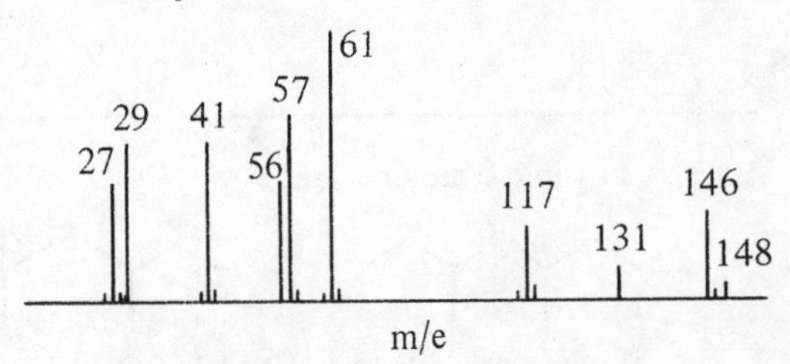


Metastable ions occur at m/e 117.5, 93.7, and 31.8. The high resolution spectrum gives m/e 146.1125.

a. What are possible formulas for the compound?

b. Give possible formulas for the fragment ions.

c. What transitions are corroborated by the metastable peaks?

**A-11** a. Formula must be $C_8H_{18}S$, ($\Omega$ = 0). The isotope pattern indicates *one* sulfur.

b. M-15 = 131 = $C_7H_{15}S^+$

M-29 = 117 = $C_6H_{13}S^+$

m/e 61 = $C_2H_5S^+$

m/e 57 = $C_4H_9{}^+$

m/e 56 = $C_4H_8{}^+$

m/e 41 = $C_3H_5{}^+$

m/e 29 = $C_2H_5{}^+$

m/e 27 = $C_2H_3{}^+$

c. $146 \xrightarrow{m^* = 93.8} 117 \xrightarrow{m^* = 31.8} 61$

$146 \xrightarrow{m^* = 117.5} 131$

**Q-12** The compound described in Q-11 could be a sulfide (R–S–R) or a thiol (R–S–H). Which of these structures is *not* consistent with the spectrum? Why? What are the major types of fragmentation which would be expected?

---

**A-12** A thiol is not consistent with the spectrum since loss of $H_2S$ (M-34) would be expected as an important fragmentation.

The important fragmentation of sulfide expected would be $\alpha$, $\beta$-cleavage and C–S cleavage.

**Q-13** Give structures for the compound described in Q-11 consistent with the spectrum. Logical steps to follow would be:

a. Explain the m/e 146 → 131 and m/e 146 → 117 transitions using partial structure.

b. Write possible structures. Attempt to substantiate or eliminate alternative structures.

---

**A-13** Possible structures are:

S or S

Logic: m/e 146 → 131 and 146 → 117 can be explained by $\alpha$, $\beta$-cleavage processes, e.g.,

$$\left[C_2H_5\text{–}\underset{\underset{R'}{|}}{CH}\text{–}\overset{\bullet +}{S}\text{–}R^2\right] \rightarrow \underset{\underset{R'}{|}}{HC}\text{=}\overset{+}{S}\text{–}R^2 + C_2H_5^{\bullet}$$

m/e 146     m/e 117

m/e 117 (odd) yields m/e 61 (odd). Thus, a complex cleavage may be occurring, i.e.,

$$\underset{\underset{R'}{|}}{HC}\text{=}\overset{+}{S}\text{–}C\text{–}C(H) \longrightarrow \underset{\underset{R'}{|}}{HC}\text{=}\overset{+}{S}H \quad \text{m/e 61}$$

(m/e 61 = $C_2H_5S$)

Thus, R′ must be $CH_3$ and $R^2$ must be $C_4H_9$, i.e.,

$$C_2H_5\underset{\underset{CH_3}{|}}{CH}\text{–}S\text{–}C_4H_9$$

(Continued over)

**Q-14** A compound gave the following mass spectrum.

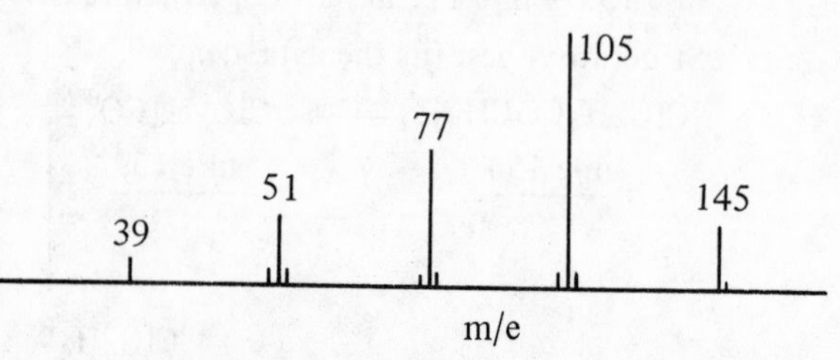


a. What conclusions can be drawn from an analysis of the molecular ion region?

b. What structural types are suggested by the fragment ions? What is the residual formula?

Four possible structures are

S S S S

A B C D

If either A or B were the correct structure, an M-43 peak (m/e 103) would be expected. m/e 103 is not present, therefore, structure C or D is indicated.

---

A-14 a. The m/e 145 peak may be the molecular ion (molecular ions are sometimes too weak to detect); if m/e 145 is the molecular ion, its *odd* value indicates the presence of an odd number of nitrogens.

The isotope pattern indicates no Cl, Br, or S (I can be eliminated because of mass). A possible molecular formula is:

$$C_9H_7ON \quad (\Omega = 7)$$

b. The presence of the 39, 50, 51, 52, and 77 peaks suggests an aromatic system, and the 105 peak is strongly suggestive of a benzoyl unit. The residual formula is $C_2H_2N$.

Q-15 Using the structural features given in A-14, suggest structures for the compound described in Q-14.

(Hint: Are all the units of unsaturation accounted for?)

---

A-15 The spectrum is purported to be of $C_6H_5COCH_2CN$. $C_6H_5CO$– accounts for 5 units of unsaturation. The residual formula $C_2H_2N$, must account for two units of unsaturation.

Possible structures for the residual formula are:

$-CH_2CN$, $-CH{=}C{=}NH$, $-C{\equiv}C{-}NH_2$,

$-N{=}C{=}CH_2$, –C(=N)–$CH_2$ (three-membered ring), etc.

These combined with $C_6H_5CO$– would give many possible structures (we have found 13).

Q-16 An ester gave the mass spectrum shown below. The high resolution spectrum gave m/e 116.0833.

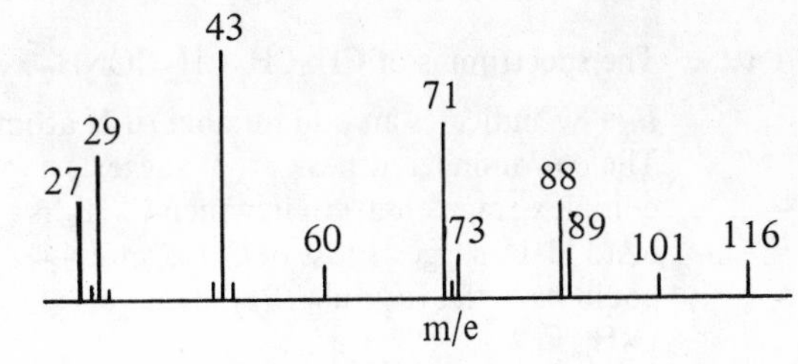


a. Suggest a formula. How many double bond equivalents does the ester have?

b. Suggest formulas for the ions of m/e 29, 43, 71.

---

A-16 a. $C_6H_{12}O_2$, which indicates one double bond equivalent. Therefore, the ester is saturated.

b. m/e 29 = $C_2H_5$

m/e 43 = $C_3H_7$ ($CH_3CO$ is also possible)

m/e 71 = $C_4H_7O$ ($C_5H_{11}$ is also possible)

Q-17 Consider the ion of m/e 88 shown in the spectrum of Q-17. What is the significance of the even m/e value?

What type of fragmentation will give m/e 88?

Postulate a tentative structure and rationalize the formation of the ions of m/e 29, 43, 71, 73, 88, and 101.

**A-17** m/e 88 is *even* and, therefore, arises by a rearrangement process – the McLafferty rearrangement.

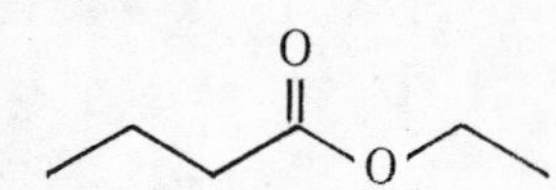


m/e 29 ($C_2H_5^+$), m/e 43 ($C_3H_7^+$), m/e 71 ($C_3H_7CO^+$), m/e 73 ($C_2H_5OCO^+$) are formed by $\alpha$-cleavage processes. m/e 88 can be formed by either of the two possible McLafferty rearrangements.

**A-18** If m/e 136 is the molecular ion, then there are no N or an even number of N atoms. There is no Cl, Br, I, or S. The formula must be $C_9H_{12}O$ ($\Omega$) = 4.

Peaks at 39, 51, 77 suggest an aromatic system.

M-18 suggests an alcohol.

Structural fragments:

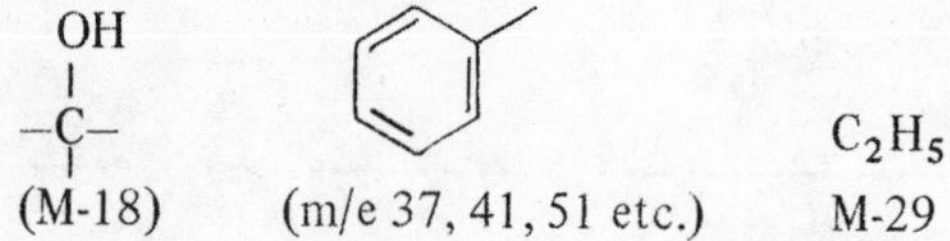


(M-18) (m/e 37, 41, 51 etc.) M-29

A structure consistent with the data is $C_6H_5CHOHC_2H_5$.

**A-19** The spectrum is of $CH_3CH_2CH_2CONH_2$.

m/e 87 indicates an odd number of N atoms. The *odd* numbered peak at 59 suggests a complex fragmentation in which $C_2H_4$ is lost. M-15 suggests loss of $CH_3$. m/e 44 could have the formula $CH_2NO$, $C_2H_4O$, or $C_2H_6N$.

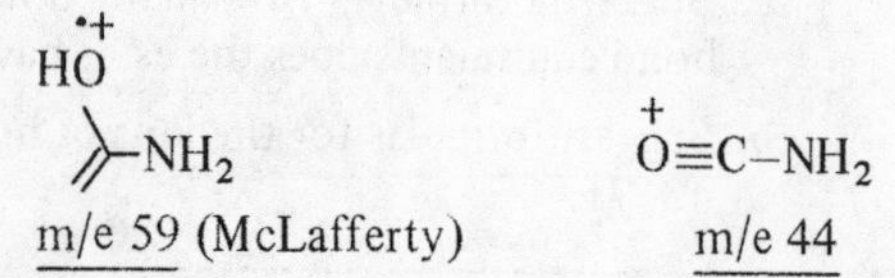


m/e 59 (McLafferty) m/e 44

**A-20** $CH_3CH_2CH_2CO_2H$

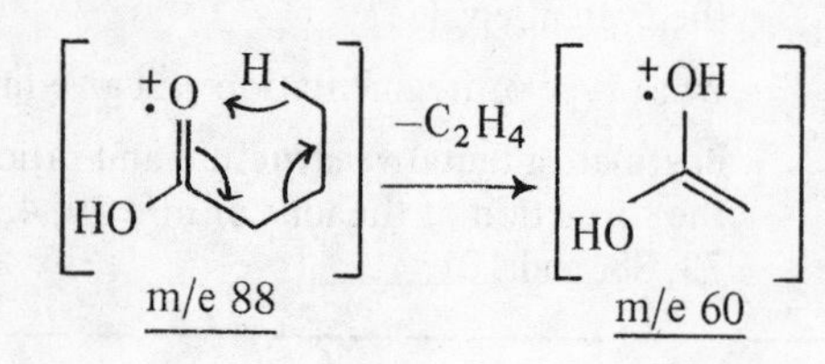


**Q-18** Suggest possible structures for the compound which affords the following mass spectrum. The high resolution spectrum gave m/e 136.0886.

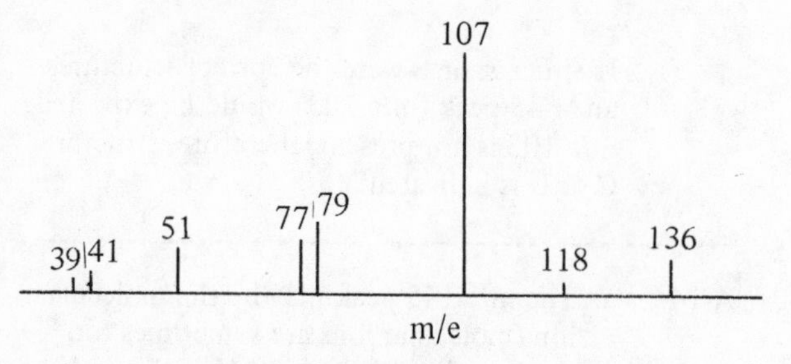


**Q-19** A solid substance known to be an amide gave the following mass spectrum. Suggest possible structures.

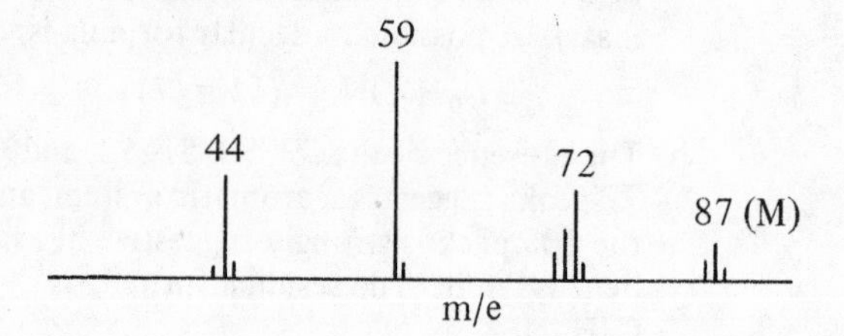


**Q-20** A carboxylic acid gave the following mass spectrum:

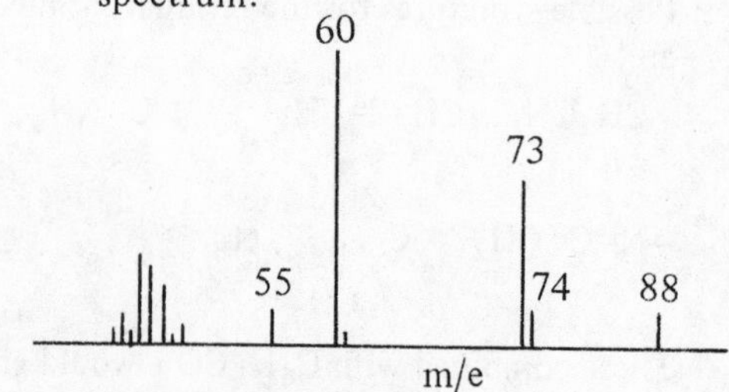


Suggest a structure.

**Q-21** Propose a structure for the compound which gives the following mass spectrum.

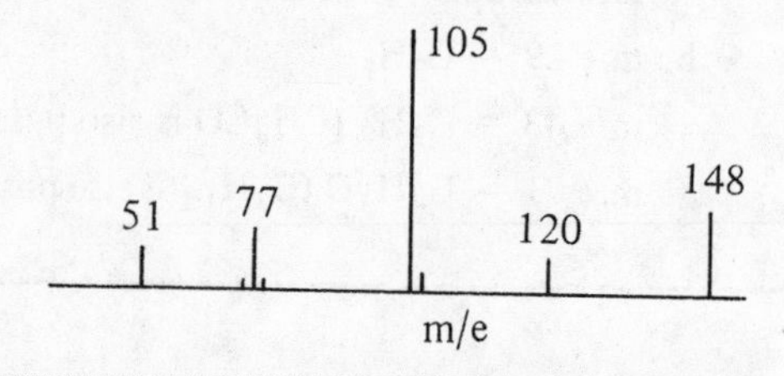

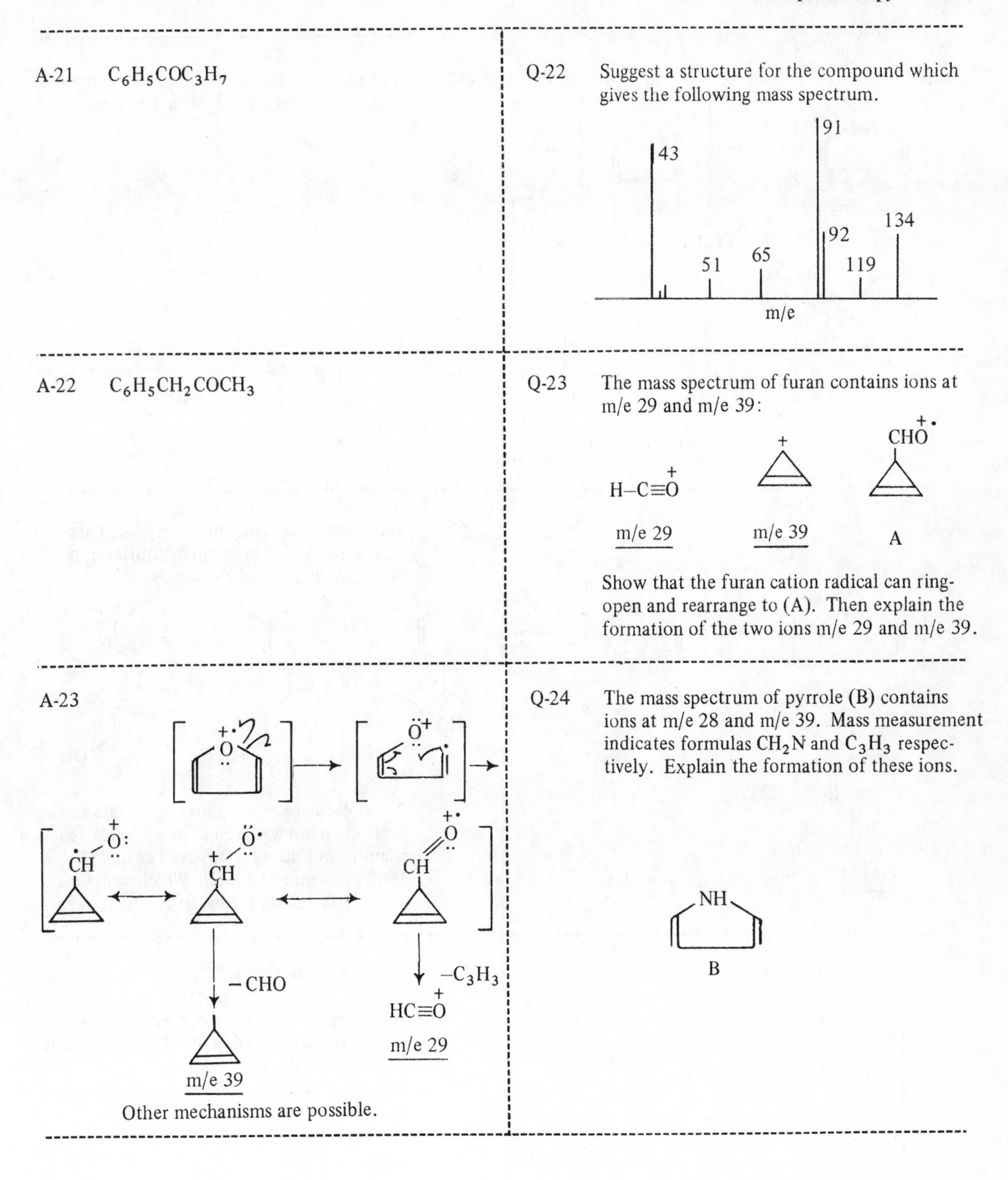

A-21 $C_6H_5COC_3H_7$

Q-22 Suggest a structure for the compound which gives the following mass spectrum.



A-22 $C_6H_5CH_2COCH_3$

Q-23 The mass spectrum of furan contains ions at m/e 29 and m/e 39:



Show that the furan cation radical can ring-open and rearrange to (A). Then explain the formation of the two ions m/e 29 and m/e 39.

A-23



Other mechanisms are possible.

Q-24 The mass spectrum of pyrrole (B) contains ions at m/e 28 and m/e 39. Mass measurement indicates formulas $CH_2N$ and $C_3H_3$ respectively. Explain the formation of these ions.

A-24

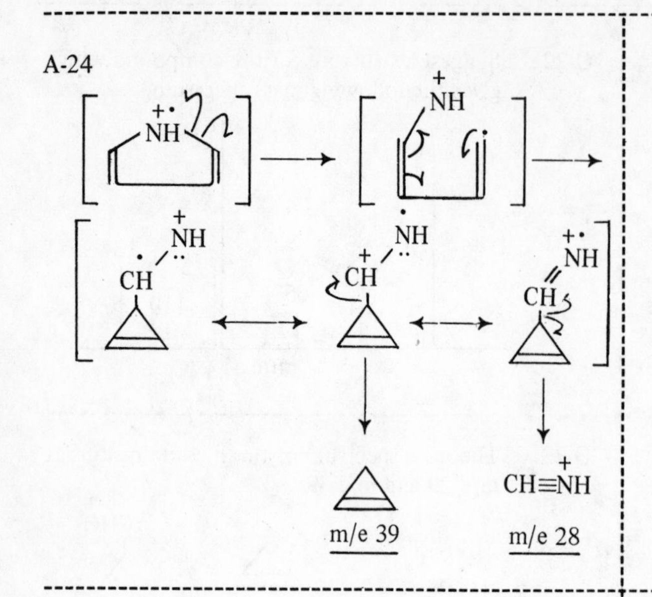


Q-25 The mass spectrum of pyrrole also contains an ion of m/e 41 ($C_2H_3N^{+\cdot}$). Explain.

A-25

NH

:NH

$-C_2H_2$

•NH

NH

m/e 41

m/e 41

Q-26 A naturally occurring monoterpenoid alcohol was isolated in microgram quantities. The structure was thought to be

HO

A

or

OH

B

or

OH

C

No molecular ion was observed at m/e 154. Instead an ion was seen at m/e 136 (M-18). Significant ions were observed at m/e 69 (50%) and m/e 71 (95%). Which structure is most probable on the basis of these data?

A-26 Structure A.

HO

m/e 69

HO

$HO^+$

m/e 71

Q-27 In order to confirm structure A (A-26) the alcohol was trimethylsilylated, and the mass spectrum of the derivative examined. A molecular ion was observed at m/e 226. The base peak in the spectrum occurred at m/e 143. Are these data in accord with the proposed structure (D)?

$(CH_3)_3SiO$

D

(The atomic weight of Si is 28)

A-27 Yes.

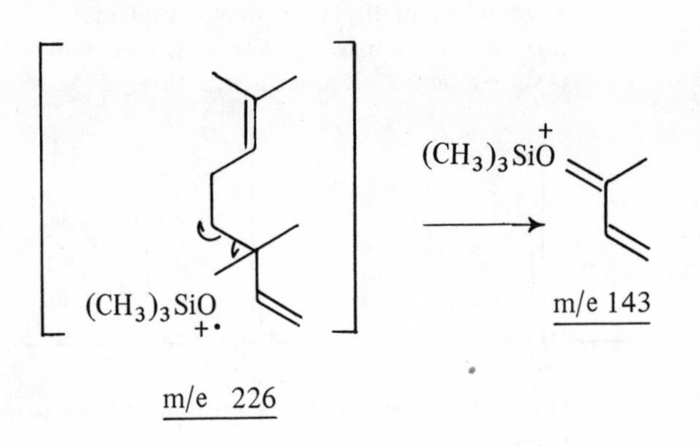


Q-28 Trimethylsilyl ethers are commonly used as alcohol derivatives in mass spectrometry. The mass spectrum of cholesterol trimethylsilyl ether is shown below. Account for the formation of the base peak (m/e 129) and the ion of m/e 329.

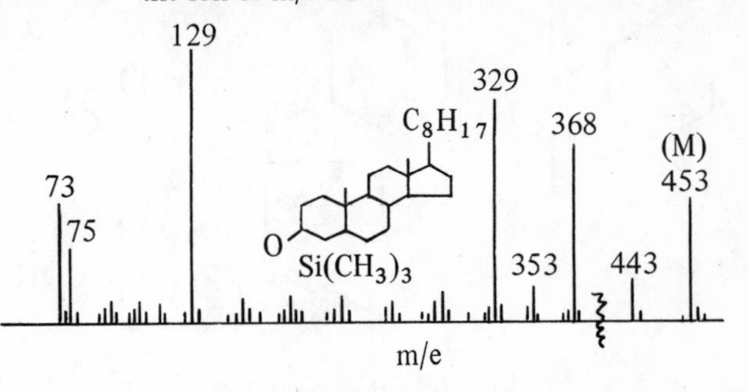


A-28

$(CH_3)_3SiO$ +• → $(CH_3)_3SiO$ + (H) →

$(CH_3)_3SiO$ + → $(CH_3)_3SiO$ + m/e 129

m/e 329 corresponds to M-129

$(CH_3)_3SiO$ + ←→ $(CH_3)_3SiO$ +• →

$(CH_3)_3SiO$ • + + m/e 329

Q-29 The base peak in the mass spectrum of cyclohexanone ethylene ketal (A) occurs at m/e 99 and corresponds to structure (B). Suggest a mechanism to account for the formation of (B).

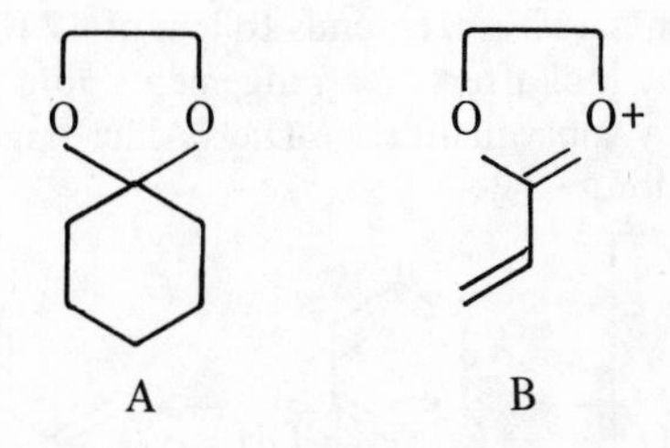

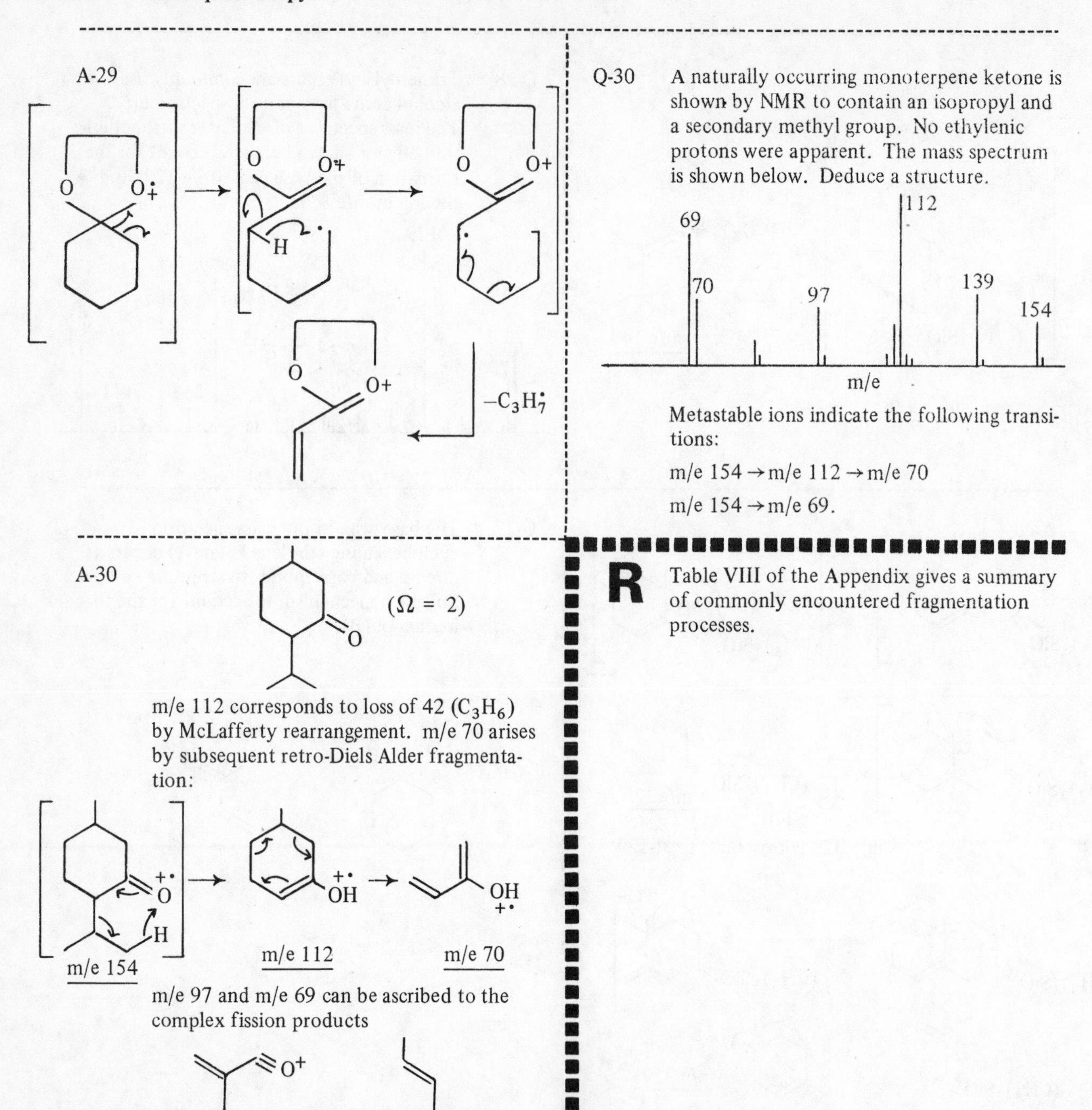


Q-30 A naturally occurring monoterpene ketone is shown by NMR to contain an isopropyl and a secondary methyl group. No ethylenic protons were apparent. The mass spectrum is shown below. Deduce a structure.

Metastable ions indicate the following transitions:

m/e 154 → m/e 112 → m/e 70

m/e 154 → m/e 69.

m/e 112 corresponds to loss of 42 ($C_3H_6$) by McLafferty rearrangement. m/e 70 arises by subsequent retro-Diels Alder fragmentation:

m/e 97 and m/e 69 can be ascribed to the complex fission products

Table VIII of the Appendix gives a summary of commonly encountered fragmentation processes.

## REFERENCES


Budzikiewicz, Herbert; Djerussi, Carl; and Williams, Dudley H. 1967. *Mass Spectrometry of Organic Compounds.* San Francisco: Holden-Day.

Campbell, Malcolm M. and Runquist, Olaf, *J. Chem. Ed.*, 49, 104(1972).

Hill, H. C. 1966. *Introduction to Mass Spectrometry.* London: Heyden and Son.

McLafferty, Fred W. 1963. *Mass Spectral Correlations.* Washington, D.C.: American Chemical Society.

——. 1966. *Interpretation of Mass Spectra.* New York: W. A. Benjamin.

# APPENDIX

**Table I**

**Rules for Diene Absorption**

| | | |
|---|---|---|
| Value assigned to parent heteroannular or open chain diene | | 217 nm |
| Homoannular diene | | 36 nm |
| Increment for | | |
| (a) each alkyl substituent or ring residue | | 5 nm |
| (b) the exocyclic nature of any double bond | | 5 nm |
| (c) a double bond extension | | 30 nm |
| (d) auxochrome –OAcyl | | 0 nm |
| –OAlkyl | | 6 nm |
| –SAlkyl | | 30 nm |
| –Cl, –Br | | 5 nm |
| $-NAlkyl_2$ | | 60 nm |
| | Total | λ calc |

(Reprinted with permission from A.I. Scott, *Interpretation of the Ultraviolet Spectra of Natural Products,* Pergamon Press, Oxford, 1964.)

## Table II

### Rules for $\alpha\beta$-Unsaturated Ketone and Aldehyde Absorption

$$\overset{\delta}{C}=\overset{\gamma}{C}-\overset{\beta}{C}=\overset{\alpha}{C}-C=O$$

**$\epsilon$ values are usually above 10 000 and increase with the length of the conjugated system.**

| | | |
|---|---|---|
| Value assigned to parent $\alpha\beta$-unsaturated six-ring or acyclic ketone | | 215 nm |
| Value assigned to parent $\alpha\beta$-unsaturated five-ring ketone | | 202 nm |
| Value assigned to parent $\alpha\beta$-unsaturated aldehyde | | 207 nm |
| Increments for | | |
| (a) a double bond extending the conjugation | | 30 nm |
| (b) each alkyl group or ring residue | $\alpha$ | 10 nm |
| | $\beta$ | 12 nm |
| | $\gamma$ and higher | 18 nm |
| (c) auxochromes | | |
| *(i)* –OH | $\alpha$ | 35 nm |
| | $\beta$ | 30 nm |
| | $\delta$ | 50 nm |
| *(ii)* –OAc | $\alpha, \beta, \delta$ | 6 nm |
| *(iii)* –OMe | $\alpha$ | 35 nm |
| | $\beta$ | 30 nm |
| | $\gamma$ | 17 nm |
| | $\delta$ | 31 nm |
| *(iv)* –SAlk | $\beta$ | 85 nm |
| *(v)* –Cl | $\alpha$ | 15 nm |
| | $\beta$ | 12 nm |
| *(vi)* –Br | $\alpha$ | 25 nm |
| | $\beta$ | 30 nm |
| *(vii)* –$NR_2$ | $\beta$ | 95 nm |
| (d) each exocyclic double bond | | 5 nm |
| (e) homoannular diene | | 39 nm |
| Total | | $\lambda_{calc}^{EtOH}$ |

(Reprinted with permission from A.I. Scott, *Interpretation of the Ultraviolet Spectra of Natural Products,* Pergamon Press, Oxford, 1964.)

## Table III

### Rules for $\alpha\beta$-Unsaturated Acid and Ester Absorption

**$\epsilon$ values are usually above 10 000**

| | |
|---|---|
| $\beta$-monosubstituted | 208 nm |
| $\alpha\beta$- or $\beta\beta$-disubstituted | 217 nm |
| $\alpha\beta\beta$-trisubstituted | 225 nm |
| Increment for | |
| (a) a double bond extending the conjugation | 30 nm |
| (b) exocyclic double bond | 5 nm |
| (c) when the double bond is endocyclic in a five- or seven-membered ring | 5 nm |
| Total | $\lambda$calc |

(Reprinted with permission from A.I. Scott, *Interpretation of the Ultraviolet Spectra of Natural Products,* Pergamon Press, Oxford, 1964.)

## Table IV

### Rules for the Principal Band of Substituted Benzene Derivatives
### $R\text{–}C_6H_4\text{–}CO\text{Z}$

| | Orientation | $\lambda_{calc}^{EtOH}$ nm |
|---|---|---|
| Parent Chromophore: | | |
| Z = alkyl or ring residue | | 246 |
| Z = H | | 250 |
| Z = OH or OAlkyl | | 230 |
| Increment for each substituent: | | |
| R = alkyl or ring residue | *o-, m-* | 3 |
| | *p-* | 10 |
| R = OH, OMe, OAlkyl | *o- m-* | 7 |
| | *p-* | 25 |
| $R = O^-$ | *o-* | 11 |
| | *m-* | 20 |
| | *p-* | 78 |
| R = Cl | *o-, m-* | 0 |
| | *p-* | 10 |
| R = Br | *o-, m-* | 2 |
| | *p-* | 15 |
| $R = NH_2$ | *o-, m-* | 13 |
| | *p-* | 58 |
| R = NHAc | *o-, m-* | 20 |
| | *p-* | 45 |
| R = NHMe | *p-* | 73 |
| $R = NMe_2$ | *o-, m-* | 20 |
| | *p-* | 85 |

(Reprinted with permission from A.I. Scott, *Interpretation of the Ultraviolet Spectra of Natural Products,* Pergamon Press, Oxford, 1964.)

# Table V

**Characteristic Infrared Group Frequencies. (Courtesy of N.B. Colthup, Stamford Research Laboratories, American Cyanamid Company, and the editor of the Journal of the Optical Society.) Overtone bands are marked 2 ν.**

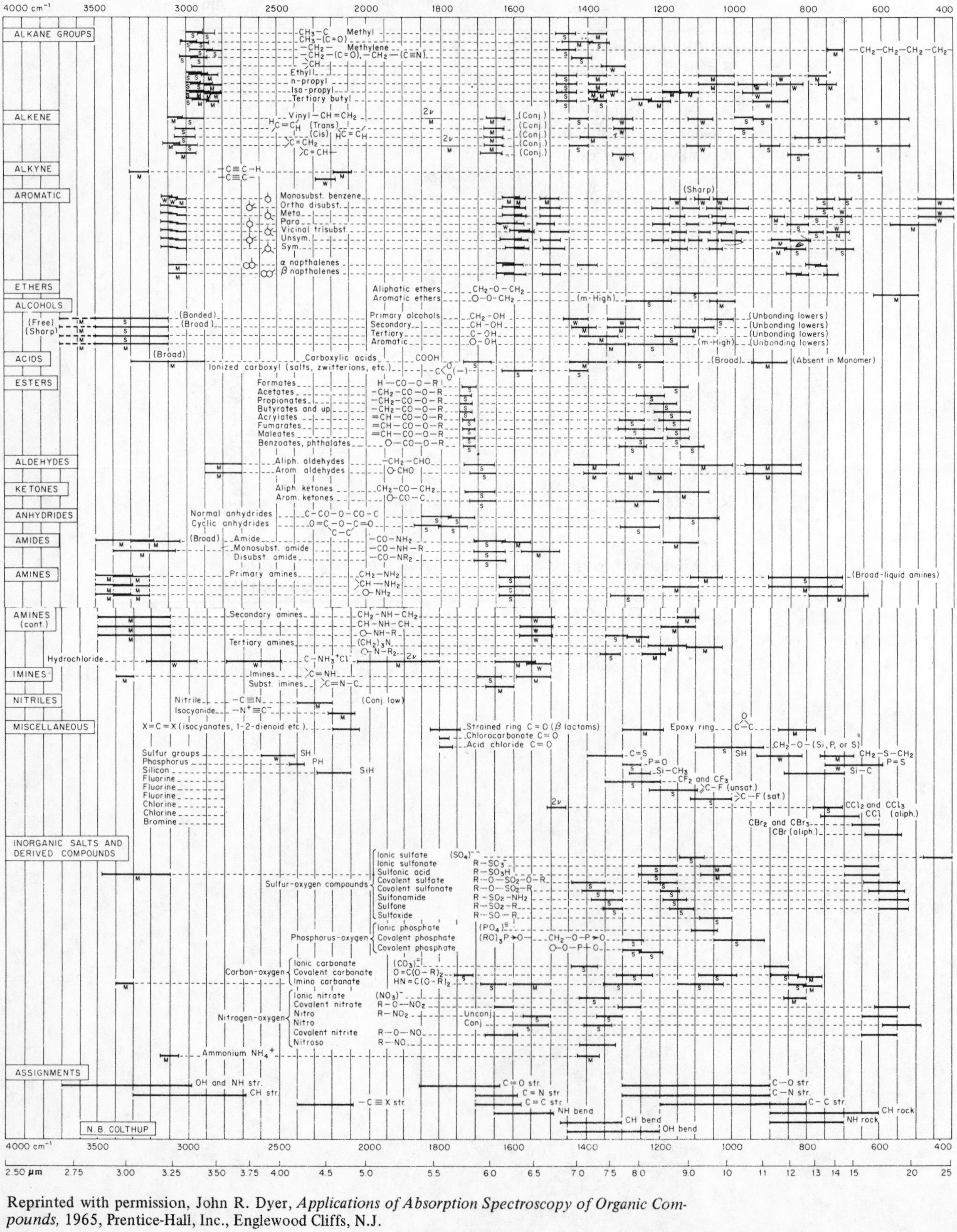

4000 cm⁻¹ 3500 3000 2500 2000 1800 1600 1400 1200 1000 800 600 400

ALKANE GROUPS
$CH_3-C$ Methyl
$CH_3-(C=O)$
$-CH_2-$ Methylene
$-CH_2-(C=O), -CH_2-(C\equiv N)$
>CH–
Ethyl
n-propyl
Iso-propyl
Tertiary butyl
$-CH_2-CH_2-CH_2-CH_2-$

ALKENE
Vinyl $-CH=CH_2$
HC=CH (Trans)
(Cis) HC=CH
>C=CH₂
>C=CH–
(Conj.) (Conj.) (Conj.) (Conj.) (Conj.)
2ν 2ν

ALKYNE
–C≡C–H
–C≡C–

AROMATIC
Monosubst. benzene
Ortho disubst.
Meta
Para
Vicinal trisubst.
Unsym.
Sym.
α napthalenes
β napthalenes
(Sharp)

ETHERS
Aliphatic ethers $CH_2-O-CH_2$
Aromatic ethers Ø–O–CH₂
(m-High)

ALCOHOLS
(Free) (Sharp)
(Bonded) (Broad)
Primary alcohols $CH_2-OH$ (Unbonding lowers)
Secondary CH–OH (Unbonding lowers)
Tertiary –C–OH (Unbonding lowers)
Aromatic Ø–OH (m-High) (Unbonding lowers)

ACIDS
(Broad)
Carboxylic acids COOH (Broad) (Absent in Monomer)
Ionized carboxyl (salts, zwitterions, etc.) $-C\langle^{O}_{O}(-)$

ESTERS
Formates H–CO–O–R
Acetates $-CH_2-CO-O-R$
Propionates $-CH_2-CO-O-R$
Butyrates and up $-CH_2-CO-O-R$
Acrylates =CH–CO–O–R
Fumarates =CH–CO–O–R
Maleates =CH–CO–O–R
Benzoates, phthalates Ø–CO–O–R

ALDEHYDES
Aliph. aldehydes $-CH_2-CHO$
Arom. aldehydes Ø–CHO

KETONES
Aliph. ketones $CH_2-CO-CH_2$
Arom. ketones Ø–CO–C

ANHYDRIDES
Normal anhydrides C–CO–O–CO–C
Cyclic anhydrides O=C–O–C=O, C–C

AMIDES
(Broad) Amide $-CO-NH_2$
Monosubst. amide –CO–NH–R
Disubst. amide $-CO-NR_2$

AMINES
Primary amines $CH_2-NH_2$
>CH–NH₂
Ø–NH₂
(Broad-liquid amines)

AMINES (cont.)
Secondary amines $CH_2-NH-CH_2$
CH–NH–CH
Ø–NH–R
Tertiary amines $(CH_2)_3N$
Ø–N–R₂
Hydrochloride $-C-NH_3^+Cl^-$ 2ν

IMINES
Imines >C=NH
Subst. imines >C=N–C

NITRILES
Nitrile –C≡N (Conj. low)
Isocyanide $-N^+\equiv C^-$

MISCELLANEOUS
X=C=X (isocyanates, 1-2-dienoid etc.)
Strained ring C=O (β lactams)
Chlorocarbonate C=O
Acid chloride C=O
Epoxy ring C–C with O
Sulfur groups SH
Phosphorus PH
Silicon SiH
Fluorine
Fluorine
Fluorine
Chlorine
Chlorine
Bromine
C=S
P=O
$Si-CH_3$
$CF_2$ and $CF_3$
>C–F (unsat.)
>C–F (sat.)
2ν
SH
$CH_2-O-$(Si, P, or S)
$CH_2-S-CH_2$
P=S
Si–C
$CCl_2$ and $CCl_3$
CCl (aliph.)
$CBr_2$ and $CBr_3$
CBr (aliph.)

INORGANIC SALTS AND DERIVED COMPOUNDS
Sulfur-oxygen compounds:
Ionic sulfate $(SO_4)^=$
Ionic sulfonate $R-SO_3^-$
Sulfonic acid $R-SO_3H$
Covalent sulfate $R-O-SO_2-O-R$
Covalent sulfonate $R-O-SO_2-R$
Sulfonamide $R-SO_2-NH_2$
Sulfone $R-SO_2-R$
Sulfoxide R–SO–R
Phosphorus-oxygen:
Ionic phosphate $(PO_4)^{\equiv}$
Covalent phosphate $(RO)_3P\rightarrow O$ $CH_2-O-P\rightarrow O$
Covalent phosphate Ø–O–P→O
Carbon-oxygen:
Ionic carbonate $(CO_3)^=$
Covalent carbonate $O=C(O-R)_2$
Imino carbonate $HN=C(O-R)_2$
Nitrogen-oxygen:
Ionic nitrate $(NO_3)^-$
Covalent nitrate $R-O-NO_2$
Nitro $R-NO_2$ Unconj.
Nitro Conj.
Covalent nitrite R–O–NO
Nitroso R–NO
Ammonium $NH_4^+$

ASSIGNMENTS
OH and NH str.
CH str.
–C≡X str.
C=O str.
C=N str.
C=C str.
NH bend
CH bend
OH bend
C–O str.
C–N str.
C–C str.
CH rock
NH rock

N.B. COLTHUP

4000 cm⁻¹ 3500 3000 2500 2000 1800 1600 1400 1200 1000 800 600 400

2.50 μm 2.75 3.00 3.25 3.50 3.75 4.00 4.5 5.0 5.5 6.0 6.5 7.0 7.5 8.0 9.0 10 11 12 13 14 15 20 25

Reprinted with permission, John R. Dyer, *Applications of Absorption Spectroscopy of Organic Compounds,* 1965, Prentice-Hall, Inc., Englewood Cliffs, N.J.

# Table VI

## Absorption Positions of Protons in Various Structural Environments

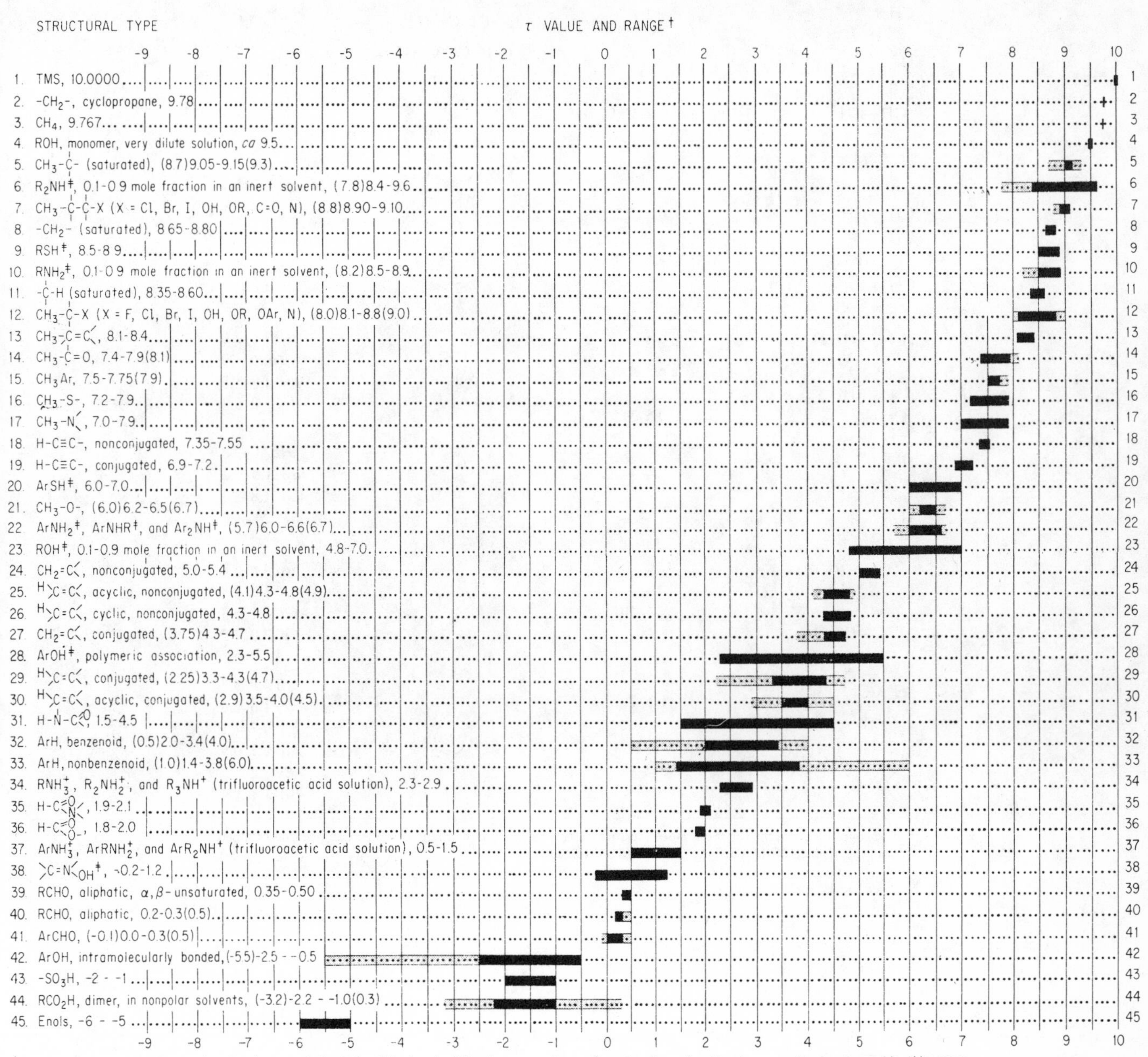

STRUCTURAL TYPE
τ VALUE AND RANGE†
-9 -8 -7 -6 -5 -4 -3 -2 -1 0 1 2 3 4 5 6 7 8 9 10
1. TMS, 10.0000
2. -CH2-, cyclopropane, 9.78
3. CH4, 9.767
4. ROH, monomer, very dilute solution, ca 9.5
5. CH3-C- (saturated), (8.7)9.05-9.15(9.3)
6. R2NH‡, 0.1-0.9 mole fraction in an inert solvent, (7.8)8.4-9.6
7. CH3-C-C-X (X = Cl, Br, I, OH, OR, C=O, N), (8.8)8.90-9.10
8. -CH2- (saturated), 8.65-8.80
9. RSH‡, 8.5-8.9
10. RNH2‡, 0.1-0.9 mole fraction in an inert solvent, (8.2)8.5-8.9
11. -C-H (saturated), 8.35-8.60
12. CH3-C-X (X = F, Cl, Br, I, OH, OR, OAr, N), (8.0)8.1-8.8(9.0)
13. CH3-C=C<, 8.1-8.4
14. CH3-C=O, 7.4-7.9(8.1)
15. CH3Ar, 7.5-7.75(7.9)
16. CH3-S-, 7.2-7.9
17. CH3-N<, 7.0-7.9
18. H-C≡C-, nonconjugated, 7.35-7.55
19. H-C≡C-, conjugated, 6.9-7.2
20. ArSH‡, 6.0-7.0
21. CH3-O-, (6.0)6.2-6.5(6.7)
22. ArNH2‡, ArNHR‡, and Ar2NH‡, (5.7)6.0-6.6(6.7)
23. ROH‡, 0.1-0.9 mole fraction in an inert solvent, 4.8-7.0
24. CH2=C<, nonconjugated, 5.0-5.4
25. H>C=C<, acyclic, nonconjugated, (4.1)4.3-4.8(4.9)
26. H>C=C<, cyclic, nonconjugated, 4.3-4.8
27. CH2=C<, conjugated, (3.75)4.3-4.7
28. ArOH‡, polymeric association, 2.3-5.5
29. H>C=C<, conjugated, (2.25)3.3-4.3(4.7)
30. H>C=C<, acyclic, conjugated, (2.9)3.5-4.0(4.5)
31. H-N-C=O, 1.5-4.5
32. ArH, benzenoid, (0.5)2.0-3.4(4.0)
33. ArH, nonbenzenoid, (1.0)1.4-3.8(6.0)
34. RNH3+, R2NH2+, and R3NH+ (trifluoroacetic acid solution), 2.3-2.9
35. H-C(=O)N<, 1.9-2.1
36. H-C(=O)O-, 1.8-2.0
37. ArNH3+, ArRNH2+, and ArR2NH+ (trifluoroacetic acid solution), 0.5-1.5
38. >C=N<OH‡, -0.2-1.2
39. RCHO, aliphatic, α,β-unsaturated, 0.35-0.50
40. RCHO, aliphatic, 0.2-0.3(0.5)
41. ArCHO, (-0.1)0.0-0.3(0.5)
42. ArOH, intramolecularly bonded, (-5.5)-2.5 - -0.5
43. -SO3H, -2 - -1
44. RCO2H, dimer, in nonpolar solvents, (-3.2)-2.2 - -1.0(0.3)
45. Enols, -6 - -5


†Normally, absorptions for the functional groups indicated will be found within the range shown. Occasionally, a functional group will absorb outside this range. Approximate limits for this are indicated by absorption values in parentheses and by shading in the figure.

‡The absorption positions of these groups are concentration-dependent and are shifted to higher τ values in more dilute solutions.

Reprinted with permission, John R. Dyer, *Applications of Absorption Spectroscopy of Organic Compounds,* 1965, Prentice-Hall, Inc., Englewood Cliffs, N.J.

## Table VII

### Spin-spin Coupling Constants

| Type | *J*, Hz | Type | *J*, Hz |
|---|---|---|---|
| $H_2$† | 280 | >C=CH—CH=C< | 9–13 |
| $CH_4$† | 12.4 | H—C≡C—H† | 9.1 |
| >C(H)(H) | 12–15 | >CH—C≡C—H | 2–3 |
| >CH—CH< | 2–9 | >CH—C(H)=O | 1–3 |
| H—C—(—C—)$_n$—C—H | ~0 | >C=C(H)—C(H)=O | 6–8 |
| $CH_3$—$CH_2$—*X* | 6.5–7.5 | | |
| $CH_3$(CH—*X*)$CH_3$ | 5.5–7.0 |  (benzene ring) | *o*- 6–9<br>*m*- 1–3<br>*p*- 0–1 |
| 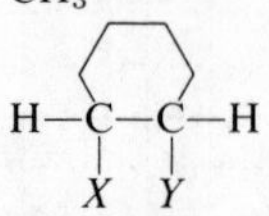 H—C—C—H (cyclohexane ring)<br>*X* *Y* | a,a 5–10<br>a,e 2–4<br>e,e 2–4 |  (ring) O | αβ 1.6–2.0<br>αβ′ 0.6–1.0<br>αα′ 1.3–1.8<br>ββ′ 3.2–3.8 |
| >C=C(H)(H) | 0.5–3 | | |
| H(>C=C<)H (cis) | 7–12 |  (ring) N—H | αβ 2.0–2.6<br>αβ′ 1.5–2.2<br>αα′ 1.8–2.3<br>ββ′ 2.8–4.0 |
| H(>C=C<)H (trans) | 13–18 | | |
| >C=C(H)—C—H | 4–10 |  (ring) S | αβ 4.6–5.8<br>αβ′ 1.0–1.8<br>αα′ 2.1–3.3<br>ββ′ 3.0–4.2 |
| >C=C(—C—H)<br>H | 0.5–2.5 | 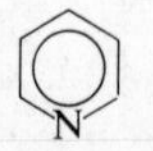 (ring) N | αβ 4.9–5.7<br>αγ 1.6–2.6<br>αβ′ 0.7–1.1<br>αα′ 0.2–0.5<br>βγ 7.2–8.5<br>ββ′ 1.4–1.9 |
| H<br>>C=C(—C—H) | ~0 | | |

† The coupling constant for these molecules, which contain only equivalent protons, has been determined from the spectra of the partially deuterated substances.

Reprinted with permission, John R. Dyer, *Applications of Absorption Spectroscopy of Organic Compounds,* 1965, Prentice-Hall, Inc., Englewood Cliffs, N.J.

## Table VIII

### Summary of Fragmentation Processes

Alkanes:

a. Simple fission of C–C bonds, most frequently at the site of branching.

$$\left[H_3\dot{C}{+}CH(CH_3)_2\right] \rightarrow H_3C\cdot + {}^{+}CH(CH_3)_2$$

$$H_3C{-}\overset{+}{C}H_2 \longrightarrow H_3C^{+} + :CH_2$$

b. Cyclic alkanes tend to lose side chains and/or extrude neutral olefinic moieties.

$$[C_6H_{11}{-}CH_3]^{+\cdot} \longrightarrow C_6H_{11}^{+} + \cdot CH_3$$

Alkenes:

a. Simple allylic cleavage (vinylic cleavage is much less frequent).

$$\left[CH_2\overset{\cdot+}{-}CH{-}CH_2{-}CH_2R\right] \rightarrow {}^{+}CH_2\,CH{=}CH_2 + \cdot CH_2R$$

b. McLafferty rearrangement (if γ H atoms are present).

$$\rightarrow + C_2H_4 \quad \text{or} \quad \rightarrow + {}^{+}CH_2\,CH_2\cdot$$

c. Retro-Diels Alder

$$\rightarrow + C_2H_4$$

Aromatic Hydrocarbons:

a. Benzylic cleavage with ring expansion to the tropylium ion.

$$\rightarrow \rightarrow + Z\cdot$$

Z = alkyl, aryl, or heteroatom

b. Vinylic cleavage.

$$\rightarrow + Z\cdot$$

c. McLafferty rearrangement (if γ H atoms are present).

$$\rightarrow + Y{=}Z$$

X, Y, Z can be almost any combination of C, O, N, or S.

d. Elimination of neutral fragments from *ortho*-disubstituted aromatic compounds.

$$\longrightarrow + Z{-}Y{-}H$$

X, Y, Z can be almost any combination of C, O, N or S.

e. Retro-Diels Alder.

$$\left[\text{(bicyclic radical cation)}\right] \longrightarrow \text{(methylenecyclohexadiene radical cation)} + C_2H_4$$

Alcohols:

a. Dehydration (thermal, prior to ionization and electron bombardment induced).

$$CH_3-\underset{H\ \ OH}{\overset{H}{C}}-CH_2 \xrightarrow[-H_2O]{\text{thermal}} CH_3CH\,CH_2 \qquad \left[CH_3\underset{H}{\overset{H}{C}}-CH_2-\overset{\overset{\cdot}{+}OH}{CH_2}\right] \longrightarrow CH_3\dot{C}HCH_2\overset{+}{C}H_2$$

(may be a 1,3-or 1,4-elimination)

1,4-dehydration product may undergo further cleavage

$$\overset{+}{\ }\!\!\diagdown\!\!\diagup\!\cdot-CH_3 \longrightarrow C_2H_4 + {}_{+}\!\diagup\!\overset{\cdot}{\ }-CH_3$$

b. $\alpha$, $\beta$-cleavage to form oxonium ions.

$$\left[\begin{matrix}\overset{+}{\cdot}OH\\ |\\ CH_3CH-CH_3\end{matrix}\right] \longrightarrow CH_3\overset{+OH}{\overset{//}{C}} + \cdot CH_3$$

c. Complex fission with H transfer in cyclic alcohols.

$$\left[\overset{\overset{\cdot}{+}OH}{\text{cyclohexanol}}\right] \longrightarrow \left[\overset{+OH}{\text{ring-opened radical}}\right] \longrightarrow \left[\overset{+OH}{\text{radical}}\right] \longrightarrow \text{(propyl radical)} + \overset{+OH}{\text{(enol)}}$$

Aliphatic Amines:

a. $\alpha$, $\beta$-fission with formation of immonium ion.

$$\left[\begin{matrix}\overset{\cdot}{+}NH_2\\ |\\ CH_3-CH-CH_3\end{matrix}\right] \longrightarrow CH_3CH{=}\overset{+}{N}H_2 + CH_3\cdot$$

Immonium ions may further cleave with transfer of H.

$$CH_3CH{=}\overset{+}{N}H-CH_2-\overset{H}{\overset{|}{C}}H_2 \longrightarrow CH_3CH{=}\overset{+}{N}H_2 + CH_2{=}CH_2$$

b. Complex fission with H transfer in hydrocarbon rings with amino substituents. (Anàlogous to complex fission of cyclic alcohols.)

Aliphatic Ethers:

a. Alkyl-oxygen fission. The charge ususally resides with the alkyl moiety.

$$\left[CH_3-\overset{\cdot +}{\underset{\cdot\cdot}{O}}-CH_3\right] \longrightarrow CH_3{}^{+} + \cdot OCH_3$$

b. $\alpha$, $\beta$-fission with oxonium ion formation.

$$\left[CH_3CH_2-\overset{\cdot +}{\underset{\cdot\cdot}{O}}-CH_2-CH_3\right] \longrightarrow CH_3CH_2-\overset{+}{O}{=}CH_2 + CH_3\cdot$$

The oxonium ion formed may undergo further cleavage with H transfer.

$$\overset{H}{CH_2}CH_2-\overset{+}{O}{=}CH_2 \longrightarrow CH_2{=}CH_2 + H\overset{+}{O}{=}CH_2$$

c. Cyclic ethers may extrude a neutral aldehyde moiety.

$$\left[\text{cyclic ether}\right]^{\bullet+} \longrightarrow CH_2{=}O + {}^{+}CH_2CH_2CH_2^{\bullet}$$

Halides:

a. Cleavage of the H-X bond.

$$\left[CH_3 - \overset{\bullet+}{X}:\right] \longrightarrow CH_3^{\bullet} + :\overset{+}{X}:$$

and

$$\left[CH_3 - \overset{\bullet+}{X}:\right] \longrightarrow CH_3^{+} + :\ddot{X}^{\bullet}$$

b. Elimination of HX. Analogous to the elimination of $H_2O$ from alcohols.

$$\left[\overset{H}{C}H_2(CH_2)_n\overset{+\dot{X}}{C}H_2\right] \longrightarrow HX + {}^{\bullet}CH_2(CH_2)_n\overset{+}{C}H_2$$

c. $\alpha, \beta$-fission with the formation of halonium ion.

$$\left[CH_3\overset{\dot{X}+}{C}H{-}CH_3\right] \longrightarrow CH_3\overset{X+}{\overset{\|}{C}} + CH_3^{\bullet}$$

d. Remote cleavage with the formation of cyclic halonium ions.

$$\left[CH_3{-}\text{ring}{-}\dot{\ddot{X}}{:}^{+}\right] \longrightarrow CH_3^{\bullet} + \text{cyclic } {}^{+}X$$

Esters:

a. $\alpha$-cleavage to form ions of the type $R^+$, $RCO^+$, $^+OR$, $^+OCOR'$ and $R'^+$.

$$\left[CH_3\overset{\dot{O}+}{\overset{/\!/}{C}}{-}OR'\right] \longrightarrow CH_3CO+ + {}^{\bullet}OR'$$

$$\quad\quad\quad\quad \hookrightarrow CO + CH_3^{+}$$

b.

$$\left[CH_3{-}\overset{\dot{O}+}{\overset{/\!/}{C}}{-}OR'\right] \longrightarrow CH_3^{+} + {}^{\bullet}OCOR'$$

$$\left[CH_3{-}\overset{+\dot{O}}{\overset{/\!/}{C}}{-}OR'\right] \longrightarrow \overset{+}{O}R' + CH_3CO^{\bullet}$$

$$\left[CH_3{-}\overset{+\dot{O}}{\overset{/\!/}{C}}{-}O{-}R'\right] \longrightarrow CH_3{-}CO_2^{\bullet} + R'^{+}$$

c. McLafferty rearrangement.

$$\left[\begin{array}{l} CH_2 \overset{H}{\frown} \dot{+}O \\ | \quad\quad\quad \| \\ CH_2 \quad\quad C \\ \quad \diagdown \quad \diagup \quad \diagdown \\ \quad\; CH_2 \quad\quad OR' \end{array}\right] \longrightarrow CH_2{=}CH_2 + CH_2{=}\overset{H\dot{O}^{+}}{\overset{|}{C}}{-}OR'$$

d. Double rearrangement of certain esters to yield protonated carboxylic acid fragments.

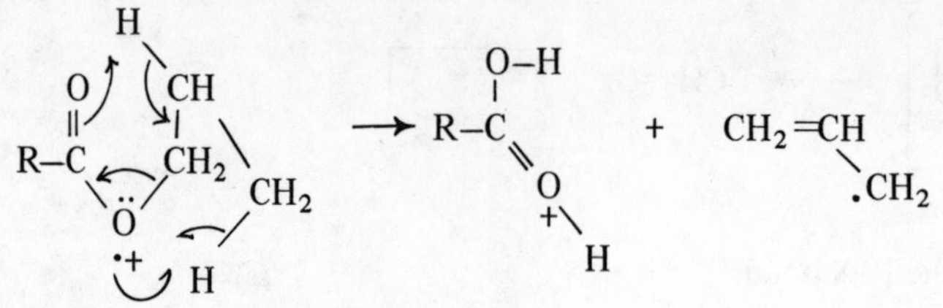

$$\text{R--C(=O)--}\ddot{\text{O}}^{\bullet+}\text{--CH}_2\text{--CH}_2\text{--CH}_2\text{--CH--H} \longrightarrow \text{R--C(--O--H)=}\overset{+}{\text{O}}\text{--H} + \text{CH}_2\text{=CH--}\dot{\text{C}}\text{H}_2$$

Aldehydes and Ketones:

a. $\alpha$-cleavage to form ions of the type $R^+$ and $RCO^+$.

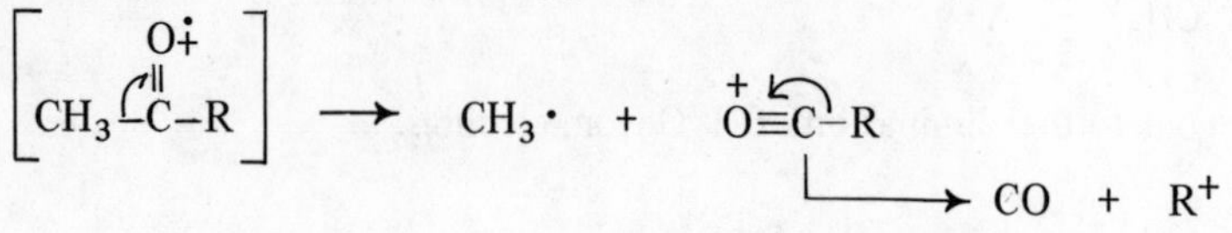

$$\left[\text{CH}_3\text{--C}(\text{=O}^{\dot{+}})\text{--R}\right] \longrightarrow \text{CH}_3\bullet + \overset{+}{\text{O}}\equiv\text{C--R} \longrightarrow \text{CO} + \text{R}^+$$

b. McLafferty rearrangement.

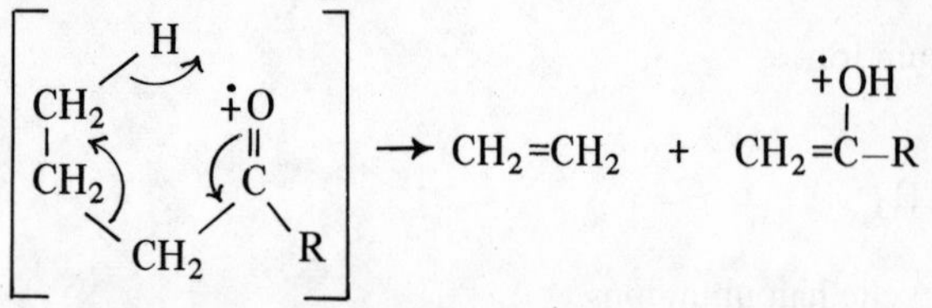

$$\left[\text{H--CH}_2\text{--CH}_2\text{--CH}_2\text{--C}(\text{=}\dot{\text{O}}^+)\text{--R}\right] \longrightarrow \text{CH}_2\text{=CH}_2 + \text{CH}_2\text{=C}(\text{--}\dot{\text{O}}^+\text{H})\text{--R}$$

c. Cyclic ketones undergo complex fission to yield neutral fragments and an oxonium ion.

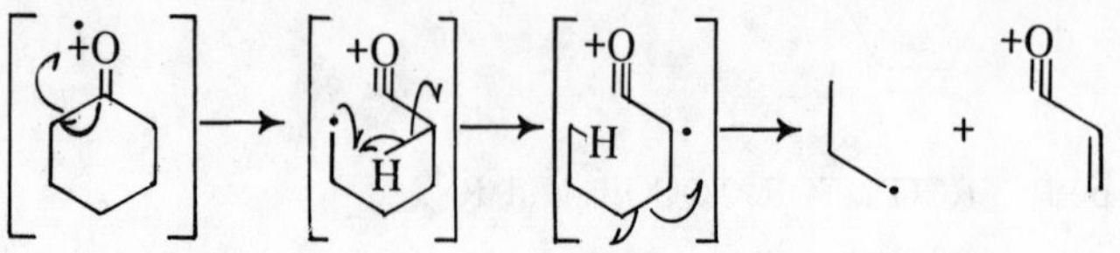

d. Bridged aromatic ketones extrude carbon monoxide.

$\longrightarrow$ + CO

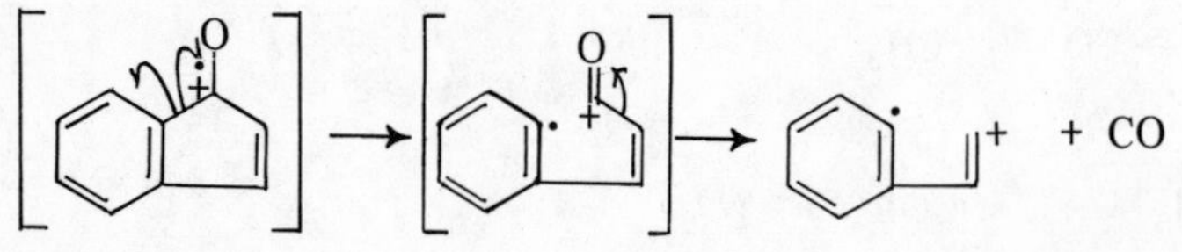

Phenols:

a. Phenols extrude carbon monoxide.

$\longrightarrow$ + CO

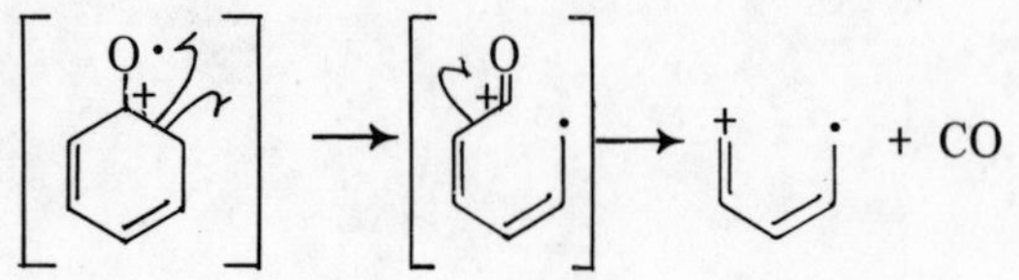

# INDEX

# Periodic Table of the Elements

| IA | IIA | IIIB | IVB | VB | VIB | VIIB | VIII | | | IB | IIB | IIIA | IVA | VA | VIA | VIIA | INERT GASES | |
|---|---|---|---|---|---|---|---|---|---|---|---|---|---|---|---|---|---|---|
| 1 H 1.0080† | | | | | | | | | | | | | | | | 1 H 1.0080 | 2 He 4.00260 | 2 |
| 3 Li 6.941† | 4 Be 9.01218 | | | | | | | | | | | 5 B 10.81 | 6 C 12.011 | 7 N 14.0067 | 8 O 15.9994† | 9 F 18.9984 | 10 Ne 20.179† | 2 8 |
| 11 Na 22.9898 | 12 Mg 24.305 | | | | | | | | | | | 13 Al 26.9815 | 14 Si 28.086† | 15 P 30.9738 | 16 S 32.06 | 17 Cl 35.453 | 18 Ar 39.948† | 2 8 8 |
| 19 K 39.102† | 20 Ca 40.08 | 21 Sc 44.9559 | 22 Ti 47.90† | 23 V 50.9414† | 24 Cr 51.996 | 25 Mn 54.9380 | 26 Fe 55.847† | 27 Co 58.9332 | 28 Ni 58.71† | 29 Cu 63.546† | 30 Zn 65.37† | 31 Ga 69.72 | 32 Ge 72.59† | 33 As 74.9216 | 34 Se 78.96† | 35 Br 79.904 | 36 Kr 83.80 | 2 8 18 8 |
| 37 Rb 85.4678† | 38 Sr 87.62 | 39 Y 88.9059 | 40 Zr 91.22 | 41 Nb 92.9064 | 42 Mo 95.94† | 43 Tc 98.9062 | 44 Ru 101.07† | 45 Rh 102.9055 | 46 Pd 106.4 | 47 Ag 107.868 | 48 Cd 112.40 | 49 In 114.82 | 50 Sn 118.69† | 51 Sb 121.75† | 52 Te 127.60† | 53 I 126.9045 | 54 Xe 131.30 | 2 8 18 18 8 |
| 55 Cs 132.9055 | 56 Ba 137.34† | 57 *La 138.9055† | 72 Hf 178.49† | 73 Ta 180.9479† | 74 W 183.85† | 75 Re 186.2 | 76 Os 190.2 | 77 Ir 192.22† | 78 Pt 195.09† | 79 Au 196.9665 | 80 Hg 200.59† | 81 Tl 204.37† | 82 Pb 207.2 | 83 Bi 208.9806 | 84 Po (210) | 85 At (210) | 86 Rn (222) | 2 8 18 32 18 8 |
| 87 Fr (223) | 88 Ra 226.0254 | 89 ′Ac (227) | 104 § (260) | 105 § (260) | | | | | | | | | | | | | | |

*Lanthanum Series

| 58 Ce 140.12 | 59 Pr 140.9077 | 60 Nd 144.24† | 61 Pm (147) | 62 Sm 150.4 | 63 Eu 151.96 | 64 Gd 157.25† | 65 Tb 158.9254 | 66 Dy 162.50† | 67 Ho 164.9303 | 68 Er 167.26† | 69 Tm 168.9342 | 70 Yb 173.04† | 71 Lu 174.97 | 2 8 18 32 9 2 |
|---|---|---|---|---|---|---|---|---|---|---|---|---|---|---|

′Actinium Series

| 90 Th 232.0381 | 91 Pa 231.0359 | 92 U 238.029 | 93 Np 237.0482 | 94 Pu (244) | 95 Am (243) | 96 Cm (247) | 97 Bk (247) | 98 Cf (251) | 99 Es (254) | 100 Fm (257) | 101 Md (258) | 102 No (255) | 103 Lr (256) | 2 8 18 32 32 9 2 |
|---|---|---|---|---|---|---|---|---|---|---|---|---|---|---|

FISHER SCIENTIFIC COMPANY
CAT. NO. 5-702-10

§The International Union for Pure and Applied Chemistry has not adopted official names or symbols for these elements.

†These weights are considered reliable to ±3 in the last place. Other weights are reliable to ±1 in the last place.

Atomic weights corrected to conform to the 1969 values of the Commission on Atomic Weights.

©Copyright 1970 by Fisher Scientific Company

# INTERNATIONAL ATOMIC WEIGHTS

Based on $C^{12}$

| *Element* | *Symbol* | *Atomic number* | *Atomic weight* |
|---|---|---|---|
| Actinium | Ac | 89 | [227]* |
| Aluminum | Al | 13 | 26.9815 |
| Americium | Am | 95 | [243]* |
| Antimony | Sb | 51 | 121.75 |
| Argon | Ar | 18 | 39.948 |
| Arsenic | As | 33 | 74.9216 |
| Astatine | At | 85 | [210]* |
| Barium | Ba | 56 | 137.34 |
| Berkelium | Bk | 97 | [247]* |
| Beryllium | Be | 4 | 9.0122 |
| Bismuth | Bi | 83 | 208.980 |
| Boron | B | 5 | 10.811 |
| Bromine | Br | 35 | 79.909 |
| Cadmium | Cd | 48 | 112.40 |
| Calcium | Ca | 20 | 40.08 |
| Californium | Cf | 98 | [247]* |
| Carbon | C | 6 | 12.01115 |
| Cerium | Ce | 58 | 140.12 |
| Cesium | Cs | 55 | 132.905 |
| Chlorine | Cl | 17 | 35.453 |
| Chromium | Cr | 24 | 51.996 |
| Cobalt | Co | 27 | 58.9332 |
| Copper | Cu | 29 | 63.54 |
| Curium | Cm | 96 | [247]* |
| Dysprosium | Dy | 66 | 162.50 |
| Einsteinium | Es | 99 | [254]* |
| Erbium | Er | 68 | 167.26 |
| Europium | Eu | 63 | 151.96 |
| Fermium | Fm | 100 | [253]* |
| Fluorine | F | 9 | 18.9984 |
| Francium | Fr | 87 | [223]* |
| Gadolinium | Gd | 64 | 157.25 |
| Gallium | Ga | 31 | 69.72 |
| Germanium | Ge | 32 | 72.59 |
| Gold | Au | 79 | 196.967 |
| Hafnium | Hf | 72 | 178.49 |
| Helium | He | 2 | 4.0026 |
| Holmium | Ho | 67 | 164.930 |
| Hydrogen | H | 1 | 1.00797 |
| Indium | In | 49 | 114.82 |
| Iodine | I | 53 | 126.9044 |
| Iridium | Ir | 77 | 192.2 |
| Iron | Fe | 26 | 55.847 |
| Krypton | Kr | 36 | 83.80 |
| Lanthanum | La | 57 | 138.91 |
| Lawrencium | Lw | 103 | [257]* |
| Lead | Pb | 82 | 207.19 |
| Lithium | Li | 3 | 6.939 |
| Lutetium | Lu | 71 | 174.97 |
| Magnesium | Mg | 12 | 24.312 |
| Manganese | Mn | 25 | 54.9380 |
| Mendelevium | Md | 101 | [256]* |
| Mercury | Hg | 80 | 200.59 |
| Molybdenum | Mo | 42 | 95.94 |
| Neodymium | Nd | 60 | 144.24 |
| Neon | Ne | 10 | 20.183 |
| Neptunium | Np | 93 | [237]* |
| Nickel | Ni | 28 | 58.71 |
| Niobium | Nb | 41 | 92.906 |
| Nitrogen | N | 7 | 14.0067 |
| Nobelium | No | 102 | [254]* |
| Osmium | Os | 76 | 190.2 |
| Oxygen | O | 8 | 15.9994 |
| Palladium | Pd | 46 | 106.4 |
| Phosphorus | P | 15 | 30.9738 |
| Platinum | Pt | 78 | 195.09 |
| Plutonium | Pu | 94 | [242]* |
| Polonium | Po | 84 | [210] |
| Potassium | K | 19 | 39.102 |
| Praseodymium | Pr | 59 | 140.907 |
| Promethium | Pm | 61 | [147]* |
| Protoactinium | Pa | 91 | [231]* |
| Radium | Ra | 88 | [226]* |
| Radon | Rn | 86 | [222]* |
| Rhenium | Re | 75 | 186.2 |
| Rhodium | Rh | 45 | 102.905 |
| Rubidium | Rb | 37 | 85.47 |
| Ruthenium | Ru | 44 | 101.07 |
| Samarium | Sm | 62 | 150.35 |
| Scandium | Sc | 21 | 44.956 |
| Selenium | Se | 34 | 78.96 |
| Silicon | Si | 14 | 28.086 |
| Silver | Ag | 47 | 107.870 |
| Sodium | Na | 11 | 22.9898 |
| Strontium | Sr | 38 | 87.62 |
| Sulfur | S | 16 | 32.064 |
| Tantalum | Ta | 73 | 180.948 |
| Technetium | Tc | 43 | [97]* |
| Tellurium | Te | 52 | 127.60 |
| Terbium | Tb | 65 | 158.924 |
| Thallium | Tl | 81 | 204.37 |
| Thorium | Th | 90 | 232.038 |
| Thulium | Tm | 69 | 168.934 |
| Tin | Sn | 50 | 118.69 |
| Titanium | Ti | 22 | 47.90 |
| Tungsten | W | 74 | 183.85 |
| Uranium | U | 92 | 238.03 |
| Vanadium | V | 23 | 50.942 |
| Xenon | Xe | 54 | 131.30 |
| Ytterbium | Yb | 70 | 173.04 |
| Yttrium | Y | 39 | 88.905 |
| Zinc | Zn | 30 | 65.37 |
| Zirconium | Zr | 40 | 91.22 |

*Mass numbers of the most stable or most abundant isotopes are shown in parentheses.

3286 1